György Darvas

Symmetry

Cultural-historical and ontological aspects of science–arts relations

The natural and man-made world in an interdisciplinary approach

Translated from the Hungarian by David Robert Evans

Birkhäuser
Basel · Boston · Berlin

Author:

György Darvas
Institute for Research Organization of the
Hungarian Academy of Sciences
and
Symmetrion
P.O. Box 994
H-1245 Budapest
Hungary
e-mail: darvasg@iif.hu

Reviewed by Szaniszló Bérczi and István Hargittai.

Library of Congress Controll Number: 2006939567

Bibliographic information published by Die Deutsche Bibliothek
Die Deutsche Bibliothek lists this publication in the Deutsche Nationalbibliografie;
detailed bibliographic data is available in the Internet at <http://dnb.ddb.de>.

ISBN 978-3-7643-7554-6 Birkhäuser Verlag AG, Basel – Boston – Berlin

Cover design: György Darvas and Tamás F. Farkas
Front cover image by Tamás F. Farkas

Part of Springer Science+Business Media
Printed on acid-free paper produced from chlorine-free pulp. TCF ∞

ISBN 3-7643-7554-X
ISBN 978-3-7643-7554-6
9 8 7 6 5 4 3 2 1

ISBN 3-7643-7555-8
ISBN 978-3-7643-7555-3 (eBook)
www.birkhauser.ch

Contents

Human Creativity

Preface

Hermann Weyl (1885–1955), who from the 1920s onwards turned the general phenomenon of symmetry into a subject of research in its own right, retired in 1951 to Europe from Princeton, where he spent his years in emigration in the company, among others, of John von Neumann (1903–1957) and Eugene P. Wigner (1902–1995). As a parting gesture, he held four lectures at the university in which he sought to summarize all there was to know about symmetry. The talks were written at the level of scientific discourse of the age, but Weyl did not address the representatives of individual disciplines — the material was accessible to those from all of the university's scholarly fields. The edited text of the lectures was published by the university in a separate volume. Weyl's *Symmetry* become a publishing sensation, being translated into some fifty languages, enjoying countless new editions, and is used in education throughout the world to this day.

Since Weyl's retirement there has been a huge upsurge in research into symmetry (and, I should add, into its absence or its violation). At almost the same time as Weyl's lectures, Buckminster Fuller patented his geodesic dome, containing hexagons, and ensuring a high level of symmetry (employing his principle of synergetics) and thereby great stability. The world first witnessed this structure in the form of the spatious dome constructed for the Montreal world exhibition, then later as the most stable sewing pattern for soccer balls, but this only really became a success following the discovery of the spherical carbon molecule fullerene, capable of stable bonding, in 1985. We know from the memoirs of J. Watson that — within a year of the publication of the Weyl volume — it was a symmetry consideration which led to the final discovery of the structure of the double helix. A year later Yang and Mills published their article about a new type of gauge invariance to describe the conservation of isotopic spin, which has become an inescapable foundation for physics ever since: no discovery in particle physics could have been made without it. During the 1950s, Eugene P. Wigner published a series of articles on the application of symmetries and conservation laws, turning these into a fundamental theory in physics, and, like some of those mentioned above, was rewarded with the Nobel prize. From this point forward, the history of physics became a series of discoveries of symmetries and symmetry breakings. Fivefold symmetry, which remained a mystery for centuries, became material reality in the form of the quasicrystals discovered in the 1980s. The concept of symmetry, as a method, became an element of heuristics, being transferred from one branch of science to the next as an idea to catalyse intellectual creativity, not to mention the interplay between arts and sciences, and the effects of the role it has played in various different cultures. We only have to think of the intellectual proximity of the eightfold way of Buddha to the classification of elementary particles and the use of the SU(3) group that describes their symmetry. Or consider the role of Japanese origami in designing structures used in spaceships, and the history of the discovery of the artificial retina, which combines branches of science formerly considered distant from one another (the theory of analogue and digital chips, ancient logic, and cerebral asymmetries). Neither have the results of symmetry research left untouched such areas of scientific research as the thermodynamics of chemical equilibria, psychology, brain research, education science, musicology and sociolinguistics.

The last half century of research into symmetry has extended our knowledge manifold. Not only has the content of this knowledge become enriched — so has the concept of symmetry itself. When in the 1990s I began holding special interdisciplinary symmetry lectures for students at the Faculty of Sciences at L. Eötvös University in Budapest, I had to confront the question of how to summarize all that we can and should know about symmetry today in a single semester — in about the same depth as Weyl described the knowledge of his day in his four lectures. Half a century ago, interpretations of symmetry were dominated by crystallography and crystallographic analogies. To be true to the proportions we see in science today, I can devote no more than two out of fifteen chapters to this approach. Over the years the lectures have developed and become more polished, and the proportions have also changed. I have been greatly helped by consultations with colleagues, with members of what was the International Symmetry Society and what is now the International Symmetry Association, and by correspondence with the authors of articles as editor of the journal *Symmetry: Culture and Science*. Also of great importance were questions and responses from students, and what I learned from examination discussions. This is what provided the material for this volume. For their very useful suggestions and proof-reading of the manuscript, I would like to express particular gratitude to Szaniszló Bérczi, as well as to László Beke, József Cseh, Gábor Gévay, István Hargittai and Ervin Hartmann. It is my pleasure to thank David Robert Evans for his great contributions to the English text.

In the course of the half century mentioned above, both the specialist literature on symmetry and the array of artistic interpretation embodied by works of art have expanded in unprecedented measure. In almost all disciplines, works have appeared discussing the symmetries and violations of symmetry in that field. As specialization has increased, so too has the number of writings discussing the points of interdependence. It was inevitable that results in one particular area based on symmetry considerations would inspire other fields of research. Interest in such work has also increased. It is no accident that D. Hofstadter's monograph *GEB* has become one of the best-read works of the last two decades. Hofstadter addresses the question of what is common in the intellectual legacy of "*G*ödel, *E*scher, *B*ach". Of course, his all-embracing work shows that all three personalities representing the main thrust of its line of thought embody the intellectual legacy of the unity of humanity, a legacy we can only truly appreciate when we make the boundaries between disciplines and art-forms, which have criss-crossed human culture in a largely artificial fashion, both transparent and traversable. It transpired that one of the key means for this could be a phenomenon, a concept, a method that is present in all of them. One such means is symmetry.

Without either denying or accidentally repeating the spirit and content of valuable earlier works dedicated to the presentation of the holistic way of thinking, I set myself the objective of writing a book which switches the perspective, putting symmetry at the focus of discussion, in the light of the scholarly knowledge we have at our disposal today. In the course of this it relies on the factual material gathered by its predecessors, keeping to its own set of proportions to present those facts in a different light and group them in an alternative way. In some chapters I have allowed my own personal observations to be expressed.

One of the book's goals is to present the unity and interdisciplinary nature of human culture. It attempts — in contrast to the division of reality into different disciplines by school education — to introduce the reader to the alternative view of the world provided by the holistic approach. In the interests of this, it emphasizes three types of possibility for bridging the split elements of that culture. First, between the various scholarly disciplines. Second, by presenting the interplay between arts and sciences. And third, with examples of ways in which the different cultures of various ages and geographical regions have influenced each other to produce new intellectual achievements.

In the course of this, the book discusses three approaches to symmetry: first, as a phenomenon; second, as a concept with varying content characterizing a group of phenomena; third, as an operation (or rather a well-defined group of operations) which is at once a method. How can symmetry operations serve as a method? They can represent a method for implementing analogies, for example. For the observation of common elements in various ages, cultures and branches of knowledge, which are invariant in the face of their differences, and for their implementation elsewhere. I would like to draw particular attention to the strengthening role of heuristics in this regard.

The discussion of symmetry as a subject in its own right gives us a particular slice of scientific endeavour, one which cannot be fitted within the framework of any traditional discipline. The mode of discussion is partly historical. In addition to the history of art, science and ideas, its historical nature also presents itself in its method. This can be seen in the way it repeatedly makes use of philosophical analyses in the course of the discussion.

In choosing my subject, I could not avoid taking certain constraints into consideration. On the one hand, adherence to the aforementioned proportions; on the other, overall length. I could not attempt a repetition of the whole of the rich literature concerning symmetry, or indeed the discussion or even mention of every single phenomenon associated with symmetry or symmetry violation. I had to select, and this selection reflects my own choice. This is what I thought it important, here and now, to say about symmetry. In considerable measure, my choice rested on my experiences noted above.

Through the chosen examples, I try to present a cross-section of the most significant areas of the interpretation and implementation of symmetry. If from time to time I succeed in evoking the feeling in the reader that I could have written more about this or about that, then the selection has achieved its goal: it has awakened associations, and connected the presented examples with knowledge the reader has from other sources. In this way it will have reconstructed, at the individual level, the bridges that symmetry has built in the collective consciousness of academics and artists between various branches of knowledge.

The first few chapters are a bit more dry in nature. Following the generalization of concepts and a historical introduction, I present some of the most successful interdisciplinary uses of the concept of symmetry. Then, tracing the path of the organization of matter from its physical structure, through the chemical, and molecules, which play a biologically important role, to living matter, the human brain, and finally to the products of human cognition and consciousness, I inspect the interdependence of art and science through the unusual lens of a series of symmetry breakings. The conceptual framework laid out at the start of the book will make it increasingly easy to become acquainted with the increasingly expanding world of this group of phenomena. Rather like a kaleidoscope, a lens, through which, on the following pages, we will observe the world, giving us a new way of seeing. I would like to emphasize this attitude as one of the book's important attributes. I would only be too pleased if its readers decide to adopt it, and put the approach they have learned and understood in these pages to use in their own respective disciplines.

Introductory chapters

Chapter 1

Symmetry, invariance, harmony

The interpretation of the concept of symmetry in everyday life, science and art

The concept of symmetry

"How nice and symmetrical," we often say or think to ourselves. The associations that the word awakens in us depend on the experiences in our past that have established its meaning for us. The term 'symmetry' can have three separate types of meaning, as a phenomenon, a concept, or an operation. The *phenomenon* is what we consider to be symmetrical on the basis of our experience or of knowledge we have learned. The *concept* is what circumscribes all such phenomena. The *operation* is what gives rise to the phenomenon or makes it possible.

The etymology of the word 'symmetry'

The word 'symmetry is a combined word of Greek origin. Its two components are the prefix *συν* [syn] and the word *μετρ(ι)ος* [metr(i)os]: *συν* + *μετρ(ι)ος*. The prefix *syn-* can appear separately as an adverb, or, in combination, as a preposition. It means together, a group, simultaneously, common, or together with. It occurs in many other words in English, like sympathy, symphony, synthesis, syntax, synonym, synod, synagogue. The 'n' at the end of 'syn' often becomes an 'm'. The adverb *μετρίως* [metrios] means in (good) measure, suitably; as the adjective *μτριος* it means measured, moderate, middling, average. *Μετριτης* [metriotes] means the correct measure, moderation, the right proportion. In Greek texts these are used in a number of figurative senses (for example, suitable, worthy, just, honourable, collected, modest, decent), and these figurative meanings were all to enrich the later content of the term symmetry that was made up of them.

The two-part verb $\sigma\upsilon\mu\mu\varepsilon\tau\rho\omega$ [simmetreo] means measure by the same standard, compare; the adjective $\sigma\mu\mu\varepsilon\tau\rho o\varsigma$ [simmetros] means of the same size, of the same content, commensurate, proportionate with something, suitable, moderated; in its adverbial form, $\sigma\upsilon\mu\mu\tau\rho\omega\varsigma$ [simmetros] means proportionately, in the right proportion, at the right time. Its meaning as the compound noun $\sigma\upsilon\mu\mu\varepsilon\tau\rho\alpha$ [simmetria] is combined measurement, measurement through comparison, good proportions, and in a figurative sense agreement, harmony.

In short, as far as its content is concerned, to the Greeks two and a half thousand years ago the word symmetry meant *the common measure of things*.

As we think of it today, symmetry belongs to the great organizing concepts. It is a comprehensive concept that appears in many areas: in our everyday lives, in science, in art. We encounter it so often that we sometimes feel as if the world were only made up of symmetrical things. But before we attribute symmetry too much significance, let me emphasize that it is by no means the only concept or phenomenon we have as a comprehensive organizing principle. We have a need for organizing principles to deal with the many phenomena around us, the many experiences we acquire, and the multitude of knowledge we deduce from these. To help us, we call upon general concepts applicable in a variety of areas and with a variety of goals. Examples of concepts playing an organizing role in our thinking are order and orderliness, harmony, hierarchy, and system. In both a wider and narrower sense, the concepts of beauty, proportion and rhythm appear as characteristics with many common features; the same can be said of (logical and aesthetic) perfection (or attempts to achieve it), analogy, invariance under change (which is very close to symmetry), and, finally — as just another in this list — the concept of symmetry.

In what does the general nature of these concepts lie? What makes them suitable to act as organizing principles? Firstly, the fact that they are applicable to a wider group of phenomena, are group concepts, useable in various different areas of life, some more than others, but always to a number of phenomena. They are at once present in everyday life, academic discourse (the search for the truth), while a good number of them

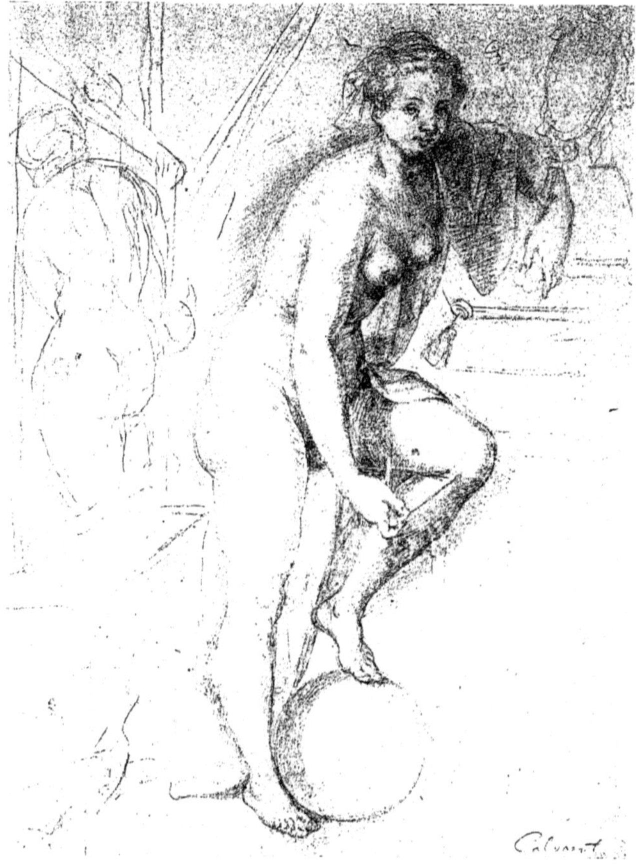

Figure 1.1. *The Allegory of Symmetry.* D. Calvaert (1540–1619), Bologna. (Graphical archive of the Museum of Fine Arts, Budapest, K.66.25.)

serve as a benchmark in arts, in aesthetics (the search for beauty) or in ethics (the search for what is right or just). They help to give us a wide interdisciplinary survey of the world.

Of the above, perhaps it is symmetry which has — by conveying analogies, ideas and methods — played a role in inspiring creativity in the greatest number of fields. As it is present in all the four areas mentioned above, it is a suitable auxiliary concept for taking us right through the dangers and beauty of interdisciplinary thought. With its help, we are given an insight into how arts and sciences, different disciplines, and various different human cultures have affected and conceptually enriched one another.

What exactly is symmetry?

Let us try to begin with the ordinary meaning of the word and built up the academically-founded concept of symmetry. In the course of this journey, we first become acquainted with the everyday interpretations of the concept of symmetry. We then consider whether we essentially see the world around us symmetric. Is symmetry a natural phenomenon? It is in the light of the answers to the above questions that we can estimate the significance of symmetry violations.

On hearing the word symmetry, the majority of people think of the simplest examples of the concept.

The best-known example of geometric symmetry is *reflection*. If we reflect a (planar) shape in a linear one (the axis of symmetry), then it appears on the far side such that the respective points of the shape and its reflection are at the same distance from the axis, albeit in opposite directions. The figure retains its shape — in mirrored form — and its size and the angles between the lines connecting its various points are also unchanged, as is its colour. Reflection changes the direction of orientation, however: left and right are swapped. We have completed an operation, that of reflection, in the course of which certain characteristics of the reflected object have changed, but some have remained the same. It is these unchanged characteristics which represent the symmetry of the original figure in relation to its reflection. If we reflect the reflected image in the same axis again, we are returned to the original shape (Figure 1.2). Taking the same reflection operation one dimension further, we can produce the same results with spatial figures by using a mirror plane instead of an axis.

The second-most mentioned symmetry operation is *rotation*. If we rotate a (planar) figure around an axis perpendicular to the plane, the figure preserves its internal characteristics and the distance of its points from the axis. Its symmetry lies in keeping these properties intact. If we complete a rotation of such an angle that after a finite number of rotations — 2, 3, 4, 5, 6, ... — the figure exactly overlaps the original, then, depending on the angle of the rotation, we can talk of 2-, 3-, 4-, 5- or 6-fold symmetry. In general, if we rotate the object by an angle of $360°/n$, we term this n-fold symmetry, where n is a natural number. We can rotate spatial figures in the same way around an axis.

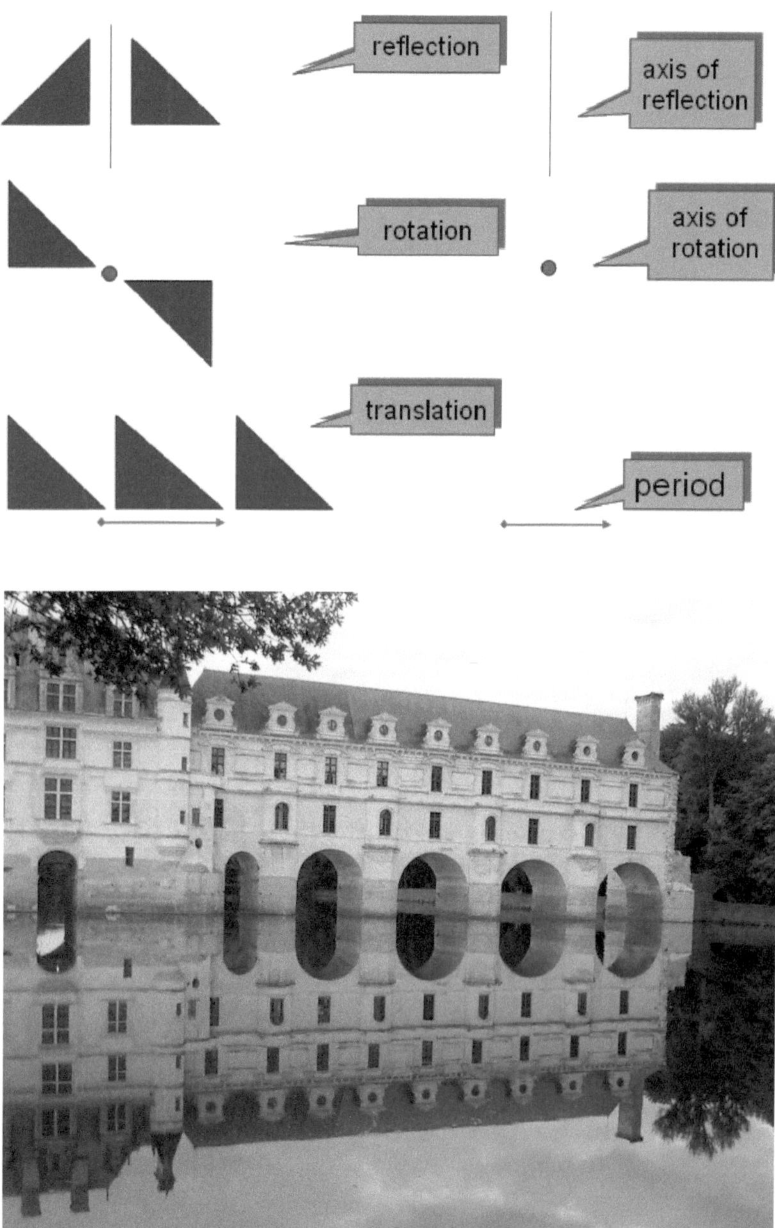

Figure 1.2. *Basic symmetry transformations, as well as the reflection and translation symmetry represented in a photo of Chenonceau Castle, France*

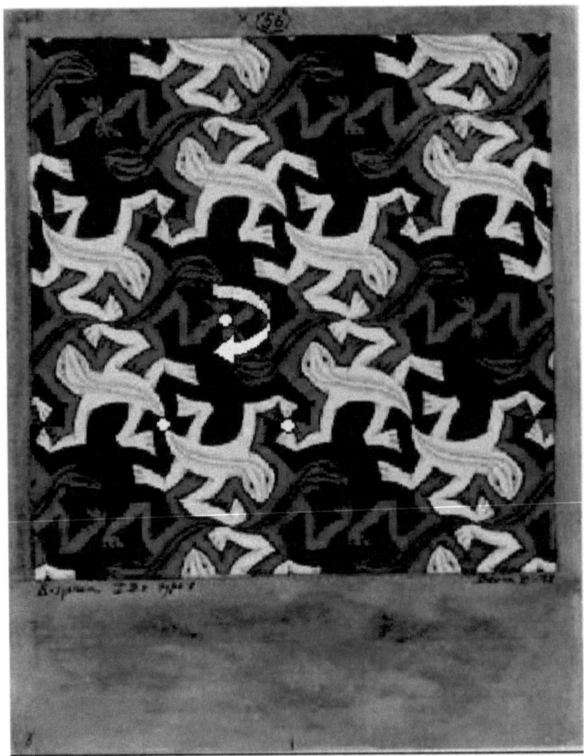

Figure 1.3. *M. C. Escher: Reptiles*

A good example of rotational symmetry is M. C. Escher's (1898–1972) mosaic of reptiles (Figure 1.3). For example, we can place an axis of rotation at the points where the reptiles' tails or right front legs touch. Rotating the reptile around these through 180° gives us another reptile of similar colour (twofold symmetry). If we turn the reptile through 60° around its left front leg, however, we overlap a reptile of a different colour. Let me draw the reader's attention to the fact that this design is also a beautiful artistic example of continuous planar layout — with figures that are congruent — for which Escher uses three reptiles of different colour but the same shape. If we are to mark the points where the possible axes of rotation would be, it becomes clear that they form a web of congruent regular triangles filling the plane of the drawing. The artist formed the congruent lizards by joining these points with similar broken lines, three by three. As seen in Figure 1.4, in a similar design, he even marked the lines of the web of triangles which assisted him.

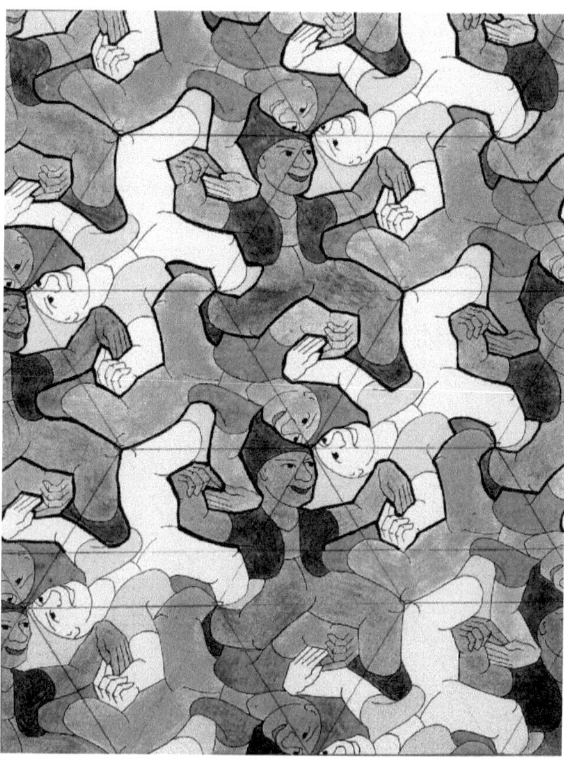

Figure 1.4. *M. C. Escher: Symmetrical Drawing 21 (1938)*(page)

The operation we have completed is a rotation around an axis. The object on which we completed it is a geometric figure. With the exception of its spatial position, almost all of its properties have remained intact, whatever we take our frame of reference to be.

Less common than the above two, but still regularly mentioned, is *translational* symmetry. If we translate a (planar) figure along a straight line with a uniform period, always in the same direction, we are given a repeated series of the given figure, in which the repeated elements are alike in every respect, and the distance between which is fixed on a periodic basis. Their symmetry lies in this uniformity. Translational symmetry is characterized by the direction of the translation and the length of the period. The symmetry operation in this instance is the straight linear shift (translation) we perform on a geometric object, as in the previous instances; in the process of this operation all of its characteristics are preserved, save its spatial location. (Figures 1.2 and 1.5).

It is in this form that symmetry most often occurs in our surroundings. This is true of the windows on most buildings, the successive carriages of trains, series of telegraph poles and electricity pylons, bars of railings, paving stones on the street, the web of squares in a school notebook, the rows of chairs in a schoolroom or theatre, the headstones in a cemetery, the frieze designs on Greek vases, the atoms in crystals, the waves of Lake Balaton, the annual recurrence of the changing seasons, the phases of the Moon, the rhythm of songs, dance movements, and the rhyme of poems (Figure 1.5).

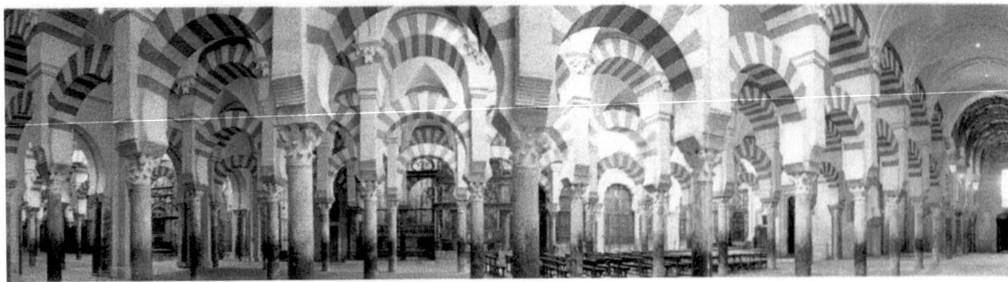

Figure 1.5. *The translational symmetry of the columns of the Mezquita in Cordoba.* (The hall of columns was built in a number of stages, between 785 and 962 A.D.)

The characteristic of translational symmetry is that it can be achieved in all three dimensions of space. The operation can occur in any direction, but only in one particular direction (dimension) in each case. Rotation requires at least two dimensions, for it requires at least one plane. To implement reflection with movement, even in the simplest instance we require all three dimensions, for the object to be reflected — even if it is just a single point — can only be fitted with its mirror image by rotating it through space, maintaining its distance, around the axis of reflection in a plane perpendicular to it. (This is true even if the starting and ending positions lie along the same line or plane, and we do not notice the path along which the reflected object has been moved in our minds.)

There are three-dimensional mirror images which cannot be made to overlap each other in space by any kind of geometric operation: our left and right hands are examples. Based on this (left- or right-) handedness, such figures are referred to as *chiral*, from the Greek.

In decorative designs (freeze designs) we often encounter *glide reflection*. In essence, this is a complex symmetry operation. The object to be reflected is first glided by one period length, then reflected in an axis parallel with the direction of gliding, and so on. The preserved characteristics are made up from the characteristics of the reflection and the glide symmetry, but each pair of elements can also be interpreted as a translational symmetry. Given that, in addition to decorative arts, glide reflection often occurs in the world of crystals, crystallography considers it to be a separate symmetry operation.

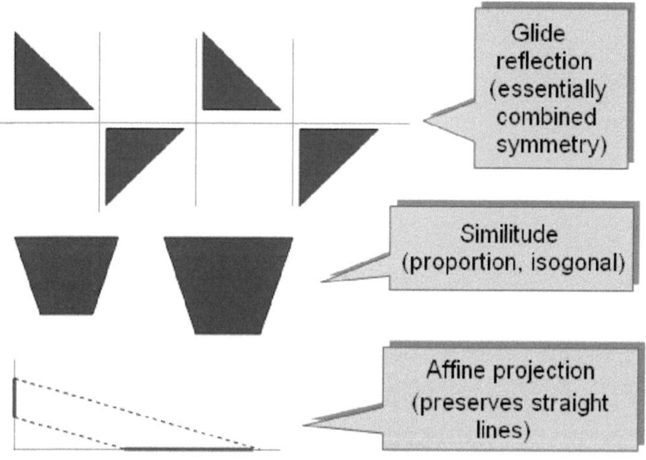

Figure 1.6. *Symmetry transformations*

Similitude is also a geometric symmetry transformation. If we enlarge or reduce a figure (perhaps rotating it at the same time) then the distances between its respective points (measurements) change, but the proportions of the distances between particular characteristic points and the angles enclosed by the lines connecting them remain unchanged. It is in maintaining its proportions and angles that the figure, similar to the original, displays symmetry. The completed geometric symmetry operation is thus enlargement-reduction (size transformation), which as before is applied to a geometric object. The measurements (and possibly the spatial orientation) are characteristics which change, while internal proportions and angles are characteristics preserved during the operation.

Affine projection preserves even fewer characteristics: it transforms a line into a line. The rays of sunshine falling on a window, for example, project the image of one edge of the frame onto the floor of the room such that its length and direction are different from that of the window edge, but remain straight. This preserved characteristic is adequate, however, for this to be regarded as symmetry.

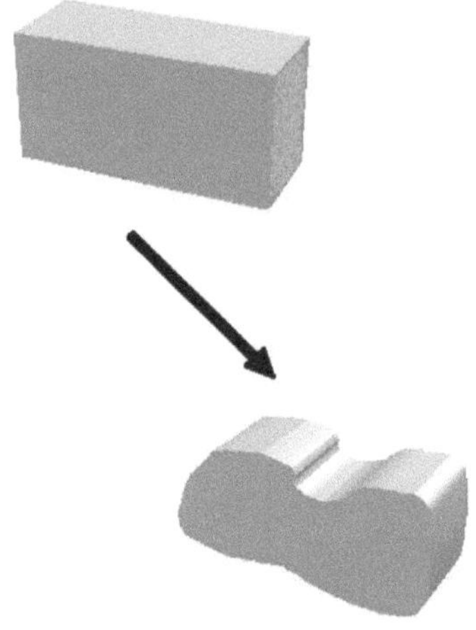

Figure 1.7. *Topological symmetry: squeezing the sponge*

When we squeeze a sponge, its shape becomes completely deformed, and its measurements change. The neighbourhood relations between the cells of the pores are unchanged, however, as are those of contact with the matter surrounding with cells. If this characteristic — the topological relations between points — is unchanged, we call this topological symmetry (Figure 1.7). Perhaps a comparison can make the essence of topological symmetry clearer. Let us imagine that we live on an estate comprised of family houses, plots, and streets. The estate is purchased by a businessman, who offers us all alternative living space elsewhere. In the new estate, the landscaping will be different, as will be the size and shape of the plots, the location of the trees and the house in a given plot, but in every direction the situation will be the same for our closest neighbours, then second-closest, third-closest, and so on, as in the place we previously lived. In a social sense, the topology of our place of residence remains unchanged.

The series of geometric symmetry operations could be continued. But these are not the only types of symmetry operations. It is also a symmetry operation if a figure is painted a different colour — if all its other characteristics are left intact. In this instance, the *object* is a geometric figure, the *operation* is the change of colour, and the *preserved characteristic* is everything apart from its colour.

Combined symmetries can also be formed from symmetries that have been learned. Examples are the various depictions of perspective in art (with one or more vanishing points, and aerial symmetry which also produces changes in colour) which combine similarity and affine projection (and colour symmetry) to preserve as many characteristic of the observed world in the picture as possible.

The aforementioned can have created the impression that the world around us is purely made up of symmetries. What's more, that everything is symmetrical. Is it really?

Is the world symmetrical?

Do we see the world as symmetrical? We cannot give a definite answer to this question — it depends on one's interpretation and point of view. Of course, in the world we find both symmetry and asymmetry. If we did not find any symmetry in it, it would not be beautiful, but if all we found in it were symmetry (permanence), it would be very dull. Let us think back to the original meaning of the word symmetry, however, which for the Ancient Greeks meant harmony and proportion. This we find both in our natural and our man-made environment. And this is what all art strives for. It is how the greats of classical art attempted to depict the human body. We can think of the statues of classical Greek culture (e.g. Pheidias, 5th century BC), the greatest artists of the Renaissance, or of the Greek Polykleitos (5th century BC), who was the first to write a book on proportions and who produced sculptures considered to be in perfect harmony, and, almost two millennia later, the studies of Leonardo da Vinci (1452–1519) or Albrecht Dürer (1471–1528) on the perfection of human proportions. The harmony of the human body also expresses the harmony of the world and the symmetry of the world-view in which man puts himself in the centre of the world (Figure 1.8). The symmetry of the (anthropomorphic) world is the embodiment of the belief in the perfection of that world.

We see ourselves (and humankind in general) as symmetrical, and this is how we try to depict ourselves, even if we know that the two sides of our face (and a number of external parts of our body) are not perfectly mirror symmetric, while inside our bodies the heart is slightly to the left of the middle line of the body, the liver rather to the right, and our digestive tract is located even more asymmetrically; indeed, we now know that the two hemispheres of our brain are different — not only in their morphology, but also in their function.

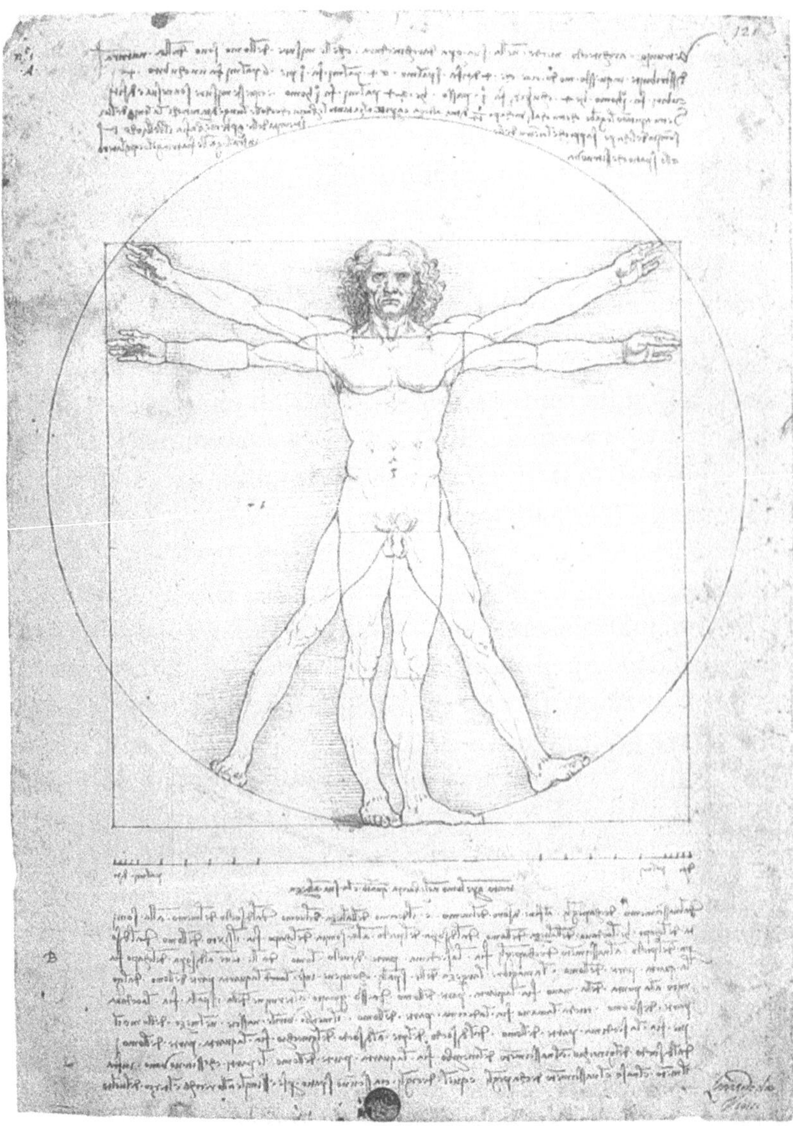

Figure 1.8. *Leonardo's drawing of the proportions of the human body*

A breaking of geometric symmetry can possibly increase the balance of another characteristic (for example, colours, tones, light and shade). Ikebana does not display geometric symmetry, for example, yet it still radiates the proportions of the different parts, and their harmony. Harmony is not lost, that is, should geometric symmetry be harmed; neither is it limited to that symmetry. Indeed, in and of itself, achieving geometric symmetry does not exhaust the possible perfection of a depiction.

Do perfect images exist at all?

If we stand facing one another, and both extend our right hands, our arms cross each other between our bodies. As our hands grasp each other, they mark a perpendicular axis of rotation between us, relative to which the two bodies demonstrate (twofold) rotational symmetry (let us ignore differences in personal characteristics for the moment). But let us step up to a mirror and extend a hand to our mirror image. In the plane of the mirror, the two hands do not cross each other, but form a straight line together. Opposite my right shoulder it is not the left shoulder of my mirror image that I see, but the right. The two bodies — the real one and the perceived one — now display mirror symmetry, with the right and left reversed. But which image should I regard as the reversed one? For I do not see the shoulders of the real-life body in front of me as reversed — I simply see its left shoulder opposite my right (Figure 1.9).

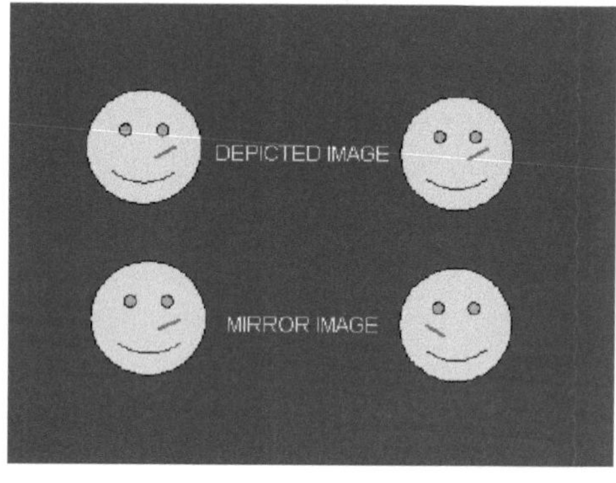

Figure 1.9. *Which is the more perfect likeness?*

Which is the more perfect likeness? The real person standing opposite, or the (perfect) mirror image seen in the mirror? On the photograph the mark on the skin is seen on the right hand side. In the mirror it is reflected onto the other side. Both pairs are symmetrical. But which is the real one? We cannot always give a clear and universal answer to this question. Should we give up our faith in the symmetry and thereby the perfection of the world? Instead of an immediate response, let us return to the earlier question of whether the world around us is symmetrical or not, and whether symmetry is a natural phenomenon at all.

From what we have seen so far, it seemed that, in some form or other, symmetry was natural. We certainly try to see it as such, and, in turn, to have the world seen as symmetrical. This is how we construct buildings, decorate things; this describes our works of art, our everyday objects and tools. The flowers in the garden are symmetrical, as are the leaves on the trees, the way animals' bodies are built, not to mention our own. Of course, each is only symmetrical in its own way: a certain symmetry violation is found in each. Even the most symmetrical building would seem lifeless without a flower in one of the windows of its nicely-arranged façade or without some windows being left open, or if they were all covered by uniform curtains, each equally straightened without being plaited. The six petals of a tulip are not laid out perfectly, and the wings of a dragonfly display a certain asymmetry. We are not shocked by these things; in fact, they only make the world more beautiful and harmonious.

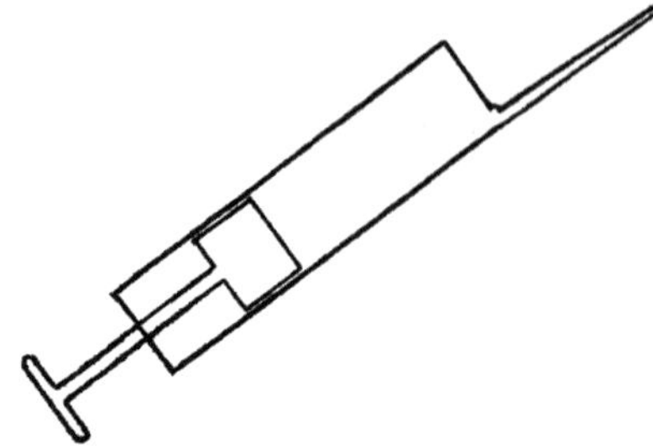

Figure 1.10. *Asymmetry is sometimes more practical: syringe with needle*

And sometimes more practical. For example, for a long time the only syringes that were produced were ones on which the needle continued the cylinder's axis of symmetry. If a doctor wanted to extract blood from the crook of our arm, it made no difference what position she held it in, the rotational symmetry meant that it would look the same in her hand. The needle did not nestle up against the skin and the blood vessels running in parallel underneath it, and so even if used with great care would penetrate the vein at an angle. Nowadays more practical asymmetrical syringes are produced for this purpose, where the needle extends from the mantle of the cylinder (Figure 1.10).

Let us put a photograph of a loved one on the table, and place an unframed double-sided mirror to it at right angles, touching the central line of the face. If we look from the left, instead of the right side of the face we see the left duplicated by the mirror. If we look from the right, on the other hand, we see the right side of the face doubled in the same way. The two faces looking out at us are not the same. Neither is identical to our loved one, not even to the photograph! Does the perfectly symmetrical loved one thus obtained seem as lovable as the original dissymmetrical one?

Whether we like it or not, the topic of symmetry-asymmetry has to some degree become part of our view of the world. We try to see the world as symmetrical, and if our experiences differ strongly from this, we are surprised, but not if the deviance is a minor one (dissymmetry). We are prone to suppose that the largely harmonically arranged world is symmetrical, or at least that it is governed by rules that strive to establish symmetry.

The astonishing asymmetry of the world

If at the outset I inspired some illusions about the perfect nature and symmetry of the world, I have only slightly dented this belief by drawing attention to minor symmetry breakings — the dissymmetry of the world — and yet, sooner or later, I will have to show that the world — if inspected more closely — can be astonishingly asymmetrical.

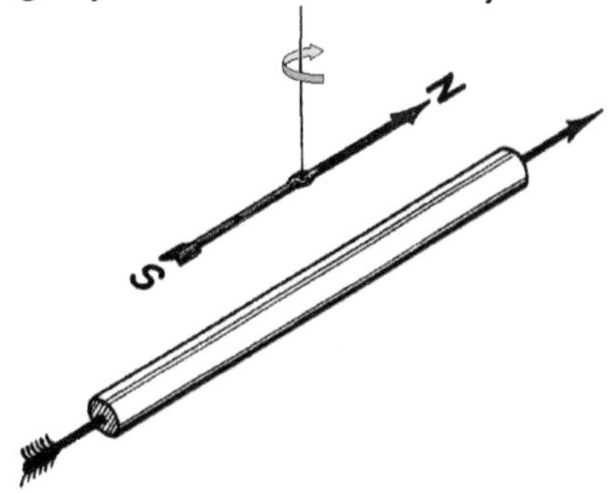

Figure 1.11. *The electricity flowing through a cylindrically symmetrical wire makes the compass deviate to the side. This physical phenomenon upsets the geometric symmetry of the system.*

Physicist (and non-physicist readers of *Symmetry* by Hermann Weyl [1885–1955]) are well aware of Ernst Mach's (1838–1916) memory of how as a schoolboy he was shocked by an experiment in which he observed that a magnetic needle placed in parallel above a wire conducting electricity would twist to the side away from the plane determined by the needle and the wire (Figure 1.11).

Looking along the plane clearly drawn by the electric cable and the compass, the arrangement appears to be symmetrical from both a geometric and physical point of view. Our faith in the isotropy of the world suggests that there should not be any particular distinguished directions in it. The two sides of the plane should be equivalent in space. In other words, there is no reason for us to think that anything should make a difference between the left and right sides of the plane. And yet, physical processes occur inside the two metals which lead out from the plane. Fixing the direction of the electric current and of the magnetization clearly determine the direction of the force generated in-between them. We will return to Mach's dilemma in relation to the interpretation of axial vectors. For the time being, let us stick to geometric transformations.

The reflection of geometric objects does not in and of itself explain the processes occurring when physical events are mirrored. The experimental layout is visibly made up of two objects. In addition, while for geometric objects we only took their geometric characteristics into account, this physics experiment has brought another property into the picture: electric charge. Like the pair of spatial directions to the left and right, electric charge also exists in two different forms, and by convention we regard one as the mirror image of the other, though this (+) ↔ (–) reflection has nothing in common with the interchange of the space directions. Or is its effect similar after all?

What, the question arises, should we consider to be the mirror image of the experimental layout? The reversal of the direction of the electric current? Or the reversal of the direction of the magnetization of the needle (switching the north and south poles)? Or do we have to reverse both for a mirrored arrangement? If we complete the reversals, will everything take place as if we were watching the original experiment in a mirror?

Spatial mirrorings and charge reflection can be achieved both separately and together. But the problem from Mach's youth did not arise from this. The explanation of the phenomenon is that the needle cannot be considered to be a one-dimensional figure. Its magnetic force is explained by the atomic-size circular flows around its axis, and when reflected spatially the directions of these is swapped. These elementary circular currents move in a given direction in a plane perpendicular to the direction of the needle — and to the electric cable originally parallel to it. Figure 1.12 shows the case in which spatial mirroring is pure, that is, when the charges are not reflected. The left hand diagram shows what happens if we look at the compass in a mirror. The right hand one shows the genuine physical situation when we rotate the elementary currents (real spatial reflec-

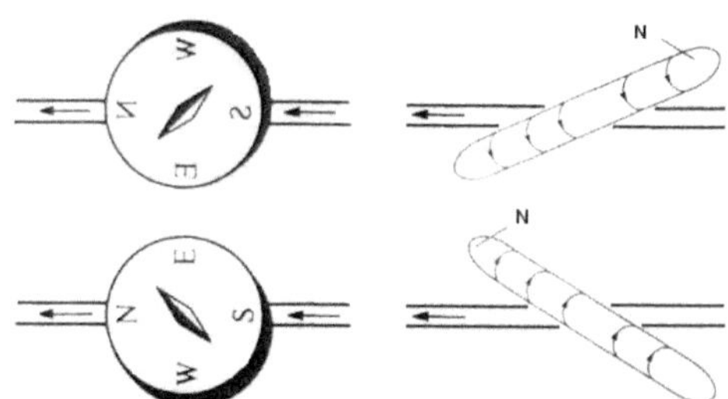

Figure 1.12. *Reflection of electric current and a compass*

tion), as a result of which the north and south poles of the compass are swapped.

At first sight, it seems as if the same thing has happened in both instances: the left hand end of the compass points upwards (to the right of the direction of the current) in the original position below, but in the upper, mirrored situation, it points downwards, i.e. to the left of the direction of the current. However, while on the left we see the north pole of the compass turn down (more precisely, to the left), on the right hand diagram, in the case of the real-life physical reflection in space, the pole of the compass has changed, and it is the (new) south pole that faces left and downwards.

Neither is the symmetry of the world simple when it comes to classical physical phenomena. The lesson of the interpretation of the experiment is that a plane without extension in the third dimension is the result of mathematical abstraction; real physical processes always occur in three-dimensional space.

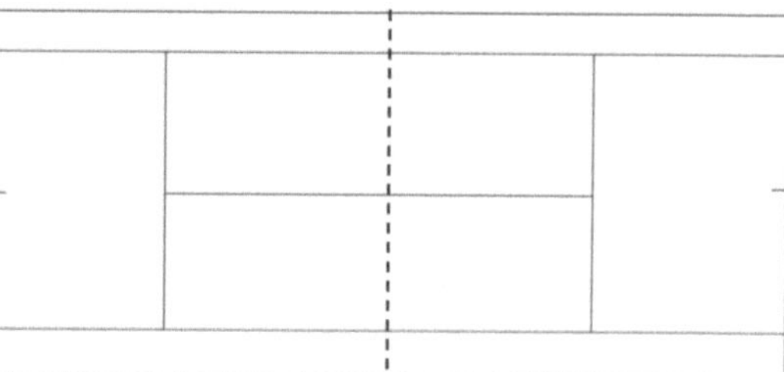

Figure 1.13. Tennis court

In the following I will provide examples of how symmetry is not natural even in instances where we would think this is self-evident. Figure 1.13 shows a tennis court.

The tennis court — we might think — is self-evidently symmetrical. It is mirror symmetric vis-à-vis the net, or rather at right angles to it, and displays twofold rotational symmetry around the court's centre. True, the court's mirror symmetry only holds until the two players step up to play the first serve. As the rules hold that the ball must be served diagonally across the court, herein on the court only displays rotational symmetry.

Why does it seem evident to us that a tennis court be symmetrical? For two reasons. First, so that the physical characteristics of the court present identical conditions for both players. Second, this is how we can satisfy the ethical requirement that the rules of the game be identical for both players. This is obvious to our current way of thinking, but was not always so self-evident. In the age of Henry VIII (1509-1547), when the ancestor of the game was developed, it was played in Hampton Court on a court that was asymmetric and in the shape of a trapezium. This gave the king the advantage, and the rules were written in his favour, too. The king played at the narrower end of the court, and on one side the ball could bounce back onto the court from the wall — naturally in the direction that was favourable to the king, while for his opponent only a backhand stroke would bounce back off this wall — while on the other side it would just leave the open court. The conditions were asymmetric from both a geometrical and ethical perspective.

Bridges are almost all symmetric from the perspective of the two banks of the rivers they span. This is even true of the Margaret Bridge in the capital of Hungary, which crosses the Danube at a point where the two banks form an angle: the bridge is bent in the middle so as to approach both banks at a right angle. The supporting structures of our bridges are symmetrical, too. We think of it as natural that a bridge be symmetrical. But this is not the case. We do not have to look far from Budapest: it is enough to think of the bridge connecting the centre of Bratislava with the southern bank of the Danube, the only cable-bearing pillar of which stands near the southern bank, asymmetrically (Figure 1.14).

The symmetry of nature is not broken by this; rather it is emphasized by it. It represents a counterpoint to the heights of the castle hill standing

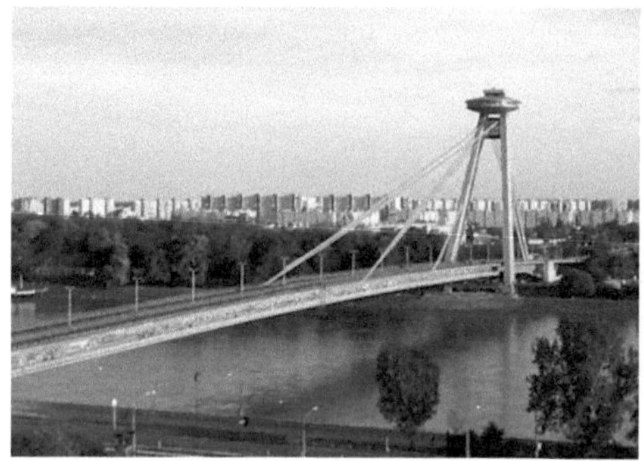

Figure 1.14. *The new Danube bridge in Bratislava*

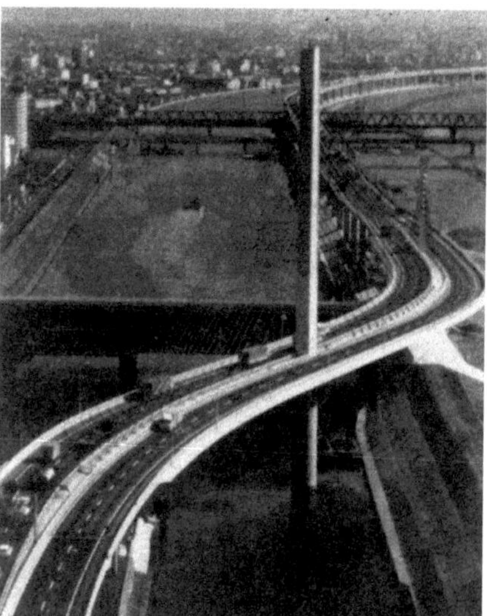

Figure 1.15. *Kyo Takenouchi's asymmetric bridge. Tokyo, Japan*

on the northern side. In Paris, not so far from the Eiffel tower, we can see a bridge crookedly spanning the Seine. The motorway leaving Tokyo towards the north follows a bridge that spans the river in asymmetrical fashion (a mirrored S shape) (Figure 1.15).

We have become acquainted with the most important geometric forms in which symmetry appears in our modern everyday lives. Before the euphoria of what a symmetrical world we live in had a chance to overcome us, we also learned that perfect symmetry does not exist. Symmetry refers to specific characteristics of the objects in question, but not to others. Following this, we also had to determine whether the world is full of astonishing asymmetries, and just as natural as symmetry is in certain phenomena, so is it far from natural in others.

The subject of this book is how our world, whether natural or man-made, is characterized by symmetry and its absence or violation. Let us first see how we can generalize our everyday concepts of symmetry so far only interpreted in geometric phenomena.

The generalization of the concept of symmetry

What did the geometric symmetries already presented have in common?

- In each instance we performed some kind of (geometrical) operation (transformation).
- In this process, *one or more* (geometrical) *characteristics of the figure* remained unchanged.
- This characteristic proved to be *invariant under the given transformation* (did not change as a result of the operation performed).

Let us generalize this in such a way that the interpretation be valid not only for geometrical operations and geometric objects, and not just for geometrical characteristics. In a generalized sense, we can speak of symmetry if

- *in the course of any kind of* (not necessarily geometrical) *transformation* (operation)
- *at least one* (not necessarily geometrical) *characteristic of*
- *the affected* (arbitrary and not necessarily geometrical) *object remains invariant* (unchanged).

The generalization, that is, took place with reference to three things:

- to any transformation,
- to any object,
- to any characteristic.

The first generalization made it possible for us not just to look for unchanged characteristics during geometrical *operations* we have learned (reflection, rotation, translation, etc.). This made it possible, for instance, for us to understand invariance under charge reflection in physics, and invariance under the swapping of colours in art. The second generalization makes our concept of symmetry capable of making any kind of *object* of science or art the subject of a symmetry operation. This paved the way, among other things, for us to use symmetry operations on the abstract objects of physics. Finally we allow the constancy of any *characteristic* to be considered as symmetry. Of the examples given so far, this was true of electric charge, but any physical quantity considered charge-like can be the object of symmetry, just as can the rhythm of a poem or the motif of a piece of music.

Related concepts

First of all, the other member of the symmetry family of concepts is *asymmetry*, used to denote the absence of symmetry. We speak of asymmetry when none of the characteristics of a given object displays symmetry.

There are instances where an object displays symmetry, but this symmetry is broken in one of its characteristics or a not too significant detail. Examples of this are a door handle, a bubble in a diamond, a freckle on a face. We refer to this as dissymmetry. We can see beautiful examples of dissymmetry in Figures 1.16, 1.17 and 1.18.

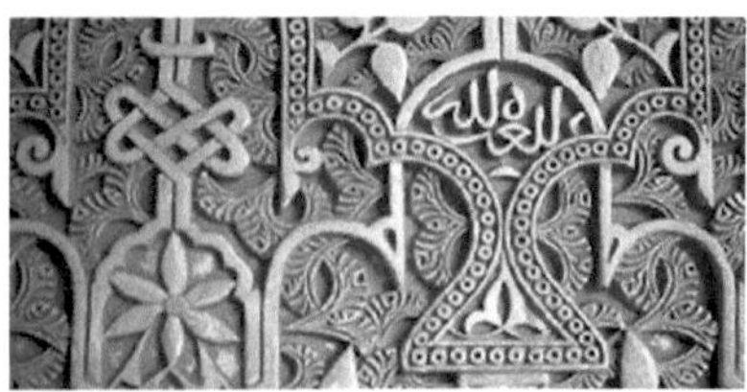

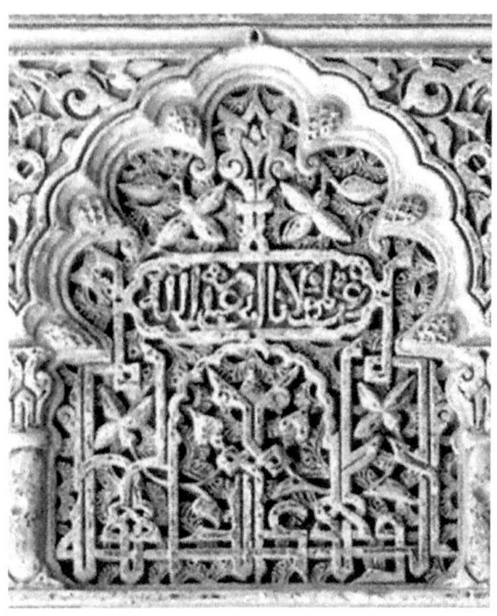

Figure 1.16. *Dissymmetrical wall details. Alhambra monastery, Granada, Spain*

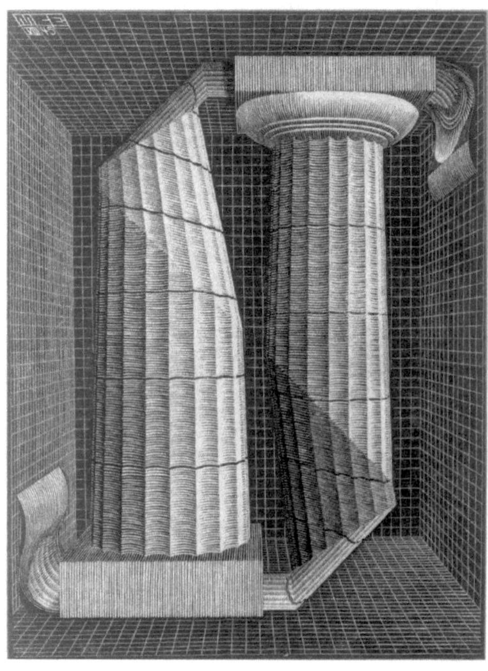

Figure 1.17. *Dissymmetrical columns. M.C. Escher: Doric columns (1945)*

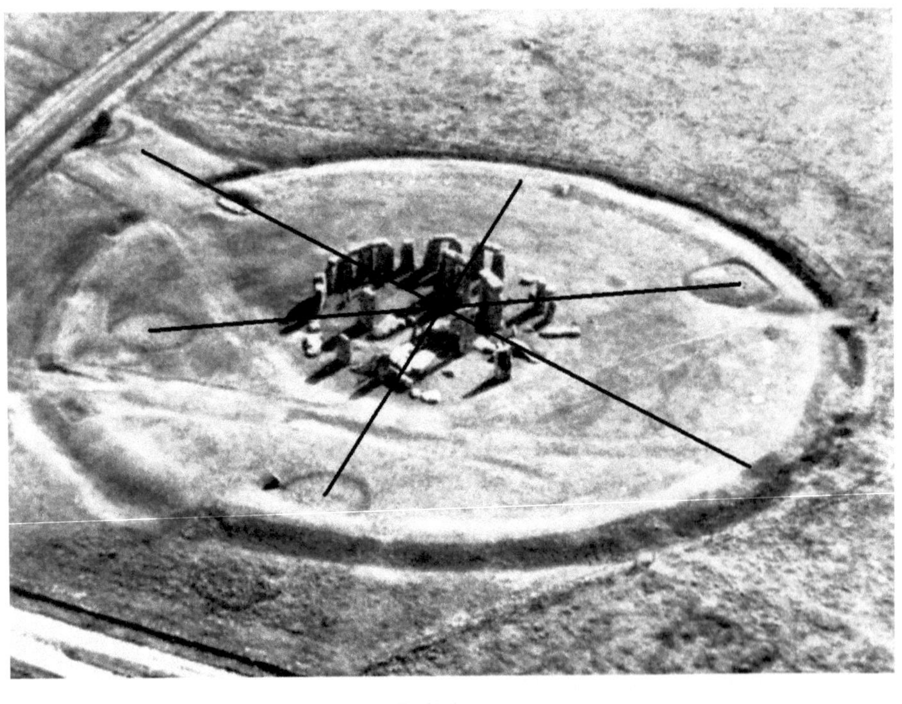

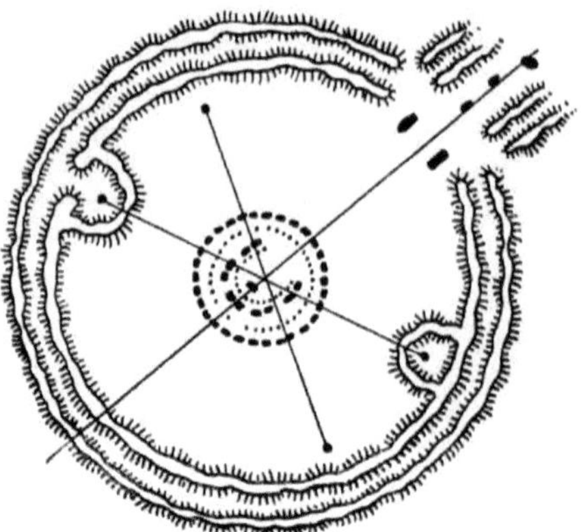

Figure 1.18. *Stonehenge's column order deviates slightly from complete symmetry (aerial photograph and ground-plan)*

One of the most beautiful symmetry-dissymmetry comparisons in world literature is given by Thomas Mann (1875–1955) in his description of the *"hexagonale Unwesen"*, the snowflakes in a snowstorm, in *The Magic Mountain*. "(...)Yet each in itself — this was the uncanny, the antiorganic, the life-denying character of them all — each of them was absolutely symmetrical, icily regular in form. They were too regular, as substance adapted to life never was to this degree — the living principle shuddered at this perfect precision, found it deathly, the very marrow of death — Hans Castorp felt he understood now the reason why the builders of antiquity purposely and secretly introduced minute variation from absolute symmetry in their columnar structures." (Transl. By H. Lowe-Porter, N.Y.: Knopf, 1927.)

The aforementioned related concepts to symmetry were defined in their current use by Pierre Curie (1859–1906). It is to him that the now famous phrase is attributed: "it is dissymmetry that makes he phenomenon" (*"c'est la dissymmétrie qui crée le phénomène"*, 1894). We can take this to mean that phenomena which are important for researchers to discover exist at the points where they encounter dissymmetry. Let me illustrate the significance of this claim with an example. The impure atoms present in miniscule measure in semiconducting crystals change the crystals' conductive characteristics to a degree measurable in macroscopic phenomena. It is the realization and implementation of this that made possible the creation first of transistors, then of the integrated circuit chips which are the soul of the apparatuses of our everyday lives. It would be hard to imagine today's world without them: the telephone would not work, and neither would radios, televisions, computers, the ATMs where we get our money, the cash registers at the supermarket where we get our food, to name but a few. All this from a handful of tiny dissymmetries.

The third related concept is that of *antisymmetry*. We talk of antisymmetry when a characteristic is preserved by being transformed into its opposite. A chessboard is antisymmetric, for example: if reflected, the white squares turn black, and vice versa. The shape of the well-known antisymmetric yin-yang (Figure 1.19) displays rotational symmetry, but if rotated 180° around its centre, black turns to white, and white to black; in terms of its colours, that is, it is antisymmetrical.

The Japanese equivalent of the antisymmetric Chinese yin-yang is the symmetrical *tomoe* (Figure 1.20). The outline with rotational symmetry

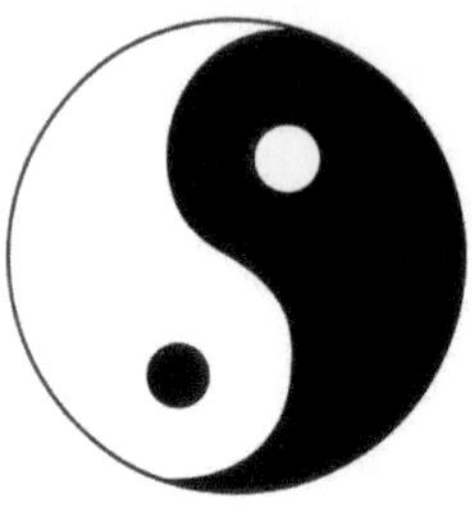

Figure 1.19. *Yin-yang*

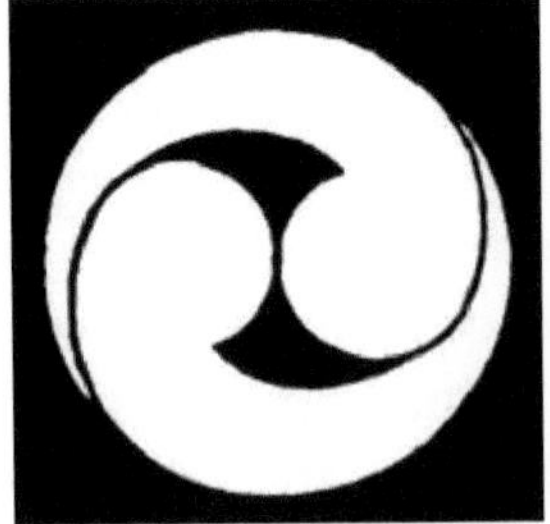

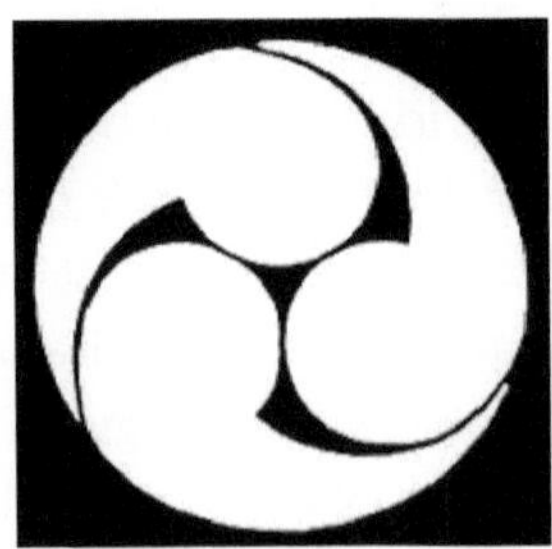

Figure 1.20. *Double and triple tomoes*

made up of identical elements makes it possible for it to be depicted in threefold and indeed manifold versions. Like its antisymmetrical cousin, it has cultic significance in its homeland.

The yin-yang symbol developed from the motif of fishes biting each other's tails. Some symmetrical motifs developed in different cultures independently of one another, and this motif is an invariant element of many human cultures. Ceramic designs similar in age and form to the Chinese plate shown in Figure 2.4 have been found in Crete, Knossos and the nearby Kamares cave, dating back to the 17th–14th centuries B.C. — when, as far as we know, there cannot have been any connection between the two cultures — and displaying two dolphins biting each other's tails, and in the Mexico City Basin from the middle prehistoric period (see Chapter 2). (According to different archaeological hypotheses, the common motifs can have dispersed from Mesopotamia in an earlier period.) In Crete, twisting whirlpool motifs were formed from these designs, partly connected on surfaces as wallpaper motifs, partly along a straight as frieze motifs. It was from these that the *meander* (the word is derived from a

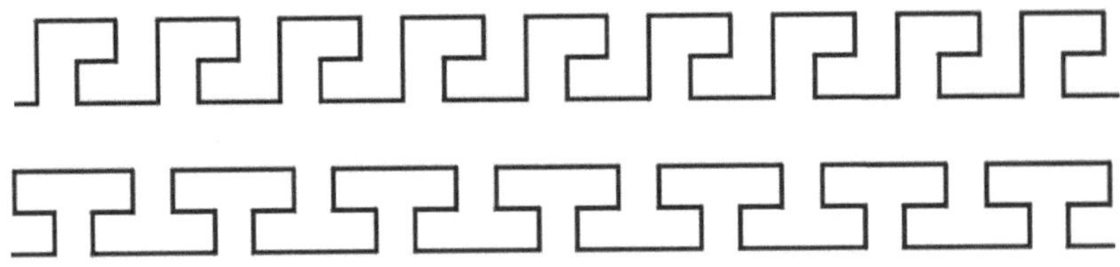

Figure 1.21. *Meander motifs*

river in Asia Minor whose course is winding) series of motifs developed, becoming increasingly simple, then rectangular.

The meander motifs belong to the classical motifs of decorative culture. (The Cretan development of dolphin and whirlpool motifs is discussed in detail in books by K. Czernohaus [1988] and G. Walberg [1987].) We encounter similar frieze motifs on Mexican bowls. A change of motifs towards the development of frieze motifs similar to the development of meander also took place in China (Figure 1.22).

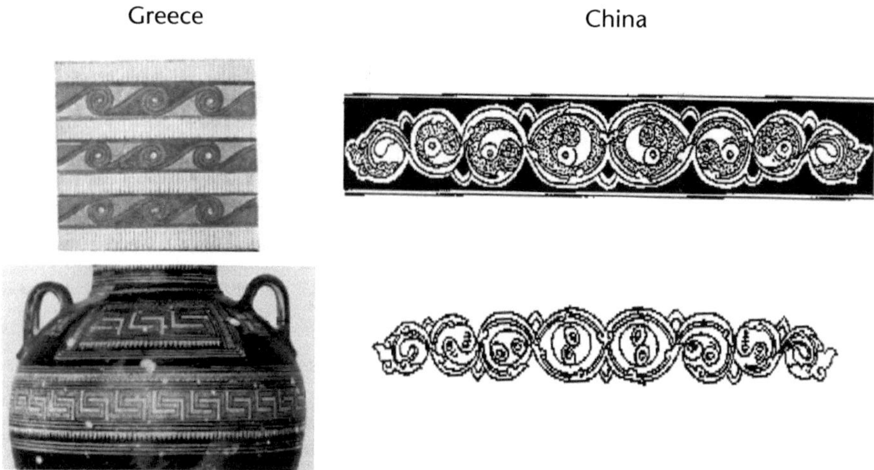

Figure 1.22. *Parallels in meander and related designs between Europe and China*

The connection between whirlpool and fish motifs also inspired Escher in the twentieth century (Figure 1.23).

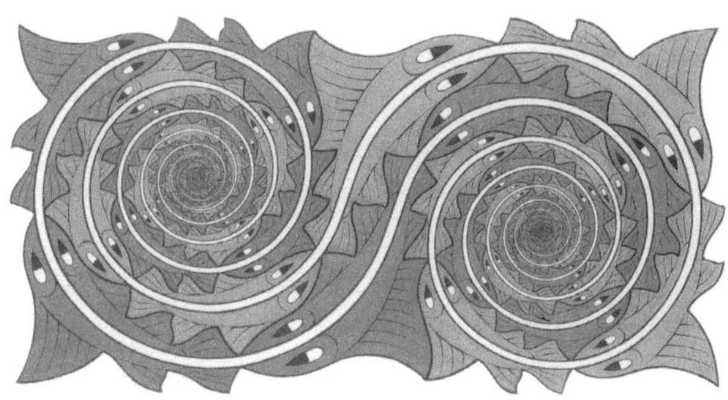

Figure 1.23.
M. C. Escher: Whirlpools (1957)

According to the dichotomic European tradition based on pairs of opposites, the main dividing line was between symmetry and asymmetry — or the lack of symmetry.

In Far-East thinking, the same mark of contrariety was drawn between symmetry and antisymmetry. In their universe of symbols, antisymmetry displayed symmetry in a *geometric* sense, but pairs of opposites in some of its *non-geometric* properties (which could be physical, e.g. black and white, hot and cold, or intellectual in nature, e.g. good and bad, wise and stupid, masculine and feminine). For example, the yin-yang (Figure 1.19) symbolized the unity of the opposites in both Tao and Confucian philosophies, originating at around the same time (6th century B.C.): rotating the figure 180° around its centre point, the geometric shape (form) turns back into itself, but the white turns to black, and the black to white. Chinese philosophical works relating to the yin-yang used antisymmetry to describe many dichotomies, the dialectics of inorganic and organic natural, social and intellectual (ethical, conceptual) substances. Here symmetry and antisymmetry appear closely united. It is through the interpretation of the dichotomies associated with them that they became part of our world-view.

Nowhere is the world perfectly symmetric or antisymmetric. Where the deviation is small in scale, we speak of dissymmetry; where it is greater, or where symmetry is hardly measurable, we refer to it as asymmetry. Can an asymmetric phenomenon be completely free of symmetry? The answer is surprising: no.

The world, while appearing to us to be asymmetric in general, can be regarded as the unity of symmetry and antisymmetry. The scholarly foundations for this claim originate not just from philosophy, but partly from the method of mathematical description applied to quantum physics: the operators describing the world's physical phenomena can always be broken down into a symmetrical and an antisymmetric component. This follows from the fact that our physical operators can mathematically be represented by matrices, and matrices possess this mathematical characteristic. It would be one-sided for us to force the approach of physics onto our view of the world. Neither is there any need. The method can be extended. Everything that can be represented by points arranged in matrix-like fashion — the grid points of a crystallized material, the pixels of a digital image, etc. — can be broken down (with the use of specialized scale transformation procedures) into symmetric and antisymmetric components along its arbitrary axis of symmetry, if the elements of the matrix used are formed such that the value of a quantifiable characteristic attributed to the given points is applied to them (e.g. the colour intensity codes of the given pixel, etc.). By the antisymmetric pair of an image point we mean the point of complementary colour (marked by the complementary colour code) that appears at the respective opposite point as reflected in or rotated around an arbitrary axis. As a result, everything that can be depicted pictorially can be broken down into symmetrical and antisymmetrical components. (This is the basis for the author's method patented in 2003.) For we really do "see", "watch" and "observe" the world with the sockets of our retinas, and our retinas forward the information, as image point codes separated by colour, to the respective part of our brain responsible for processing and interpreting images.

The two parts of the picture artificially divided into symmetrical and antisymmetrical components reproduce the original when projected onto

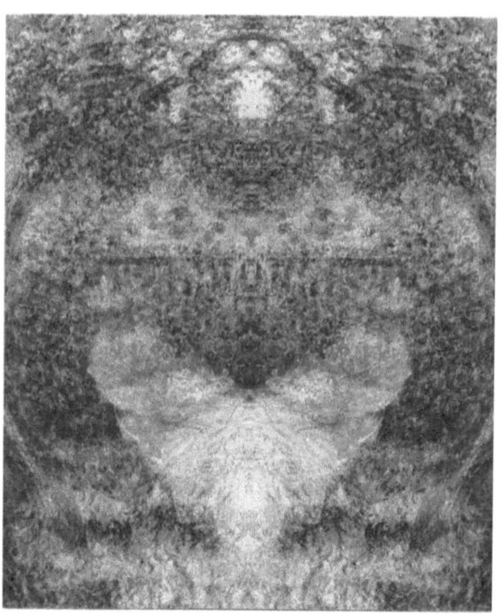

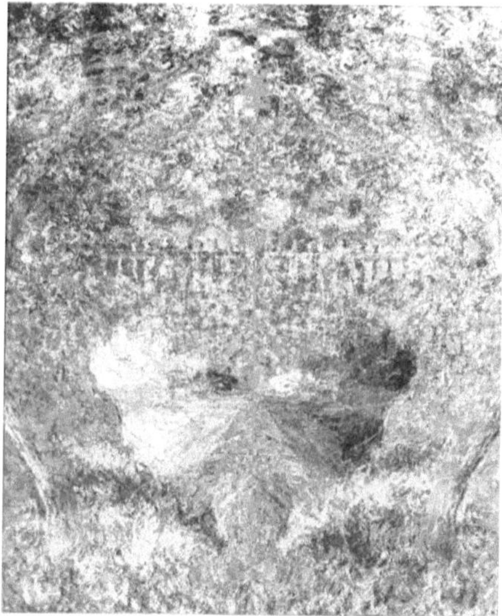

Figure 1.24. Symmetrical image **Figure 1.25.** Antisymmetrical image
on the two sides of the perpendicular mirror axis, the colour of the respective mirror points is
identical *complementary*

one another. The antisymmetric image component displays the deviations from symmetry of the original image. This is the basis for the technological applications made possible by this method.

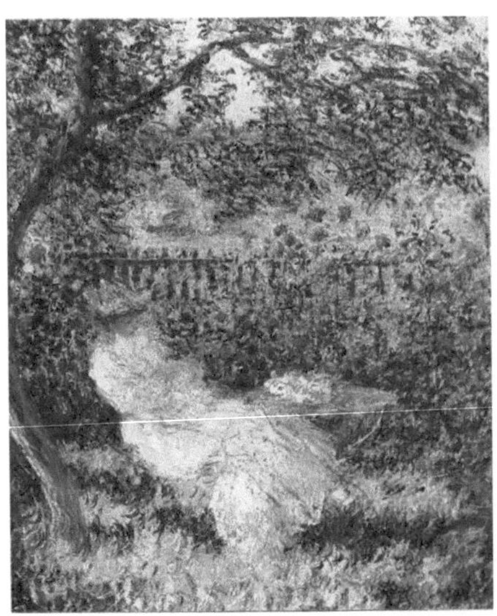

Figure 1.26. *C. Monet: Alice Hoschedé in the Garden* (1881)

If we copy the symmetrical and antisymmetrical images seen in Figure 1.24 and 1.25 onto transparent material and place one on top of the other, we are given a reproduction of the original. In Figure 1.26 we can see which (asymmetric) work of art we split into two image components, each in itself hard to identify with the original. The method can be performed with any picture: all images can be assembled from a symmetrical and an antisymmetrical component.

To summarize, we can say that according to our present knowledge asymmetry is the synthesis of symmetry and antisymmetry.

Mathematic description of symmetries

Both geometric symmetries and invariances under other types of change are phenomena that can be dealt with using a mathematically unified theory. The more unified a description we seek for characterizing symmetry (one with which we can mathematically model as many or even all of the theoretically possible symmetries in existence), the higher the level of abstraction required. For this reason, I apologize in advance to readers less versed in mathematics for briefly burdening them with a few concepts which exceed that learned at high school, but which, for those more at home with the exact sciences, will remain at the elementary level.

The concept of a *group* has proved to be the most useful tool for the mathematical description of symmetries. A group is an algebraic structure, and is defined as follows:

(1) A *group* is a set of *elements*
(2) which have an *operation* applied to them, which is subject to:
(3) the group *axioms*:
   (3.1) closure, i.e. the operation cannot bring a result outside the group;
   (3.2) associativity, i.e. operations can be grouped together;
   (3.3) the existence of an identity element (neutral element);
   (3.4) the existence of an inverse element.

In order to show what this all means and how it can be used to describe symmetries, we will formulate first in mathematical form, then in words, before finally applying them to simple examples.

The formalized group axioms:

(1) $\exists\, \boldsymbol{G}$, a (non-empty) set, where

(2) $\forall a, b \in \boldsymbol{G}2$ ordered pair of elements has an element $a \circ b = c$

unambiguously assigned to $\boldsymbol{G}$ for which

(3.1) $a, b \in \boldsymbol{G} \rightarrow a \circ b, b \circ a \in \boldsymbol{G}$

(3.2) $a, b, c \in \boldsymbol{G} \rightarrow (a \circ b) \circ c = a \circ (b \circ c)$

(3.3) $\exists e, \rightarrow \forall a \in \boldsymbol{G} \rightarrow e \circ a = a \circ e = a$

(3.4) $\forall a \in \boldsymbol{G} \rightarrow \exists$ an inverse element $a^{-1} \in \boldsymbol{G} \rightarrow a \circ a^{-1} = a^{-1} \circ a = e$

(Here $\exists$ is taken to mean "there exists", the meaning of $\forall$ is "all", $\in$ denotes "is an element of the following", while a^{-1} denotes the inverse of the element a.)

Put in words: anything can be an element of a group. The elements of a group can be numbers, but can also be certain operations or indeed anything else. This ensures that symmetry can be interpreted in any set of phenomena. The operation defined for the elements of the group (which we have symbolically denoted with "$\circ$", to differentiate it from the symbols for all well-known mathematical operations) can also be anything (not only basic arithmetical operations) that satisfies the requirements of the four group axioms.

The first group axiom states that if we apply the group operation to two elements of the group, thus creating a third element, this must also be an element of the group: the operation cannot result in an element outside the group. The second states that if we apply the operation twice in a row to three elements all grouped together, it will lead to the same result on both occasions. Associativity does not imply commutativity! We do not require of the elements of the group that the equation $a \circ b = b \circ a$ be universally true. Commutativity is not a group axiom! The third axiom states that there exists a neutral (or identity) element which with the group operation turns an element into itself. Finally, the requirement of the existence of an inverse element means that for all elements there exists another element such that application of the group operation to them results in the identity element.

In our first example, let our elements be the integers. The set of integers generates a group for the operation of addition, for it satisfies the group axioms:

(1) the sum of two integers is an integer,

(2) addition is associative: $(a + b) + c = a + (b + c)$,

(3) there exists a neutral element, 0, for which $a + 0 = 0 + a = a$, where a is any element of the group,

(4) $(-a)$ is the inverse element for a, because $a + (-a) = (-a) + a = 0$.

Similarly, it can be seen that the *set of rational numbers* also represents a group with regard to addition. This group expresses the fact that we can perform translations along a number line that will be independent of where along the line the starting point was chosen. The group of rational numbers is the same group that in geometry characterizes invariance to translation.

For our second example, we show that the *rational numbers other than zero* represent a group with regard to multiplication. The identity element is 1, while the inverse of an element is its reciprocal. The group axioms are satisfied as follows:

(1) $a * b$ is also a rational number,

(2) $(a * b) * c = a * (b * c)$,

(3) $a * 1 = 1 * a = a$,

(4) $a * \frac{1}{a} = \frac{1}{a} * a = 1$.

We can depict the groups with matrices — generally with finite elements — which depict the projection of the group's elements onto one another according to the defined group operation. As an example, let us take the group of *rotations on a plane.* From high school mathematics we can remember that within a planar coordinate system the rotation of a point around the origo by the angle α is described by the matrix

$$\begin{bmatrix} \cos\alpha & -\sin\alpha \\ \sin\alpha & \cos\alpha \end{bmatrix}.$$

Similarly, a rotation by the angle β is

$$\begin{bmatrix} \cos\beta & -\sin\beta \\ \sin\beta & \cos\beta \end{bmatrix}.$$

If we perform the two rotations in succession, then the given point will be rotated by the angle $(\alpha+\beta)$. Symbolically, we denote the implementation of the two rotations in succession as the operation "$\oplus$". Let the first rotation we perform be of angle α, and the second of angle β. Then:

$$\text{Second rotation } (\beta) \oplus \text{ First rotation } (\alpha) = \text{Resulting rotation } (\alpha+\beta).$$

We know that in group operations we cannot — in the absence of commutativity — change the order of the elements. We notice that we wrote the rotation with angle β first and the rotation with angle α afterwards. By convention, the order in which we perform operations is first the one directly before the equals sign, then the one before it. The fixed order states that we implement the rotation with angle α first, and this rotated point will then be acted upon in the next step by the rotation with angle β. Applying the operations to the matrices of the individual rotations

$$\text{Second rotation } (\beta) \oplus \text{First rotation}(\alpha) = \text{Resulting rotation}(\alpha+\beta).$$

$$\begin{bmatrix} \cos\beta & -\sin\beta \\ \sin\beta & \cos\beta \end{bmatrix} \oplus \begin{bmatrix} \cos\alpha & -\sin\alpha \\ \sin\alpha & \cos\alpha \end{bmatrix} = \begin{bmatrix} \cos(\alpha\beta) & -\sin(\alpha\beta) \\ \sin(\alpha\beta) & \cos(\alpha\beta) \end{bmatrix}$$

and taking into consideration the way in which the circular functions of the angle sums are identical, we see that on the right hand side of the equation we get the matrix of the rotation through angle $(\alpha+\beta)$ from the two matrixes on the left hand side by using the algebraic multiplication of matrices ("$*$") as the group operation denoted by the symbol "$\oplus$". It takes elementary arithmetic to see that planar rotations do in fact represent a group, because they satisfy the group axioms, in which the identity element is rotation by an angle of 0 (which is equivalent to the identity matrix $\begin{bmatrix} 1 & 0 \\ 0 & 1 \end{bmatrix} = \begin{bmatrix} \cos 0 & -\sin 0 \\ \sin 0 & \cos 0 \end{bmatrix}$), while the inverse element is a "counter"-rotation through the angle ($-\alpha$).

Dipping into the rich literature on symmetry we can hardly avoid concepts describing or at least relating to group theory. To make this easier, we now touch upon a few special concepts we can often come across. Depending on the number of group elements, we speak of *finite* or *infinite* groups. In instances in which the group operation proves to be commutative, we refer to the group as an *Abel* (1802–1829) *group*. If there is a correspondence between every element of $\boldsymbol{G}$ and one of the elements of $\boldsymbol{G'}$, we call this *homomorphism* between the two groups. If this equivalence is mutual, we call this *isomorphism*.

Let us take a square as our next example. Intuitively we think of a square as naturally symmetrical, but why is it so, mathematically speaking? We can show eight basic operations which bring a planar square into equivalence with itself, such that we cannot distinguish the figures before and after the operation. These operations are as follows: reflection in the lines drawn between the midpoints of opposite sides (2), reflection in the lines drawn between opposite vertices (2), and planar rotations around the geometric centre around 90°, 180°, 270° and 360° (4). To these we can also add the operation of leaving the square untouched (neutral element). We consider the inverse operation to be any of the above operations in reverse. We can see that, together, these operations represent a group, because they satisfy the group axioms. They form the symmetry group for the square. We can characterize the square and its respective symmetry group with the smallest number of basic operations with which all of its reflexive projections can be represented. Together, we term these operations the *generators of the* (square as a shape and its associated) *group*. We refer to the matrices which describe the individual operations (in this particular instance the reflections and rotations through the given angles, i.e. the elements of the symmetry group) as the *representation of the group*.

Symmetries and invariances under transformations appear in the world in the most varied forms. There are common criteria that group them together in a single set of phenomena. These include less universal ones and those valid in a wider sense. We treat the latter as laws. Certain laws of symmetry apply to a group of phenomena, for example crystallography (Th. Hahn, 1985), the world of molecules (D. Avnir, 1998), physics, or more generally the inorganic natural sciences (J. Rosen, 1995), while others hold true for the whole course of the evolution of matter (Sz. Bérczi, 1980; Gy. Darvas, 1998). The Swiss mathematician Andreas Speiser (1885–1970), who provides a detailed group theory discussion of ornamental symmetries in his book *Die Theorie Der Gruppen von Endlicher Ordnung*, argues that the presentation of the five regular bodies was the greatest achievement of deductive geometry in the system created by the Greeks and canonized in the *Elements* of Euclid (c. 325 BC–270 BC).

Is symmetry a mathematical concept?

In the light of the exact conceptual and mathematical apparatus outlined above for its examination, the reader might well assume that symmetry is indeed a mathematical concept. But this suspicion is eliminated if we consider the broad sphere of phenomena in which symmetry appears, and whose rich store of treasures we are going to become acquainted with in the following chapters. The diversity and richness of these phenomena is not narrowed by describing them by similar means. The head, the feet, the kidneys or the heart do not lose any of their peculiarity just because we might treat diseases relating to all of them with medicines, most of which are in tablet form and distributed in rectangular boxes. The common form disguises differing active ingredients and the treatment of differing diseases, but the medicines do also have an invariant characteristic, namely the objective of healing. Naturally doctors also use medicines on themselves. Mathematics is one area — indeed a very rich area — in which symmetry phenomena appear, but only one of many.

Before continuing, let us make a brief and not necessary comprehensive survey of how various academic disciples discuss symmetry and how some artistic fields present it.

- In *physics*, the permanence of the laws of nature and their connection to symmetry principles, variational principles; the list of conservation laws, study of spontaneous symmetry breakings, and the symmetries and asymmetries of the structure of matter;

- in *earth sciences* and *astronomy*, the investigation of periodical events and the preparation of calendars;
- in *crystallography*, regular filling of space, examination of spatial arrangements, study of crystal distortions (dislocations), quasi-periodic arrangements;
- in *chemistry*, the form of molecules, with particular attention to the spatial structure of more complicated ring and chain molecules, and macromolecules (including the examination of enantimoer pairs);
- in *biology*, the structure of DNA and RNA molecules, that of protein, morphological and functional symmetries in living creatures, the relationship between evolution and symmetry;
- in *psychology*, the regularities of the conscious development of the individual, personality tests, as well as brain asymmetry and its wide-ranging consequences;
- in *music*, the rules of rhythm, melody, and the architecture of the musical work;
- in *literature*, rhyme, meter, the structure of the literary work;
- the movements of *dance*;
- in *fine art*, proportions, perspective, and harmony of proportion and colour;
- in *applied art*, the harmony of the proportions between the object and the person using it, the harmony of form and function;
- proportions and form in *architecture*;
- in *ethics*, virtuous behaviour, finding moderation, the middle way;
- in *logic* and *philosophy*, the form and order of thought, and the role of symmetry in developing a world-view;
- in *economics*, balance, asymmetric decision-making in game theory, simulation.

Chapter 2

Historical Survey

The appearance of the concept of symmetry; its substantive expansion and applications

The first chapter acquainted us with the concept of symmetry. We saw what significant substantive changes the concept itself has undergone over the millennia. We now survey the study and depiction of the phenomenon over the ages.

Like the concept, the phenomenon of symmetry has been an important bridge between the arts and the sciences. It is no accident that symmetry played a key role in exactly those periods which we regard as the golden ages of human culture. I use this term to refer to the periods in which art and science flowered at about the same time, when works of particular significance emerged from both areas. There have not been many such eras. We acknowledge perhaps three such special periods in the history of European culture: the Greek golden age (approximately the end of the 6th century and the 5th century BC); the Renaissance, particularly the Quattrocento and Cinquecento (fifteenth-sixteenth centuries AD); finally, we can perhaps dispassionately include the present era (twentieth century and early twenty-first century AD), though this will only be decided for certain by posterity. Symmetry emerged at an early stage of human development, at the same time as culture, and entered the symbol system almost immediately. The mirror symmetric dolphin motif, for example, which we see in Figure 2.2, existed earlier in its rotationally symmetric version, closely associated with the whirlpool motif that emerged from it. As we have mentioned, according to a number of archaeological schools of thought, this process took place in distant cultures, independently of one another. The bowl with a whirlpool motif in Figure 2.1, for example, was found in China. The interdependence between symmetry and human culture can be linked to the development of *homo sapiens* as a unified species. The separation of left-handedness and right-handedness occurred in the very earliest stage of development into human beings. We can regard the lateralization of the human brain in the early Stone Age as a completed process of development, and after the motor centre the speech centre also found its dominant position in the left hand temple lobe, and it appears certain that the functional separation of the two halves of the brain also took place — which can be traced back to anatomical (morphological) differences. We can also see, at this early stage, man's instinctive attraction to symmetry. What, exactly, is the nature of the symmetry under discussion? In their first approach, our ancestors were probably looking for harmony. The form and subsequently decoration of their everyday utensils attest to this; harmony originally had a cultic function, and only later became in itself an objective satisfying an aesthetic need. With time, this harmony demanded order and orderliness, and so figures displaying translational (repetitional), mirror and rotational symmetry became elements of the decoration. The symbol system has common elements to be found in all cultures. It appears that these developed in different cultures in parallel, independently from each other, and so we have every right to suppose that they represent some common element in human culture. As our example chosen from four thousand years ago shows, we can find almost exactly the same decorative motifs in the Far East, in Mesopotamia, the cultures of the Mediterranean, and — from not much later — the American continent.

On the bowl shown in Figure 2.1 (Yang-shao culture, c. 2000 BC) we see motifs very similar to bowls found in Crete from the same period, or ones found in Mexico from a few hundred years later. Centuries before the development of the silk road or the discovery of America, acceptable hypotheses state that, without the conditions required for transportation, the cultures of these distant parts of the globe cannot have influenced one another, though there are suggestions that even in the preceding couple of millennia, common motifs can have spread from Mesopotamia in various directions. In Figure 2.2 we see a fresco, which is very similar to the floor decoration found at the same place.

Similar patterns had previously appeared on bowls in Egypt and Crete. The appearance of a dolphin as a symbol would not in itself be surprising for a people living by the sea. What is surprising, however, is the similarity in the utilization of the fish motifs in China just as in America (Figure 2.3). The Megarón dolphin motif is widespread in the Greek world in its rotational symmetry as well as mirror symmetry form.

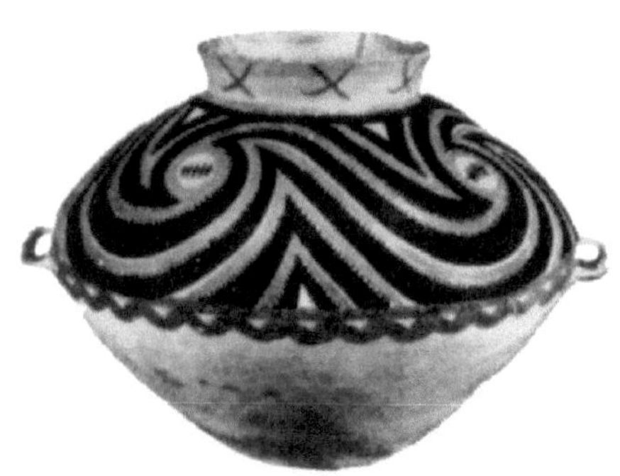

Figure 2.1. *Neolithic bowl.* Yang-shao culture, c. 2000 BC

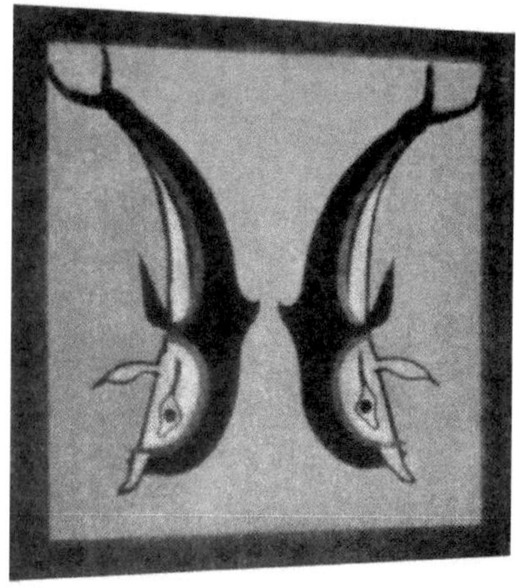

Figure 2.2. *Fresco,* Megarón, Tiryns, c. 1200 BC

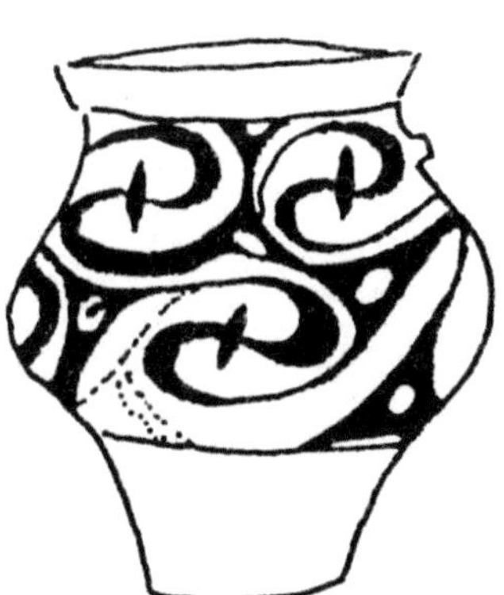

Figure 2.3. *The appearance of the dolphin motif in the Tripolis culture in North Africa c. 4000–3500 BC*

It is almost the same two dolphins biting into each other's tails that appear on the relief design on a bronze plate in ancient China as on a painted ceramic Cretan plate from the 17th–16th century BC, or a ceramic plate found in the Basin of Mexico from the middle prehistoric period between the 10th and 6th centuries BC (Figures 2.4, 2.5, 2.6).

These symmetrical symbols gradually went on to pursue their own independent paths. In China, the dolphins biting each other's tails turned into a decoration that increasingly filled the circular plate, then finally became the symbol of the yin-yang. The antisymmetrical design of the yin-yang became a symbol for pairs of complementary opposites, accom-

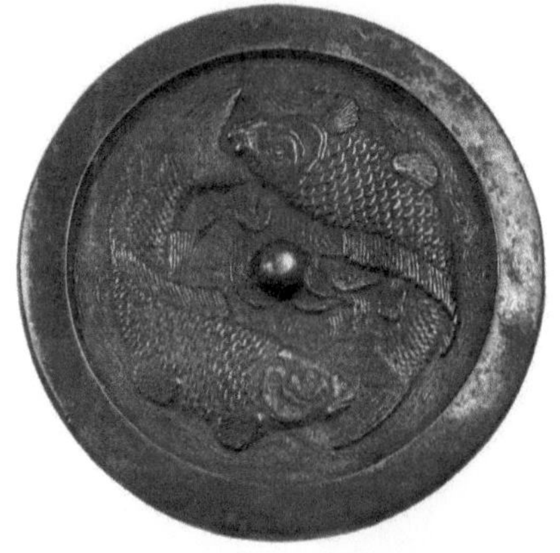

Figure 2.4. *Ancient Chinese dish, with two fish biting each other's tails*

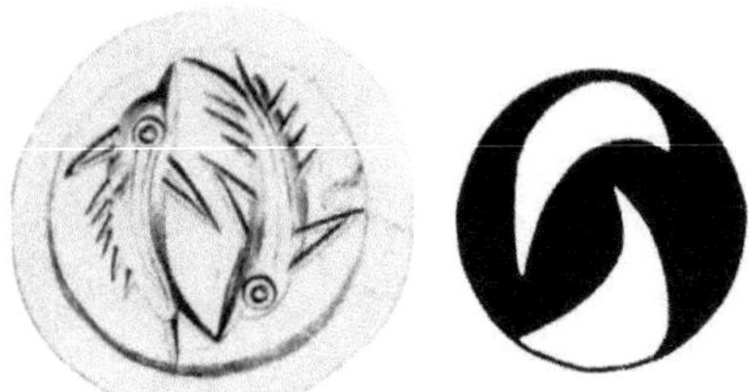

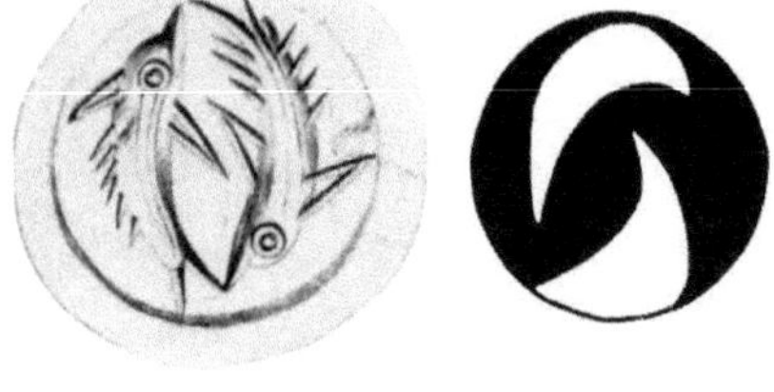

Figure 2.5. *Design of dolphins biting each other's tails, as it appears on the decoration of Cretan ceramic dishes from the 17th–14th centuries BC (K. Czernohaus) and in a stylized version (G. Walberg, right)*

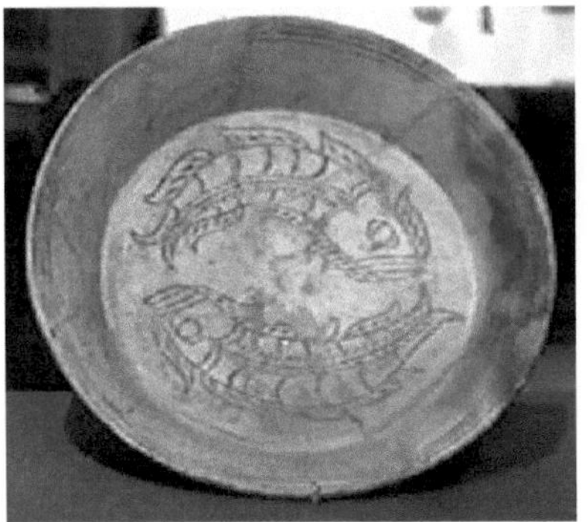

Figure 2.6. *Ceramic plate with two fish biting each other's tails. Basin of Mexico, 10th–6th century BC*

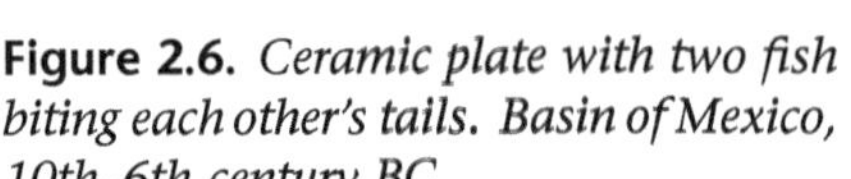

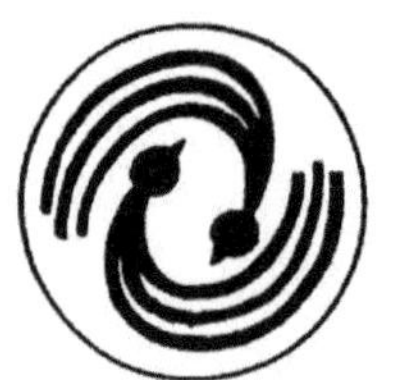

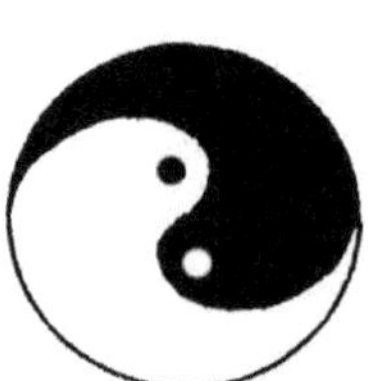

Figure 2.7. *Some stages in the development, according to Chinese sources, of the fishes biting each other's tails into the yin-yang*

panied to this day by countless cultic, philosophical and other scholarly explanations (Figure 2.7).

Much the same motif has existed and continues to exist in Japanese culture, in a slightly different, symmetrical form; this is the tomoe, mentioned above. The tomoe does not fill the circular space without leaving gaps, meaning that the background is given the opposite colour, and that antisymmetry is replaced by symmetry. This made it possible for the tomoe to appear in Japanese religious symbolism and later in family coats of arms not only in twofold, but also threefold and indeed multifold symmetrical versions (Figure 2.8).

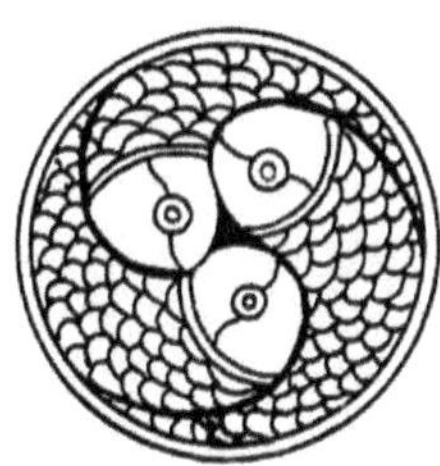

Figure 2.8. *The Chinese predecessor of the threefold tomoe*

In Greece, the dolphins biting each other's tails have become wound up into a ball, and these balls have been linked together to produce frieze designs and designs covering surfaces in groups of three. Linked together in series, the balls later became rectangular, before being reduced to the classical meander motif (Figure 1.20). Surprisingly, the association of this symbol with a ball-shaped frieze design is also found in Mexico. We first see it in the form of two fish twisted into a ball and filling a plate in an antisymmetric and continuous fashion (c. 850 BC, Chalchihuites), then connected in a series (Xicalcoliuhqui), and afterwards in the Teotihuacan culture in the period from 200 BC to 800 AD, for example on the frieze to be found on the lower part of statue of Chalchiuhtlicue, the goddess of corn and water. Large numbers of glazed, painted clay bowls with a circular-arc "meander" design running around them have been found in Mexico, and equally in Greece, and the same types of design can be seen on Etruscan gravestones. The development of the meander motif is an interesting example of how the dominant role played in decorative art by rotational symmetry was slowly taken over by translation symmetry. Some stages in the development of this series can be seen in Figure 2.9, from Kamares in Crete (after Walberg).

The few examples given illustrate how certain rotation and translation symmetry motifs are invariant elements of human culture as it developed in a variety of locations and eras. It is mirror symmetry that has achieved the most impressive cultic career, however. Mirror symmetry is also an

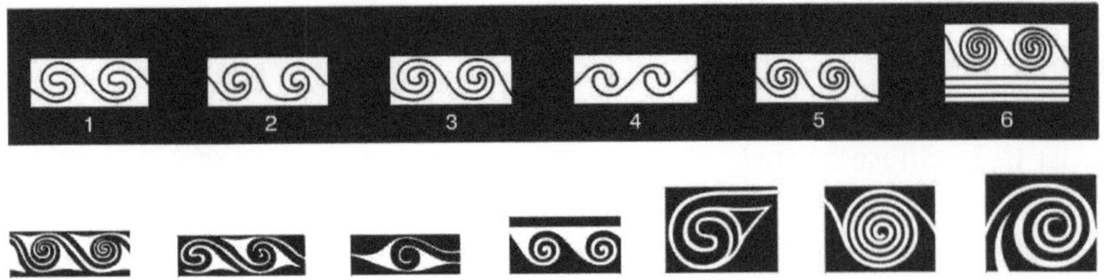

Figure 2.9.

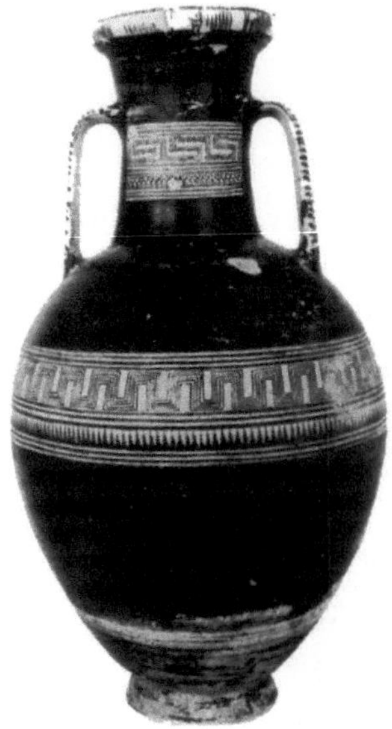

Figure 2.10. *Greek vase with meander designs from the early geometric period*

ancient symbol, which became sacred over time, one that was raised from an everyday decorative element to a royal symbol. Even Weyl makes mention of the role of two-sided or mirror symmetry in the symbols of rulers in antiquity (Figure 2.11). In Sumerian insignias (ruling class symbols), for example, it appears alongside translation symmetry. It is worth noting the symmetry violations. On the upper row on the silver bowl seen in the figure, the goats of the lower row are replaced by deer; indeed, even within a row, the lions are not symmetrical.

This small-scale violation of the symmetries is an example of the concept of dissymmetry that we have already explored. Dissymmetry is not only a dependable accompaniment to art; this is also a sign that the phe-

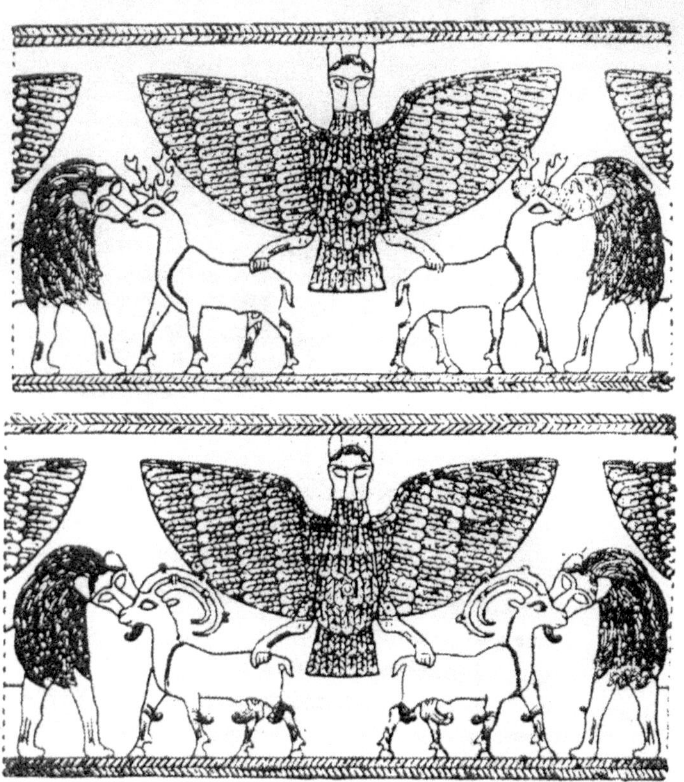

Figure 2.11. *Sumerian decoration on silver bowl of King Entemena.* Lagash, around 2700 BC

nomenon is an irremovable, integral part of nature. The inevitability of dissymmetry is also shown in the Sumerian picture, also referred to by Weyl, in Figure 2.12.

The two eagle-headed people stand facing one another almost symmetrically. But not entirely. Both of them are reaching toward the grapes with their right hand; in reality only rotational symmetry can superimpose one onto the other, but not on a two-dimensional plane. With their wings, the artist has done the opposite: to preserve the appearance of mirror symmetry, one of them has the left wing pointed downwards, and the other the right. Let us think back to the dilemma posed by the concept of symmetry. Which symmetry is more "true to life": if we offer our hand to the mirror,

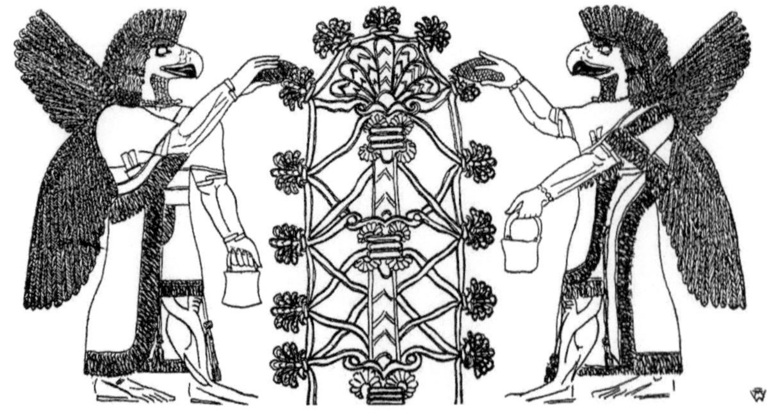

Figure 2.12.

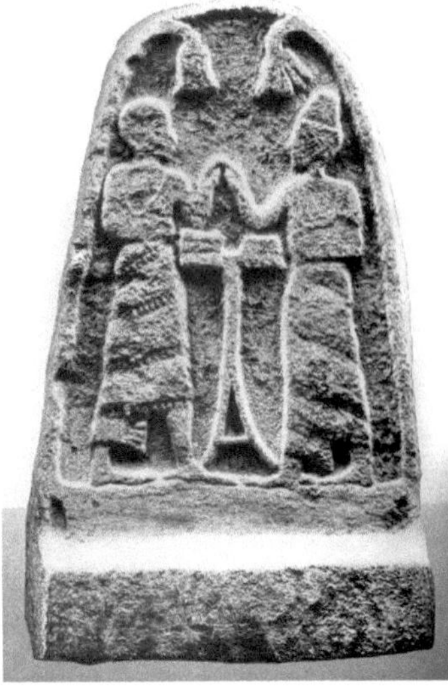

Figure 2.13. *Settling a contract.* Ugaritic stele, 14th–13th Century BC

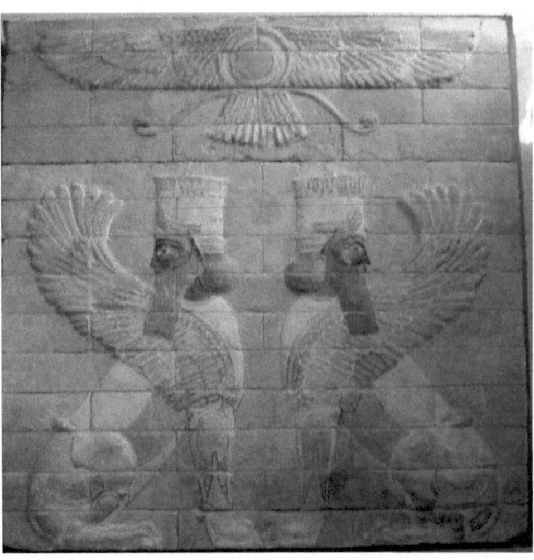

Figure 2.14. *Glazed sphinxes from the palace of Dareios at Susa.* Persia, c. 490 BC

or to an acquaintance? In the former instance, symmetry means reflection in a plane; in the latter, rotation around an axis. In two dimensions, we cannot simultaneously portray the spatial difference between left and right and preserve symmetry at the same time. Just as we cannot make

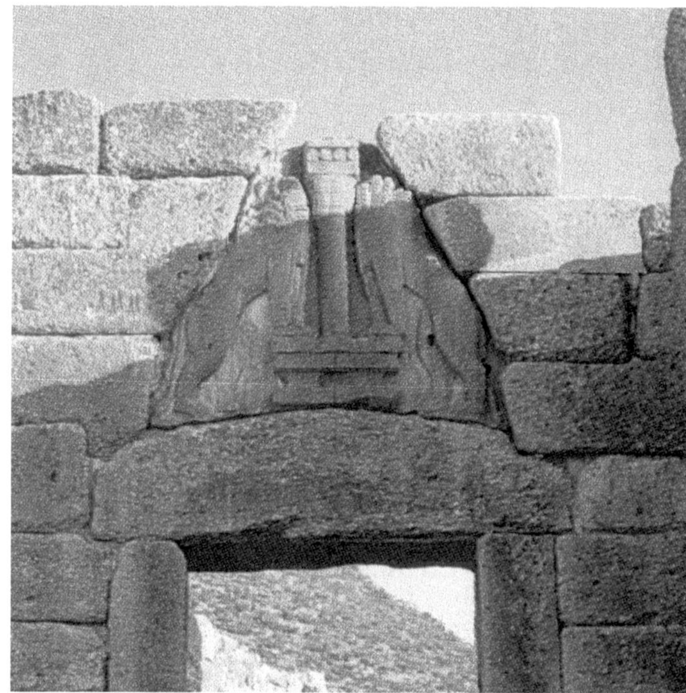

Figure 2.15. *The lions on the gate of the citadel of Mycenae*

our two hands overlap spatially. The creator of the Hittite stone column from Ugarit (today northern Syria) seen in Figure 2.13 solved this problem of depiction by making one of the contracting partners offer his left hand, and the other his right.

Even in modern times, common and recurring elements in coats of arms have included symmetrically-placed animal figures and eagles with their wings stretched out symmetrically (perhaps, for the sake of symmetry, even with two heads), for example in the coats of arms of the Habsburg rulers or the Russian tsars. These depictions present order, beauty and perfection all at once. Mirror symmetry displays perfection derived from God, the legitimacy of power — as seen on the enamelled sphinxes from the era of Dareios (ruled 522–486 BC), seen in Figure 2.14, or in essentially the same motif from Mycenae, from 750 years previously, shown in Figure 2.15. We will return to the role played in Egyptian culture by this symmetry when we deal with the golden section.

Mirror symmetry as representing divine perfection appears in the first written mention of symmetry, in the Bible. In the Book of Kings, chapter one, we read of the construction of the temple of King Solomon (c. 970–930 BC).

"6:23 And within the oracle he made two cherubims of olive tree, each ten cubits high.
6:24 And five cubits was the one wing of the cherub, and five cubits the other wing of the cherub: from the uttermost part of the one wing unto the uttermost part of the other were ten cubits.
6:25 And the other cherub was ten cubits: both the cherubims were of one measure and one size.
6:26 The height of the one cherub was ten cubits, and so was it of the other cherub.
6:27 And he set the cherubims within the inner house: and they stretched forth the wings of the cherubims, so that the wing of the one touched the one wall, and the wing of the other cherub touched the other wall; and their wings touched one another in the midst of the house."

The intellectual world of antiquity was accompanied throughout by the search for perfection and a faith in the perfection of the world.

"Hidden harmony is stronger than explicit," wrote Heraclitus (c. 535–475 BC) at the beginning of the fifth century BC. And what did harmony mean for Heraclitus and his age? Essentially it meant what we now term symmetry. What is more, it referred both to what we now call scientific and artistic approaches. According to ancient thought, harmony pervades nature and its laws, at once expressing artistic harmony, proportion, and beauty. Translations from two thousand years later referred to this as symmetry. Explicit harmony referred to that which we feel, we see, to that which is external. Hidden harmony was rational, that which we grasp with our reason, what we understand, that is, what is internal. One is connected to form, the other to content.

In classical Greek philosophy, the concept of symmetry is to be found in the work of all those philosophers who synthesized the thinking of earlier periods.

Plato attributes to Socrates (469–399 BC) the words (which will be interpreted in detail later) which suggest that symmetry is a golden middle road between beauty and truth, and this thought continues throughout the moral narrative. In one of the dialogues attributed to him, Socrates teaches his student about the relationship between truth and symmetry (or proportionality, which for them was almost the same thing) as follows:

"*Socrates*: And do you think that truth is akin to measure and proportion or to disproportion?
Adimantus: To proportion." (Plato, The Republic, VI, 486d)

Symmetry plays the measured, average, connecting role between the beautiful and the true, sometimes augmented by the good — not just for Aristotle (384–322 BC), soon after Socrates, but in later European philosophy and literature, and even in Far Eastern philosophies. It was in the work of Plato (427?–c. 347 BC) that faith in the perfection of the world came to fruition. Plato sought the most perfect bodies. In his *Timaeus* — probably based on earlier mathematical works — he describes the five perfect bodies, all made up purely of congruent planes and edges with equal angles between them, and which have no alternatives in existence (Figure 2.16). The semiregular, so-called Archimedean bodies, which can be obtained by truncating those of Plato, at best only come close to perfection. We will deal with their properties later.

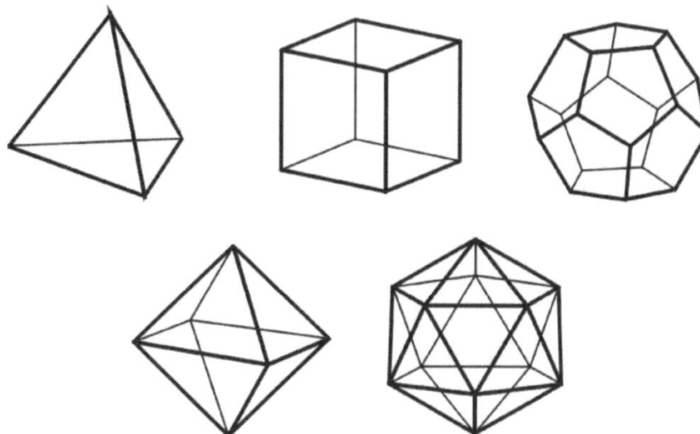

Figure 2.16. *Plato's regular bodies: the tetrahedron, the cube, the octahedron, the dodecahedron and the icosahedron*

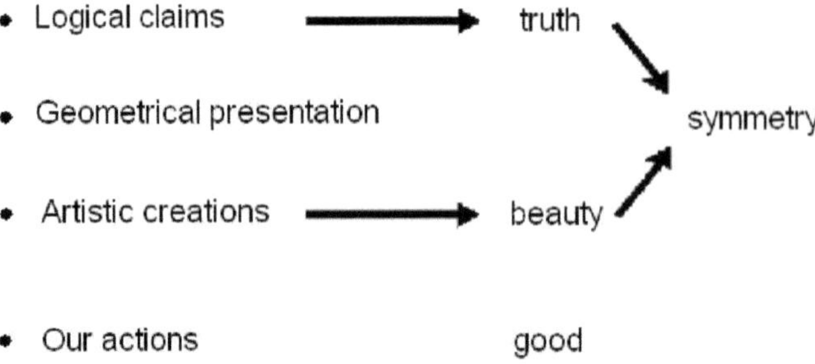

Figure 2.17. *Perfection in the thought system of Aristotle*

The perfection that Plato demonstrates with geometrical bodies is expressed in their philosophy (their love of wisdom). It was in the work of Aristotle that this philosophy became a system, and it is his name with which we associate the unification of the ancient Classical Greek view of the world.

For Aristotle, perfection is the form and content of our thoughts. (We need only think of Heraclitean harmony.) Our logical claims are the embodiment of the truth. Our artistic creations are the embodiment of beauty. The middle-ground between the two is represented by geometrical presentation and symmetry (just as it did for Socrates and Plato). The measure of perfection of our actions is the pursuit of good (ethics). It is finding the right degree or proportion, the happy medium, that is harmony itself.

We can also find this pursuit of perfect harmony in the numerology of the Pythagoreans. They looked for harmony in the proportions of integers, which they measured by the proportion of the length of the strings on a harp (Figure 2.18). This led them to the conclusion that musical harmony and the harmony of numbers are connected. And as they presumed the entire universe to be harmonious, the proportions of spheres, and hence of the spheres of the cosmos, had also to be like musical harmony, for the cosmos meant nothing other than the harmonious order of the world.

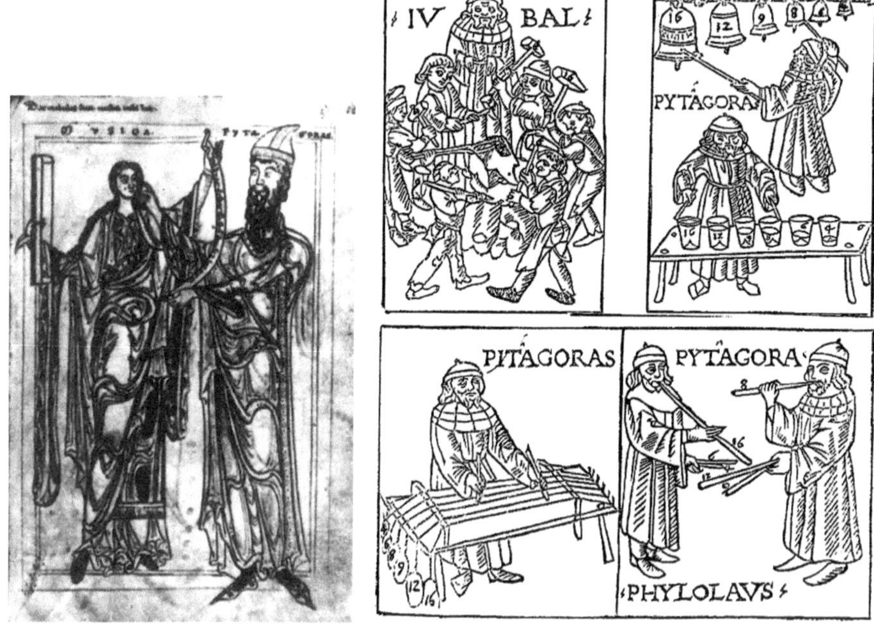

Figure 2.18. *Pythagoras (c. 560–c. 480 BC) and music.* Miniature, c. 1200

Figure 2.19. *Gaffurio (1480) Opera teorica della disciplina musicale*

This later became the foundation for the Orpheus cult, which survived throughout the Middle Ages (Figure 2.19).

It was Herodotus (c. 484–425 BC) who collected the existing knowledge of antiquity into a unified worldview. In addition to historical data, his work preserved much else of the ancient world and the worldview of antiquity. Although he did not prepare a map of the areas known at the time, Herodotus did give exact enough a description of these areas for them to be drawn (Figure 2.20). In his geographical worldview, the Danube and the Nile flow in mirror symmetric fashion respective to the Mediterranean Sea and its continuation, the Black Sea. According to his information, both rivers first flow from west to east, then the Danube goes from north to south and the Nile from south to north before pouring into the sea. Along the bank of the Nile, in the Nubian desert, we find oases that are the same number of walking days apart (translation symmetry), and he describes

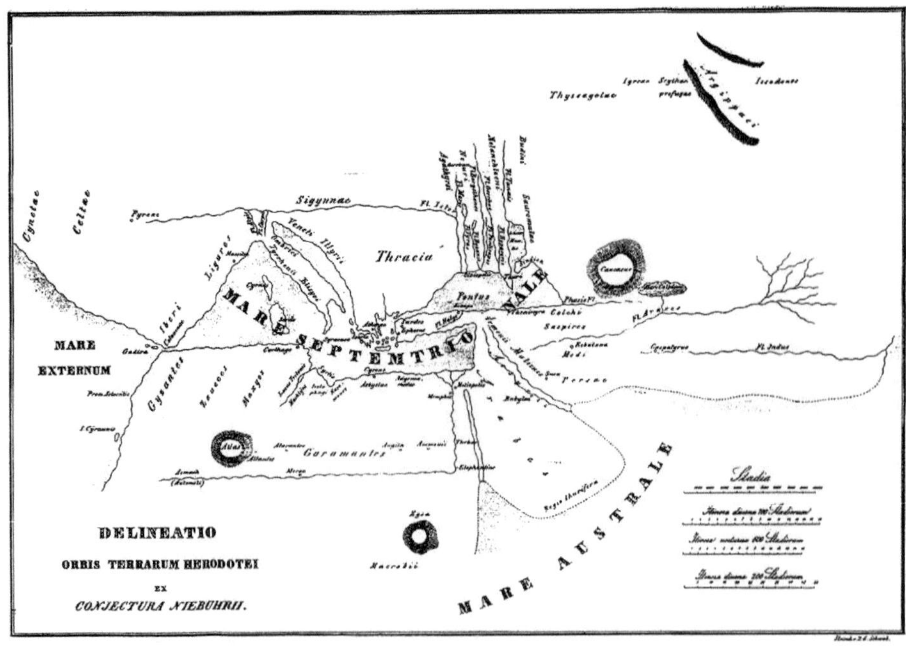

Figure 2.20. *The world as Herodotus saw it*

the temporally periodic flooding of the Nile. He does mention that in the Danube's case there are no periodic changes to the flow of water, but he explains this with an unexpected twist, a symmetry assumption. This is so, he writes, because the Sun is higher above the basin of the Nile, while its rays reach the Danube basin from the South at a flatter angle, meaning there is more rain all year round. But let us suppose, Herodotus continues, that the Sun shone symmetrically to what we experience, from the North! Then the flow of the Nile would be more even, and that of the Danube periodic, and the harmony of the world would be restored at once. Today we would likely smile at this. But we can see that his deduction has a rational basis. For a thinker in ancient Greece, faith in the symmetry of the world was stronger than what was actually experienced. When a decision had to be made between a faith in symmetry and empirical experience, the advantage was given to the former. This is how symmetry was able to become a principle.

As early as the Hellenistic age, symmetry had become the subject of a separate branch of study. As we have mentioned, the sculptor Polykleitos wrote a book about proportions and their application in sculpture. Pliny the Elder (AD 23–79) and Vitruvius (c. 90–20 BC) mention the painter and sculptor Euphranon, who wrote a book about symmetry. Vitruvius' work *Ten Books about Architecture*, written in the second half of the first century BC, is not only noted for being the first (surviving) theoretical discussion of symmetry, and did not only become well-known for its theory of proportions, but because it is a link between antiquity and the Renaissance.

After this, symmetry was not a subject for writings of the Latin cultural world for almost a millennium and a half. Naturally, it was used in works of art, architecture, and here and there in other genres, e.g. the Iberian Latin poet of late antiquity, Prudentius (348–405 or later) used symmetry in the structural construction of his hymns, but aesthetic or scholarly works did not deal with it.

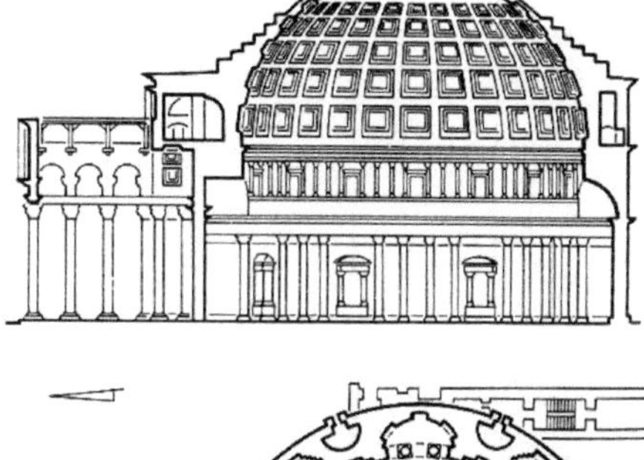

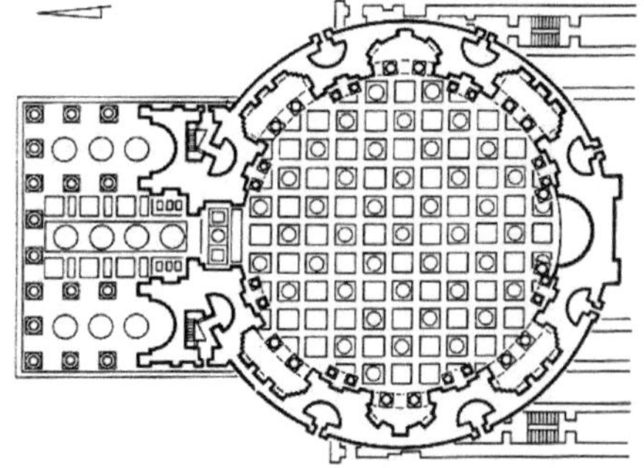

Figure 2.21. *A nice example of architectural proportions: the Pantheon in Rome*

In the middle ages the concept lived on in the sense of today's harmony, rhythm and proportion. The word symmetry was not mentioned, and was not translated into European languages. It survived in the way it was applied, including in a principally creative sense in Arabian (Moorish) architecture, decorative arts, and not least mathematics. In self-evident fashion rather than for theoretical reasons, it was used by Romanesque (Figure 2.22) and Gothic art (Figure 2.23), poetry (rhyme), and it was in this era that the Pisan mathematician L. Fibonacci lived and worked (c. 1170–1250), whose significance was only properly appreciated long after his death.

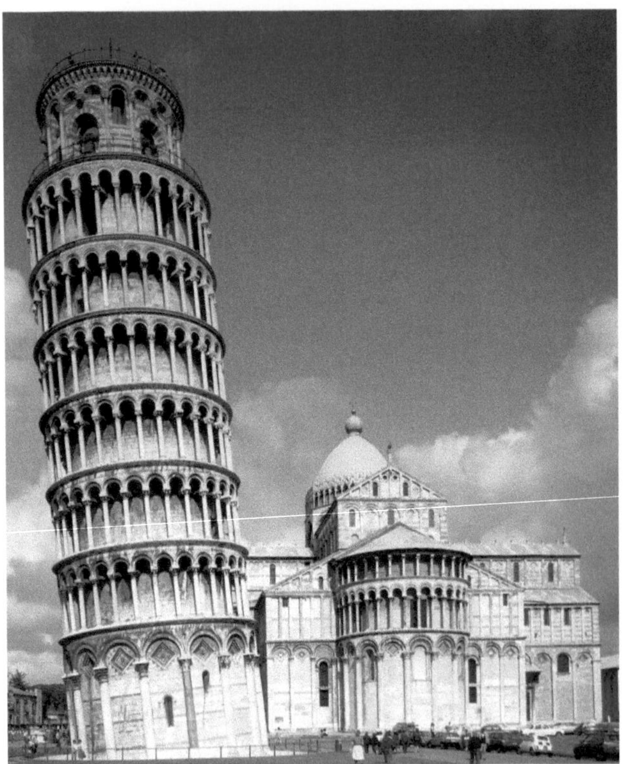

Figure 2.22. *The Romanesque duomo of Pisa (1063) and its bell-tower (1174)*

It was the Renaissance that was to rediscover Vitruvius. He was referred to on a continuous basis, and from the end of the fifteenth century a whole series of translations of his works was to appear (Figure 2.23).

Petrarch (1304–1374) mentions that symmetry is a Greek word which has no Latin equivalent. The Renaissance reinterpreted and modernized the term symmetry for its own ends, putting the emphasis more on its *proportione* sense. It was in this period that the theory of art came into being as a discipline.

The word symmetry was rehabilitated a century later by Leon Battista Alberti (1404–1472). In all probability, Alberti was the first to use the term symmetria in his works on aesthetics, and initially only did so with regard to painting. He wrote his book on painting in two, slightly distinct, versions: in 1435 in Latin, then in the following year in Italian (Vulgare). Human proportions were initially referred to by the term *misura* (measure), then in later works by the Latin word *proportione* (proportion). Unlike his

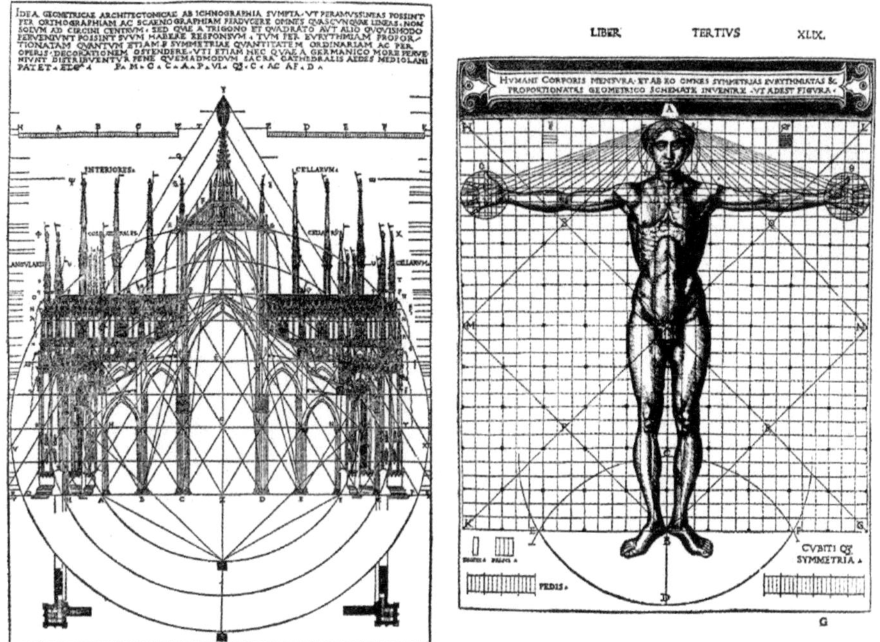

Figure 2.23. *The wood-engraved illustration of the Gothic duomo of Milan used in the translation of Vitruvius (1521) by C. Cesariano (1483–1543). The depiction of Vitruvius'* related proportions of the human body and architecture *demonstrates the transition from the Gothic to the Renaissance*

predecessors, Alberti uses the term *symmetry* as an independent concept, in both its geometric and aesthetic senses. But he did not translate the term into Italian, and in the Italian version of his work *De pictura*, entitled *Della pittura*, we find the word *misura*.

It was Lorenzo Ghiberti (1378–1455) who, in his work *Commentarii* (1455), established the concept of symmetry, and who introduced the use of the word *proporzione*, using the two as synonyms, and who indisputably connected proportionality with beauty. By the end of the century, the word *proporzione* was unambiguously the Italian equivalent of symmetry. It becomes indispensable in all works concerned with art, starting with *Perspective* by Piero della Francesca (c. 1420–1492), through L. Pacioli (c. 1445–c. 1514), Leonardo and Dürer to the comprehensive synthesis of C. Ripa (c. 1555–1622).

Leonardo da Vinci was the greatest polymath of the Renaissance, whose achievements were equally prodigious in the arts and the sciences. In accordance with his humanist principles, it is man who is at the centre of his works; in his theory of proportions, the benchmark is the person. For him, then, symmetry is associated with the proportions of the human body (see Figure 1.6). Few of his paintings — in which symmetry is a decisive compositional element — survive, but many more of his sketches do. In these, Platonic regular and Archimedean semi-regular bodies appear just as much as the outlines preserving the harmony of the proportions of the human body in geometrical form. He illustrated Luca Pacioli's book *De Divine Proportione* (The Divine Proportion, 1509), in which, among other things, we can discover the structure (Figure 2.24, bottom right) of *fullerene* (see Chapter 8), which would enjoy meteoric success half a millennium later.

Translations of Vitruvius and other Italian Renaissance works on aesthetics appeared one after the next in modern European languages. This is how the concept of symmetry was introduced into a growing number of countries. For Albrecht Dürer, in early sixteenth-century Germany, it was already a natural concept. Dürer was also a multifaceted Renaissance personality, who concerned himself with painting, graphic design, goldsmithery, physics (geometric optics), geometry, and even military engineering. He was one of the first to look for the solution to the mystery of plane tiling with pentagons. His results relating to the theory of proportions and the applications of symmetry were published in his works *Instruction How to Measure with Compass and Straight Edge* (1525) and *Four Books on Human Proportion* (1528). He is associated with the creation of one of the first golden proportion compasses, the middle of the three endpoints of which always divides the distance between the two edges according to the proportions of the golden section — however it is opened. Accordingly, if we put the two outer arms next to the figure's legs and the top of its head respectively, the middle point is usually at the person's navel (Figure 2.25).

Neither was the classic meaning of symmetry relating to harmony and beauty to be forgotten. Shakespeare's Sonnet LIV — which we will discuss in Chapter 14 — extols the connection between beauty and truth, as do successors like John Keats (1795–1821) or Attila József (1905–1937).

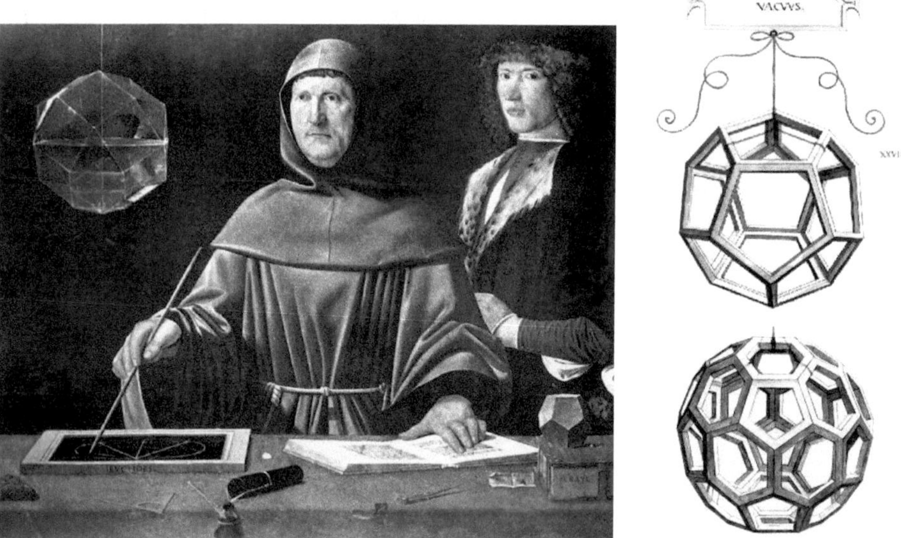

Figure 2.24. *A likeness of Luca Pacioli, and the dodecahedron and the truncated icosahedron in Leonardo's drawing in Pacioli's book* Divine Proportion. *Pacioli's portrait was painted by Jacopo de Barbari (1440–1516) in 1495, a year after the publication of his work* Summa de aritmetica geometria proportioni et proportionalita, *which summarized the work of his predecessors from antiquity to the end of the fifteenth century. On the picture, Pacioli is proving on of the theorems of Euclid, with a dodecahedron in front of him, and a rhombic cuboctahedron half-filled with water hanging next to him. According to some sources, the figure alongside him is his mentor, Prince G. d'Urbino, but it also bears a strong likeness to Albrecht Dürer's self-portrait from three years later.*

The Renaissance also influenced the scholarly thinking of areas of Europe further away from Italy. The Copernican Turn also fits the applications of symmetry. Plato was already striving to create a picture of a perfect world. Perfect meaning simpler, more symmetrical. It is this turn — the choice of the description that is more symmetrical — that Copernicus (c. 1473–1543) makes. His worldview is, in point of fact, a choice between a simpler and a more complicated description.

From the ancient Egyptians onwards, the geocentric and heliocentric views of the world were equally well-known, and survived in parallel in scholarly attitudes. Both premises can be used to describe the discernible apparent movement of celestial bodies. The geocentric worldview suited

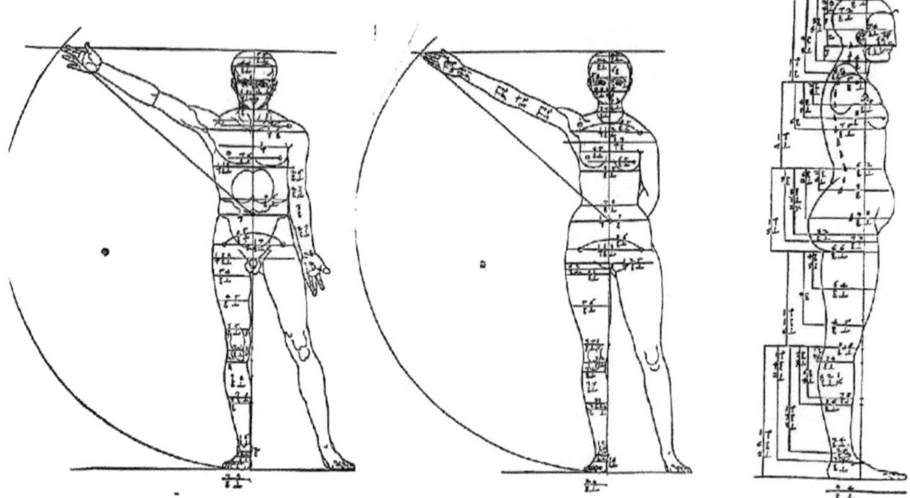

Figure 2.25. *Dürer's proportions of the human body*

the simple attitude: I am in the middle, and the celestial bodies move around on the surface of the concentric spheres around me. True, they follow paths that are more than a little complex to describe — the planets trace cycloids, for example — but, for want of anything better, they can be reconciled with empirical experience. While the heliocentric worldview throws the Earth — and with it man — from the centre of the universe, it makes possible much simpler paths, descriptions and explanations of the movement of the planets. In return for sacrificing our privileged position, we are given a picture of a more perfect and more symmetrical world. This can also be reconciled with experience. Copernicus, that is to say, opted for symmetry. It was easier to establish the laws governing the movement of the planets for these paths presumed to be more symmetrical. These discoveries required a genuine late Renaissance figure able to combine the culture of the Renaissance, the fledgling science of experimentation and observation of the seventeenth century, and a faith in the harmony of the ancients.

Relying on the extensive list of empirical observations of Tycho Brahe (1546–1601), Johannes Kepler (1571–1630) established the laws of planetary motion. His laws emerge from a faith in the harmony of the world,

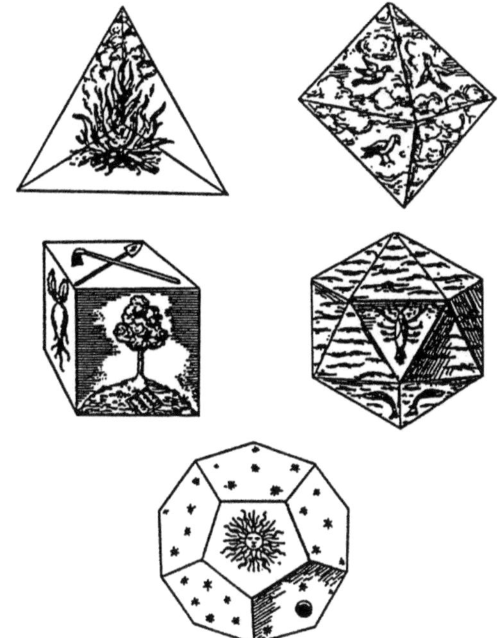

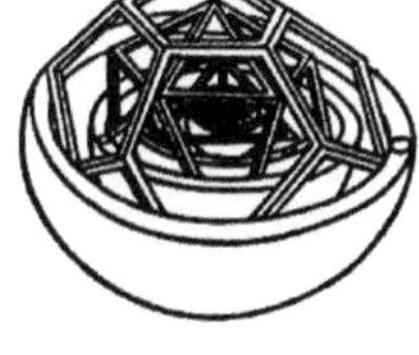

Figure 2.26. *The spheres drawn around Plato's regular bodies, in Kepler's* Mysterium Cosmographicum (1596)

Figure 2.27. *The harmony of the universe and the Platonic regular bodies*

into which even the elliptical path — whose two foci do not in this instance play an equal role — only fits by departing from the elemental symmetrical instinct. His thought is well illustrated by the titles of his books: *Mysterium Cosmographicum* (1596), *Astronomia nova seu Physica coelestis* (1609, first and second laws), *De harmonice mundi* (1619, third law). Kepler was so attached to his picture of the perfect world that in his previous system the paths of the planets were located around nested Platonic regular bodies and on the surface of the inscribed and circumscribed spheres (Figure 2.26).

The proportions of the rays of these spheres and of the planetary paths appear compatible with one another, not just with the periods of orbit raised to the respective power. Following Plato, he also makes the harmony of the world more perfect by identifying the five perfect bodies with the four primary elements and (for the body seen as the most mystical and perfect, the dodecahedron) the universe — the universal whole (Figure 2.27).

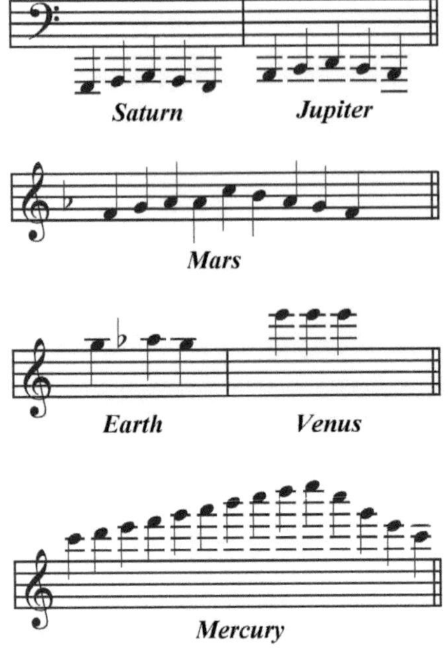

Figure 2.28. *Kepler's music for the spheres (see Pythagoreans)*

Kepler also attributed musical distances to the distances between the respective spherical shells (Figure 2.28), and thus, led by the teaching of the Pythagoreans, composed the music for the spheres. In the twentieth century, this inspired the German composer Paul Hindemith (1895–1963) to compose the opera *Harmonices mundi* (1959) and the *Kepler symphony*.

At the end of the Renaissance, just like in the twilight years of the Greek golden age, the beginning of the seventeenth century saw the paths of science and the arts again separate. For some three hundred years, the implementation of symmetry was also characterized by this separation. At the same time, the seventeenth century was when modern science was born. This was when Galilei expounded his relativity principle in which he proclaimed the equivalence of the various inertia systems. This period marked the beginning of experimental science. The new role of science was described in the *Novum Organum* (1620) of Francis Bacon (1561–1626). This was when research into natural laws began: Descartes (1595–1650), Newton (1642–1727), whose Philosophiae naturalis principia mathematica was published in 1687, and Leibniz (1646–1716). As rationality became all-powerful, so the arts were expunged from scholarly thinking. In the space of four decades, Kepler's search for harmony became an outdated method. It was in this century that the first scientific laws were established. It is customary to refer to it as the century of scientific laws, in contrast to the subsequent one, which was to be the century of symmetry principles.

But for the moment let us stay in the seventeenth century, with the programme of Francis Bacon. The scientific laws then stated in exact form and the road to their discovery presuppose the repeatability and reproducibility of the experiments (events), irrespective of their time and place. Further, they declare the natural laws thus found and stated to be independent of time and place. If we recall the generalized definition of the concept of symmetry, do we not see a similarity? Let us think through what change occurs. An experiment conducted at a different time and place; a law discovered at a different time and place. What remains unaltered when the time and place are changed? The result of the repeated experiment, and the law we thereby prove. To this day it is this symmetry that is the highest criterion of scientific endeavour.

Science became the subject of pure rationality, and proof one of its tools. Emotional motivation, impressions and intuition no longer had any place in science, and were left to art. By the middle of the century, Kepler's method was no longer defensible.

It is the Baroque and the French *classique* that become the characteristic artistic trends of the era. Overblown decoration and billowing tunes do not exactly bear the classic hallmarks of order. So did symmetry disappear under the Baroque? Not at all — in fact, it was the bringer of new things.

It was at that time that ellipsis appeared as a new element in architecture; it can clearly be seen in the ground-plan for the church of Borromini (1599–1667) (Figure 2.29). Helical columns appear as a characteristic Baroque decorative element. The Baroque is characterized by symmetrical façades of buildings and French gardens trimmed in orderly fashion. In music, it is enough to think of the fugues of J. S. Bach (1685–1750). The most that the asymmetric ornaments appearing in or layered upon certain elements can do to the symmetry of the Baroque is to make it dissymmetric.

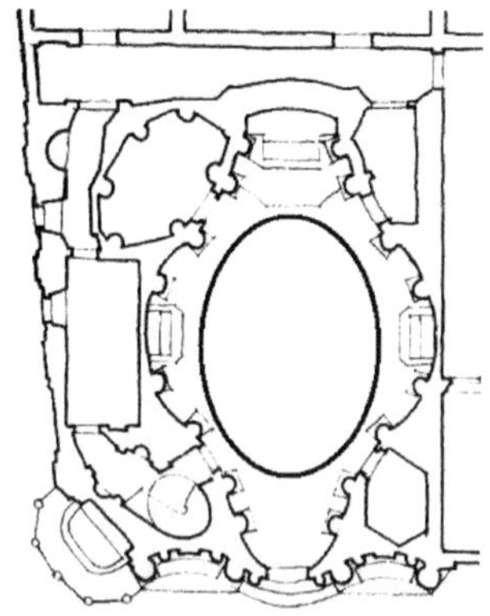

Figure 2.29. *Borromini: the ground-plan for the Church of S. Carlo alle Quattro Fontane in Rome (1638–41)*

The eighteenth century is often referred to as the century of mechanics. In a sense this view is justified, especially if we think of philosophy or the Industrial Revolution. Staying within the sciences, the search for natural laws was replaced by that for variational principles, which were essentially symmetry principles. These symmetry principles sought to find the optimal (minimal) distance, work, time, etc., from among the transitions between two physical states.

If, in the light of our knowledge of physics from the twentieth century, we think back to its roots, we have to say that — as far as its mode of description is concerned — the path toward modern physics can be traced without interruption from physics based on variational/symmetry principles. If we begin with Newtonian physics, on the other hand, we have to draw a caesura between classical and modern theories, not to mention change our mode of description. From this it also follows that the physics of the century of mechanics was not the direct result and continuation of the physics of the seventeenth century.

We have seen that both were founded on symmetry considerations. The invariance of experiments and laws under temporal and spatial changes is not, however, equivalent to the mathematically-expressed symmetry of the optimal variation, neither in terms of form nor of the content of the invariance. We will see that in modern physics it is variational principles that lead to the elaboration of the conservation laws that play such an important predictive role. These were not at first built into physics from theory, however, but on a gradual, empirical basis.

The nineteenth century enriched science with further symmetry principles. It was around this time that the most significant conservation laws were elaborated, principally that of the conservation of energy. Why do conserved quantities represent symmetries? Because they are invariant under any change; whatever might happen to the closed system in which they are measured, their quantity is not affected by the passage of time. In the nineteenth century, physics came to include charges from electricity and magnetism, as fundamentally new types of physical quantity, as well as the laws governing them. The laws of electricity were originally established using the analogy of the laws of mechanics. They measured and described actions of forces. Coulomb's (1736–1806) Law took gravitational force as its model. It was expanded, however, to include charge symmetries: the positive and negative electric charges in many respects act as mirror images of each other.

The Maxwell equations, which unite the laws of electricity and magnetism, at first sight seem distinctly asymmetric. Yet it appears that James

C. Maxwell (1831–1879) saw the symmetry hidden in them in advance. In 1871, in his lecture to the Royal Society, he announced that "Mathematics loves symmetry above all". He was proved right. After his death it transpired that his laws, as he had laid them out, were from the outset relativistically invariant (their form does not change under the Lorentz transformation); indeed, in their relativistic form, they can be written in a fashion that is visibly symmetrical.

Let me draw the reader's attention to how Maxwell's sentence quoted above inverts Heraclitus' statement about hidden and explicit harmony. For him, symmetry is connected to mathematics, that is to explicit harmony. Maxwell's result shows how mathematical modern (natural) science has become (for this is the source of its exactness). And today we tend to regard as science only that of which exactness, that is explicit harmony, is required. Leaping forward half a century, let me quote Hermann Weyl, who, in line with Maxwell, said the following: "The mathematical laws governing nature are the source of symmetry in nature". He, too, saw the explicit harmony seen in mathematics as the source of symmetry or universal harmony.

The examination of crystals developed alongside the progress of physics. It is no surprise that, after centuries of attempts by alchemists, in the era of scientific orderliness research into the structure of material should have turned to materials with regular structures — like metals. The beginnings of the study of crystal structures date back to the seventeenth century. Kepler investigated the form of snowflakes, without success. After examining mineral crystals, N. Steno (1638–1686) then C. Huygens (1629–1695) discovered one of the regularities of the angle of refraction for layers of crystal. The breakthrough was brought at the end of the eighteenth century by the works of R.-J. Haüy (1743–1822), who introduced the molecule-like explanation similar to the concept of elemental unity of Democritus (460–370 BC). Haüy was also successful in applying the geometry of three-dimensional space; then, as the culmination of his oeuvre, he published the law of crystallization in 1815, which he called the *law of symmetry*. This became the basis for further research. In the nineteenth century, the study of the form of crystals essentially followed two parallel but interdependent threads. Geometrical crystallography tried in a theoretical way to classify the possible cells that fill space on a continuous basis.

The other school of thought considered the experimental observation of real crystals to be more important. L. M. Frankenheim and J. F. C. Hessel identified the thirty-two crystal classes on the basis of mathematical symmetry considerations. Two decades later, A. Bravais (1811–1863), a researcher into minerals and crystals, also discovered the existence of the thirty-two crystal classes by studying the fourteen possible space lattices (Frankenheim distinguished fifteen of them) and the twenty-three possible symmetries of the polyhedra that can be placed at the nodes of the lattice (certain scientists at the time imagined molecules as being polyhedra). The advance is associated with the names of two mathematicians, C. Jordan (1838–1922) and A. Schoenflies (1853–1928), and three crystallographers, L. Sohncke (1842–1897), E. Fedorov (1853–1919), and in part to Barlow. The task was to find out how many ways the possible point group symmetries could be located in the possible space lattices. In the 1860s, Jordan applied the concept of a group as acquired from Galois' manuscript (see below) to the description of the symmetries of the crystals. By the end of the 1870s, Sohncke continued along this path to the point of classifying the crystallographical space groups, distinguishing 65 so-called groups of motion. The end result, the identification of the 230 space groups, was achieved almost independently by Russian mineralogist E. Federov (1890) and German mathematician A. Schoenflies (F. Klein's student, 1891), followed soon after by W. Barlow (1845–1934) of Great Britain. The problems of continuous geometrical space-filling using a single type of space element and the structure of natural crystals simultaneously found a definitive solution. This created a closed system, that until recently was unassailable, and a unified description which became the bible for every crystallographer.

In 1894, almost immediately after the recognition and systematization of the 230 crystallographical space groups, Pierre Curie introduced the concept of symmetry to physics. He was also responsible for defining the terminology of symmetry, asymmetry and dissymmetry. Curie's statement that "it is dissymmetry that makes the phenomenon" came into general use some fifty years later, in the physics and material science of the twentieth century. Its essence is that it is where the seemingly perfect symmetry is distorted that the physicist is needed. Let us just think of how radically a trillionth or less of dirt or perhaps crystal defect (dislocation) can transform a crystal's characteristics. As we noted in the previous chapter, the functioning of semiconductors is an example of something based on this property.

In addition to those already mentioned, the introduction of symmetry properties as an organizing principle in science was also thanks to the work of H. Poincaré (1854–1912) and later É. Cartan (1869–1951). There is, however, another important trend in the history of science that we must discuss in this regard. None of these results could have been achieved without the development of group theory. This story originally began on another thread of science, and only after many decades did it transpire that it could be applied to crystallography (then later to many other areas of physics!). The early history of group theory is the most romantic chapter in the history of science, and volumes have been written about it. In addition to giving us the most efficient tool for describing symmetries, it is good example of how continuous problems can be treated and solved with discrete mathematical methods, and further of how — naturally with symmetry considerations — ideas and methods can be transferred from one area of science to another, however distant the two might appear.

The story began in a forest next to Paris in 1832. A respectable citizen heading towards the city picked up a student fatally injured in a duel and took him to hospital in his carriage. Évariste Galois (1811–1832) was a mathematics student at the Sorbonne. His last request before his death was that one of his manuscripts reach the great mathematician of the age, A. L. Cauchy (1789–1857). The document suggested a general method for determining the roots of higher-order equations on the complex plane. Cauchy probably did not ascribe particular significance to the document, but thankfully, as a man of precision, he did not throw it away. After Cauchy's death, the aforementioned Camille Jordan found it in a file while sorting his papers. Jordan recognized the significance of the concept of groups and applied it to the description of the infinitesimally small translations and rotations in crystallography (1868–69). It was around this time that two students studying in Germany, Felix Klein (1849–1925) and the Norwegian Sophus Lie (1842–1899), decided to complement their mathematical studies in Paris, under Jordan's auspices. It was here that they became acquainted with group theory, and devoted themselves to it for life. Their Parisian adventures also came to an abrupt, and romantic, though not tragic, end. The Franco-Prussian War broke out a few months after they arrived, and Klein, a German, was forced to return home from the hostile country. After a misunderstanding that led an excursion to end in imprisonment at Fontainebleau prison, the tall and blonde Lie could not stay much longer, either. The method they had learned remained in their possession, however, and, separated from their master, they had no option but to uncover its secrets on their own. It was thanks to the work they later produced that group theory was to emerge as the essential (and to this day considered all-powerful) tool for the description and study of symmetries.

Although they were to be separated from one another, they remained friends for life. They set to the task with incredible passion, enriching the theory very productively. It was in the following year, 1872, that Klein elaborated his well-known so-called Erlangen programme: "Geometry is the science which investigates the invariances under a group of transformations". It was an immediately arresting and cheekily image-breaking claim, particularly from the pen of a youth in his twenties. But Klein and Lie succeeded in realizing their programme.

Group theory grew in leaps and bounds, its applications expanded, particularly in physics. Generations of students grew up learning this new method. The next milestone is associated with Emmy Noether (1882–1935), a student of D. Hilbert (1862–1943), or rather the two theorems of group theory she published in 1918, in which she displayed the mutually unambiguous correspondence between symmetries and the conservation laws. The pioneering mathematician's election to the position of professor was rejected by the conservative University of Göttingen with the excuse that there was no women's lavatory in the senate building. Even before the appearance of the methods of quantum mechanical description, Noether's theorems opened the way for physics to discuss both, to deliberate the classical conservation laws and the so-called gauge invariances.

As early as the year in which Noether's theorems were published, Hermann Weyl attempted to prove the conservation of electric charge based on a gauge invariance examination, which he, alongside Wolfgang Pauli (1900–1958), was later to put into a more precise form. In addition to a number of his achievements in applications in mathematics and physics, Weyl's work is relevant to this study because it was he who made symmetry, in its modern interdisciplinary sense, the subject of a separate field of research. Thanks to his work, the gauge theories (spatial coordinate-dependent conservation laws) opened a new class of symmetries and made possible the explanation of a whole series of physical phenomena (not just the conservation of electric charge, but also charge-like and similarly-behaving physical quantities, and then the conservation of the so-called isotopic spin, the classification of elementary and sub-elementary particles, later the approximate (non-exact) symmetries of particles playing a role in weak interaction, and finally quantum chromodynamics).

The physics of the second half of the twentieth century was essentially a series of researches into symmetry breakings followed by the discovery of new symmetries. As a result of Noether's theorems, no theoretical debate could avoid the question of symmetries and their violations.

The description of symmetries and their related mathematical theories had almost no applications in physics where Weyl did not make a contribution. In 1928, three years after the appearance of quantum mechanics, he published his book *Group Theory and Quantum Mechanics*. At the same time as him, Eugene P. Wigner (1902–1995), who by then was working in Berlin, wrote his book *Group Theory and its Applications to the Quantum Mechanics of Atomic Spectra* while on a family vacation in Alsógöd, Hungary. Wigner's work would be connected to symmetry for the rest of his life. So much so, that although his extraordinary intellectual output would contribute to a number of chapters in physics, he was awarded the Nobel Prize in 1963 primarily for his development of the theory of symmetries and its applications in physics. Wigner is associated with the classification of physical symmetries, namely the division into geometric and dynamic invariances. For a period of their lives, Weyl's and Wigner's careers were connected: they both spent their years of emigration away from Nazism in Princeton, New Jersey; Wigner for the rest of his life, Weyl until his retirement in 1951. This was when the latter held his four famous lectures on symmetry. These lectures summarized knowledge of symmetry at the time, and the book in which they were published has since become a classic, translated into a number of languages.

Countless discoveries and discoverers could be mentioned from the history of the last half century of research into physical symmetries, all of which are ground-breaking in their own right. It is a thankless task to highlight just a few of them, but let us have a glance at the various discoveries of particle physics.

The isotopic spin gauge invariance of Chen Ning Yang (1922–) and Robert L. Mills (1927–1999) opened a new chapter in the history of physical symmetries in 1954, which was to be completed by Gerald t'Hooft (1946–) in 1971. So too did parity violation, anticipated by Tsung Dao Lee (1926–) and Yang, and demonstrated by Chien Shung Wu (1912–1997). After the violation of particular symmetries (like parity [P], time reflection [T], and charge conjugation [C]), attention turned to the conservation of so-called combined symmetries. This is how the so-called law of conservation of CPT symmetry was born — the work of Val L. Fitch (1923–) and James W. Cronin (1931–) — which states that the three properties together remain invariant, even if separately they are violated.

The next significant step was the discovery of so-called $SU(3)$ symmetry (1961), which is associated with the names of Murray Gell-Mann (1929–) and Yuval Ne'eman (1925–). Their discovery led to the classification of the elementary particles that were known at the time, to the prediction of which particles were missing, then to the establishment of the quark model, and later to quantum chromodynamics.

Here we should mention the first so-called unification theory, which explained the violation of certain symmetries, and the electroweak theory, the work of Steven Weinberg (1933–), Abdus Salam (1926–1996) and Sheldon L. Glashow (1932-), which presented a unified discussion of electromagnetic and weak interaction. This then paved the way for the so-called standard model of microphysics — to this day the theory that, to the best of our knowledge, most exactly describes the structure of the matter of microparticles — and for further unification theories, and the search for a unified supersymmetrical theory.

Particle physics has become the science of symmetries and symmetry breakings.

This explains why we have made a long but not unmerited description of physical symmetries, but symmetry considerations have also had a productive effect on other disciplines. At the end of the last chapter we saw a number of examples of these, and now we will highlight some other events, significant from the point of view of the history of science, that we will discuss later, in the thematic chapters.

The works published by Haeckel (1834–1919) from the 1860s to the 1880s brought an attitudinal change to biology. The first volume was

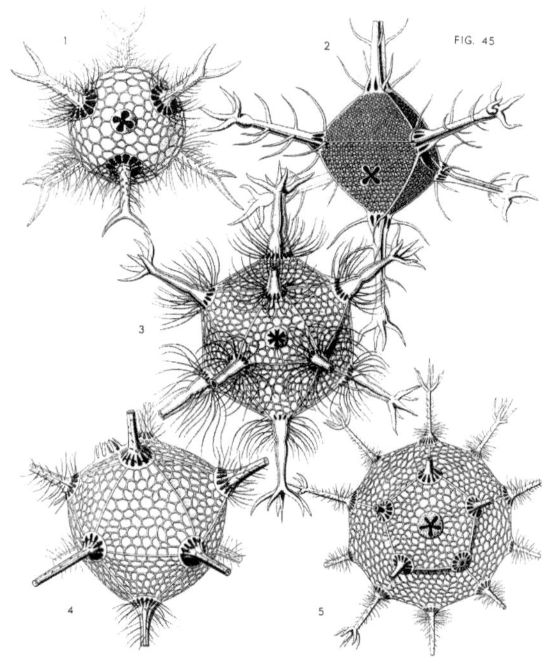

Figure 2.30. *Drawing of a few radiolaria from a page of Haeckel's Challenger Monograph*

published in 1866 with the title *Morphological Symmetries in the Animal Kingdom*, which was followed by several volumes of material describing the many animal species he discovered on his travels (Figure 2.30).

In 1917 D'Arcy Thompson (1860–1948) first published his work *Growth and Form*, which enjoyed many further editions, in which the morphological changes observed in the development of living beings are followed with the help of symmetry transformations. Thompson's work cannot be regarded as merely a specialized biological textbook: the original scientific insight with which it applies physical experiments to the study of the development of symmetries, in Weyl's respectful words, "combines profound knowledge in geometry, physics and biology with humanistic erudition" (Figure 2.31).

Since the first half of the nineteenth century and the discovery of organic molecules, the discovery of the boundary between the living and the inanimate has excited scientists. The experiment of Louis Pasteur (1822–1895) on enantiomer molecule pairs in 1848, during which he separated the chiral molecules of tartaric acid from raceme solution, was a demon-

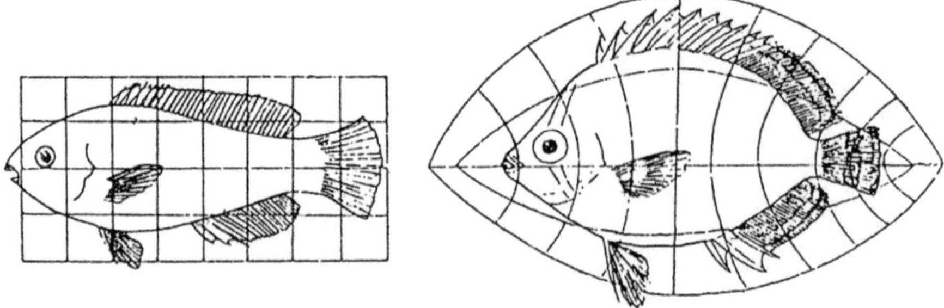

Figure 2.31. *One of D'Arcy Thompson's drawings of the metamorphosis of living creatures (519–520. Scarus sp. → Pomacanthus)*

stration of biological selection: living bacteria selected from the left- and right-handed molecules of the inanimate material.

The final precise determination of the spatial structure of the double helix of DNA (1953), discovered by James Watson (1928–) and Francis Crick (1916–2004), was the result of a symmetry hypothesis, paying attention to the symmetries of the molecules present in it.

The study of chemical equilibrium and non-equilibrium processes was given new momentum partly by the irreversible thermodynamic research of Lars Onsager (1903–1976) and partly by the results of Ilya Prigogine (1917–2003) concerning the generalization of non-equilibrium thermodynamics (for both chemical processes and processes beyond chemistry and biology).

In addition to many other results, the most visible application in psychology was presented by the Rorschach (1884–1922) tests. The research of Jean Piaget (1896–1980) put the development of a child's brain and its thinking into a new light. His results were reinforced by recent discoveries relating to the asymmetry of the brain, which, *inter alia*, extended our previous knowledge of the dependence on cerebral hemisphere of the speech centre, motor centre and left- and right-handedness.

In the twentieth century, art and science found each other again. A new relationship developed between them. They had a much more direct effect on one another than before; they inspired each other. As well as the symmetry to be found in works of art, technology and science became built into artistic education. The principles of the Bauhaus (1919–1933) and later institutions operating in the same spirit represented a particular breakthrough in this. Twentieth-century art, among other things, developed a new affinity for space: new dimensions were considered to be real, non-Euclidean spaces were granted rights, and the search for order amidst chaos was given new impetus.

As a conclusion to this historical introduction, we must mention two discoveries in material science from the last two decades in which artistic influences have played a role, and which we will discuss in greater detail in the thematic chapters. One is the discovery of quasicrystals (Dan Shechtman et al., 1983), the very notion and then acceptance of which was made possible by the unperiodic tiling of a plane showing fivefold symmetry by Roger Penrose (1931–). A spatial structure with fivefold symmetry — even locally — had previously been unthinkable to crystallographers (Figure 2.32).

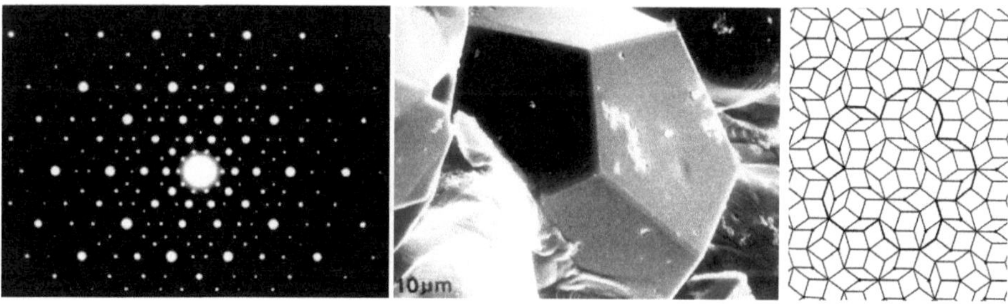

Figure 2.32. *The diffraction picture of a quasicrystal, its electron microscope image, and a Penrose tiling*

The other discovery was that of fullerene, the work of R. Smalley (1943–), H. Kroto (1939–) et al. in 1985, which displayed the connection of carbon atoms into sphere-like molecules, as a chemical realization of the synergetic principle of Buckminster Fuller (1895–1983), following with the spatial directions of chemical bonds the optimal load distribution of the supporting framework he designed as an architectural construction (Figure 2.33).

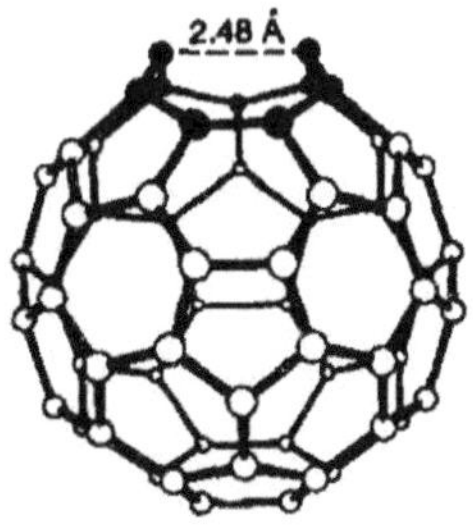

Figure 2.33. *The structure of the 60 carbon atomic fullerene molecule*

Chapter 3

Symmetry in geometrical decorative art

Frieze patterns, wallpaper patterns, space groups

The universality of decorative elements

In the preceding chapters I showed how there are certain common elements in humanity's cultural history which can equally be found in European, Middle and Far Eastern, African, American and Australian aboriginal cultures. Their origins are ancient. These decorative elements already appeared in periods when — we assume — these cultures existed independently, without influencing one another. The silk route, for example, only began to 'operate' in the 14th century BC at the earliest. We are not suggesting that neighbouring peoples did not establish contact with each other before this, but that larger items displaying decorative elements can only have been taken to far-away lands when the required means of transportation became available. There are many possible explanations for similar decorative elements in different places, one of which can be from humanity's natural environment.

Examples of decorative elements appearing in more than one place are the archetypes of rotational symmetric forms, like the dolphins and fishes previously discussed, or the mirror symmetric two-headed eagles that became symbols of power. Non-figurative motifs were increasingly abstracted from depictions of flora and fauna, and periodically repeated ribbon and tiling patterns developed. Similar decorations have been preserved for posterity, whether painted, carved, engraved or made up of mosaic, on friezes, larger surfaces, bowls and other utensils, from a variety of cultures.

These include motifs that occur in a large number of cultures, that can almost be termed universal. Other decorative elements are characteristic of particular peoples and tribes. When cultures meet, the various national elements can influence one another and become layered upon one another. Archaeologists and historians can provide a lot of assistance in researching the history of a particular people, and in determining the origins and affiliation of individual relics that are discovered. The repetition of the patterns (their symmetry elements) display common features everywhere. This shows us that while repeated and transformed elements that appear in the patterns may be varied, the symmetries themselves are nevertheless universal basic phenomena of human culture. We address these common features below.

To take an example, the original motifs of Hungarian decorative culture, the artistic roots of the ancient Uralic communities, were complemented during migration of the tribes with elements from the Steppe, from Turkistan, and from Persia. This is why Hungarian decorative art partly displays similarities to the decoration of far-away lands like Iran (Persia) and the motif world of the Turks. The elements that have influenced Hungarian accessories culture also made their way in the other direction, from Central Asia to the Far East, thus establishing very distant, indirect, influences. In the same way, of the decorated objects left to us from later periods, it is the embroidery, principally cross-stitch, that displays similarity to the textile motifs and embroidery of the other peoples of East-Central Europe.

Classification of motifs

First of all, we organize the decorative motifs according to the number of dimensions they exist in. One-dimensional motifs — which lie along a single line, usually a band (ribbon) — we call frieze patterns. For two-dimensional motifs — which can continuously be laid on a larger surface, usually a plane (without any gap or overlap) — we normally use the specialist term wallpaper pattern (and refer to their regularities as wallpaper groups), and more rarely call them tilings, tessellations or plane patterns. In three dimensions it is technically harder for us to present decorative motifs. (True, architecture divides space into cells, or rooms, but it is mostly in glass palaces that this could become visible, and for the pattern of their repetition to be appreciated. We cover the architectural "decoration" of space at the beginning of Chapter 6.) We can admire the 3-D structures created by nature, however. We call these crystal structures. Their classification and the discovery of the regularity of their symmetries succeeded in uncovering their secrets, and in producing new, man-made crystals. Given the technological opportunities now afforded us, they are as much a part of our decorative arts as are frieze and wallpaper patterns.

Some motifs that display repetitions (symmetries) go further than the one-dimensional structure of friezes, but do not exploit the potential of the two dimensions of a plane. Similarly, there are shapes with exceed the two dimensions of the plane, but which do not make use of the three dimensions offered by space. We refer to these as structures of non-integer (fractional) dimensions, or as fractals.

We normally take two considerations into account when classifying motifs — regardless of their dimension.

One is the basic element, or design, that is repeated. A planar example is a crocheted tablecloth composed of repetitions of the same motif; a spatial one is the elementary cell. We usually circumscribe the basic design of the motif with the help of a concept borrowed from crystallography, point groups. The other is the order of repetition: the potential operations and transformations with which one of the elementary motifs can be brought into coverage with the others. These potential operations determine a lattice on the plane or in space. The lattices are represented by the angles given by the edges of the cells, and the length of the edges. Together, we term these defining data of the lattice as the lattice parameters. Again we use a specialist term from crystallography, space lattice, to refer to them (even if the lattices in question are only two-dimensional). Together, the point groups and the space lattices determine the space group structure of a pattern.

The basic motif (point group) of a design can be completely asymmetrical — displaying single-fold rotational symmetry, which is transposed onto itself with a 360° rotation — but with translations and reflections a symmetrical pattern can be made out of it. It can be a floral or human motif, a human figure, or even a complete scene — let us just think of the friezes running around Greek amphorae — but it can also be an abstract design, like a snail's trail.

In other instances the basic motif can display a number of symmetries. If we take a single point as the basic motif, for example, this possesses all possible symmetries, for it can be rotated around itself in any number of ways, can be reflected in any axis that goes through it: all of these operations will transform it onto itself. A basic plane motif can be a figure both with mirror symmetry and n-fold rotational symmetry.

with identical, regular elements, leaving no gaps:

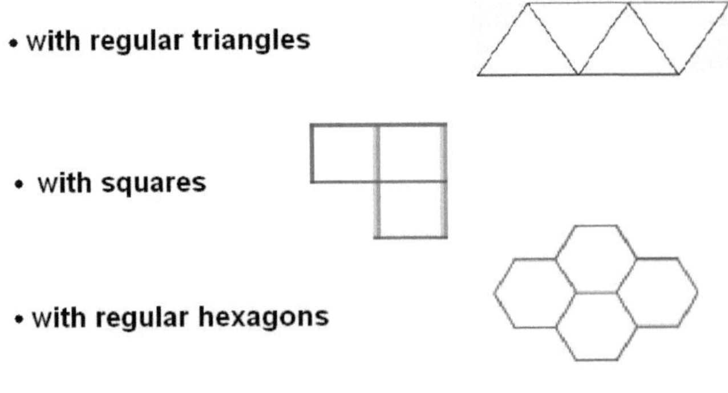

with irregular polygons, or with differing elements

Figure 3.1. *Filling the plane without gaps*

Many operations can be applied to a basic motif which can build a pattern. We can place a point group at the vertices of plane and space lattices of various shapes, or at the centre-points of their edges, faces and cells. We can make one overlap the other using some kind of symmetry operation (translation, reflection, rotation or glide reflection). The basic motif must satisfy two conditions, however: whether it is a plane or space lattice, we require that it should fill the plane or the space without gaps, in such a way that it be made up of *congruent cells* (of one given type).

These two conditions present certain restrictions. If we want to be very strict, we could also require that we fill the plane in continuous fashion with *regular*, congruent *polygons*. (The edges and angles of regular polygons are all equal.) Only three shapes are capable of covering the plane in this way: the regular triangle, the square, and the regular hexagon (Figure 3.1).

Though mathematically it is easy to prove, it nevertheless remained a mystery for centuries why the plane cannot continuously be filled with regular pentagons or polygons with more than six edges, while this is possible with irregular pentagons, or why, for example, the surface of a sphere can be covered continuously with regular pentagons, but not with only regular hexagons. These mysteries, which have inspired many artists and scientists alike, we will return to in a later chapter.

Table 3.1.

Lattice	Basic cell	System	Symmetry
oblique	parallelogram	parallelogrammatic	1, 2
primitive rectangular	rectangle	rectangular	1*m*, 2*mm*
centred rectangular	rectangle	rectangular	1*m*, 2*mm*
square	square	square	4, 4*mm*
hexagonal	60° rhombus	hexagonal	3, 3*m*, 6, 6*mm**

**Of the earlier examples, tiling with regular triangles and hexagons are included here as special cases of the 60° rhombus cell. This means no distinction from the point of view of their symmetry groups. Both a regular hexagon and 60° rhombus can be composed of regular triangles*

Similar symmetry problems are raised by the tiling of the plane with circles with identical radii. What is the densest possible arrangement? Or in the case of space: what is the densest possible filling of a given volume with identical spheres? (We consider the latter as Kepler's problem.) While the planar problem can be traced back to regular polygons on the basis of their inscribed circles, in space it is certain that the densest arrangement is not given by polyhedra inscribed inside the spheres or by spheres inscribed inside the polyhedra. In general, problems of this nature can be traced back to variational problems, where of the potential variations it is the most symmetrical solution that gives the extreme value (the minimal or maximal use of space, for example). To give a simple example, in borderline instances the inequality between the arithmetic and geometric mean transforms into equality (an extreme value) when the elements in question are equal; among rectangles of the same perimeter, to give another example, the one with the largest area is that with edges of equal length, i.e. the most symmetrical rectangle, the square.

To return to our original problem, if we drop the condition that the plane be filled with a mosaic lattice consisting purely of regular polygons, this gives us five different types of mosaic lattice. This is shown in Table 3.1.

The aforementioned conditions represent similar restrictions in the three-dimensional context. Of the regular bodies, the space can be filled without gaps only with cubes, or with a combination of tetrahedra and octahedra. If we relax the regularity condition, however, keeping only the continuity and congruence, we are still only left with fourteen possible space lattices. These are surprisingly small figures! From the study of crystals, we know that nature does not allow us to relax any further conditions.

We should note that so far we have only paid attention to the geometrical characteristics of the potential motifs. If we include further possible characteristics in the scope of our examination, we may find further variations. One example is colouring. What we lose thanks to prescribing the condition of perfection (continuous filling with the same shape) we can comfortably regain with other characteristics. It is enough for us to consider that we have made no condition with regard to the point groups that can be used, so in principle we can place anything on the vertices of the lattice, especially if we only try to apply translation symmetries (for this makes no condition on the point groups placed on the lattice points). Our inventory of attainable patterns is thus infinitely rich.

Those operations which transform particular lattice vertices onto themselves create a group. We became acquainted with the concept of a group in the first chapter. We saw two examples. Now we will see how groups can be used to describe symmetry operations.

We described the two-dimensional rotational group in Chapter 1 with the help of the matrix of rotations. Space rotations can be created on the same basis. We showed that rotations of arbitrary angles represent a group, for they do not produce results outside the limit of the group, successive rotations can be associated, there exists an identity element (rotation through 0°), and an inverse element (rotation in the opposite direction through an angle with the same absolute value).

We saw that the set of integers forms a group with regard to addition. Translations are characterized by the same group. Two different translations along a given distance are both translations, successive translations can be associated, there is an identity element, that is the translation by a distance of 0, and an inverse element, a translation in the opposite direction but by the same distance.

Reflection is a group whose operation transforms the two objects (*a* and its mirror image) onto one another. There is a neutral element, that which does not affect either of them, and an inverse element, which enacts the reflection in the opposite direction. If we perform this twice, we are given the neutral element, for we are returned to the original object.

A given object can have numerous symmetry characteristics. In such case, more than one symmetry operation can be carried out on the same object. In the case of numerous symmetry operations, their symmetry groups are "added together", i.e. the matrices representing their groups are multiplied together.

Crystallography uses a variety of notations to characterize symmetry operations. Our goal is not to list these in detail, but if we wish to see the order in them, we must make use of at least one of these notations. We can characterize the symmetries of an object with the notation of the operations which make up its symmetry group. We use the following notation:

m	mirroring in an axis
g	glide reflection (translation + reflection)
1	translation by one unit
2, 3, 4, 6	2-, 3-, 4-, 6-fold rotation (by *n*-fold rotation we mean that the figure is transformed onto its original self after *n* rotations through angle $360°/n$)
p	primitive cell (which only has objects on the vertices of the lattice)
c	centred lattice (which contains an object in the centre of one of its lattice element)

One-dimensional decorative elements: friezes

We can arrange a given object (in crystallography, point groups) along a straight line in seven different ways. Frieze groups are composed of particular point groups and the operations (transformations) applied to them. Whatever basic motif we place on the point group, the basic unit can be transformed in the following seven ways. To put it another way, taking all possible ways into account, the following seven frieze groups can be produced from the applicable group operations (symmetry transformations) used, i.e. translation, reflection, glide reflection and half-turn. These can be seen in Figure 3.2, together with notation in line with the above conventions.

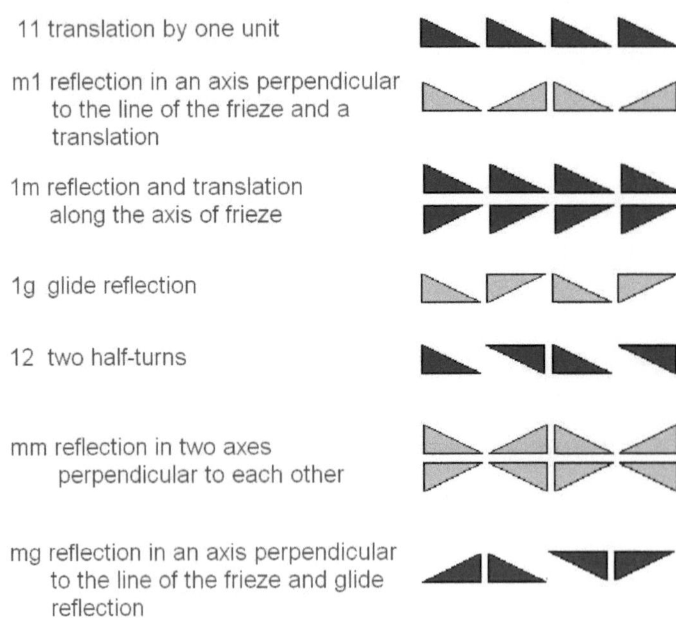

Figure 3.2. *The seven frieze groups*

It can be proved that there are seven, and only seven possible frieze groups. If we use two colours to decorate the basic motif (point group) of the friezes, we can count 17 possible frieze groups. If we place frieze motifs along both sides of a ribbon, we call this a ribbon group. With a single colour, 31 ribbon groups can be established. To illustrate what a rich variety of one-dimensional patterns can be created with help of the seven frieze groups, we will provide some examples on the basis of A.V. Shubnikov's illustrations (Figures 3.3, 3.4, 3.5, 3.6, 3.7, 3.8).

Group 1 is the simplest: translation by one unit of the basic motif.

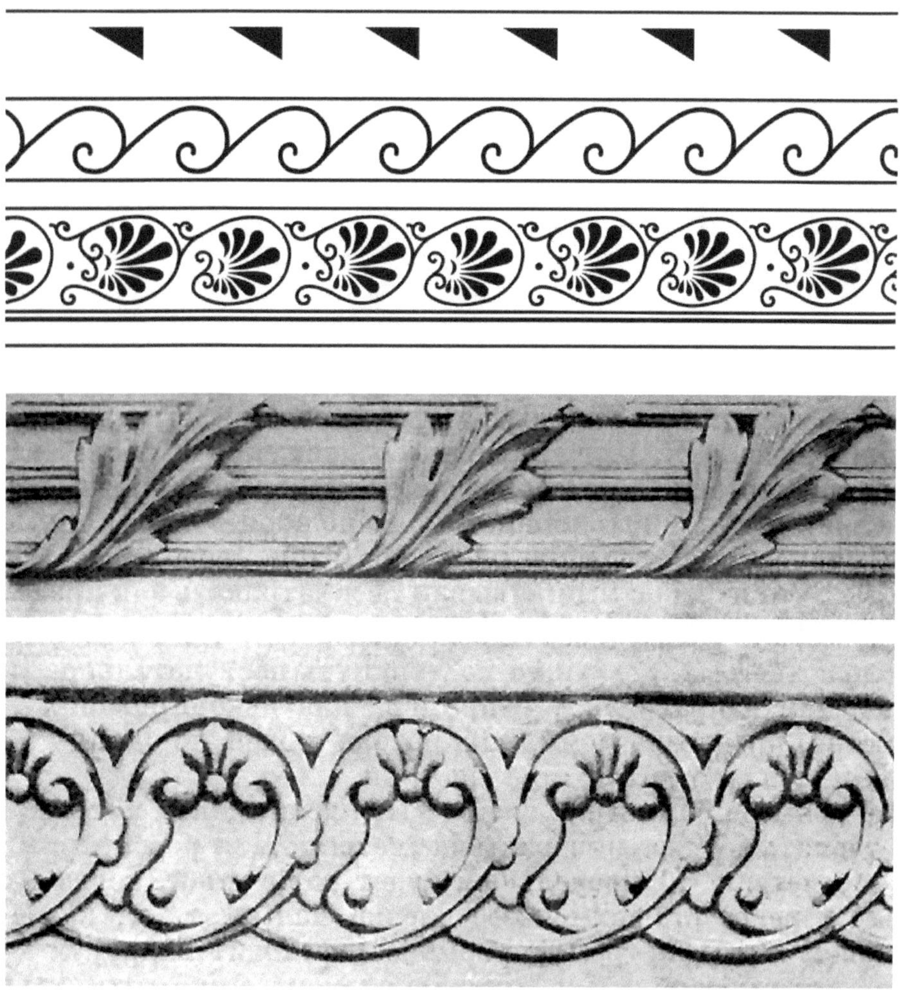

Figure 3.3. *Examples of frieze group* 11

In Group 2, we reflect the basic motif in an axis perpendicular to the longitudinal axis of the frieze (and translate by one unit).

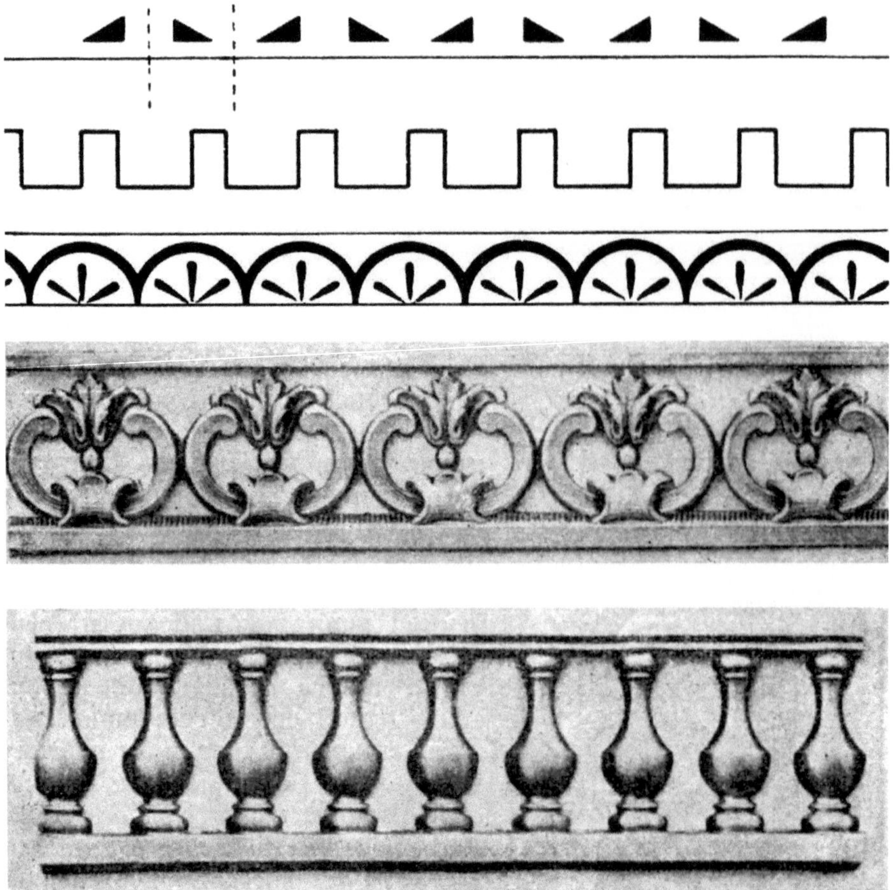

Figure 3.4. *Examples of frieze group* m1

In Group 3, after executing the reflection in the line of the frieze, we move the reflected element and its mirror image together along by one unit.

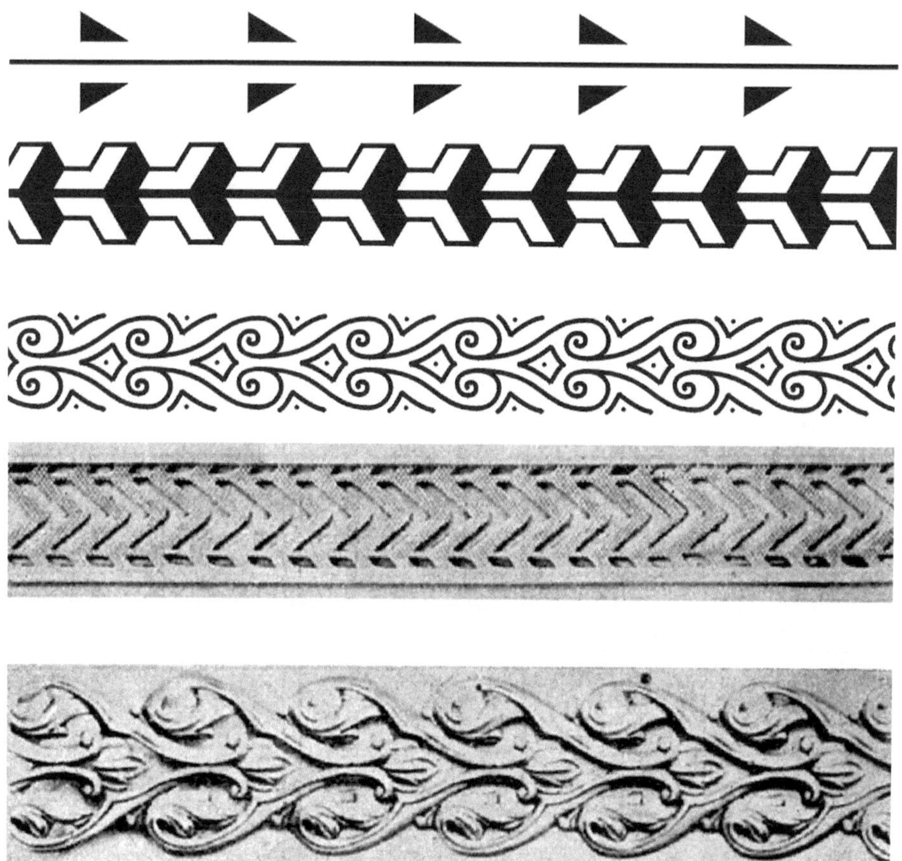

Figure 3.5. *Examples of frieze group* 1m

Group 4 is the example of glide reflection. It is made up on two operations (symmetry transformations): we reflect a basic motif in the axis of the frieze, then perform a translation by one unit. Although glide reflection is a combined symmetry transformation made up of two operations, in line with the traditions of crystallography it is considered as an independent basic symmetry operation.

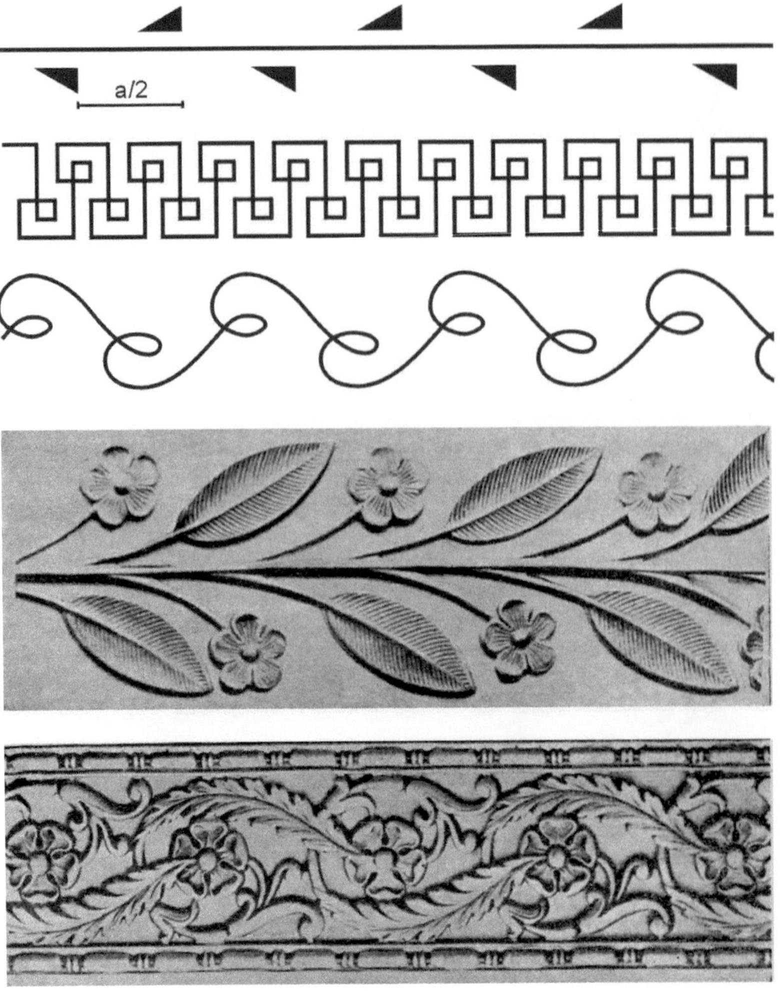

Figure 3.6. *Examples of frieze group* 1g

Group 5 consists of successive half-turns (rotations by 180°), always translated along by one unit. A characteristic example of this group is the already much mentioned meander motif so commonly found in classic Greek friezes.

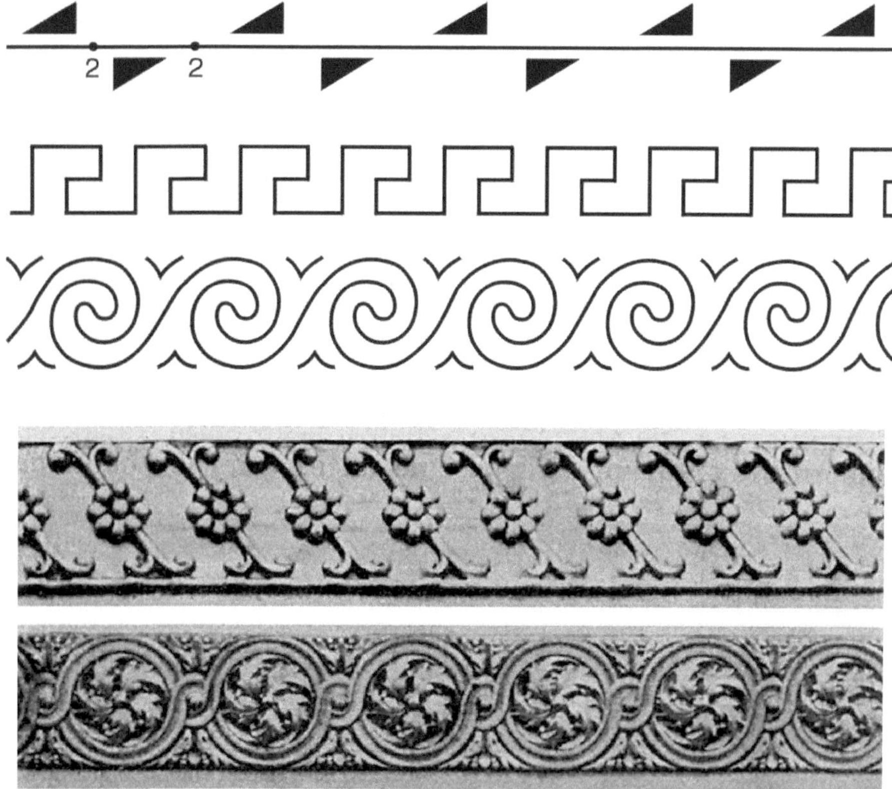

Figure 3.7. *Examples of frieze group* 12

Group 6 consists of reflection in two axes perpendicular to one another.

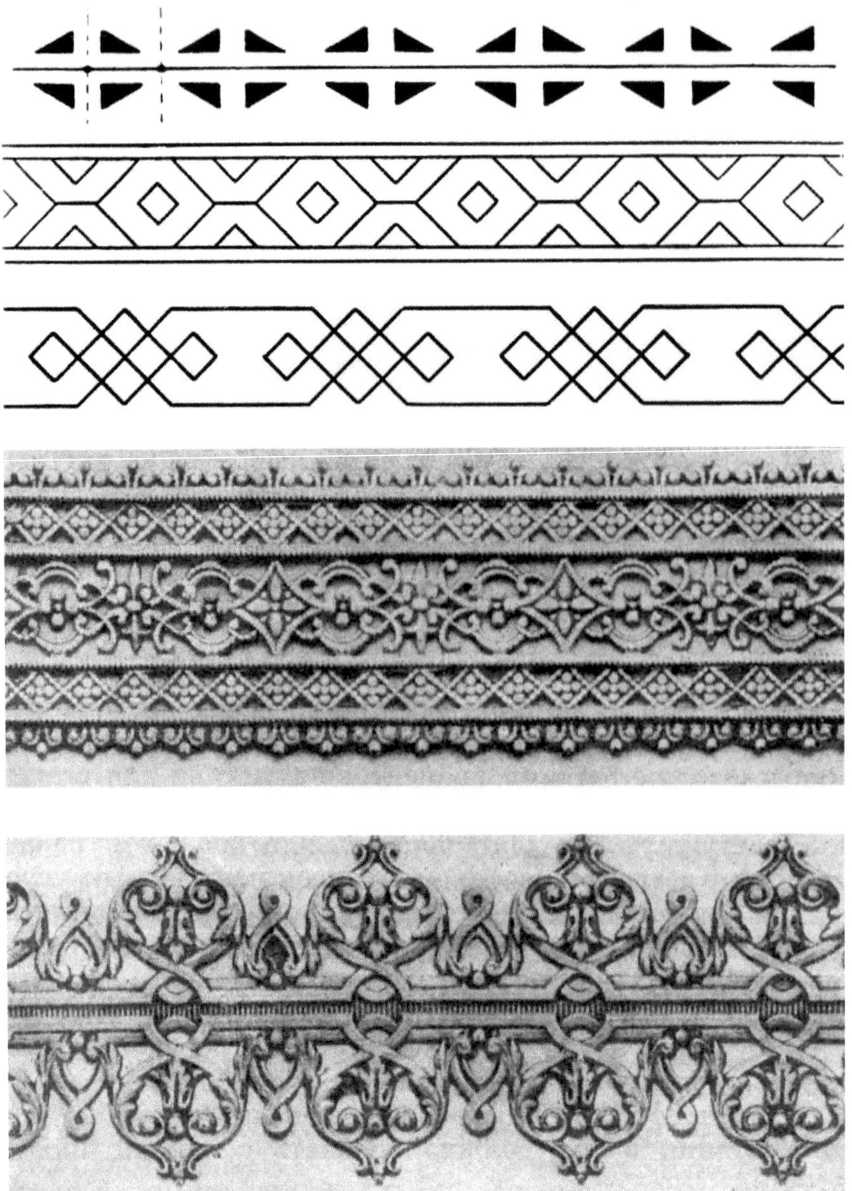

Figure 3.8. *Examples of frieze group* mm

Group 7 is a combination of reflection in an axis perpendicular to the axis of the frieze and a glide reflection.

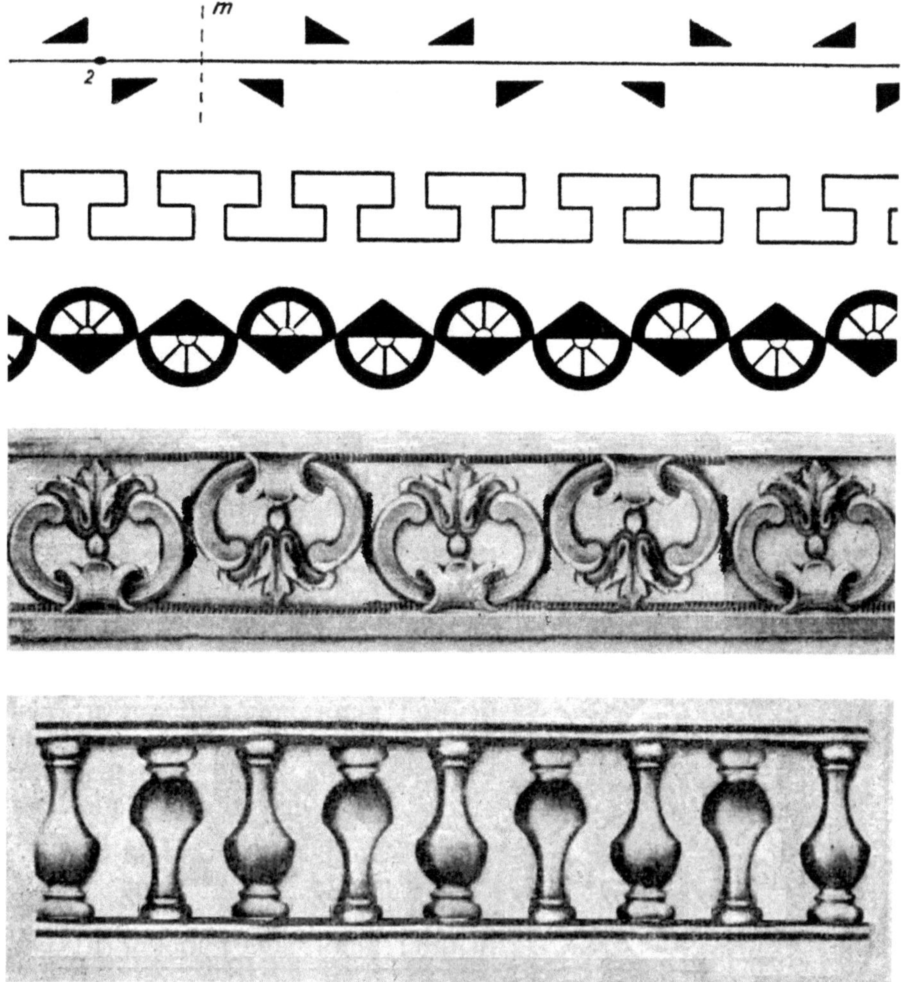

Figure 3.9. *Examples of frieze group* mg

Finally, to introduce the seven basic friezes, we present characteristic examples from the drawing of footprints in the sand (Figure 3.10), and a selection of elements of Hungarian decorative art from the age of the Magyar conquest of Hungary (9th-10th c.), as seen in natural phenomena in Szaniszló Bérczi's collection (Figure 3.11).

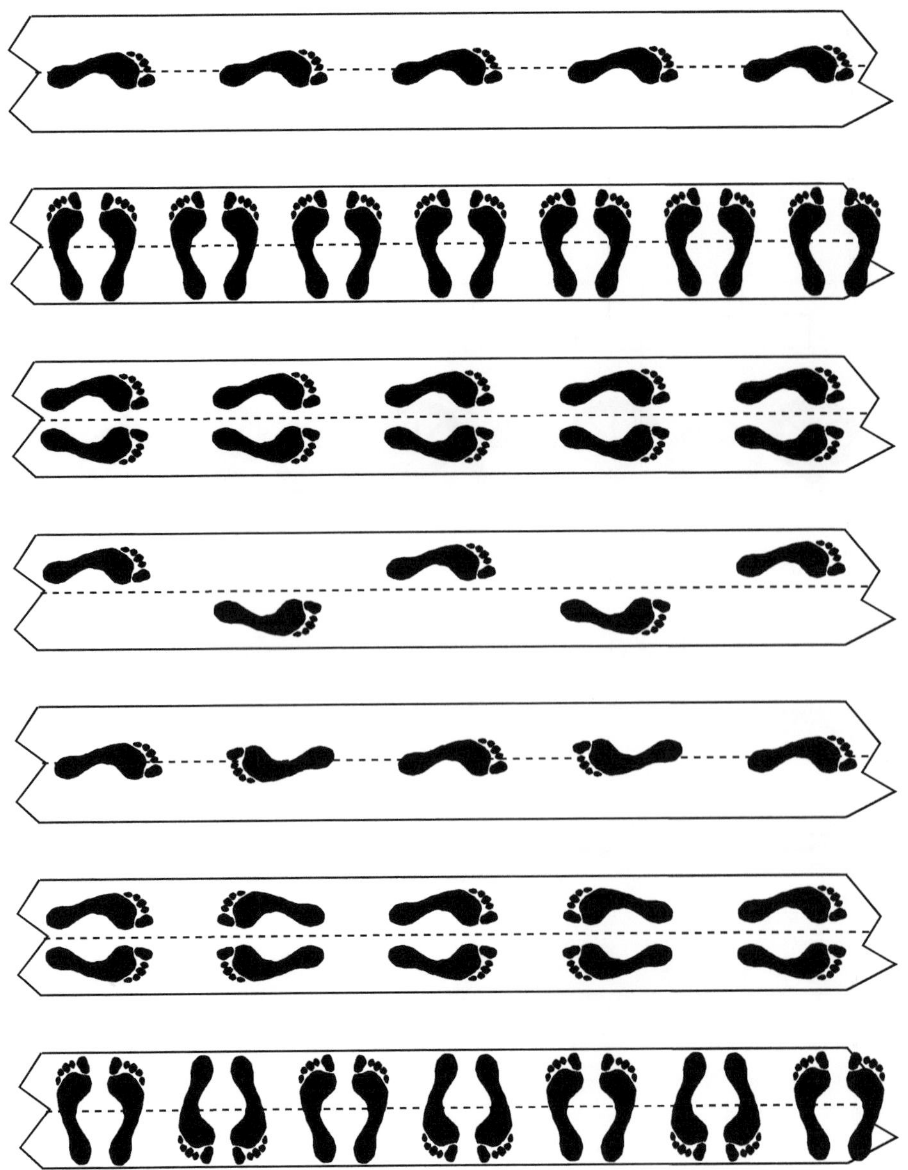

Figure 3.10. *The seven basic friezes illustrated with footprints*

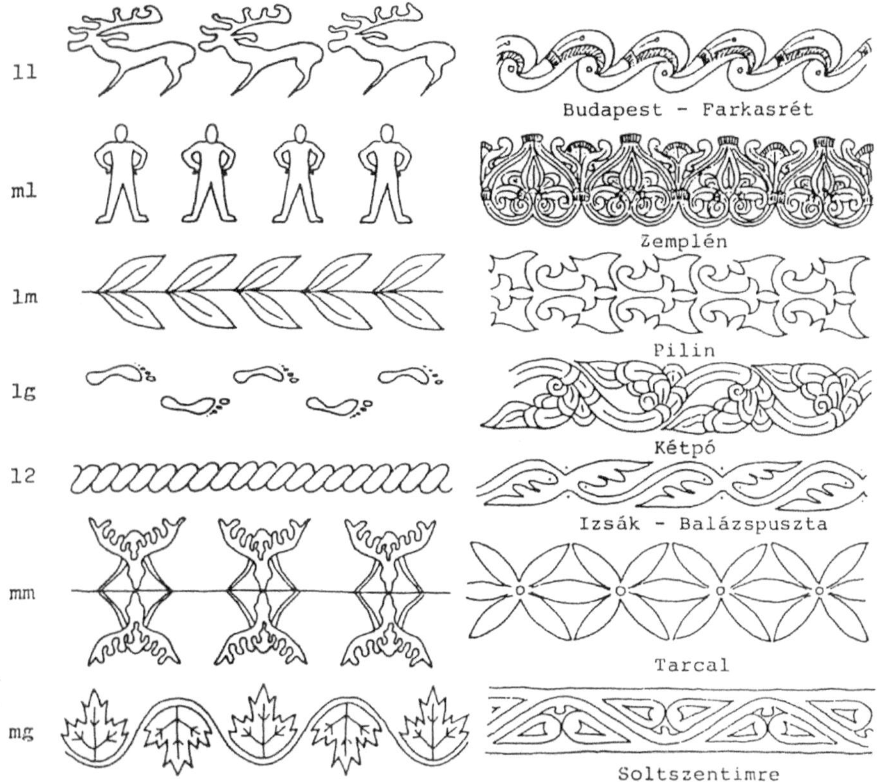

Figure 3.11. *The seven basic friezes: on the left, in natural phenomena, and on the right, in elements of Hungarian decorative art from the time of the Hungarian conquest*

It was using these regularities that ribbon and frieze designs and running building and vase decorations were created. (The visitor can view the best "exhibition" of one-dimensional decorative art in its historical context along the corridor connecting Heathrow airport with the underground station leading to central London.)

Tiling the plane

Filling the plane without gaps is often referred to as tiling or as wallpapering. The point of it is to fill the theoretically infinite planar surface in a continuous fashion with a single (uniform) repeated element. The big question is how many ways we can achieve this. For classification, we use the knowledge gained when we looked at friezes.

We can place an infinitely large number of basic motifs in the place of the individual point groups. There is a limit, however, to how many types of lattice we can use to cover the plane with infinite repetitions, and to the symmetries displayed by the point groups we place in the lattices in order for them to remain, together with the given lattice, invariant under a symmetry operation. The following example illustrates why the last condition represents a genuine limitation. We cannot place a point group with threefold symmetry onto the points of a square lattice, which is symmetrical with regard to reflection in its edges, while satisfying the condition for mirror symmetry, for the mirror symmetry of the pattern would not be preserved during reflection. It would not form a group, as reflection would produce an element which falls outside the group, which contradicts the closure condition.

On the basis of this, we can interpret the previously introduced concept of *point group* more precisely. In an exact sense, we consider a point group to be the sum of the symmetry operations that can be applied at one point of a planar figure or a spatial body (the term lattice cell applies to both) and under which the given figure or body (cell) is invariant.

We can also make our definitions more precise as concerns lattices. We give the name space lattice (Bravais lattice) to those lattices (even on a plane) which are required by the point group operations applied to the lattice points. What do we mean by required by them? We mean that only those lattices come into consideration that are of a symmetry that the given point group is capable of producing (on the basis of its own symmetries). The only lattices which can be attributed to a given point group (in other words, the only lattices which can be realised in nature) are those which possess the symmetries of that point group.

We can allow four kinds of plane transformation on the lattices: translation, rotation, reflection and glide reflection (Figures 3.12, 3.13, 3.14, 3.15).

Bearing all this in mind, in two dimensions five space lattices (Bravais lattices) are possible. It can easily be seen that no other lattice can be considered. These were listed in Table 3.1.

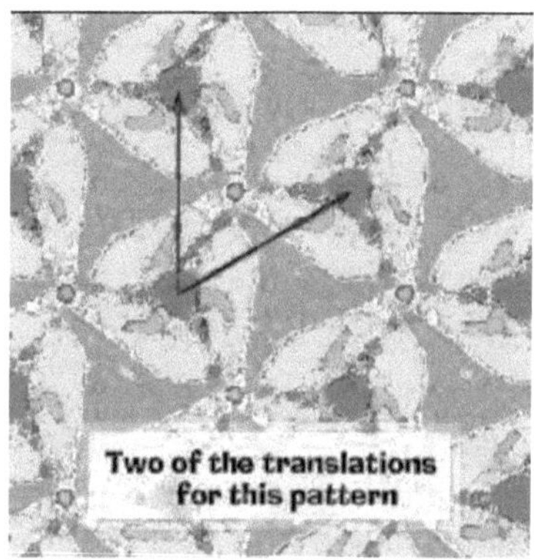

Figure 3.12. *Translation. We have marked two possible translations on the figure.*

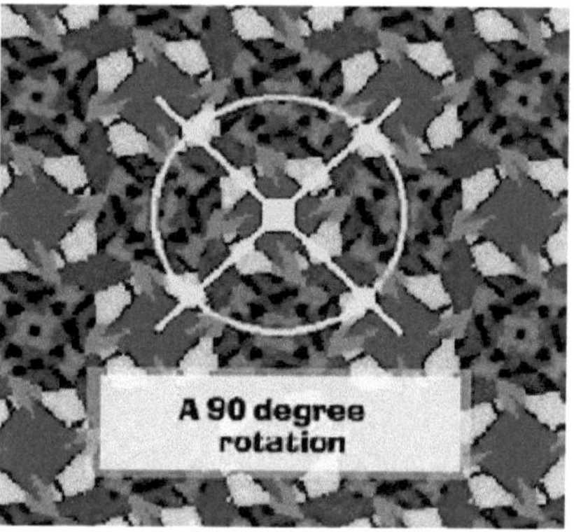

Figure 3.13. *Rotation. We have marked a rotation through 90° on the figure.*

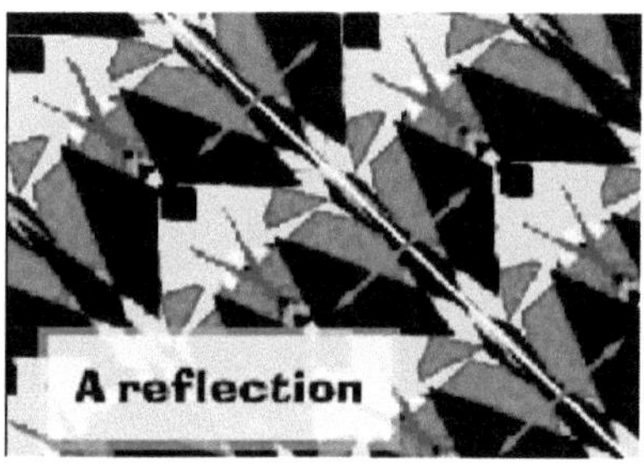

Figure 3.14. *Reflection. We have marked the axis of a reflection on the figure.*

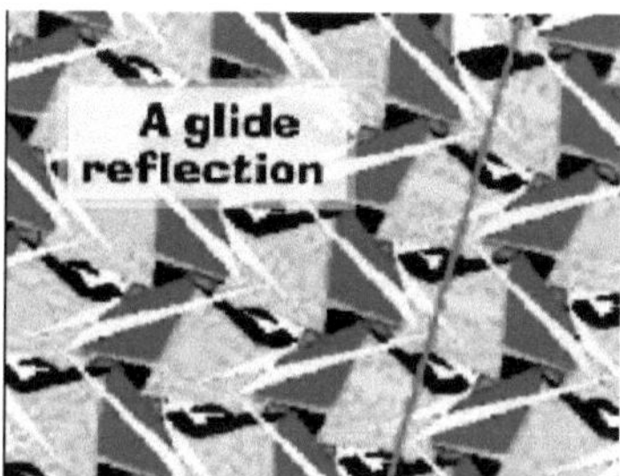

Figure 3.15. *Glide reflection. We have marked the glide reflection on the figure with an axis, and marked the place of the glide-reflected elements with arrows*

One characteristic of point groups is that they can be considered as being fixed on a single point. So we can exclude translation and glide reflection. So, in the case of point groups, only the following symmetry operations can be taken into account:

- 1-, 2-, 3-, 4- or 6-fold rotation around a point, or
- reflection in a straight (symbol: m)

Using the above operations both individually and together in two dimensions, the following ten point groups are possible (we can easily see that it is only these):

- 1, 2, 1*m*, 2*mm*, 4, 4*mm*, 3, 3*m*, 6, 6*mm*.

The ten possible two-dimensional point groups are displayed schematically in Figure 3.16, marking the symmetry operations used, and the points of equivalence.

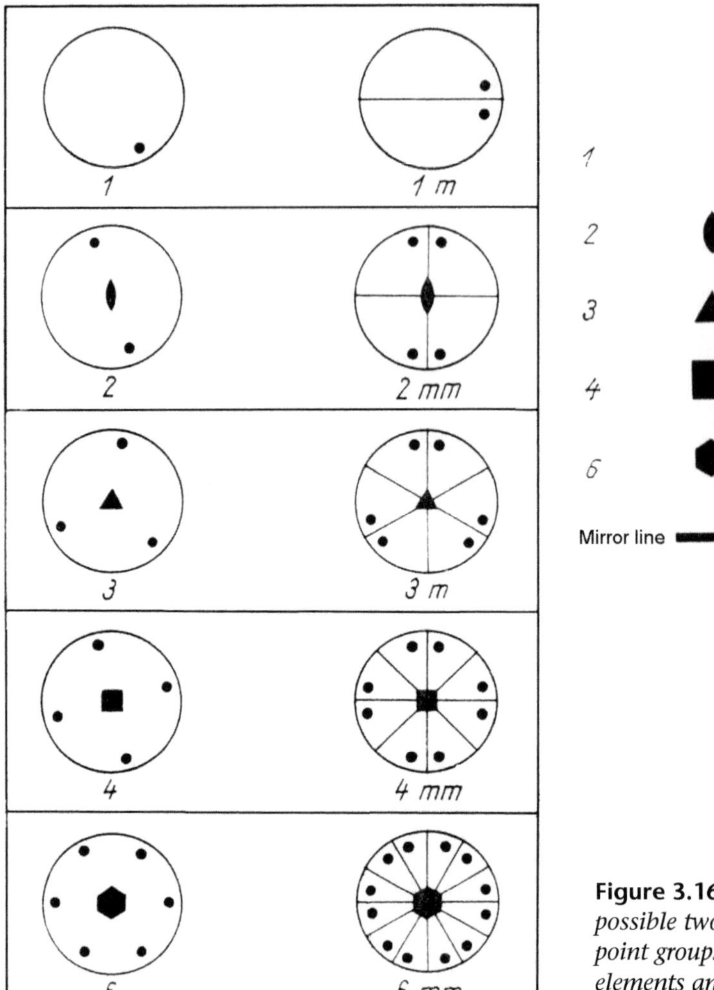

Figure 3.16. *The ten possible two-dimensional point groups, the symmetry elements and the equivalent points*

Comparing that said about space lattices and point groups, the ten point groups that can be placed on the lattice points of the five space lattices possible in two dimensions give a total of seventeen possibilities for tiling the plane. We call these wallpaper groups. It is such a small number because we saw that the symmetries of particular point groups make it impossible to place them on the point of any lattice. It can be proved that there are seventeen and only seventeen plane transformation groups that can exist.

To conclude, in 2D we have the following possibilities:

- 10 2D point groups
- 5 2D space lattices (Bravais lattices)
- 17 2D plane groups (wallpaper groups).

The wallpaper groups are summarized in Table 3.2.

Table 3.2. *Symbols and characteristics of the 17 wallpaper groups*

Symmetry group number	International Union of Crystallography symbol	Lattice type	Rotation order	Axis of reflection
1	*p*1	parallelogrammatic	none	none
2	*p*2	parallelogrammatic	2	none
3	*pm*	rectangle	none	parallel
4	*pg*	rectangle	none	none
5	*cm*	rectangle	none	parallel
6	*pmm*	rectangle	2	90°
7	*pmg*	rectangle	2	parallel
8	*pgg*	rectangle	2	none
9	*cmm*	rectangle	2	90°
10	*p*4	square	4	none
11	*p*4*m*	square	4*	45°
12	*p*4*g*	square	4**	90°
13	*p*3	hexagon	3	none
14	*p*3*m*1	hexagon	3*	30°
15	*p*31*m*	hexagon	3**	60°
16	*p*6	hexagon	6	none
17	*p*6*m*	hexagon	6	30°

* *All centres of rotation lie on reflection axes.*
** *Not all centres of rotation lie on reflection axes.*

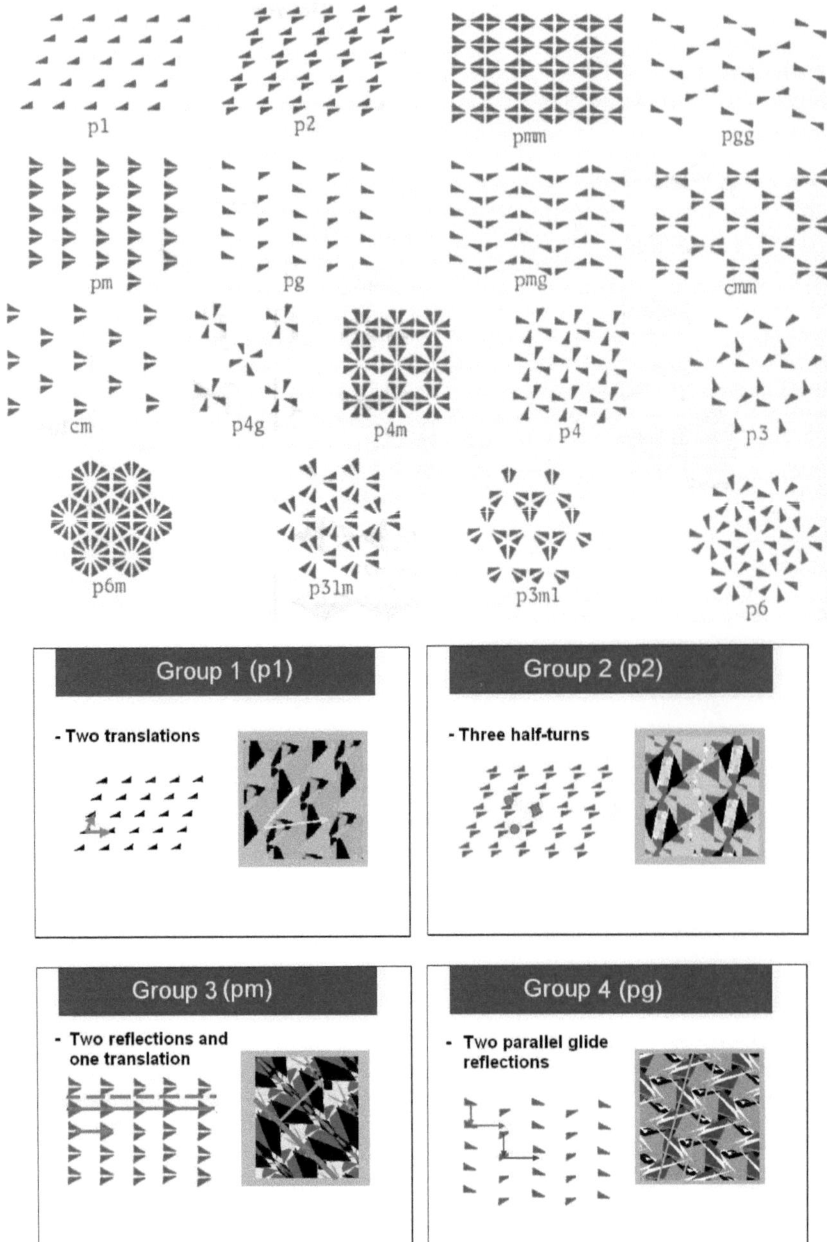

Figure 3.17. *A presentation of the 17 wallpaper groups*

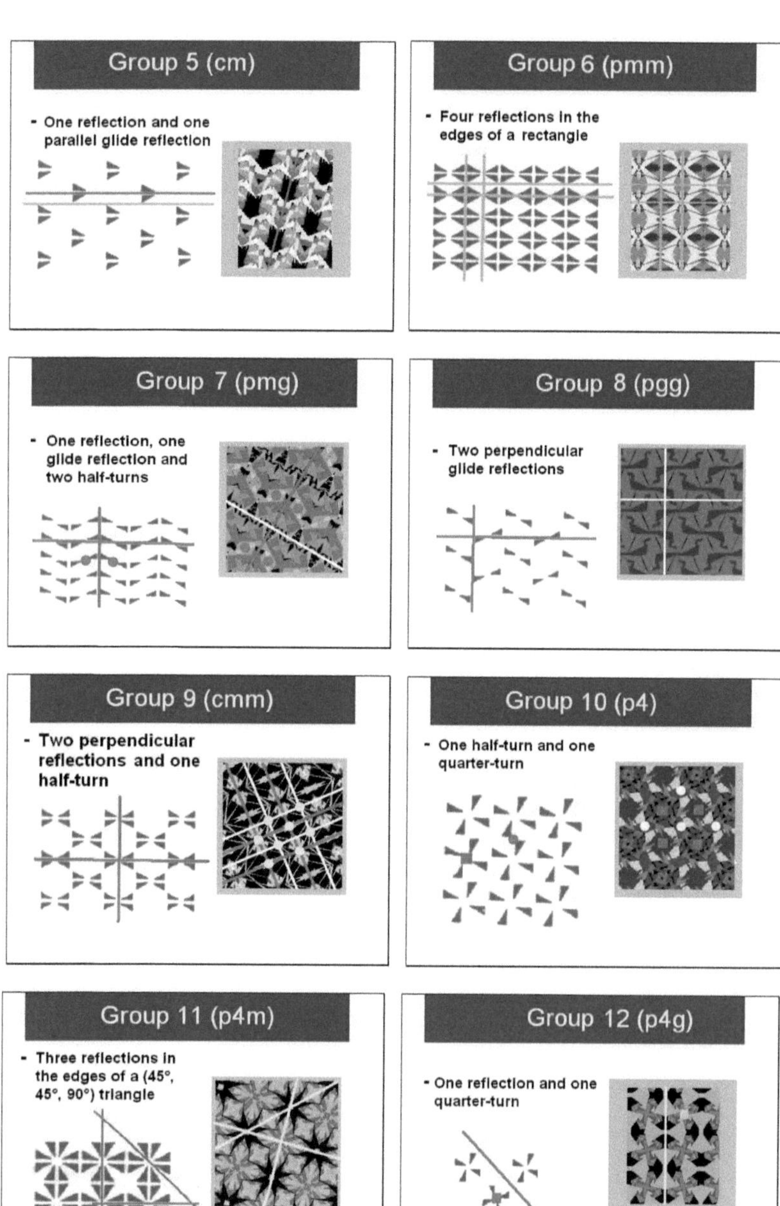

Figure 3.17. (cont.)

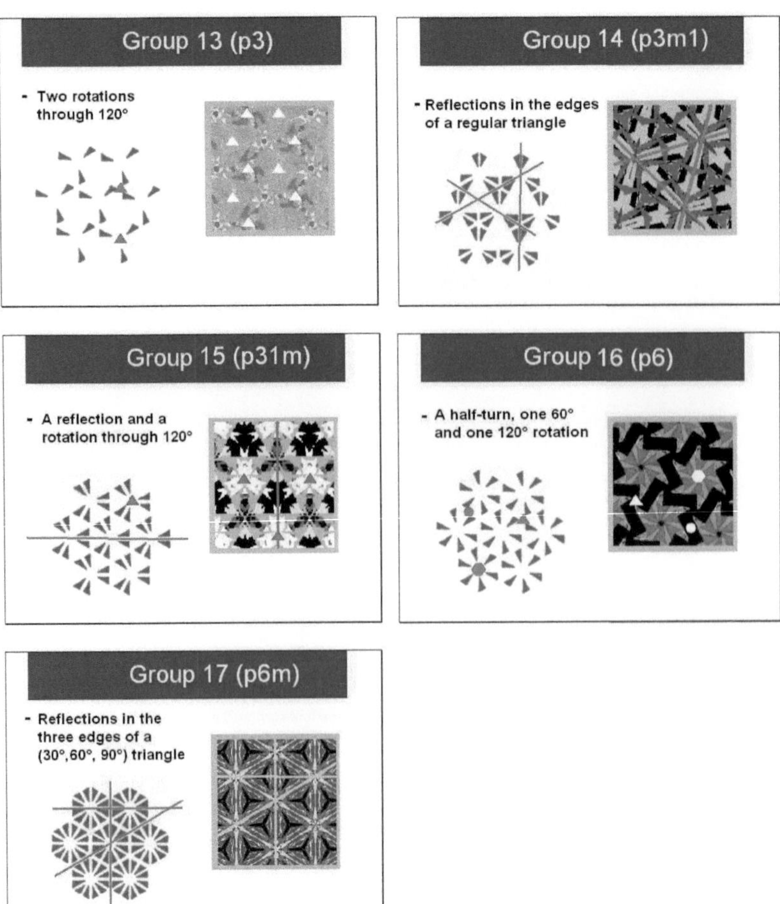

Figure 3.17. (cont.)

It can be shown that with two colours there are 48 wallpaper groups. It is evident that with a few colours even a single basic motif can be turned into a very rich world of pattern designs.

One of the most elegant examples of the multifarious opportunities for filling a plane is the collection of coloured tile mosaics covering the walls of rooms in the Alhambra palace in Granada. Among the mosaics in the various rooms, we find examples of each of the seventeen wallpaper groups within a single building (Figure 3.18).

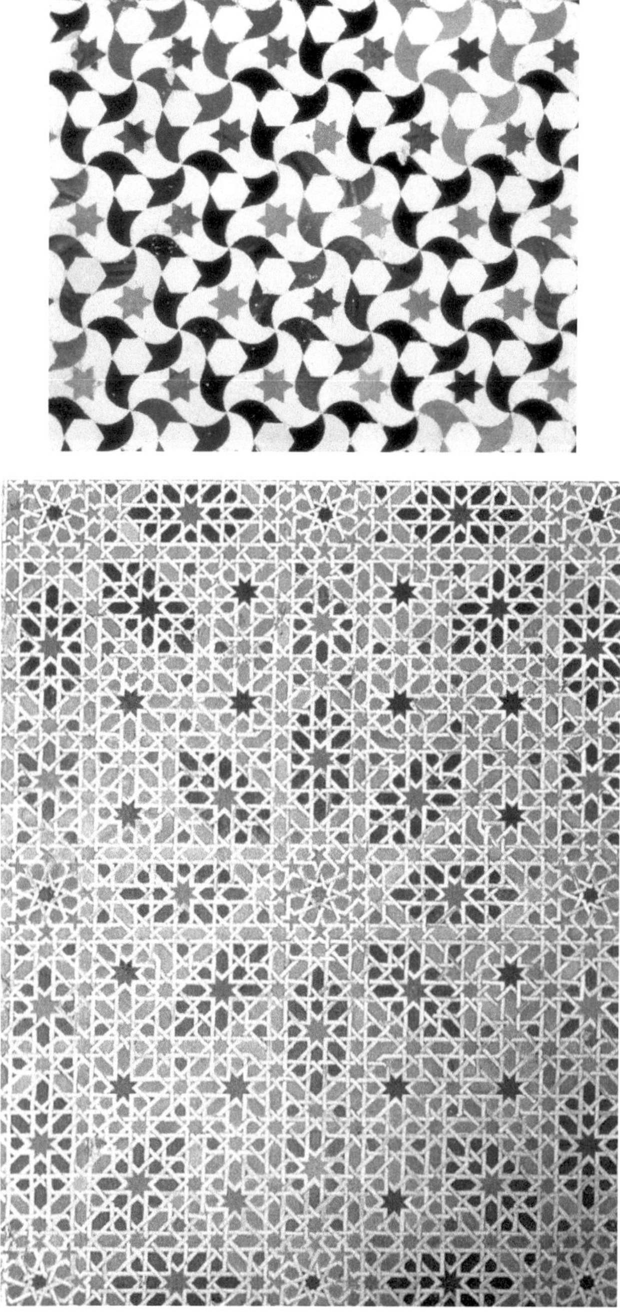

Figure 3.18. *Details from the tile mosaics of the Alhambra*

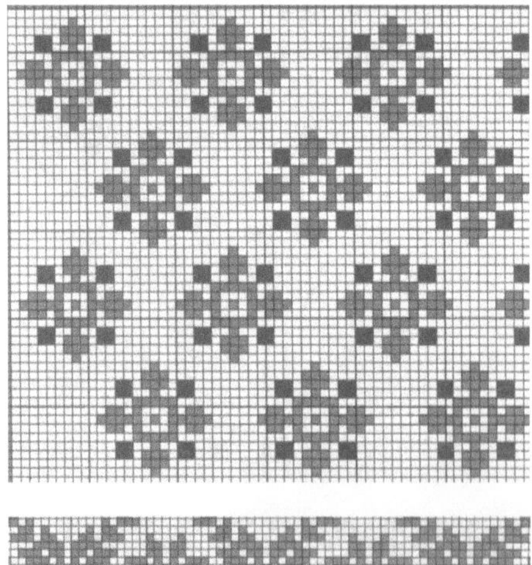

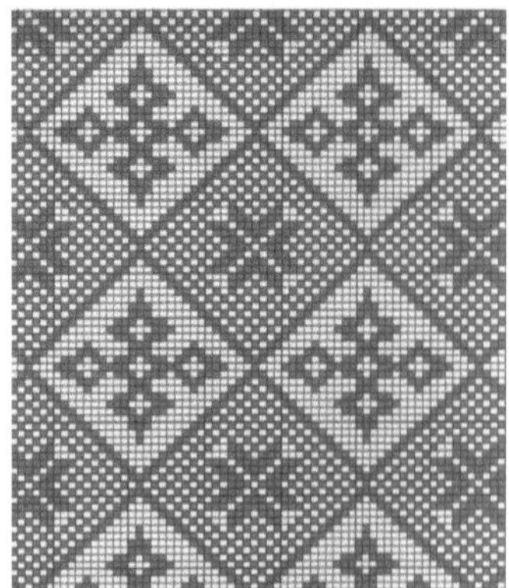

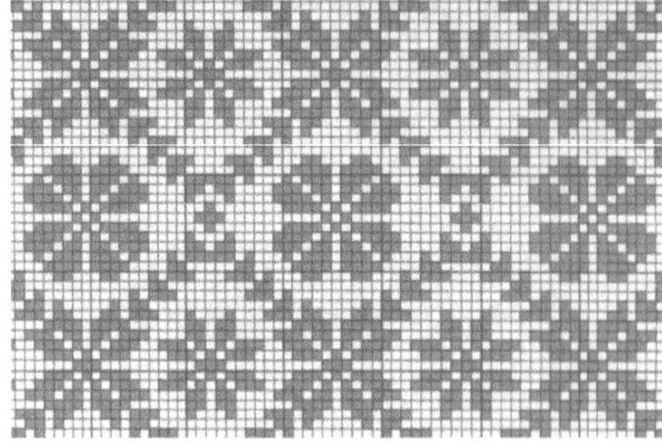

Figure 3.19. *Cross stitch embroideries from Hungarian folk art. (1) Somogy motif (2) Kalotaszeg motif (3) Baranya motif*

Another field that provides rich opportunity for coloured decoration of the plane is weaving and embroidery (Figure 3.19). The very nature of looms dictates that parallel lines can be sewn, with others placed at right angles to them, and so hexagonal space lattices cannot really come into consideration (though examples of weaving with three threads are occasionally to be found in W. Endrei's collection). Thus this set of applications only usually features the first twelve wallpaper groups. However, the triangular and hexagonal rotation groups can be found amongst the designs of lace-makers, and those of crochet-work.

The richness and beauty of plane tiling has been exemplified throughout the history of decorative art, from antiquity to the present day, by floor mosaics, parquet floors, tiles, and other repeated (wallpaper) designs filling a (plane or curved) surface. We could present thousands, even tens of thousands, of eye-catching examples. Our task, however, is not to inspire wonder at how many things in the world are symmetrical, but rather to show how systematic they are. So we leave it to readers to compare their own experiences with what they read about classification.

Periodic filling of space

The continuous filling of space with point groups to be placed in congruent elementary cells is a fundamental task of crystallography.

In order to determine the number of ways of possible space-filling, attention has to be paid to:

- cells of different shapes,
- point groups that can be fitted in the cells,
- the position of the point groups within the cell.

On the basis of this, in three dimensions there are a total of:

- 32 possible 3D point groups,
- 14 possible space lattices (Bravais lattices), and
- 230 possible space groups.

The various different crystals can be classified on this basis. We could hurry to predict the outcome on the basis of what we saw on the plane, but let us not be hasty. The world of crystals is unparalleled in terms not only of its richness, but of the cultural historical associations of the thinking connected with it. For this reason we devote a separate chapter to it.

Chapter 4

The golden section

Despite all rational explanations, a number of enigmatic questions were raised by the subject of tiling a plane and filling a space, the answers to which have been pursued for centuries, just like the secret of "magic numbers", behind which further mystical explanations were expected as to why these numbers are as special in nature as they are. The most perfect *planar* shape (that with an infinite number of symmetries) is the circle. The sphere is a *spatial* body with similar characteristics. A special proportion, π, is attributed to both. Another mathematical proportion, the base number for the natural logarithm, e, is attributed to the perfectly swirling line, the logarithmic spiral. These numbers are made all the more mystical by the fact that, as delineators of proportion, they are independent of the scale chosen, that is they remain the same even when measured in different units — and it is in this scale invariance that their symmetry lies.

An even more exciting experience than these irrational numbers is that there are certain natural numbers that nature gives special treatment. The Pythagoreans were the first to look for the secret of the proportions between the lengths of the strings that vibrated the most harmoniously, which they claimed to find in the ratio between the smallest natural numbers, and to which they gave priority over physical explanations. We have seen that the harmony they sought meant the same to them as what symmetry generally means to someone in our age.

The first example of the mysterious questions mentioned is why, of the regular planar figures, tiling the plane is only possible with triangles, squares and hexagons. And why can a sphere only be covered with regular triangles, squares and pentagons?

Why can filling space without gaps only be achieved with the cube (or the tetrahedron and octahedron combined)?

In inanimate nature, perfect symmetry is not attained with fivefold symmetry. How is it possible that in living nature we so often encounter fivefold and even higher-order rotation symmetries?

In the next three figures, we can see various fivefold symmetries that are found in nature. Among flora, for example, the common sunflower

(Figure 4.1), belonging to the Malvaceae, has five axis of reflection in addition to its fivefold rotational symmetry. In contrast, the flower of the evergreen myrtle (Figure 4.2) "only" displays rotational symmetry.

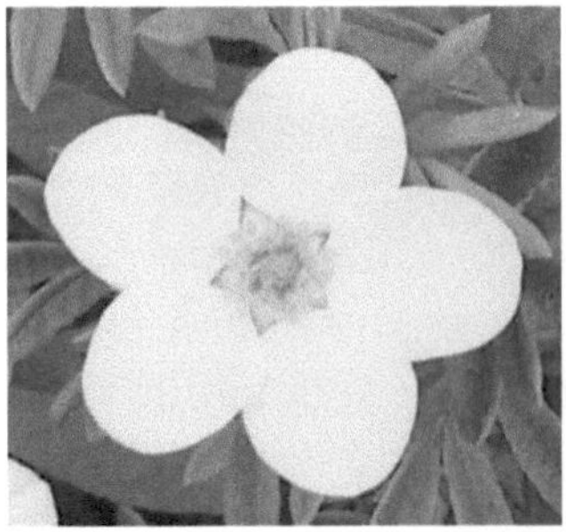

Figure 4.1. Common sunflower

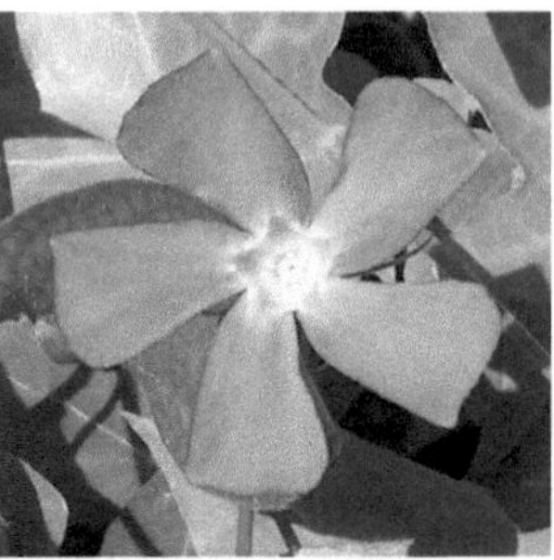

Figure 4.2. Evergreen myrtle

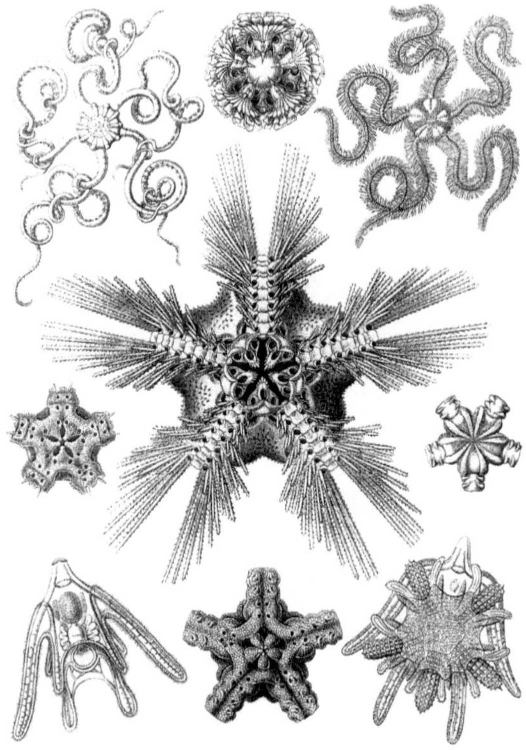

Figure 4.3. Echinoderm (*sub-order Ophidia*) displaying five-fold symmetry, which during its ontogenesis goes through an interesting metamorphosis: its larvae display bilateral symmetry

In the last decades of the nineteenth century, based on his own collections, Haeckel described and drew the morphological symmetry for many thousands of animal species in a number of volumes, including an enormous number that displayed fivefold symmetry. We have him to thank for the echinoderm in Figure 4.3 and the *Discomedusa* with eightfold symmetry in Figure 4.4.

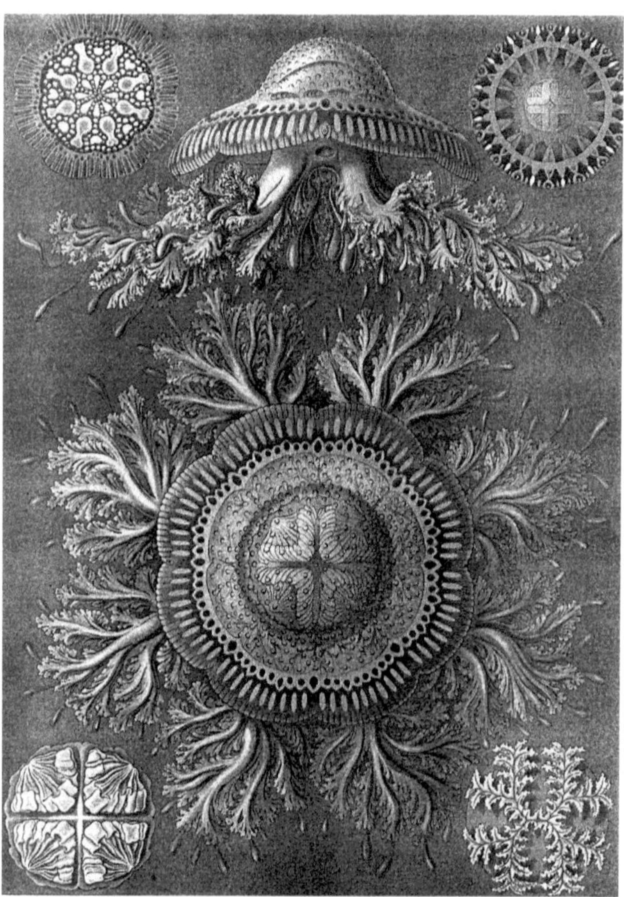

Figure 4.4. Eightfold symmetry in the animal kingdom: the Discomedusa

In general, we can observe that symmetries in living matter are related to members of the number sequence 2, 3, 5, 8, 13, ... etc.

What laws could govern the arrangement of living material that deviate from the organization observed in inanimate nature? This order is governed by another mysterious (?) proportion. First let us inspect the mathematical background to this order.

The proportion considered to be perfect

The geometrical properties of the golden rectangle

Let us draw a rectangle whose edges are such that if we use a straight line at right angles to its longer edge to divide it into a square and a rectangle, the rectangle thus produced is similar to the original one (Figure 4.5). Not all rectangles can be divided in this way: only those whose edges are in a given proportion to one another.

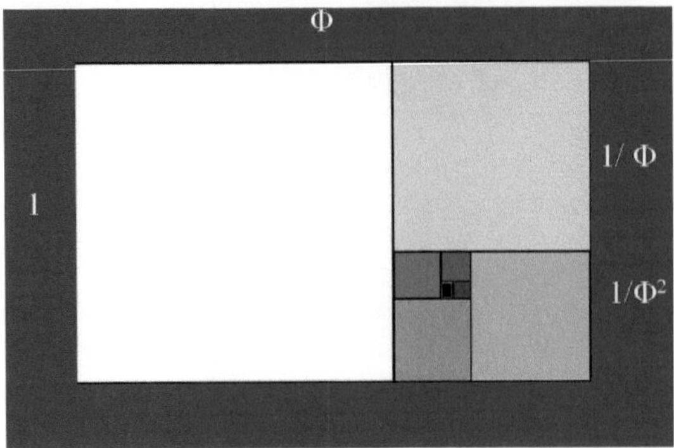

Figure 4.5. Golden rectangle

This particular rectangle has interesting properties, however. For example, as a result of the fact that the rectangle produced by the division is similar to the original, its edges are in the same proportion to one another as those of the original rectangle. So if we cut a square out of it we are again given a rectangle similar to the original one, which has the same properties, and so on.

The edge proportions of a rectangle with these properties have since antiquity been regarded as the perfect proportion. We denote this proportion with the symbol Φ; the rectangle with these proportions later came to be known as the golden rectangle.

The edges of the smaller rectangle after division of the golden rectangle are $1/\Phi$ of the edges of the original. The shorter edge of the individual golden rectangles is to the longer edge as the longer edge is to the sum of the two edges. If we denote the edges of the original rectangle as a and b, then we can express this as an equation:

$$b : a = a : (a + b)$$

From which

$$a^2 - ab - b^2 = 0$$

if $b = 1$ unit, then

$$a^2 - a - 1 = 0,$$

and as in this instance we assume

$$a = \Phi$$

we can write the equation as follows:

$$\Phi^2 - \Phi - 1 = 0$$

For the time being, let us keep to the geometric interpretation of Φ. Φ can be constructed with a compass and a straight-edged ruler. Let us start from the positive root ($\frac{1 \pm \sqrt{5}}{2}$) of the second-order equation given for Φ. For a section, let us measure out two units, and one unit at right angles. Let us extend the hypotenuse of the resulting right-angle triangle ($\sqrt{5}$) by one unit, and divide the whole section in two. We get precisely Φ (Figure 4.6).

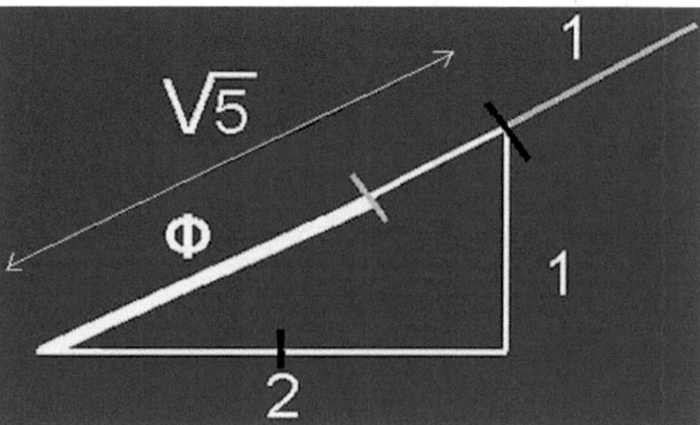

Figure 4.6. Construction of Φ

The trigonometric properties of the golden rectangle can be read from Figure 4.7. Let us take the shorter edge of the golden rectangle as one unit, meaning that its longer edge is Φ. Let φ be the angle between the diagonal

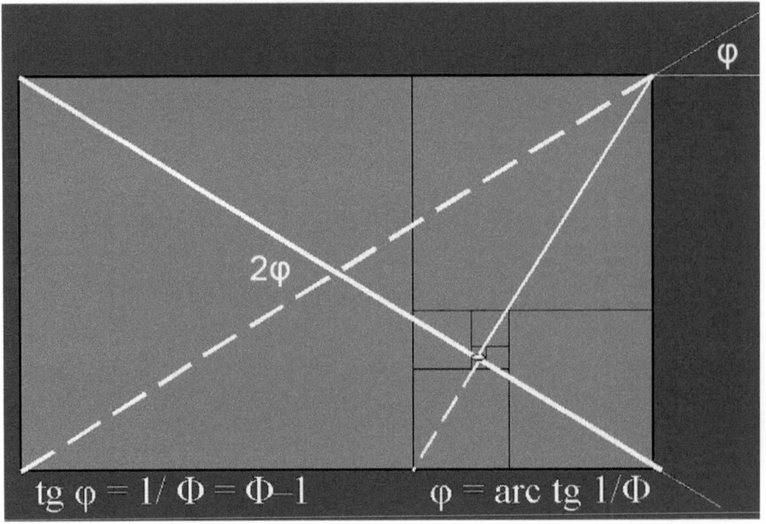

Figure 4.7. *The trigonometrical properties of the golden rectangle*

of the golden rectangle and its longer edge. Let us look for the relationship between angle φ and the proportion Φ of the edges.

We can see from the figure that

$$\mathrm{tg}\varphi = \frac{1}{\Phi} = \Phi - 1.$$

From this,

$$\varphi = arctg\frac{1}{\Phi} \quad \text{and} \quad 2\varphi = 2arctg\frac{1}{\Phi}$$

Let us use the following well-known relationship between the tangents of double angles: ($\mathrm{tg}2\varphi = \frac{2\mathrm{tg}\varphi}{1-\mathrm{tg}^2\varphi}$)

$$\mathrm{tg}2\varphi = \frac{2\frac{1}{\Phi}}{1-\left(\frac{1}{\Phi}\right)^2} = \frac{2\left(\frac{1}{\Phi}\right)}{\frac{\Phi^2-1}{\Phi^2}} = \frac{2\left(\frac{1}{\Phi}\right)}{\frac{\Phi}{\Phi^2}} = 2$$

$$\mathrm{tg}2\varphi = 2 \rightarrow 2\varphi = arctg2 \rightarrow \varphi = \frac{1}{2}arctg2$$

This gives us an interesting result: the angle φ can be expressed as a function of the smallest natural number other than 1.

Meanwhile we also produced φ above as a function of Φ: ($tg\varphi = {}^{1}/_{\Phi}$). Replacing the expression for φ gives us

$$\frac{1}{\Phi} = \mathrm{tg}\frac{1}{2}arctg2,$$

with which we have expressed Φ as a function of 2.

To summarize, we can express both the angle φ and the proportion of edges Φ with the help of simple natural number '2':

$$\varphi = \frac{1}{2}arctg2$$

$$\frac{1}{\Phi} = \mathrm{tg}\frac{1}{2}arctg2$$

The two equations hold for all golden rectangles. The latter expression is a geometric way of producing the golden section.

The algebraic characteristics of the golden rectangle

To determine the size of Φ, we previously produced a simple algebraic expression: $\Phi^2 - \Phi - 1 = 0$. Rearranging, $\Phi^2 = \Phi + 1$. From this can be seen the first interesting characteristic of Φ: its square is exactly one unit larger than itself. Rearranging again, $\Phi^2 - \Phi = 1$, which gives us $\Phi - 1 = 1/\Phi$. And this shows that the reciprocal of Φ is exactly one unit larger than itself.

Solving the second-order equation, we get two roots:

$$\Phi_1 = \frac{1+\sqrt{5}}{2} \quad \text{and} \quad \Phi_2 = \frac{1-\sqrt{5}}{2}$$

$$\Phi_1 = 1{,}618\ldots \qquad \Phi_2 = -0{,}618\ldots$$

The fractional part of the two roots is the same infinite, unrepeated decimal fraction. The following relationship holds between the roots:

$$\Phi_1 - 1 = \frac{1}{\Phi_1} = |\Phi_2|$$

Many volumes of specialist literature deal with the increasingly interesting properties of Φ. Anyone who is willing to take the trouble, and who likes playing with the interesting characteristics of numbers, can easily discover further properties of this kind. Here we mention a few of them to give a taste.

$$
\begin{aligned}
\Phi^0 &= 0 + 1 && \\
\Phi^1 &= 0 + 1\Phi && \\
\Phi^2 &= 1 + 1\Phi &= \;& \Phi^0 + \Phi^1 \\
\Phi^3 &= 1 + 2\Phi &= \;& \Phi^1 + \Phi^2 \\
\Phi^4 &= 2 + 3\Phi &= \;& \Phi^2 + \Phi^3 \\
\Phi^5 &= 3 + 5\Phi &= \;& \Phi^3 + \Phi^4 \\
\Phi^6 &= 5 + 8\Phi &= \;& \Phi^4 + \Phi^5 \\
\ldots & && \\
\Phi^n &= a_{n-1} + a_n\Phi &= \;& \Phi^{n-2} + \Phi^{n-1} = (1 + \Phi)\Phi^{n-2} = \Phi^2\Phi^{n-2} = \Phi^n
\end{aligned}
$$

The sequence of numbers to be found in the columns, 1, 1, 2, 3, 5, 8, 13, 21, ... a_n, ... , we call the Fibonacci numbers. They are named after Leonardo Fibonacci (Pisano), the first person known to have written them down. They are formed as follows:

$$a_{n-2} + a_{n-1} = a_n$$

That is, any given member of the sequence is the sum of the two previous members. By convention we begin the series with the number 1 twice.

By keeping to this convention, we reach the interesting result that the ratios of subsequent members of the sequence approach a clearly-defined boundary value, and this boundary value is none other that the proportion of the golden section we just discussed:

$$\lim_{n\to\infty} \left(\frac{a_n}{a_{n-1}} \right) = \Phi$$

This is the relationship which can help to uncover the connection between the natural numbers appearing in the symmetries of the living world and the "magic" decimal fraction we encountered as the proportion in the golden section. This relationship expresses the harmony between the geometry and algebra of nature.

One interesting characteristic of the sequence of fractions approaching Φ is that they can be put in a very appealing symmetric form using chain fractions and only the digit 1:

$$1 = \frac{1}{1}$$

$$1 + 1 = \frac{2}{1}$$

$$1 + \frac{1}{1+1} = \frac{3}{2}$$

$$1 + \cfrac{1}{1 + \cfrac{1}{1+1}} = \frac{5}{3}$$

$$1 + \cfrac{1}{1 + \cfrac{1}{1 + \cfrac{1}{1+1}}} = \frac{8}{5}$$

$$1 + \cfrac{1}{1 + \cfrac{1}{1 + \cfrac{1}{1 + \cfrac{1}{1+1}}}} = \frac{13}{8}$$

etc.

Another interesting characteristic to mention is that the square of any member of the sequence deviates in its absolute value by one from the product of the preceding and subsequent numbers in the sequence, while the sign of this difference swaps from one to the next, being –1 for members of the sequence whose position is even-numbered, and 1 for those whose place in the series is an odd ordinal number:

$$\left|a_i^2 - a_{i-1}a_{i+1}\right| = 1$$

On further examination, we can reach various expressions that create a connection between particular members of the sequence and the set of the preceding members. By using these, we can generate individual members of the sequence according to certain rules, with the help of the preceding members. For this reason the *sequence* of Fibonacci numbers is also sometimes referred to as a *series*. For example, it is easy to show that the $(n + 2)$th member of the sequence can be generated from the sum of the first n members, as follows:

$$\sum_{i=1}^{n} a_i = a_{n+2} - 1,$$

or the product of the nth and the $(n+1)$th members from the sum of the squares of the first n members:

$$\sum_{i=1}^{n} a_i^2 = a_n a_{n+1}.$$

Finally, let us look at another special generation of the Fibonacci number sequence. If we write down the well-known mirror-symmetric Pascal (1623–1662) triangle, made up of so-called binomial coefficients, and slice through it with parallel lines at the appropriate angle, the sum of the numbers found along the lines gives us the Fibonacci sequence:

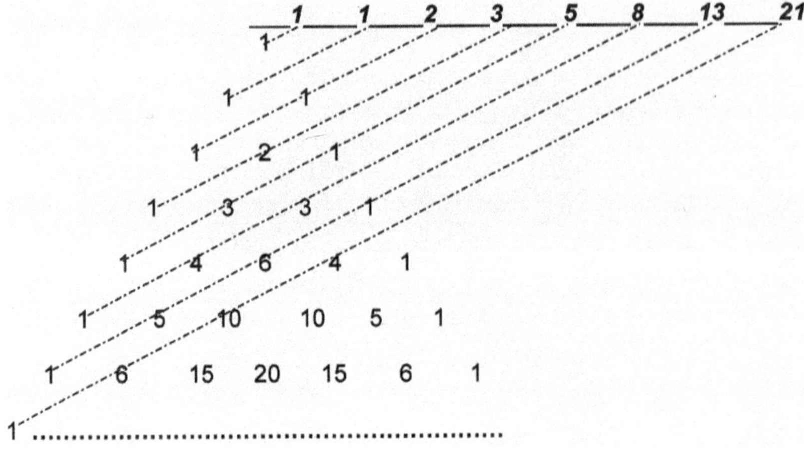

The relation between the golden section and Fibonacci numbers has shown us a connection between certain algebraic properties and their geometric equivalents, on the one hand, and a sequence of numbers made up of natural numbers and a "magic" irrational number, the proportion of the golden section, on the other. Just as the proportions of the most perfect planar shape, the circle, and the most perfect spatial body, the sphere, can be characterized with an irrational number, π, so too we can characterize the beauty to be found in nature, the orderliness of the development of living material, and the proportions of the units delimited by the human body, together with the proportions of artistic beauty based on proportions observed in nature, with the proportion Φ of the golden section. We will introduce these examples and relationships of the application of Φ in the next chapter.

Interdisciplinary applications

Chapter 5

Fibonacci numbers in nature

Phyllotaxis in the living world. The harmony of the built environment

The example of the reproduction of rabbits (or any other animal that is quick to breed) is well-known from most people's secondary school studies. If we begin with a pregnant rabbit, it produces another rabbit, and we have two rabbits. In the first generation, the little rabbit will not breed, so at the next stage it will only be the old rabbit that gives birth again, giving us three rabbits. In the following generation, this little rabbit will not yet breed, but we have two rabbits from the previous generation that will, which adds two rabbits to the existing three rabbits, giving us a total of five rabbits. At the next stage, the two new rabbits will not yet breed, but the older three will, which together with the existing five rabbits gives us eight rabbits in all. In the following generation, five rabbits will produce new rabbits, and there will be eight plus five, that is thirteen rabbits, of eight will breed, giving us thirteen plus eight, that is twenty-one rabbits, and so on, until this system is not upset by some external factor. The number of new-born rabbits will always equal the cumulative number of rabbits from two generations before. As we began with the second member ($a_2 = 1$) of the sequence of numbers, there is no need for us to subtract the 1 ($a_1 = 1$) on the right hand side of the equation $\sum_{i=1}^{n} a_i = a_{n+2} - 1$.

In addition to the breeding of rabbits, there are many other phenomena that follow similar rules. In plants, the Fibonacci numbers appear as a spatial order. The properties of this sequence are displayed on the cylindrical surface of the stalks and roots of plants. The branching (bifurcation) of the stalks of plants follows a similar reproduction rule to that for rabbits. In each "generation", the number of new branches equals the subsequent member of the Fibonacci sequence (Figure 5.1).

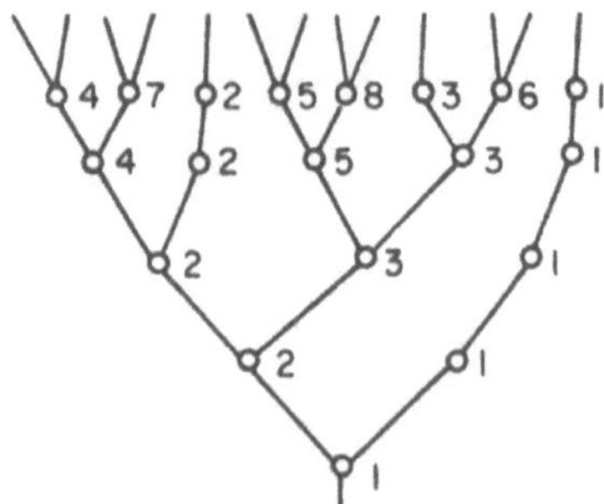

Figure 5.1. *The order of the branching of a plant's stalks follows the Fibonacci sequence*

Cylindrical surfaces

In plants, macroscopically observable results are the consequence of microscopic phenomena. The cells at the end of stalks and roots follow similar reproductive rules, which appear in macroscopically observable formulae. New cells are divided along the meeting points of two spiral bands moving in opposite directions and twisted around the cylindrical surfaces (stalks, roots) culminating in cones (after Roger Jean). The ripening of seeds and the growth of flower petals also follow rules similar to that governing the reproduction of cells (Figure 5.2).

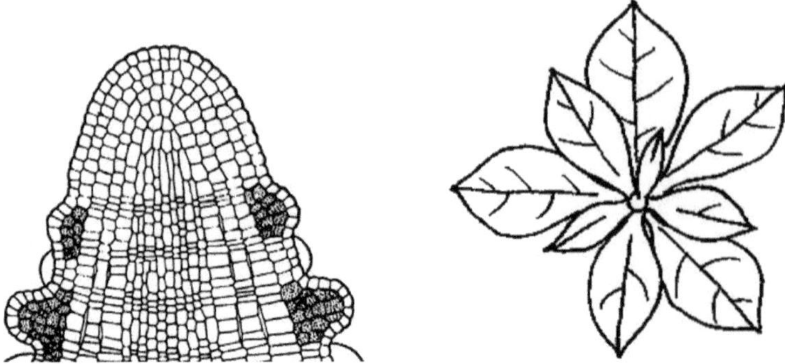

Figure 5.2. *The order of the division of cells on the stalk and the resulting growth of leaves in directions with golden angle divergence*

Vortices

Yet it would be a mistake to believe that development with growth based on the regularities of the Fibonacci sequence is only characteristic of living material. The vortices that we observe in inanimate nature, like star clouds, galaxies and (with different conditions) material flows in general follow this same order (Figure 5.3).

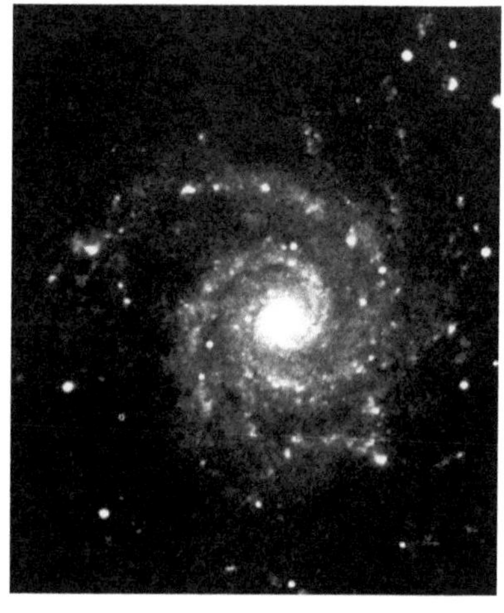

Figure 5.3. *Galaxy. Vortex phenomena in inanimate nature*

Vortex phenomena are very common in nature. Rather closer to home than galaxies, this describes the behaviour of the water flowing out of the bath, a ('swirling') river whose flow has been altered, a whirlwind or tornado, or equally the less destructive cloud formations and cyclones that typify our weather as seen on satellite images. The same path is drawn by the hands of a skater performing a pirouette by drawing in her arms and speeding up, or by sugar being stirred in coffee.

Let us return to the division of the golden rectangle we see in Figure 4.7. We can observe that the squares converge on a point, a point which always lies on the diagonals of the successive decreasing rectangles, every second of which coincide, as they are at right angles to one another. Indeed with similar triangles we can show that this point divides the diagonals according to the golden ratio, that is in a proportion of $\frac{\Phi}{\sqrt{5}}$ to $1 - \frac{\Phi}{\sqrt{5}}$. Drawing arcs in each of the squares, these will produce a good approximation of a spiral with decreasing radius which ends in this point as its centre-point. We could equally say that it spirals outwards from this point. We call this continuously "untwisting" curve which begins at this point a logarithmic spiral, characterized by the fact that the angle between its tangent at any point and the line drawn between this point and the centre-point remains constant. This angle ϑ, at the meeting point of the line drawn from the centre-point to the golden rectangle and the logarithmic spiral, can be determined using the sine theorem on the triangle circumscribed by the diag-

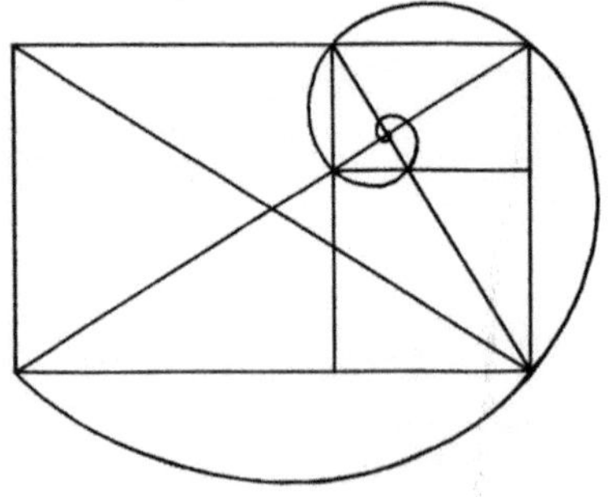

Figure 5.4. *Logarithmic spiral drawn around a golden rectangle (after Arthur Loeb)*

onals and the edges of the rectangles, and can be calculated as a function of φ and Φ — that is, as we have seen, "with the help of the number 2". Its path can be characterized by the following equation in the (r, ϑ) system of polar coordinates based on the centre-point of the logarithmic spiral: $r = ae^{k\vartheta}$, where r is the radius drawn to the given point, and a and k are pre-defined parameters (Figure 5.4).

The role of Fibonacci numbers in the living world

One of the best-known logarithmic spirals found in the living world is the pattern of seeds on a sunflower (Figure 5.5).

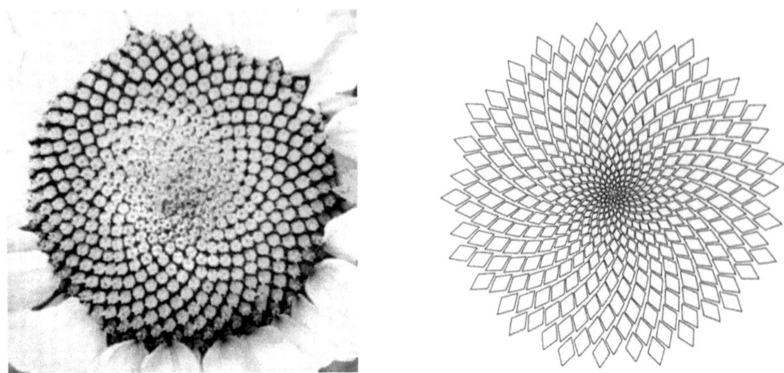

Figure 5.5. *Logarithmic spirals in the natural world. The sunflower*

The Fibonacci sequence appears in the pattern of the sunflower seeds in two ways. We can see that the individual seeds lie along two spirals. The location of every seed is the point of intersection of an arm curving to the left and another curving to the right. The number of spiral arms on a sunflower that curve to the left does not equal that of those curving to the right. However, these two numbers are always two successive members of the Fibonacci sequence.

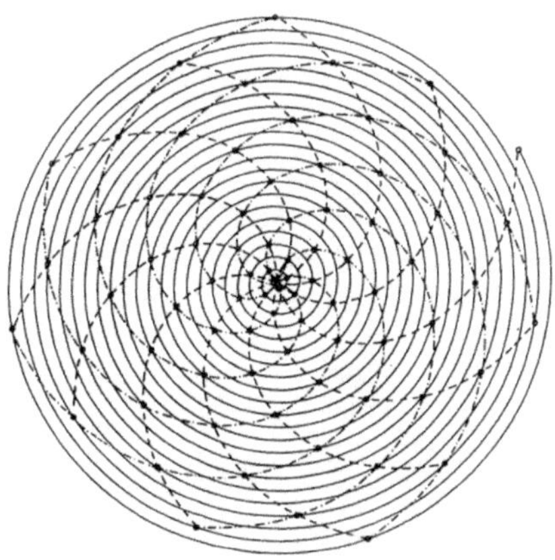

Figure 5.6. *The spiral positioning of the seeds on a sunflower*

We can mark the sunflower spiral seen in Figure 5.6 with the number pair (13L, 8R), referring to the fact that it is made up of thirteen arms curving to the left and eight arms curving to the right. If we count the seeds along two particular intersecting arms from the centre until the point of intersection, we find that the seed whose number on one arm (e.g. the left) is a Fibonacci number (e.g. 21) will be a neighbouring Fibonacci number (e.g. 13) on the other arm. While externally the sunflower seems to be mirror-symmetric, it turns out not to be so, for the number of spiral arms to the left and to the right are not equal.

What we have observed in a planar case with the disc of the sunflower, we can see wrapped (almost) cylindrically onto an ovoid surface in the case of the scales on a pine cone or a pineapple (Figure 5.7). The spiral sym-

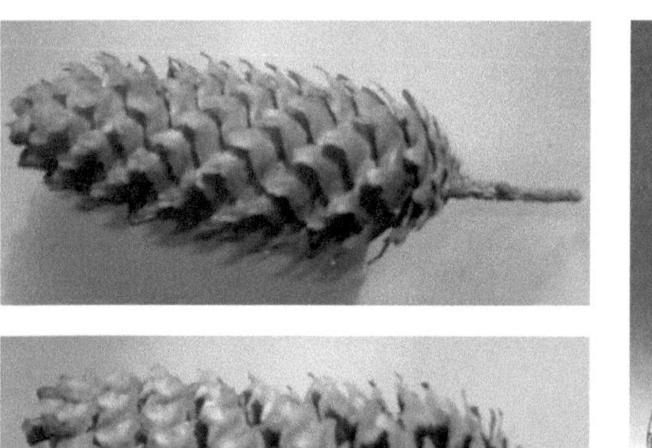

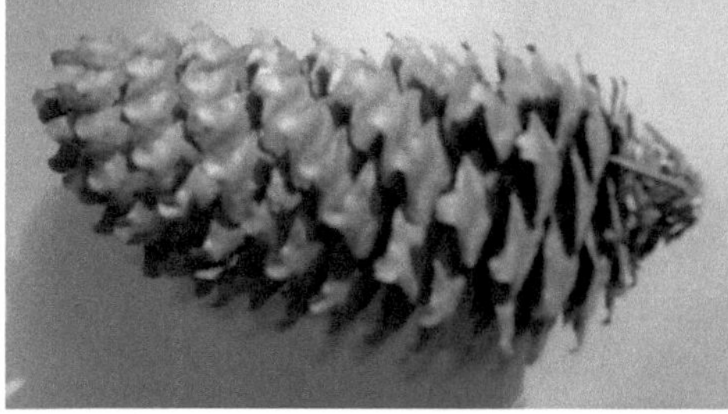

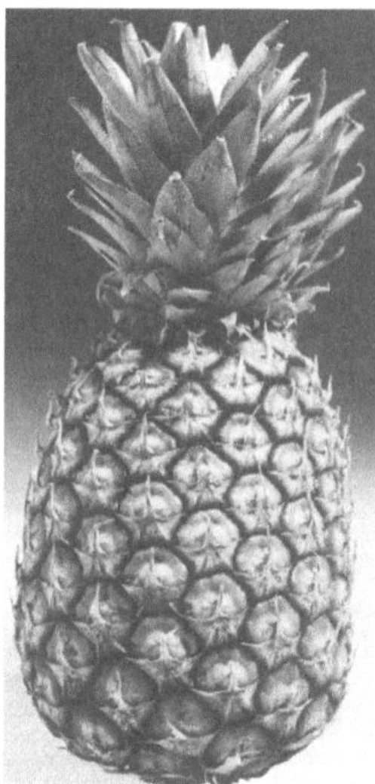

Figure 5.7. *Pinecones, pineapple scales*

metries lying on the plane (constant tangential angle, radial translation, and rotation around the centre-point) are accompanied by a translation at right angles to the plane. If the radius does not increase, but stays constant, the spiral extended into space is transformed into a helix (which we often colloquially, but incorrectly, refer to as a spiral, e.g. spiral-bound notebook).

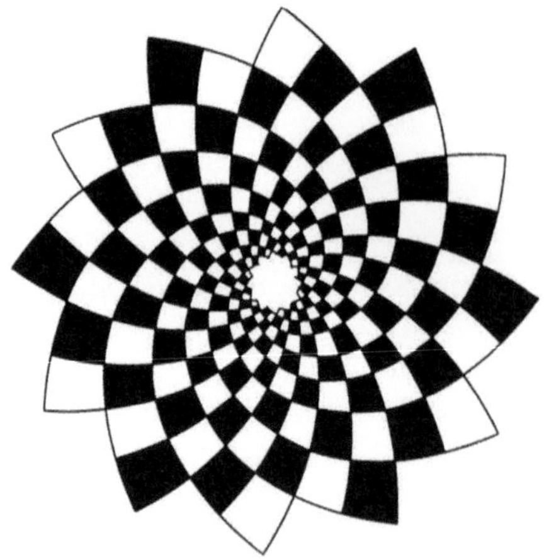

Figure 5.8. *Spiral similarity symmetry*

To return to the plane, the increasing elements (cells) lying on the individual spiral arms as they twist outwards display symmetries of similarity, combined with the rotational symmetries of the spiral arms. This is clearly visible in Figure 5.8, which contains squares in alternating black and white.

Let us construct a spiral that confirms to the smallest Fibonacci number, that is from a single arm. Here we only have a spiral arm in one particular direction, which has no other arms to intersect. If we place similar cells on this lone spiral arm that increase in size, we build a snail's shell. Just as nature does.

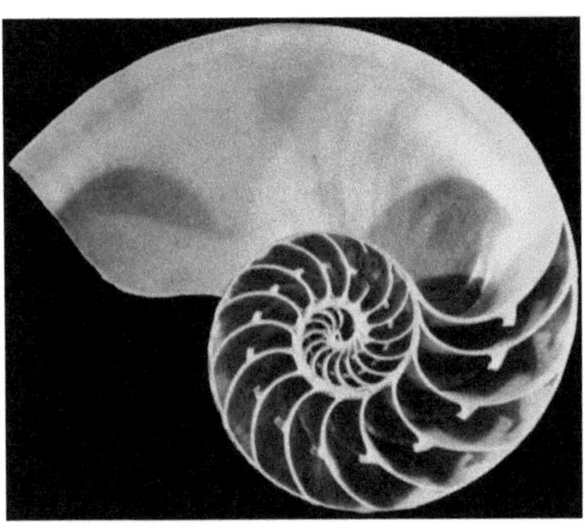

Figure 5.9. *A snail's shell (Nautilus shell)*

Figure 5.9 shows a snail extending on the plane. Despite the fact that the rich variability of spiral arms we saw for sunflowers and pine cones is here limited to a single arm, we can still build an infinite number of different snails by varying the shape of the cells and their angle from the plane. This richness is displayed in Figure 5.10.

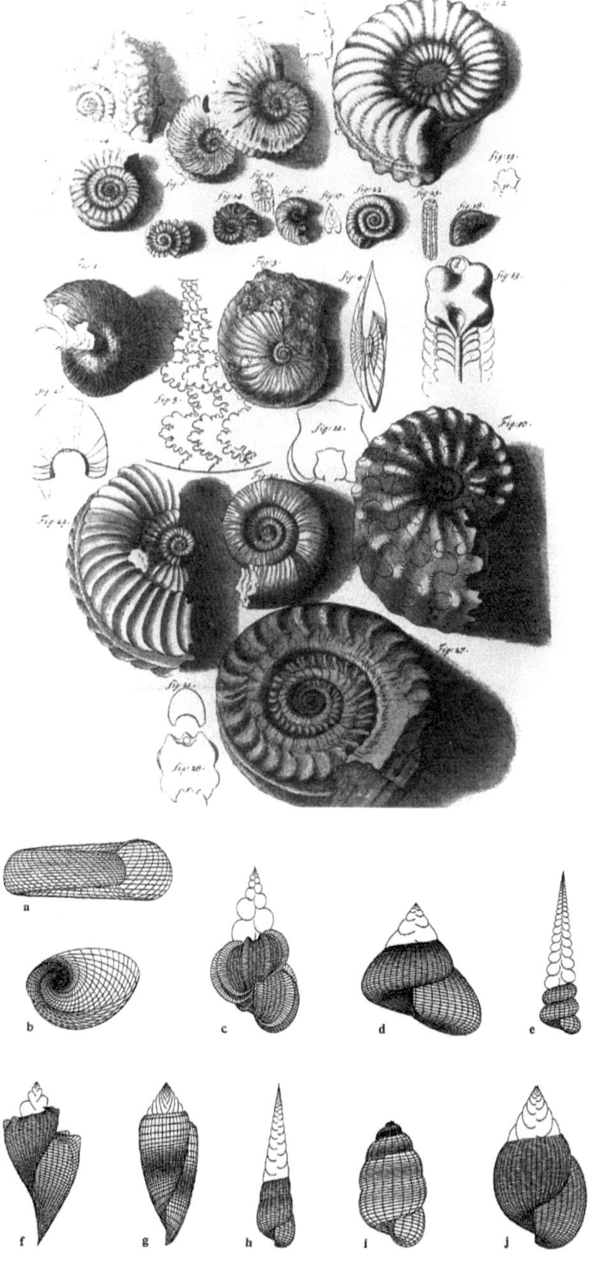

Figure 5.10. *Snails' shells (after R. Hooke and M. Cortie)*

In Figure 5.11, on the other hand, we have marked the parameters with which the shape of the cell, the similarity of different cells and the angle of the spiral arm can be described (M. Cortie). These allow the snail's shell to be described in a mathematically precise way.

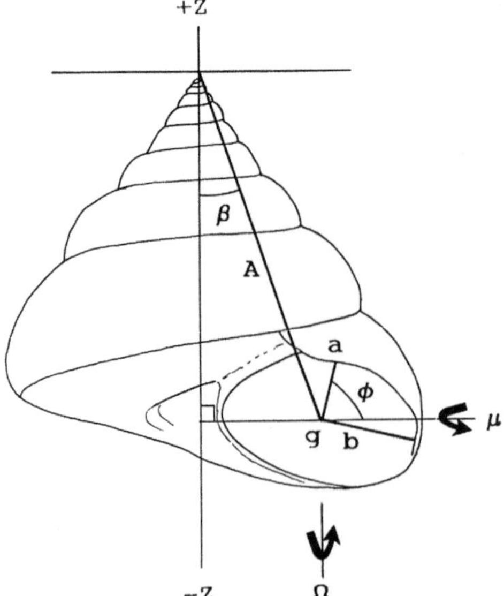

Figure 5.11. *The parameters of a snail's shell*

Chirality in the living world

We might think that nature is as likely to build snail's shells twisting to the left as to the right. Any French chefs will tell you that it does not. The vast majority of snail's shells twist to the right. So rare are those with a spiral twisting to the left, which cannot be taken out of their shell in a single motion with the special spoons designed for right hand snails, that they are simply thrown away as rejects. Aside from very rare specimens that have spontaneous symmetry violations, only two snail species, *Physa* and *Clausilia*, are specifically twisted to the left. It is even more rare for there to be roughly equal numbers of specimens with left- and right-twisting spirals within the same species: one such example is a Cuban species of snail that lives in trees, *Liguus poeyanus*. The mystery of the difference in the numbers of snails with spirals twisting to the left and to the right, that is the phenomenon of chirality, leads us to a separate class of natural phenomena, which we will deal with in a separate chapter.

Self-similar fractal structures

We can observe spiral structures in flora other than the sunflower. An everyday and very distinctive example are broccoli flowers. Individual flowers are built up of their elements in a similar way to the snail, then these flowers themselves, as elements, build upon themselves in snail-like fashion. Unlike snails, however, broccoli flowers do not fill space continuously. They create a space that is more than two dimensions, but less than three dimensions — one of so-called 'fractional' dimensions. We call such structures one of fractal dimensions, or simply fractals. Fractals (structures with fractional dimensions) do not necessarily have to display any symmetries, but broccoli does. In addition to its fractal structure, a spiral structure can also be observed on it. Indeed, as we pointed out, this construction takes place at a number of levels (Figure 5.12).

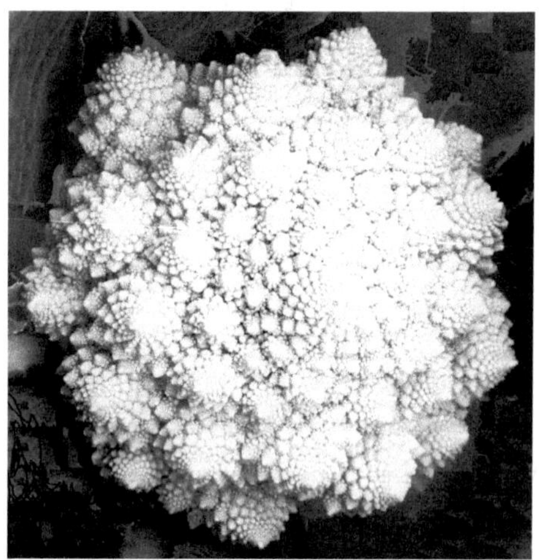

Figure 5.12. *Broccoli flowers. In addition to the spiral structure, we can see a self-similar fractal arrangement*

Let us take a look at this structural construction projected onto a plane, in the following four successive, mathematically plotted, idealized figures. This figure is more than just a line (1D), but does not fill the entire plane, that is it is less than two-dimensional. Let us enlarge a part of the first broccoli-like planar spiral, then a detail from this, and finally a single element ('flower') of the original 'flower'. We find that this single flower, whether looked at in whole or in part, is — apart from its size — a perfectly faithful copy of the original flower. Theoretically we could continue this selection of a smaller element *ad nauseam*; in practice, we know there are limits to this. The same can be said of individual star systems, galaxies,

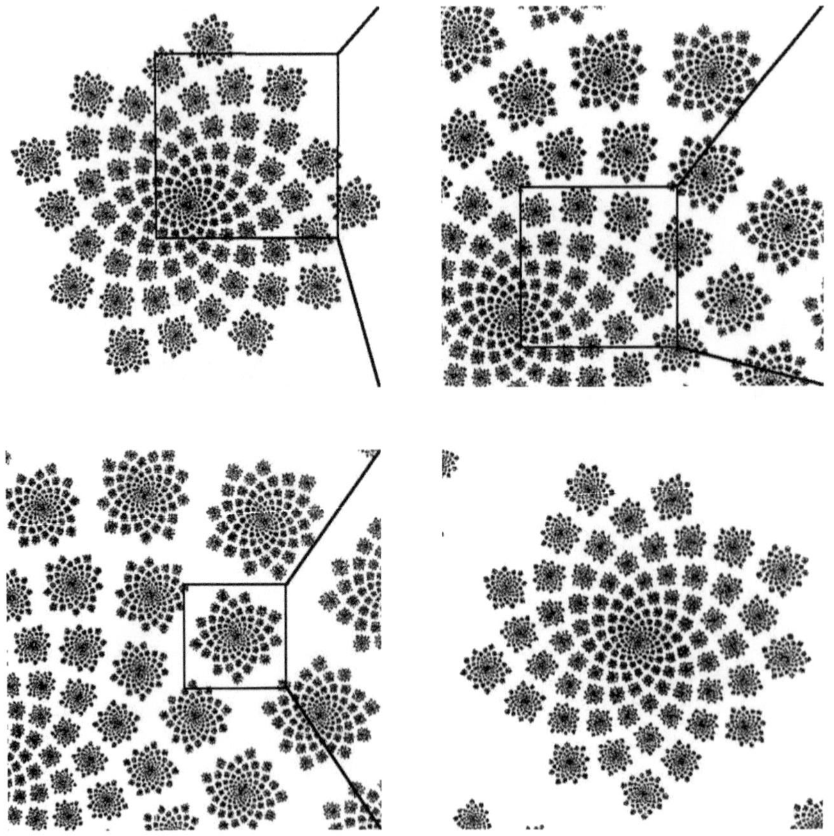

Figure 5.13. *Self-repeating broken-dimensional (fractal) structure*

extragalaxies: they are embedded in each other in the same way. We see that a structure — in this case the structure of broccoli flowers — repeats on a small scale the structure of the larger unit (Figure 5.13). This phenomenon can be continued for a few levels. We call this class of fractals self-similar fractured-dimensional structures. Their beauty is served by the similarity symmetry in which a part is similar to the whole.

Phyllotaxis

If we observe the arrangements of the leaves around the stalks of most plants, we find rotational symmetry, perhaps combined with translation.

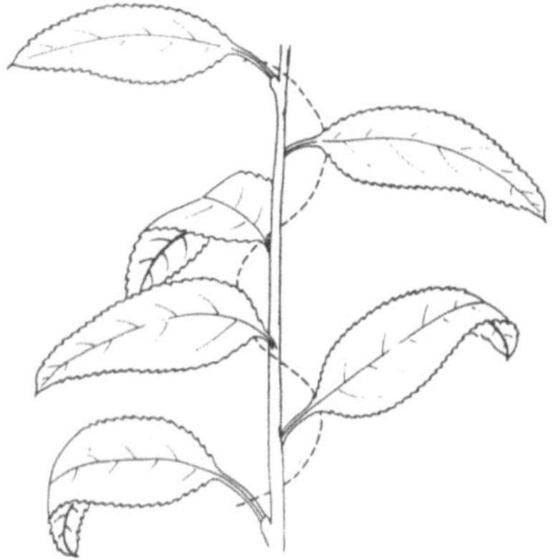

Figure 5.14. *Position of leaves (phyllotaxis). The spiral arrangement of cherry leaves. The phyllotaxis proportion is 2:5, i.e. five leaves represent two full rotations*

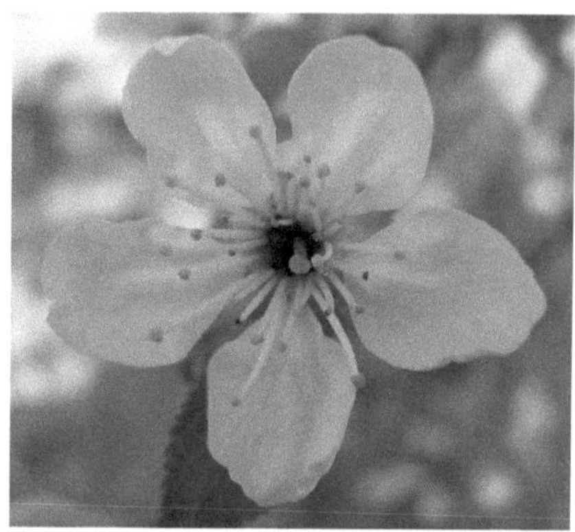

Figure 5.15. *Cherry flower. Petals arranged in a single plane, with fivefold rotation and mirror symmetry*

After Charles Bonnet (1720–1793), we refer to this order of arrangement around stalks as *phyllotaxis* (1754). We saw in the example of the sunflower, then that of the snail's shells, that during the rotational symmetries of living creatures nature often distinguishes a direction of rotation, or at least that the two directions are not equivalent. In the arrangement of leaves, the order of the rotational symmetry always corresponds to some Fibonacci number (Figures 5.14 and 5.15). We should not be confused by the fact that the flowers of the tulip, for example, have six petals. If we inspect them closely at their base, we see that there are two circles of three petals, one above the other, with each petals turned through an angle of $2\pi/6$ compared to the previous one.

The fractions characteristic of the screw-like arrangement of the leaves are often members of the sequence 1/1, 1/2, 2/3, 3/5, 5/8, 8/13, 13/21, 21/34, ..., which can easily be recognized as made up of the Fibonacci numbers and as the limits to which the reciprocal of the golden number ($1/\Phi$) tend. The points at which they emerge from the stalk can be seen as the intersections of ribbons made up of cells and spirally running up the stalk as a cylinder, in two directions. As it reaches the tip of the branch or root (possibly fruit), this cylinder can turn into a cone (as, for example, is macroscopically visible in the case of the cones grown on a pine), while in certain cases the cone flattens to such a degree that it can effectively be considered a plane (as in the case of the sunflower). View from at right angles to the plane of the leaves, the arrangement of the leaves is projected onto a single plane even when translated, and this is how the order of their rotation is best observed (Figure 5.2). What we can observe in the order of the macroscopically visible leaves, petals, seeds and scales is repeated in the details, as in the case of the broccoli flowers, in the order of the reproduction of the surface cells, only observable under a microscope.

The order of phyllotaxis, hardly surprisingly, displays regularities similar in a number of ways to those we presented in the case of reproduction. Why would the division and reproduction of these cells follow any order other than that designed by the laws of nature for the breeding of rabbits? We can illustrate the reproduction of cells and of the protuberances and stalk bifurcations in an interesting fashion with simply mathematical schemes. In the interest of observing the common origin of the floral lattices characterized by Fibonacci numbers, let me quote Szaniszló Bérczi's first addendum for the Hungarian edition of H. Weyl's book *Symmetry*, and the accompanying figure (Figure 5.16).

"Let us number the repeated elements located on the cylindrical stalk of the plant one by one, according to height: this gives us a numbered surface mosaic lattice. If we peel this off the plant (metaphorically) and stretch it from a lattice of rhombuses (or hexagons) into a lattice of squares, we are essentially given a square-netted ribbon twisted around a cylinder. (The "scale" deformation does not affect that lattice structure, only helps simpler depiction.) We can cut this ribbon up in two ways: we can make it into a ribbon twisting to the left or to the right. The two types of ribbon that can be peeled off a plant will always be a Fibonacci number in width, and there is only one case in which they can be of the same width [the first two Fibonacci numbers, 1 and 1, coincide — G. D.]: in cases of width being 1 + 1, similar to the symmetry of wheat ears. In the figure we display the ribbons that can be peeled off various plants, together with their pairs, and list them in an arrangement — referring to a genetic or development relationship — in which, starting with the simplest 1 + 1 (one scale in width to the right and to the left), a surface lattice (in both mirror-symmetric variants) with a ribbon pair accompanying any Fibonacci pair can be induced from one type of gliding operation. During this gliding operation the partial ribbons must be slid along by one unit (lattice constant) alongside one another within one of the pair of ribbons. It is by alternating this gliding between the ribbon twisting to the left and that to the right that we achieve one of the two (mirror-symmetric) "family

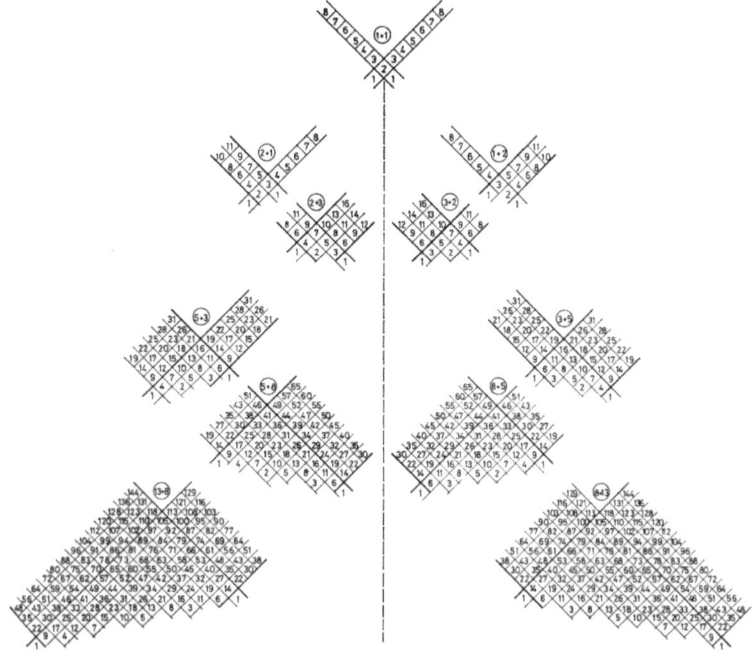

Figure 5.16. *The common origins of floral lattices that can be characterized with Fibonacci numbers*

trees" presented in the figure. We get the evolutionary sequence on the left if, on the initial 1 + 1 lattice, we first apply the gliding operation on the right-twisting ribbon; in the opposite case, we get the evolutionary sequence on the right."

(Szaniszló Bérczi: Addendum 1. In H. Weyl: Szimmetria. Budapest, 1982, Gondolat. pp. 206–208.)

Let us observe the lattice order described in the blossoming of rose petals. Let us start with the rotational symmetry of the plant organ. Then let us imagine the lattice order made of similar elements on the surface of the rotation body. In Figure 5.17 we can see the arrangement of the lattice order into Fibonacci-numbered ribbons from an axis-oriented view from above. After this, in Figure 5.18, we can follow the shape of the future rose petals being drawn along the logarithmic spirals. In Figure 5.19 we see the grown roses according to the phyllotaxis (arrangement of petals) dictated by the Fibonacci numbers (after Szaniszló Bérczi).

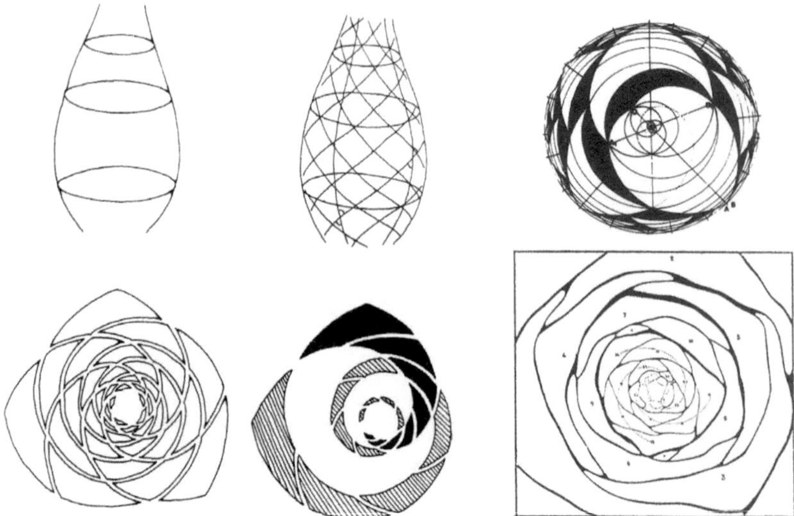

Figure 5.17. *(1) The rotation symmetry of the plant organ, (2) Lattice order made up of similar elements on the surface of the rotation body; (3) The arrangement of the lattice order in Fibonacci-numbered ribbons*

Figure 5.18. *The shape of the future rose petals becomes visible along logarithmic spirals. The developed rose petals according to the phyllotaxis (leaf arrangement) demanded by the Fibonacci numbers*

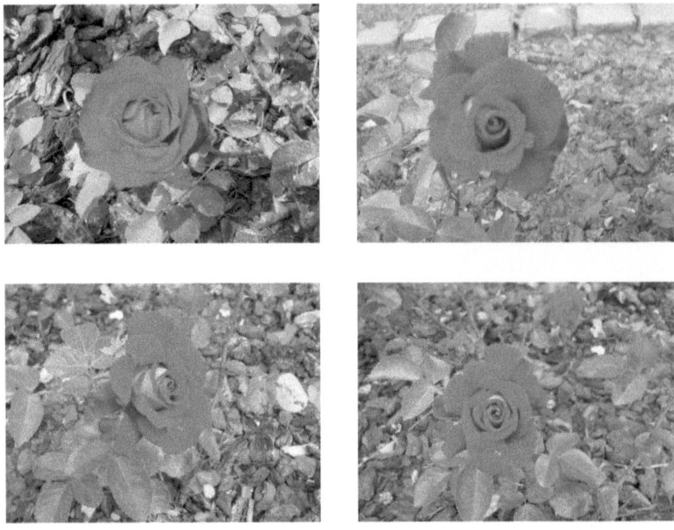

Figure 5.19. *Rose petals that have blossomed*

The Fibonacci order and the golden section in the man-made world

Man has used the symmetry spotted in roses, other flowers, and green plants, and the defining regularities of beauty created by nature to decorate his artificial environment. It is with the use of these proportions and rules that the wonders of decorative art were produced. Good examples of the structure of proportions and symmetries seen in plants are Michelangelo's stone mosaic on the Capitolium in Rome (Figure 5.20) and, projected on a concave surface, the Renaissance dome seen from below in Figure 5.22.

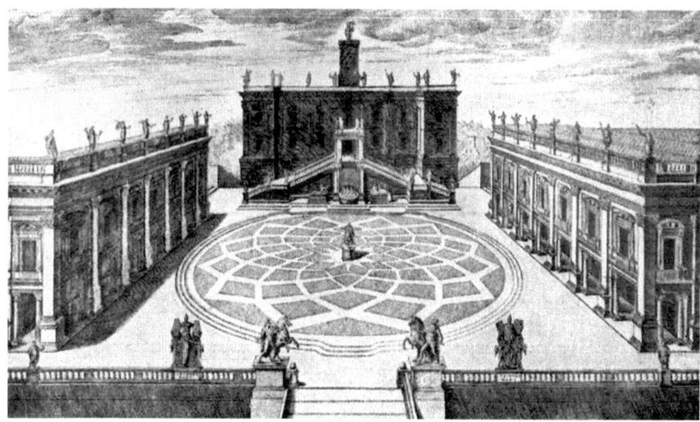

Figure 5.20. *Michelangelo's mosaic at the Capitolium in Rome* (Du Perac's engraving, 1569) ...

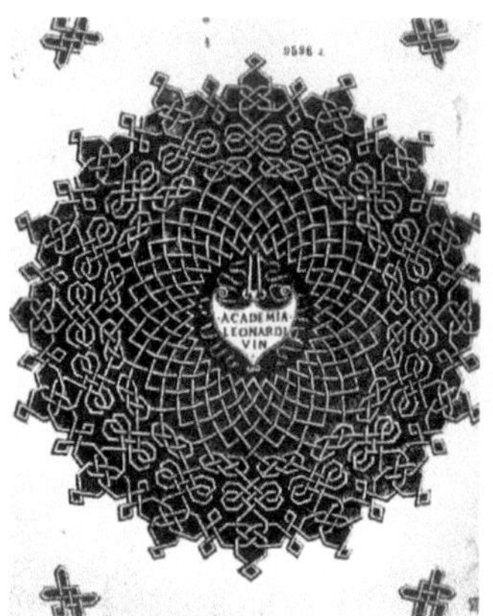

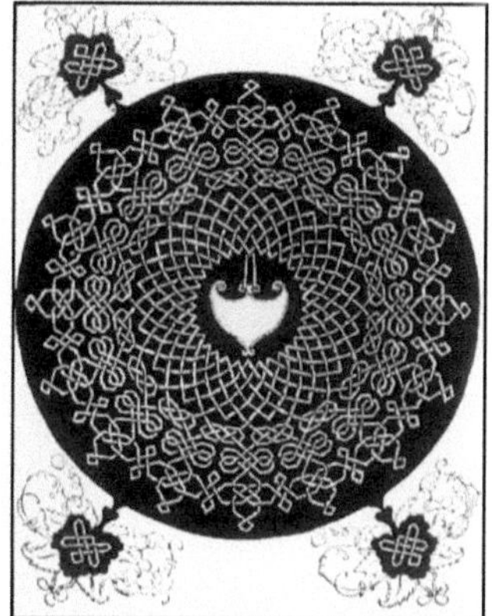

Figure 5.21. ... *and its forerunners at the centre of Leonardo's engraving* (Accademia Vinciana, c. 1495), *and a design by Dürer* (1507)

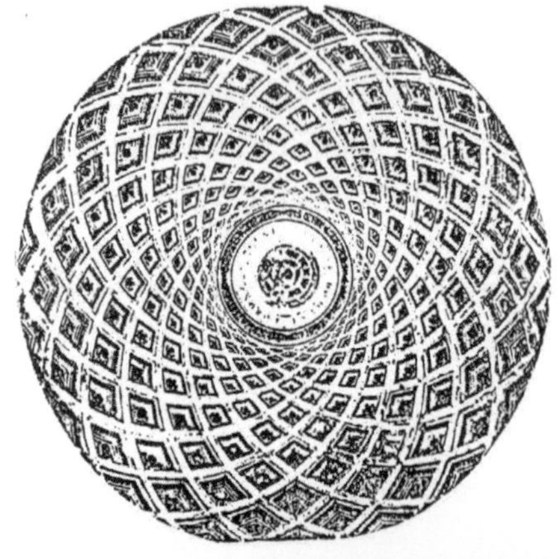

Figure 5.22. *Renaissance dome seen from below*

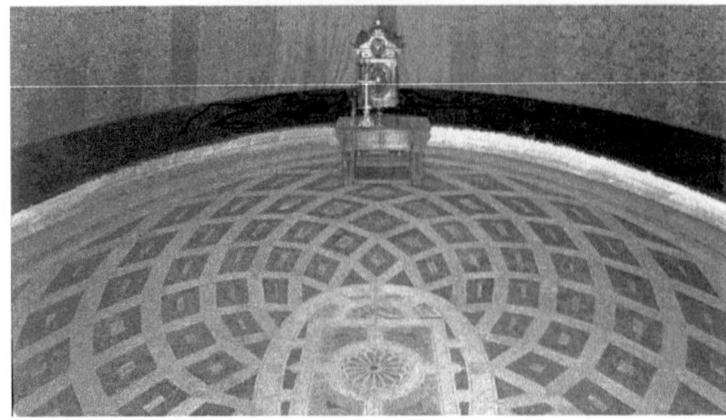

Figure 5.23. *The floor mosaic in the basilica in Seville*

As the art of the creation of internal space, architecture is itself capable of embodying perfect proportions. We can wonder at beautiful geometrically constructed examples of the proportion Φ of the golden section in architectural masterpieces.

All three spatial extensions of the Parthenon on the Acropolis in Athens are related to one another with the proportion $\sqrt{5}$, and the edges of its tympanons according to Φ (Figure 5.24). The architecture of later periods also made conscious use of the proportion regarded as perfect; we illustrate these common proportions by comparing certain marked measurements of the Parthenon and Saint Peter's Basilica in Rome (Figure 5.25).

The proportions of the golden section in various buildings are examined by Y. Shevelyov (1924–), V. I. and G. N. Korobko, as well as Oleg Bodnar, mainly following the architectural applications of the dynamic symmetries of J. Hambidge (1867–1924), or on the basis of Shevelyov's studies of architectural proportions. Here I would take one example from their

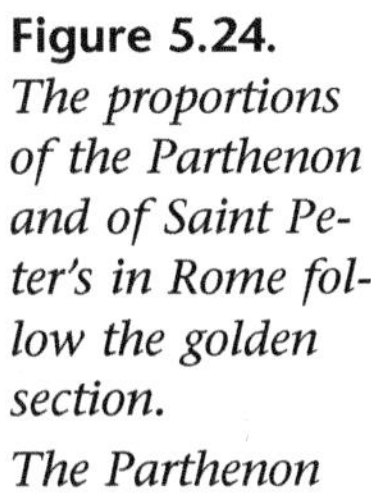

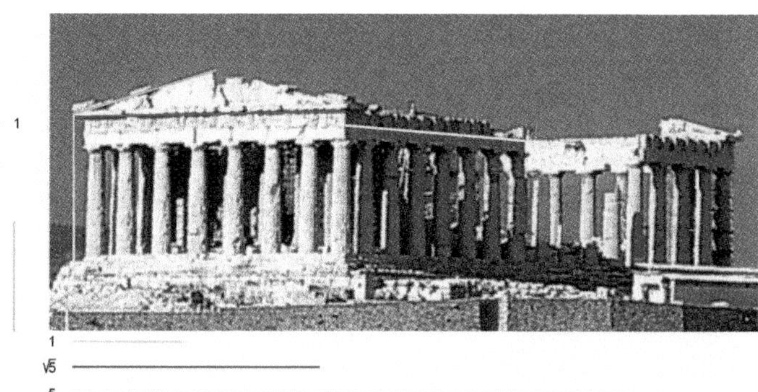

Figure 5.24.
The proportions of the Parthenon and of Saint Peter's in Rome follow the golden section.
The Parthenon on the Acropolis

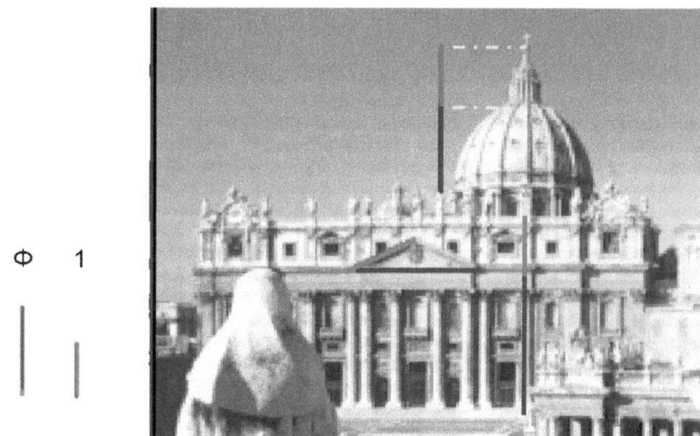

Figure 5.25.
Saint Peter's in Rome

work (Korobko, the proportions of the basilica in Ulm, Figure 5.26) which displays the whole system of the utilization of the golden proportion in the relationship between the sizes of the building's individual architectural elements. We see similar proportions (in G. Dóczi's book) in the masterpieces of sculpture (Figure 5.27).

We can also see eye-catching examples of consciously applied proportions in garden architecture. They well illustrate the creation of harmony between nature and the built environment, and show how pleasure is generated, and increased, if the formed elements of nature are fitted around a man-made construction (building, or group of sculptures) in the right proportions.

The design art of the modern age has made conscious use of the proportions of particular works of art in fixtures. Twentieth-century design has projected all the proportions that were natural in the spatial creation of a building onto the world of fixtures and furniture both inside and outside

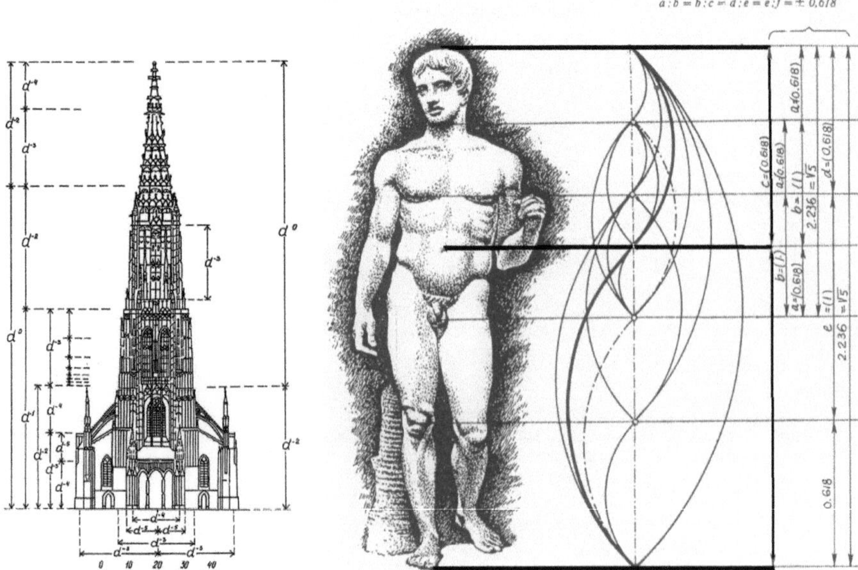

Figure 5.26. *The proportions of the Gothic minster in Ulm*

Figure 5.27. *Polykleitos: The Lance Bearer*

it. There were two schools that put this programme most firmly on their agenda. One was associated with the *Bauhaus* movement in Germany, where, together with the founder Walter Gropius, a number of Hungarian professors played a pioneering role in establishing founding principles and later in teaching and research work (the best-known being László Moholy-Nagy and Marcell Breuer). The other school is associated with Le Corbusier (1887–1965). The modulor system he elaborated (Figure 5.31) — following in the footsteps of Leonardo, Luca Pacioli, Dürer, etc. — prescribed that the proportions of the objects to be designed correspond to the proportions of the human body (Figure 5.32 and Figure 5.33). These proportions are nothing other than those of the Fibonacci numbers, and thereby they

→

Figures 5.28–5.30. *The Baroque gardens of the Castle of Versailles (La Nôtre, 1661). Alongside the symmetry, there are also a number of clearly visible dissymmetric elements. La Nôtre's garden designs must have been the inspiration for the gardens of Schönbrunn Palace next to Vienna (1759, contemporary picture) and the plan of the city of Karlsruhe (1715), still evident today*

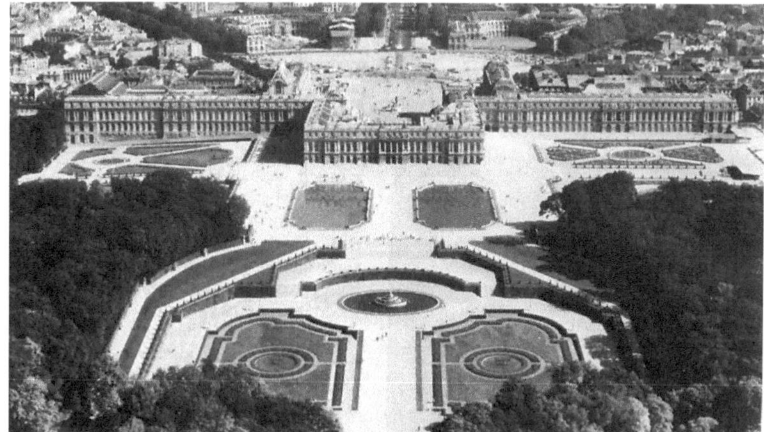

Figure 5.28.

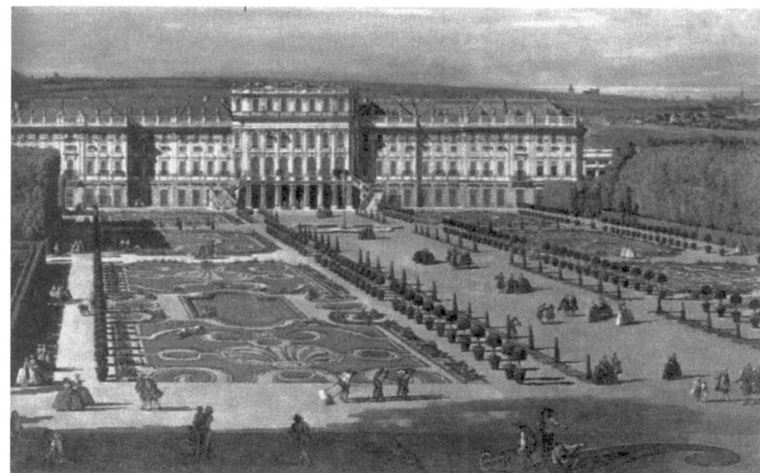

Figure 5.29.

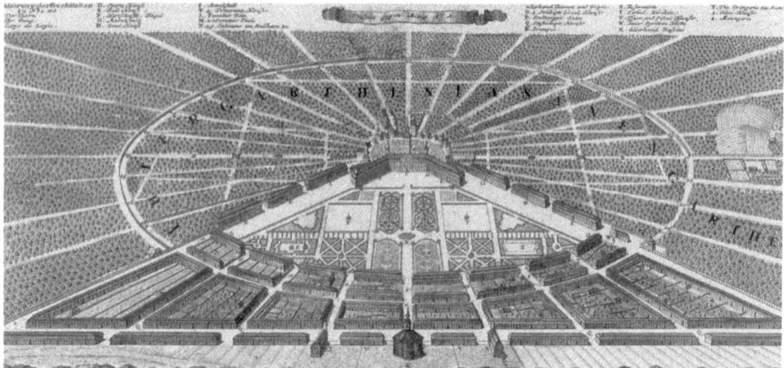

Figure 5.30.

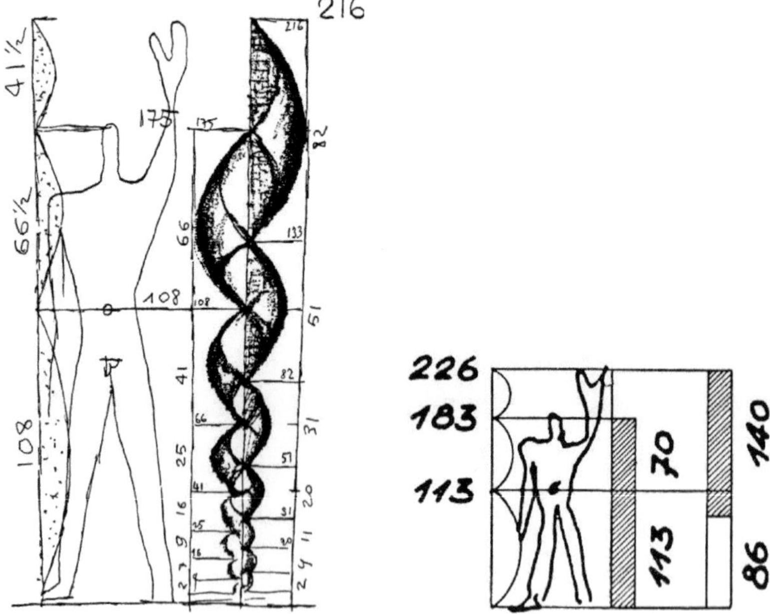

Figure 5.31. *Le Corbusier's modulor*

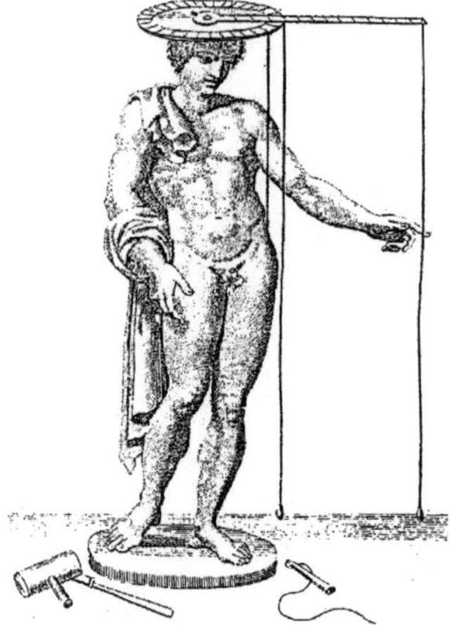

Figure 5.32. *L.B. Alberti Della Statua (1435)*

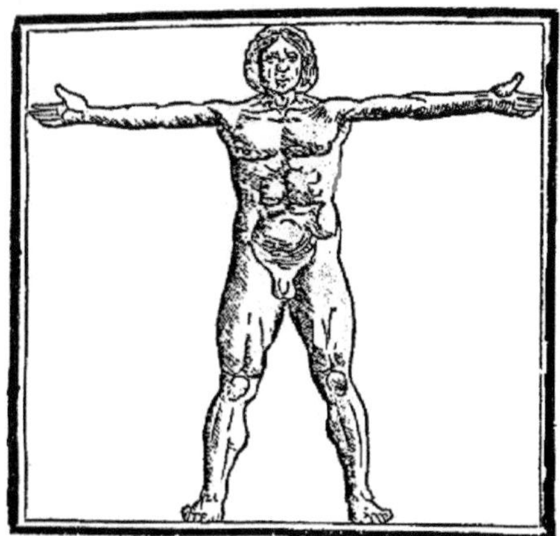

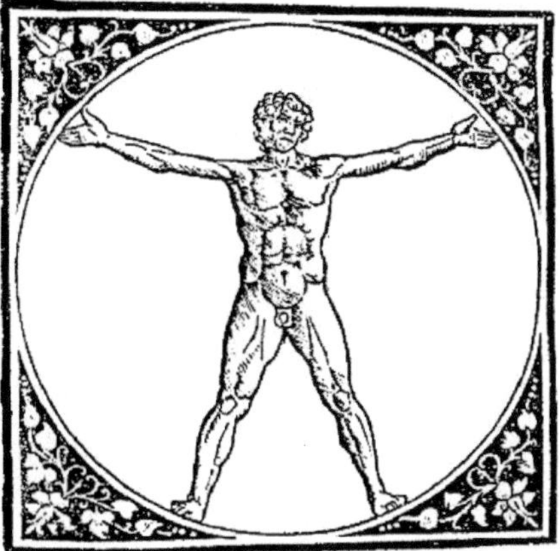

Figure 5.33. *The proportions of the human body in the 1511 Vitruvius edition by F. Giocondo*

followed the proportions of the golden section. The system, elaborated at the end of the 1920s, set the appropriate proportions for the height of chairs, tables, wardrobes, indeed for the rooms themselves, and for their proportion to one another.

It was the search for perfect proportions that led the urban planning ideas of Le Corbusier and a few of his contemporaries, like Roger Anger. With regard to the mathematics of the application of the modulor, Le Corbusier outlines the scheme of spiral architecture, which on the one hand he calls organically growing, and on the other hand refers to as organic. It was this concept that he realized (1939) in his plans for a museum to be built in Philippeville (Figure 5.34).

This is how, following the Renaissance concept of the ideal city, the so-called galaxy plan (by R. Anger) was born in the 1960s for a city built in spiral arms, Auroville, near Pondicherry in India. He planned main roads as spiral arms leading out from the public buildings located around the centre of the city, with residential areas in the expanding space between these, meaning an increasingly loose urban structure the further out one went from the centre, and guaranteeing a human scale and proportions consistent with continuous urban development (Figure 5.35; for the ground-plan, see Chapter 15).

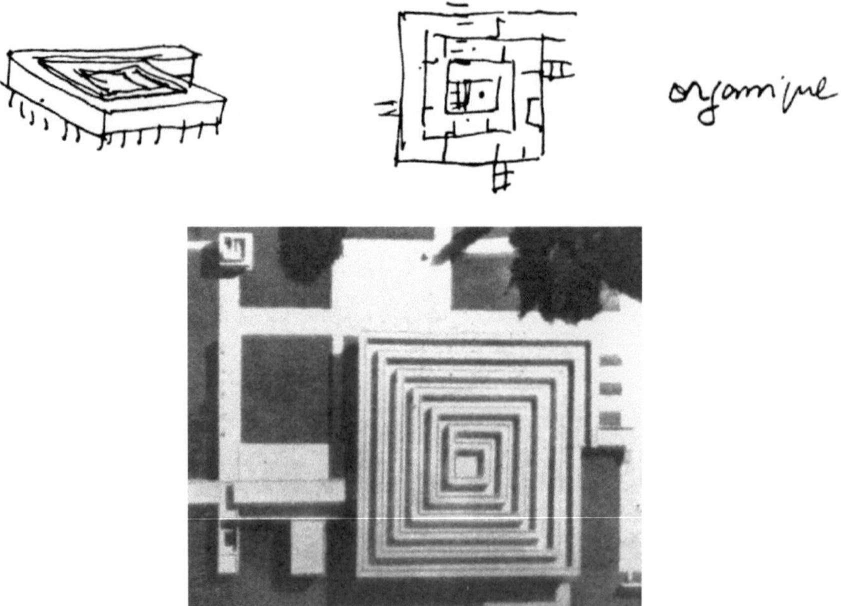

Figure 5.34. *Le Corbusier's museum design for Philippeville*

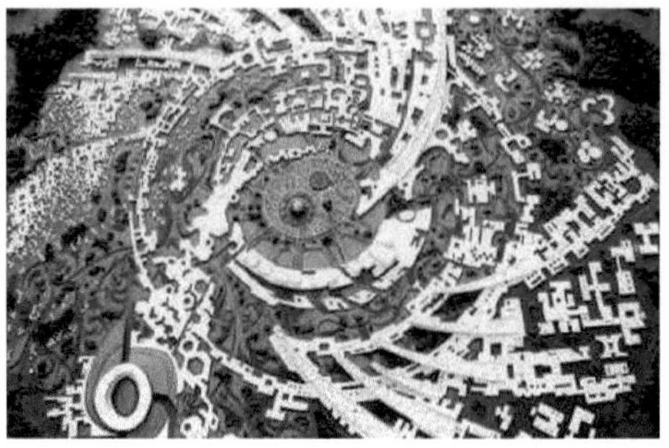

Figure 5.35. *Bird's-eye view of Auroville*

Chapter 6

Perfection and beauty

From Plato to crystals

In Chapter 3 we encountered frieze and ribbon designs that can be placed in one dimension, and wallpaper groups, suitable for filling the plane or surfaces, that can be used for tiling or tessellation. Since ancient times, the possible repeating patterns in both of these dimensions have been used to decorate our environment. We completed our train of thought by saying that we would deal with the possible repeated close filling of space without gaps in a separate chapter.

The periodic filling of space can, for example, mean the placing of the bricks in a wall alongside and on top of each other, or the arrangement of the rooms in a multi-storey building, divided by walls and ceilings. The proportions of their similarity or difference are hidden from us by their intransparency. In the absence of a fairy-tale glass palace, we can observe such structures in soap bubbles or the head of a glass of beer. Their application in decoration at best becomes possible through the use of transparent material. Practical disadvantages set the limit for this, however. Despite all this, we cannot say that spatial symmetries possess fewer aesthetic treasures for us than one- or two-dimensional ones. Even if we cannot artificially decorate our environment with repeated spatial cells in a transparent and enjoyable manner, nature does this for us.

The crystals appearing in inanimate nature present us with the most varied richness of beauty. The most beautiful examples of crystals found in nature or extracted from it can equally be awe-inspiring pieces in museums or jewels symmetrically ground by a jeweller's hand. A smaller proportion of crystals allow an insight into their secrets through their transparency. Examples are a few jewels, perhaps the most visual of which is the diamond, which reflects colourless light. The beauty of most, however, becomes visible in the way that the submicroscopic regularity of their internal structure is displayed projected on the surface in macroscopic size. Their repeated regular atomic structure makes possible the reflection of light of uniform colour, angle and strength. Their surface is shaped by splitting along planar surfaces with regular angles between them and which correspond to their microscopic crystal structure. It presents itself inside out. The regular geometric structure of its elemental cells is multiplied macroscopically by this repetition. This beauty can be increased deliberately if the jewel is polished parallel to planes that follow the regularity of the faces of the elementary crystal cell. Anyone who has seen raw and polished diamonds alongside each other can tell the difference. For with optimal polishing, the light reflecting in the cut edges, and the light travelling perpendicularly and without breaking through the faces of the elementary cells inside the crystal, can be multiplied due to multiple reflections and by keeping loss through scattering to a minimum. This is why we see a particularly strong gleam if we look from a certain angle. It is on the basis of this that the expert eye of the jeweller decides that a jewel is genuine.

In 1669, Danish physician and naturalist Nicolaus Steno identified the law of constancy of interfacial angles, which holds that the angles between the faces of a crystal, as a convex polyhedron, are constant, irrespective of the crystal's shape and size. The crystal is a body whose external form reflects the order of its internal construction at a higher level of organization. We can observe this in the case of sugar or kitchen salt crystals, which are less hard than jewels, but are brittle, and which are split of their own accord in parallel with the crystal faces of their elementary cells. Steno's discovery was later further developed by Haüy.

An interesting experiment demonstrates the way in which a crystal presents itself to the outside world. W. Kleber established the correspondence principle of crystal growth and dissolution in 1932. If we dissolve a crystal sphere in a solution, the sphere takes on the shape of a polyhedron, which is delimited by planar surfaces, as it diminishes in size. Similarly: if the crystal is grown on a spherical shape like a seed, we get a complementary polyhedron form. (If, for example, the growth form is a cube, the solution form is an octahedron). If in the centre of the crystal we further stretch a spherical cavity with solution, the growing cavity will take on the same polyhedron form as the crystal sphere growing in solution. If we grow the crystal inwards, towards the cavity which was originally spherical, it will fill up the cavity in such a way that the decreasing cavity takes on the same polyhedron form as a dissolving crystal sphere. That is to say, the correspondence principle states that the growth of the convex crystal sphere is in accordance with the dissolving of the concave sphere; the growth of the concave crystal sphere (when the cavity becomes smaller), on the other hand, corresponds to the solution of the convex sphere.

The discovery that the crystal displays its internal structure outwards — macroscopically — is much older than the experimental proof of the material's atomic structure. This made it possible for the classification of crystals mentioned in the historical chapter to begin as early as the eighteenth century. It became possible to study the properties we now know to originate at the molecular level at a larger size.

The road from the search for the most perfect forms to the uncovering of the secrets of crystals took several millennia. The scholarly study and classification of the beauty of crystals appearing in nature came much later than the search for the beauty and perfection that lies in geometrical forms in general. Filling space without gaps and determining the symmetries of point groups, then relating this to the structure of matter, was the result of a lengthy process.

Even in the earliest times, there was no doubt that the most perfect body was the sphere. Today we would say that the sphere has infinitely many axes of rotation and planes of symmetry. From the spherical body, we can reach a polyhedron delimited by planar facets by truncating spherical caps, then spherical slices. By removing these slices in regular fashion, we can obtain polyhedra with rotation symmetry. The question was which of the bodies delimited by facets were those which best approximated the perfection of a sphere. Which are those which, during the process of truncation, preserve the most of the sphere's symmetries? It would seem obvious to look for this perfection among the bodies with properties that are distinguished in a particular fashion. *Prisms, antiprisms, regular bodies* and *semiregular bodies* have distinguished properties. Also with distinguished properties are *kaleidoscopes* and *bodies suitable for filling space without gaps*. Of all these, the ones that expect the title of the most "perfect" or the most "beautiful" are those with the most symmetries. Of course, complete perfection does not exist. The sphere mentioned above as the model of perfection is not suitable for filling space without gaps, for example.

All *faceted bodies* are subject to one common rule. The very fact that there is a rule common to all of them is enough to shroud these geometrical shapes in a certain mystery. This law was discovered in the eighteenth century by Euler (1707–1783). His theorem states that, for any body delimited by edges and vertices, the total number of its vertices and faces is two more than the number of its edges. We can really be convinced of the guiding role of this theorem if we think of the fact that the bodies it applies to can be cut in two, chopped along their facets, turned into one another, fitted together along their faces, without Euler's theorem losing its force for them: it is equally applicable to every single example of the bodies cut in two or stuck together.

Of the bodies with special properties, let us first take the *prisms*. Prisms are bodies that have two coincident parallel faces connected by parallelograms (Figure 6.1).

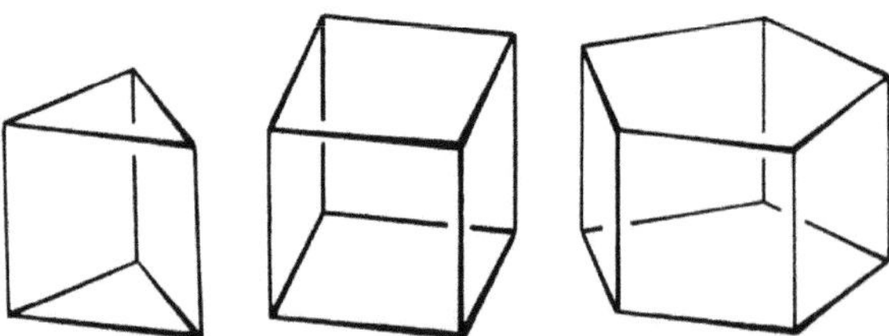

Figure 6.1. *Prisms*

Those prisms in which the parallelograms are mirror surfaces, and which generate repeated mirror images, we call kaleidoscopes. The kaleidoscope was one of my favourite childhood toys. I would never tire of admiring the infinitely rich world of colours and shapes created by the rotation of this toy made up of a few colour glass fragments, and in particular their symmetry. Kaleidoscopes were patented by David Brewster (1781–1868) in 1817. Their name derives from the combination of the Greek *kalos* (beautiful), *eidos* (shape) and *skopien* (view). Examples of kaleidoscopes are prisms delimited by regular polygonal planes whose angles are $180/n$ degrees (where n is an integer). The geometrical discussion and classification of kaleidoscopes — which took place on the basis of the classification of the symmetries of shapes with a fixed mirror plane — is associated with the work of Fedorov, after the earlier work of Möbius (1790–1868). The prismatic kaleidoscopes placed in a cylinder are prisms delimited at both their ends by polygons for which all the connecting parallelogram faces are made of mirrors (Figure 6.2). Their angle can be 90°, 60°, 45° or 30°,

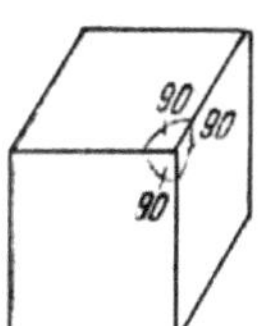

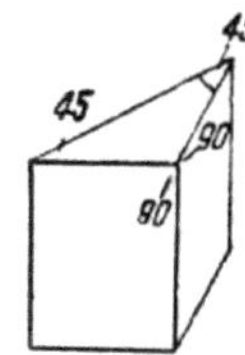

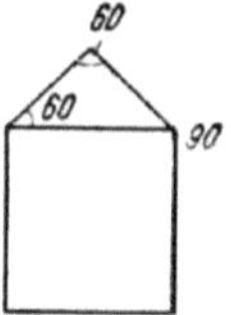

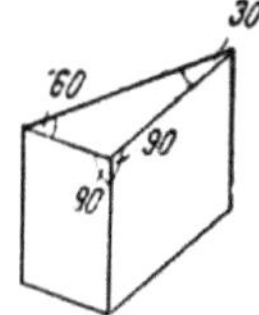

Figure 6.2. *Prismatic kaleidoscopes*

but not smaller. The possible distribution of angles are (60°, 60°, 60°), (90°, 60°, 30°), (90°, 45°, 45°) and (90°, 90°, 90°, 90°).

We cannot produce a kaleidoscope with more mirror faces if it is to fit into a cylinder. The reason for this is that each pair of mirrors must enclose a kaleidoscope angle of $(180/n)$ degrees. At the same time, all four prismatic kaleidoscopes display the characteristic of the Möbius-Fedorov kaleidoscope, in that they must have vertices in which two faces with a right-angled vertex meet one with a vertex of $(180/n)$ degrees (Figure 6.3).

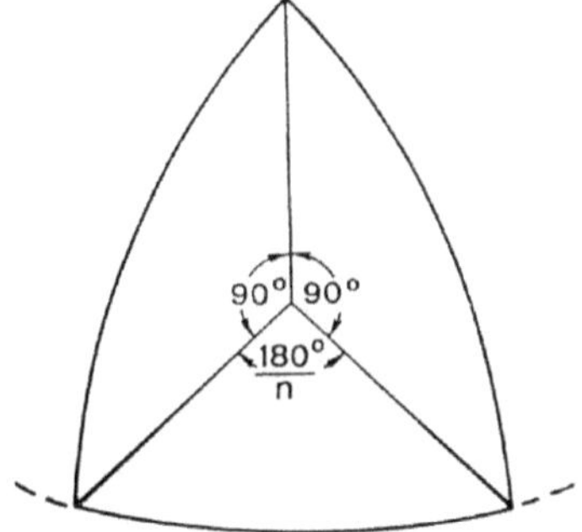

Figure 6.3. *The angles of Fedorov-type kaleidoscopes*

We must mention another class of kaleidoscopes, which are not prismatic types, but which also produce repeated mirror images. These are the so-called tetrahedral kaleidoscopes. Fedorov showed that in a tetrahedral kaleidoscope the angles of the planes meeting at a vertex can, with a little rounding-up, be the following: (55°, 55°, 70°), (55°, 45°, 35°) and (37°, 32°, 21°). By dividing up a cube (into 48 congruent parts), then putting the parts together, three other tetrahedral kaleidoscopes can be produced, as seen in Figure 6.4.

These can be produced by cutting up a cube, as shown in Figure 6.5 (*a*,*b*).

Of the three possible tetrahedral kaleidoscopes that can be produced by the congruent division of the cube, the one on the left is shown here by part (b) of the figure. On this, we can easily recognize two vertices between faces formed by the angles characteristic of prismatic kaleidoscopes (90°, 90°, 180°/4). We get the middle one by extending the previous one with

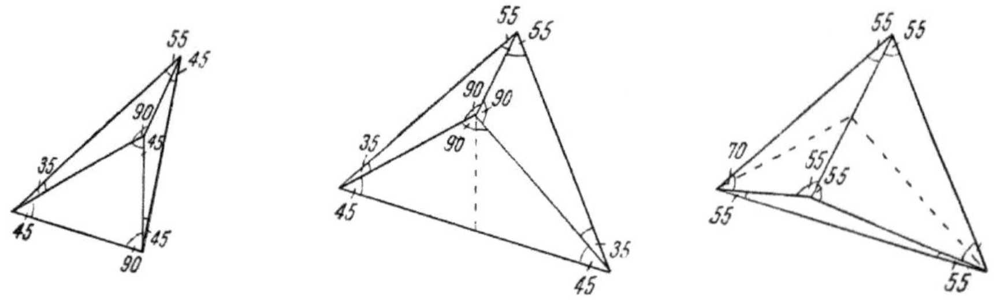

Figure 6.4. *Tetrahedral kaleidoscopes*

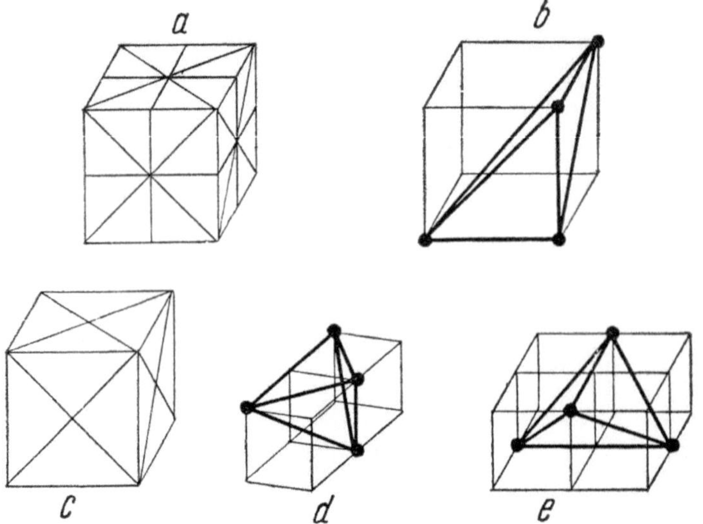

Figure 6.5. *Producing kaleidoscopes by breaking up a cube*

its mirror image reflected in one of the faces characterised by the angles (90°, 45°, 45°). The projection of the mirror plane is marked on one of the faces with a dotted line. In the course of this reflection, we lose one of the Fedorov vertices; the other is replaced by a Fedorov vertex delimited by the face angles (90°, 90°, 180°/2). The third one (on the right hand side) is reached in a similar fashion, by extending its own reflection in the face with angles (90°, 45°, 45°). The projection of the mirror plane is marked with dotted lines on two faces. The tetrahedral kaleidoscope this produces is entirely delimited by facets with the angles (55°, 55°, 70°). By placing rods in the tetrahedral kaleidoscopes that can be produced by dividing up the cube, we are given a broken series of self-returning lines of their mirror images, which, depending on the position of the rods, produce the image of regular or semiregular polyhedra.

We have not used the kaleidoscopic application of the threesome of angles (37°, 32°, 21°). Although Fedorov drew attention to this opportunity, its detailed discussion and implementation was the work of H. S. M. Coxeter (1907–2003) in the 1960s, followed, with some adjustments, by N. and C. Schwabe in the 1980s. This is the only kaleidoscope in which the sum of the angles meeting at a vertex is 90°. This means it can be folded from a square-shaped mirror. It can be constructed as described in Figure 6.6. It

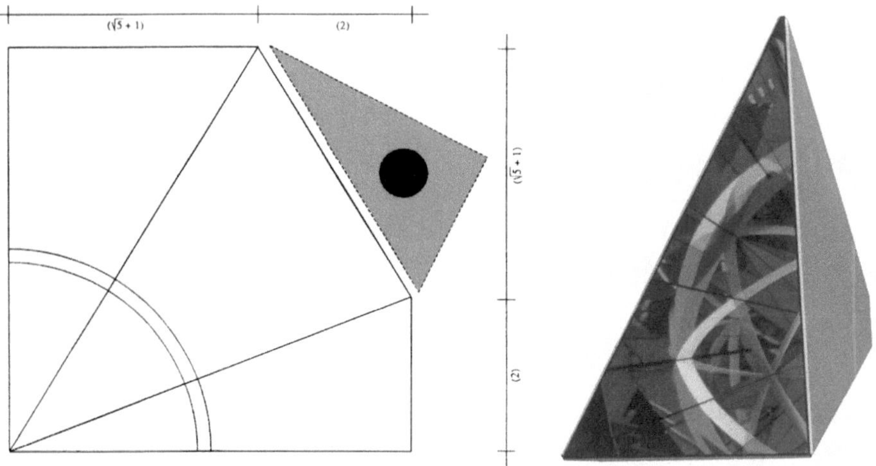

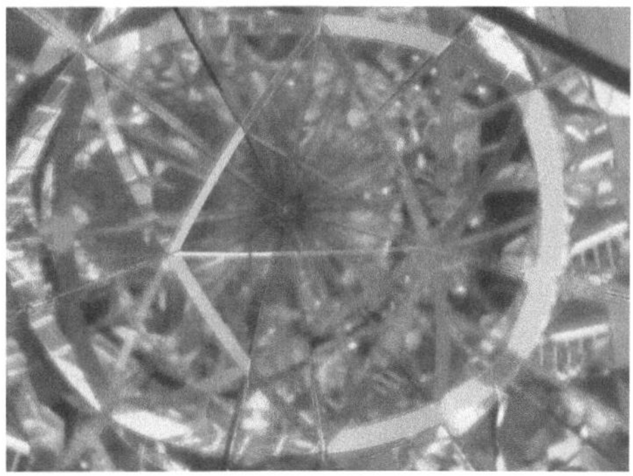

Figure 6.6. *The golden section kaleidoscope: (a) construction from a square; (b) side view; the so-called Kepler star can be seen in the centre (C. Schwabe, 1986); (c) images of the polyhedral shapes produced by passing light through the curved slits in the mirrors of the kaleidoscope (G. Darvas, 2003)*

was the Schwabes who showed that, placing the three angles in the right order (the largest in the centre), at one vertex of the square the stems at the angles would divide the opposite sides of the square according to the golden section. For this reason, their kaleidoscope is also referred to as the golden section kaleidoscope. This division makes it possible for the triangle which can be cut at the points of division marked in the figure to be used as the fourth facet of the tetrahedral kaleidoscope produced by folding, and thereby to close it. So let us divide the sides of a square according to the golden section (with a proportion of $1{:}\Phi$), then connect the division points with the opposite vertex. Finally, let us fold it into a tetrahedron along the connecting lines thus generated. (We can make our kaleidoscope more interesting if we make it possible to light it along the marked circular arc, or if we close our tetrahedron with a triangle cut out of the square, turned around according to the figure and cutting a spy-hole in the middle of it.) By putting rods at the appropriate angle in the so-called golden section kaleidoscope thus produced, the reflections are capable of producing the image of any of the five regular polyhedra.

Bodies with special characteristics that deserve mention alongside the prisms are the *antiprisms*. The antiprisms are polyhedra which also have two coincident parallel faces, but these are connected by triangles (Figure 6.7).

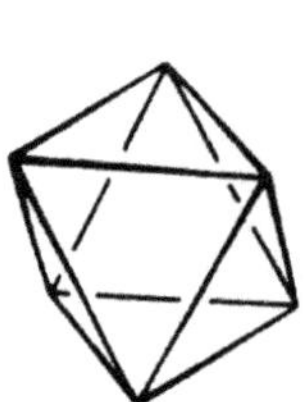
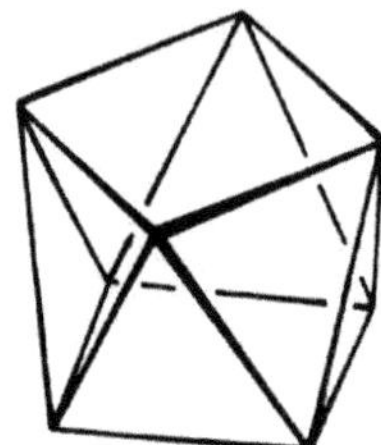
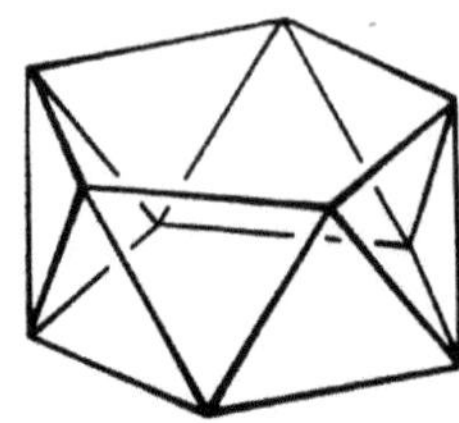

Figure 6.7. *Antiprisms*

Among polyhedra, the bodies considered the most regular are the so-called Platonic bodies (Figure 6.8). The faces of these are coincident regular polygons, their edges are equal, their vertices are coincident and regular (the angles between edges and faces are all equal). As there are no more properties of a polyhedron which could be made uniform, we regard them to be perfect bodies. Their faces, edges and angles display the maximum symmetries.

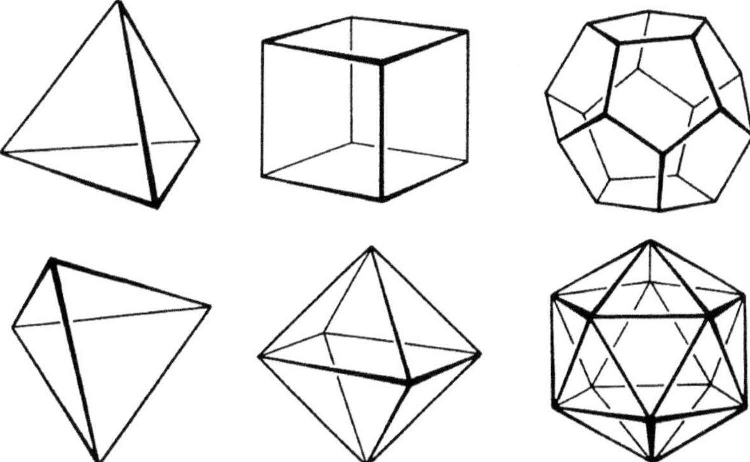

Figure 6.8. *The five Platonic bodies as dual pairs. (The tetrahedron appears twice as its own dual.)*

The best known of these is the cube. The reason we use a dice in gambling is that, because of its symmetries, it is equally likely to fall on any of its sides. It guarantees equal chances for all players. In truth, any regular body satisfies this condition of falling on any side with the same probability, not just the six-sided cube (i.e. hexahedron) that we in contemporary Europe are accustomed to call a dice in this context. Etymologically, the noun *dice* does not even refer to the cube. (It is the plural of the noun *die*, here meaning a surface with a relieved design forming one of the facets of a polyhedron.) In principle, any of the five regular polyhedra can serve as a dice. There is evidence to suggest that in the Italy of old, dodecahedra were used in games, while in Etruscan cultures they can have had religious significance (Figure 6.9a).

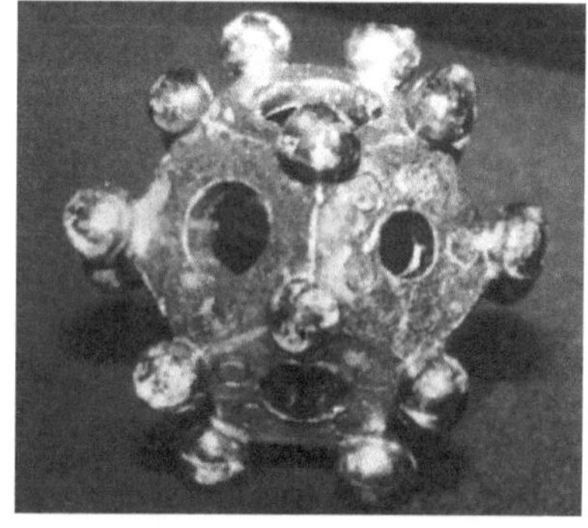

Figure 6.9a. *Etruscan dodecahedron*

In Japan, for example — where the number five is considered a lucky mascot — a dodecahedron delimited by regular pentagons is still used for

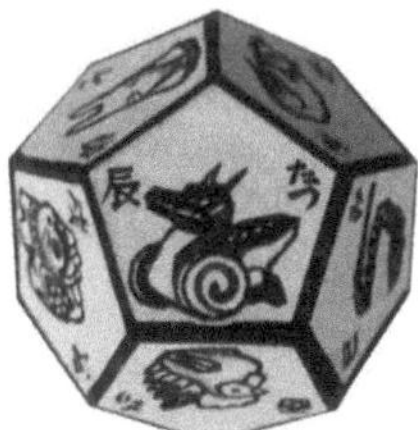

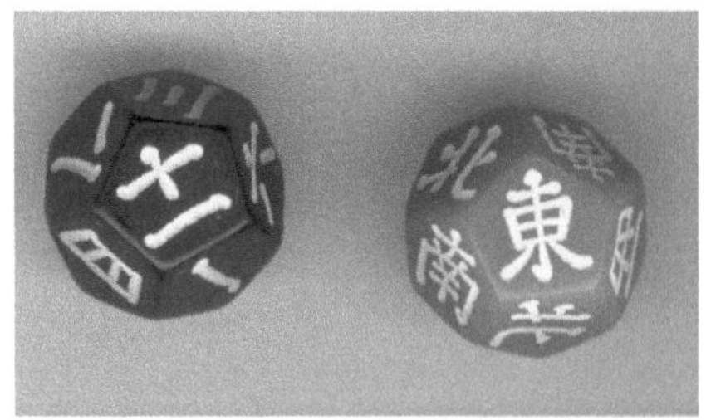

Figure 6.9b. *Japanese dodecahedral dice (after K. Miyazaki)*

this purpose to this day. Sometimes it is customary to write the digits from one to twelve on its faces, sometimes the names of the twelve months, as shown in the right hand picture in Figure 6.9b.

Plato — and probably others before him — already knew that were exactly five and only five perfect bodies. It is a surprisingly small number, but if we consider how many conditions a regular body has to satisfy, perhaps this is not so startling, after all. The less there is of something, the more respect we have for it, the more valuable it is.

The five perfect bodies are the regular *tetrahedron, cube, octahedron, dodecahedron* and *icosahedron.* These five regular bodies can be grouped into three classes, according to their duals. The cube and the octahedron are the duals of one another. Connecting the centre-points of the faces of the cube gives us a regular octahedron, and vice versa. Similarly, connecting the centre-points of the faces of a dodecahedron gives us a regular icosahedron, and vice versa. They are also the duals of one another. Neither is the tetrahedron left on its own: by connecting the centre-points of the faces of a regular tetrahedron, we are given another regular tetrahedron, that is the tetrahedron is its own dual. Whichever of their axes of rotation we rotate the various regular bodies around, we are returned to the original state with the same number of rotations as for the dual. The rotation symmetry of the various regular bodies and their duals are described by the same group. The symmetries of the regular bodies and their duals, pair by pair, are easy to read in Table 6.1, which lists their faces, vertices and edges.

Certain symmetry properties of the various bodies can immediately be made obvious by placing marks. Let us use the so-called Steiner symbols for this purpose: in parentheses, we list the regular polygons meeting in

Name	Boundary faces	Number of faces	Number of vertices	Number of edges
Tetrahedron	triangle	4	4	6
Cube	square	6	8	12
Octahedron	triangle	8	6	12
Dodecahedron	pentagon	12	20	30
Icosahedron	triangle	20	12	30

Table 6.1. *The regular polyhedra*

a vertex, giving the number of their edges (vertices). For example, the symbol of the prisms and antiprisms just mentioned is (k, 4, 4) and (k, 3, 3, 3) where k is the number of angles in the polygon of the face, the two 4s refer to the two parallelograms which meet at the vertex of a polygon, and the three 3s to the three triangles which meet at the vertex of a polygon. Using these symbols, we present the Platonic bodies in Table 6.2.

Name	Steiner symbol	Faces meeting at one vertex
Tetrahedron	(3, 3, 3)	3 triangles
Cube	(4, 4, 4)	3 squares
Octahedron	(3, 3, 3, 3)	4 triangles
Dodecahedron	(5, 5, 5)	3 pentagons
Icosahedron	(3, 3, 3, 3, 3)	5 triangles

Table 6.2. The properties of the regular bodies

The dual properties of the various pairs can be found in this table. Regular bodies representing each other's duals can be produced from one another by truncation. By cutting along the planes that can be laid between the centre-points of the faces, we are given the dual of the body in question.

Let us apply this truncation, like a milling machine, in small steps, with the so-called cell automata method. We should begin the slicing near one of the vertices, in parallel to the plane laid between the centre-points of neighbouring faces, but always only slicing off a thin layer. In Figure 6.10 we show certain stages of the series of truncations of the regular octahedron, finishing with the cube. By cutting off increasingly large squares from the vertices of the octahedron, we are given a body made

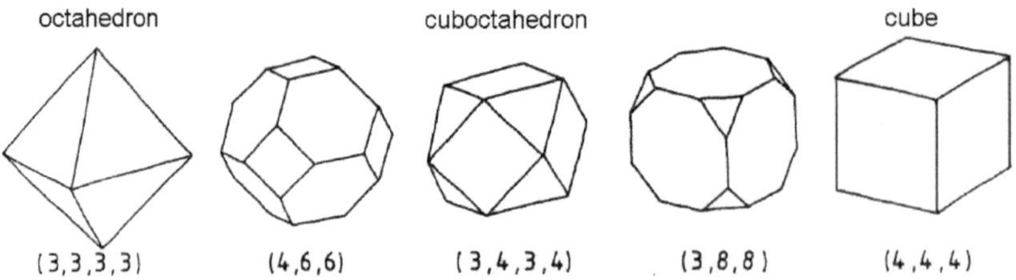

Figure 6.10. *The series of simple truncations of the octahedron*

up of 6 squares and 8 hexagons. The edges of the squares become longer and longer; the edges of the hexagons coinciding with those of the octahedron become shorter, while those bordering the squares produced by truncation increase in length, until they are equal. This gives us a body whose edges are all of equal length. Continuing the truncation, the edges of the squares grow, while the edges of the hexagons coinciding with the original edges of the octahedron decrease, until they disappear altogether, and the hexagons each turn into regular triangles. The edges of the so-called cuboctahedron thus produced are all equal in length. Continuing the truncation in parallel with the existing faces, the length of the edges of the triangles decreases while the squares turn into octagons, the edges of which, alternating, increase or decrease until they are equal. We are again given a body whose edges are all of the same length. Finally, continuing the truncation, the triangles disappear, together with every other edge of the octagons, which are thereby turned into 6 squares. This is how, by the end of the series of truncations, we reach the cube. We can equally imagine the series of truncations going backwards. We can execute a similar series of truncations from dodecahedron to icosahedron and vice versa, as well as from tetrahedron to tetrahedron.

During the series of truncations, there were situations in which we were given an intermediate body whose edges were all equal in length, whose faces were two types of regular polygon, and whose vertices coincided, but were not regular. We call such bodies semiregular (Archimedean) bodies. The Archimedean bodies can partly be produced from the regular bodies with the latter simple truncation, partly with a complex or screw truncation. The full set of the semiregular bodies (which also include those which are delimited by three types of polygonal face) can be described with Steiner symbols, as follows:

(3, 3, 3, 3, 5), (3, 4, 5, 4), (3, 5, 3, 5), (5, 6, 6),
(3, 3, 3, 3, 4), (3, 4, 4, 4), (3, 4, 3, 4), (4, 6, 6),
(3, 6, 6), (3, 8, 8), (3, 10, 10), (4, 6, 8), (4, 6, 10)

The way to make regular polyhedra delimited by planar surfaces approach the ideal body, the sphere, is to project their edges on the surface of the sphere that can be traced around them. In Figure 6.11, we show the projection of the octahedron onto a sphere.

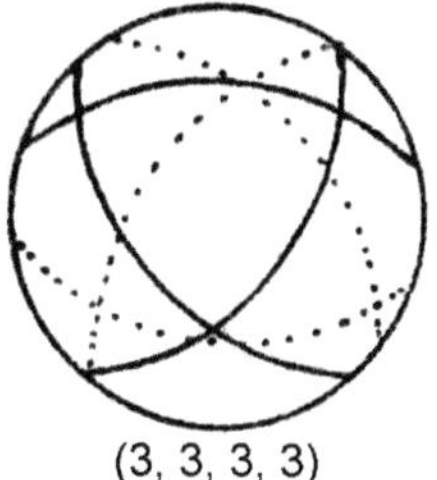

Figure 6.11. *The projection of an octahedron onto a sphere*

It can be seen that the surface thus produced preserves all of the symmetry properties of the polyhedron, and can be described using the Steiner symbols of the original. The simple series of truncations of the regular (and in intermediate stages semiregular) bodies projected onto the surface of a sphere (after S. Bérczi) can be seen in Figure 6.12. It can easily be seen from the bottom row, for example, that the structure (5, 6, 6), better-known as the stitching pattern for soccer balls, comprised of twelve regular pentagons and twenty regular hexagons, a structure known as Fullerene, can be produced from the truncation series for a dodecahedron or icosahedron.

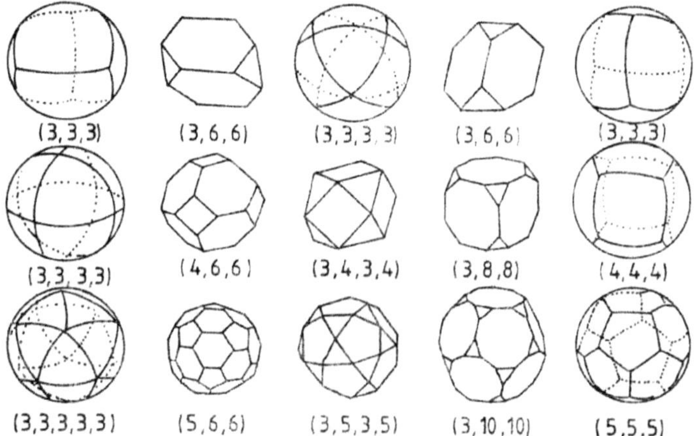

Figure 6.12. *The series of simple truncations of the regular (and intermediate semiregular) bodies*

Of the regular bodies, it is only the cube which is suitable for filling space without gaps. There are of course other bodies with this property, such as the rectangular prism, certain prisms delimited by polygons that fill the plane continuously, the rhomboid and the parallelepiped, and indeed all parallelohedra. Of the latter, we distinguish five types, namely those delimited by (1) 8 hexagons and 6 parallelograms, (2) 4 hexagons and 8 parallelograms, (3) 12 parallelograms, (4) 3-, 4- or 6-sided prisms, and (5) 6 parallelograms.

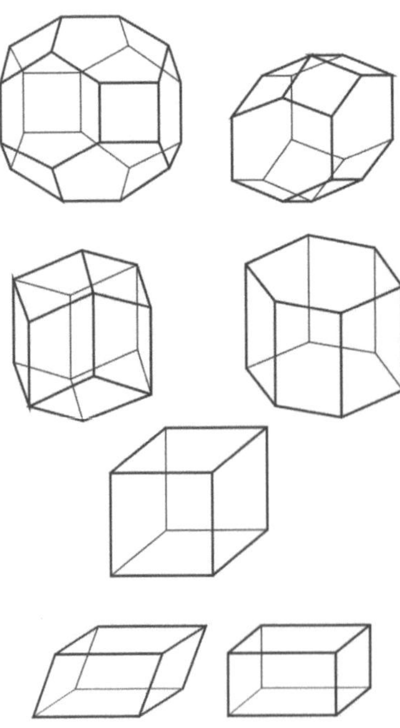

Figure 6.13. *Parallelohedra*

From the splitting of crystals, the French mineralogist Haüy deduced that crystals are built from elementary parallelepipeds. It was by replacing the latter with their vertices that his fellow countryman Bravais would later introduce the concept of the point lattice. It is the bodies capable of filling space without gaps that represent the set of potential cells that are suitable for giving us crystal lattices when point groups, groups of atoms and molecules are placed at their particular points. Because of the key role they play in natural phenomena, we need to deal with them in rather greater detail.

The symmetries of crystals are made up of two things. First, of the symmetries (shapes) of the imagined polyhedra (i.e. cells delimited by planar surfaces) which fill the space without gaps when placed alongside each other, and at particular points of which we can place the atoms (ideally considered to be point-like) that make crystals, or compounds of atoms; second, of the symmetry groups of the atoms (points) that can be placed at particular points of the cells. The symmetries of these two things, namely the cells and the point groups, can be examined separately, and if put together they produce the symmetries of crystals.

An elementary cell has the property that some fixed point, or a number of fixed points, can be associated with its symmetries. Such a fixed point or accumulation of fixed points can be the centre-point of a rotation or centred reflection, the axis of a rotation or reflection, or perhaps a mirror plane. These points (or axes, or points of a plane) remain invariant under the symmetry transformations.

The symmetries we met in the previous chapter (e.g. the similarity symmetry of the snail's shell) were all, with the exception of the operation of reflection, described by a group of repetitions of a single operation. Such a group will only be finite if the operation is a rotation through an angle $2\pi/n$, where n is an integer. For in this instance the rotated figure will be transformed onto itself after at most n rotations. The transformation of planar rotations in a rectangular system of coordinates (x, y) placed with any direction at the centre-point of the rotation is described by the matrix seen in the first chapter,

$$\begin{bmatrix} \cos\alpha & -\sin\alpha \\ \sin\alpha & \cos\alpha \end{bmatrix},$$

where α is the angle of rotation. We have seen that planar rotations form a group: the angles of two successive rotations accumulate, and the group axioms are satisfied.

Spatial rotations can similarly be composed of two successive rotations around two different axes. For this reason, they can be described by two angles and a [3x3] matrix. We can easily see, for example (by inserting numerical values for the trigonometric functions), that a rotation on the plane (x, y) by 90° can be characterized by the matrix

$$\begin{bmatrix} 0 & -1 & 0 \\ 1 & 0 & 0 \\ 0 & 0 & 1 \end{bmatrix}$$

and a rotation by 60° by the matrix

$$\begin{bmatrix} \frac{1}{2} & -\frac{\sqrt{3}}{2} & 0 \\ \frac{\sqrt{3}}{2} & \frac{1}{2} & 0 \\ 0 & 0 & 1 \end{bmatrix}.$$

Attention must be paid to one thing in the course of successive symmetry transformations, however. In defining a group, we did not require commutativity, that is the interchangeability of the order of operations. Our caution now proves justified. Let us place a book in front of ourselves. Let us rotate it by 90° around its spine (the process can be followed in Figure 11.3). Then let us rotate it again by 90° around its shorter right hand edge. Let us make a note of the position. Then let us place the book back in its original position in front of us, and let us rotate in first by 90° around its shorter right hand edge, in the same direction as before. Next, we rotate it around its spine, again by 90° and in the same direction as before. The book is now in a different position than before. The order of the two rotations are not interchangeable. Marking the two rotations as S_1 and S_2, the final result is $S_1S_2 \neq S_2S_1$. This property of rotations has significant consequences, and — particularly in physics — can engender surprise at the nature of the world in those not aware of it. In such instances, we are not justified in immediately announcing the breaking of symmetry.

We have seen that the groups of rotations on the plane and in space can be characterized by the above matrices. We can describe reflections in a similar fashion. Let us take a point (for example the vertex of a cell) and also a rectangular system of coordinates with any origo not identical to this point (this could be another vertex of the cell, for example) and any direction. If we reflect the point in the xy plane of the system of coordinates, its x and y coordinates will not change, but its z coordinate will change sign. The reflection, then, can be described by multiplication by the following matrix:

$$\begin{bmatrix} 1 & 0 & 0 \\ 0 & 1 & 0 \\ 0 & 0 & -1 \end{bmatrix}.$$

Reflection in the xz or yz planes can be described similarly, with a (−1) instead of a 1 in the second or first element of the diagonal, respectively.

We can describe the reflection of a point in an axis in a similar way. With reflection in axis x, the x coordinate of the point remains unchanged, while its y and z coordinates swap signs. The matrix for reflection in the x axis is

$$\begin{bmatrix} 1 & 0 & 0 \\ 0 & -1 & 0 \\ 0 & 0 & -1 \end{bmatrix}.$$

In the case of reflection in the y axis, the 1 is the second digit along the diagonal; for the z axis it is the third. We can immediately see that in the case of reflection in a centre-point, along the diagonal of the $[3 \times 3]$ matrix there is a -1 in all three places. We normally refer to this operation as inversion.

As we have already observed in one and in two dimensions, in space it can also be shown that transformations leading to symmetry (reflections, rotations, inversions and rotoinversions) can all be traced back to a finite number of operations. In addition to the identity matrix

$$\begin{bmatrix} 1 & 0 & 0 \\ 0 & 1 & 0 \\ 0 & 0 & 1 \end{bmatrix},$$

spatial symmetry transformations can be described by means of fourteen generating matrices.

$$M_0 = \begin{bmatrix} 1 & 0 & 0 \\ 0 & 1 & 0 \\ 0 & 0 & 1 \end{bmatrix}$$ identity, identity element

$$M_1 = \begin{bmatrix} -1 & 0 & 0 \\ 0 & -1 & 0 \\ 0 & 0 & -1 \end{bmatrix}$$ inversion

$$M_2 = \begin{bmatrix} -1 & 0 & 0 \\ 0 & -1 & 0 \\ 0 & 0 & 1 \end{bmatrix}$$ twofold rotation around z axis

$$M_3 = \begin{bmatrix} 1 & 0 & 0 \\ 0 & 1 & 0 \\ 0 & 0 & -1 \end{bmatrix}$$ reflection in plane xy

$$M_4 = \begin{bmatrix} 1 & 0 & 0 \\ 0 & -1 & 0 \\ 0 & 0 & -1 \end{bmatrix}$$ twofold rotation around x axis

$$M_5 = \begin{bmatrix} -1 & 0 & 0 \\ 0 & 1 & 0 \\ 0 & 0 & 1 \end{bmatrix}$$ reflection in plane yz

$$M_6 = \begin{bmatrix} 1 & 0 & 0 \\ 0 & -1 & 0 \\ 0 & 0 & 1 \end{bmatrix}$$ reflection in plane xz

$$M_7 = \begin{bmatrix} 0 & -1 & 0 \\ 1 & 0 & 0 \\ 0 & 0 & 1 \end{bmatrix}$$ fourfold rotation around z axis

$$M_8 = \begin{bmatrix} 0 & -1 & 0 \\ 1 & 0 & 0 \\ 0 & 0 & -1 \end{bmatrix}$$ fourfold rotoinversion around z axis

$$M_9 = \begin{bmatrix} -\frac{1}{2} & -\frac{\sqrt{3}}{2} & 0 \\ \frac{\sqrt{3}}{2} & -\frac{1}{2} & 0 \\ 0 & 0 & 1 \end{bmatrix}$$ threefold rotation around z axis

$$M_{10} = \begin{bmatrix} \frac{1}{2} & -\frac{\sqrt{3}}{2} & 0 \\ \frac{\sqrt{3}}{2} & \frac{1}{2} & 0 \\ 0 & 0 & -1 \end{bmatrix}$$ threefold rotoinversion around z axis

$$M_{11} = \begin{bmatrix} \frac{1}{2} & -\frac{\sqrt{3}}{2} & 0 \\ \frac{\sqrt{3}}{2} & \frac{1}{2} & 0 \\ 0 & 0 & 1 \end{bmatrix}$$ sixfold rotation around z axis

$$M_{12} = \begin{bmatrix} -\frac{1}{2} & -\frac{\sqrt{3}}{2} & 0 \\ \frac{\sqrt{3}}{2} & -\frac{1}{2} & 0 \\ 0 & 0 & -1 \end{bmatrix}$$ sixfold rotoinversion around z axis

$$M_{13} = \begin{bmatrix} 0 & 0 & 1 \\ 1 & 0 & 0 \\ 0 & 1 & 0 \end{bmatrix}$$ threefold rotation around [1 1 1] axis

$$M_{14} = \begin{bmatrix} 0 & -1 & 0 \\ 0 & 0 & -1 \\ -1 & 0 & 0 \end{bmatrix}$$ threefold rotoinversion around [1 1 1] axis

Table 6.3. *The 14 matrices generating the spatial symmetry transformations* (after Ervin Hartmann [1935–] and Vladimir A. Koptsik [1924–2005])

How can we characterize the reflection when we define the system of coordinates with a different direction, i.e. if the axes of reflection are not parallel to the coordinate axes? There are two things we can do. One — with the help of the transformation with two parameters mentioned in the previous paragraph — is to rotate the system of coordinates in such a way that its axes coincide with those of the reflection. In this case, we can write down the reflection of our chosen point in the rotated system of coordinates. The second option is to rotate the given point — in the opposite direction — with the same angle as that of the rotation between the two coordinate systems. This gives us the original reflection matrix again. The reflection of the point cannot depend on the choice of system of coordinates. We can also see that reflection in a centre-point is always interchangeable with a rotation: the result does not depend on the order of the operations. (And in all cases the distance of the point from the origo remains unchanged.) This independence is a symmetry property. In general, we can say that if some property is a symmetry with regard to an operation (e.g. reflection with regard to the choice of systems of coordinates pointing in different directions), then the two operations (in this case reflection in a given axis and the rotation of the axis of symmetry) are interchangeable.

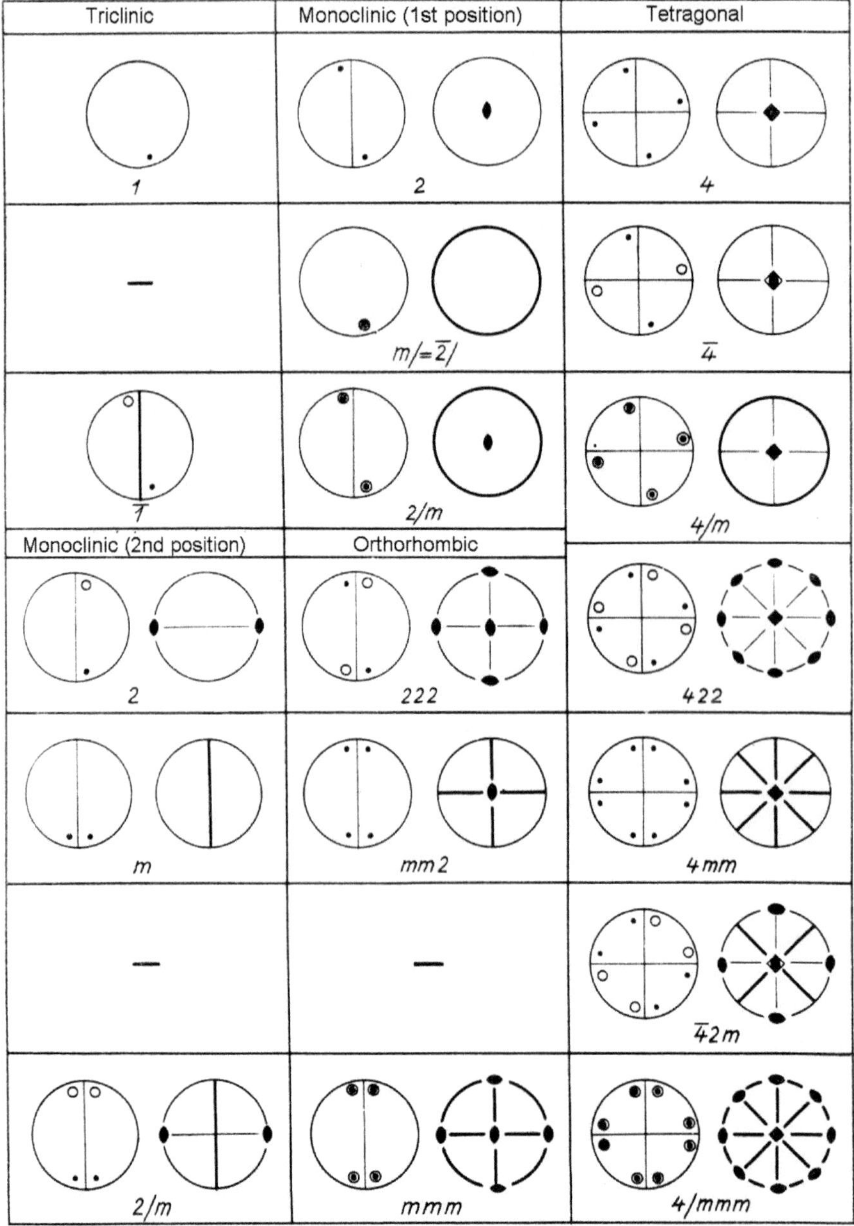

Figure 6.14. *The stereograms of the three-dimensional point groups*

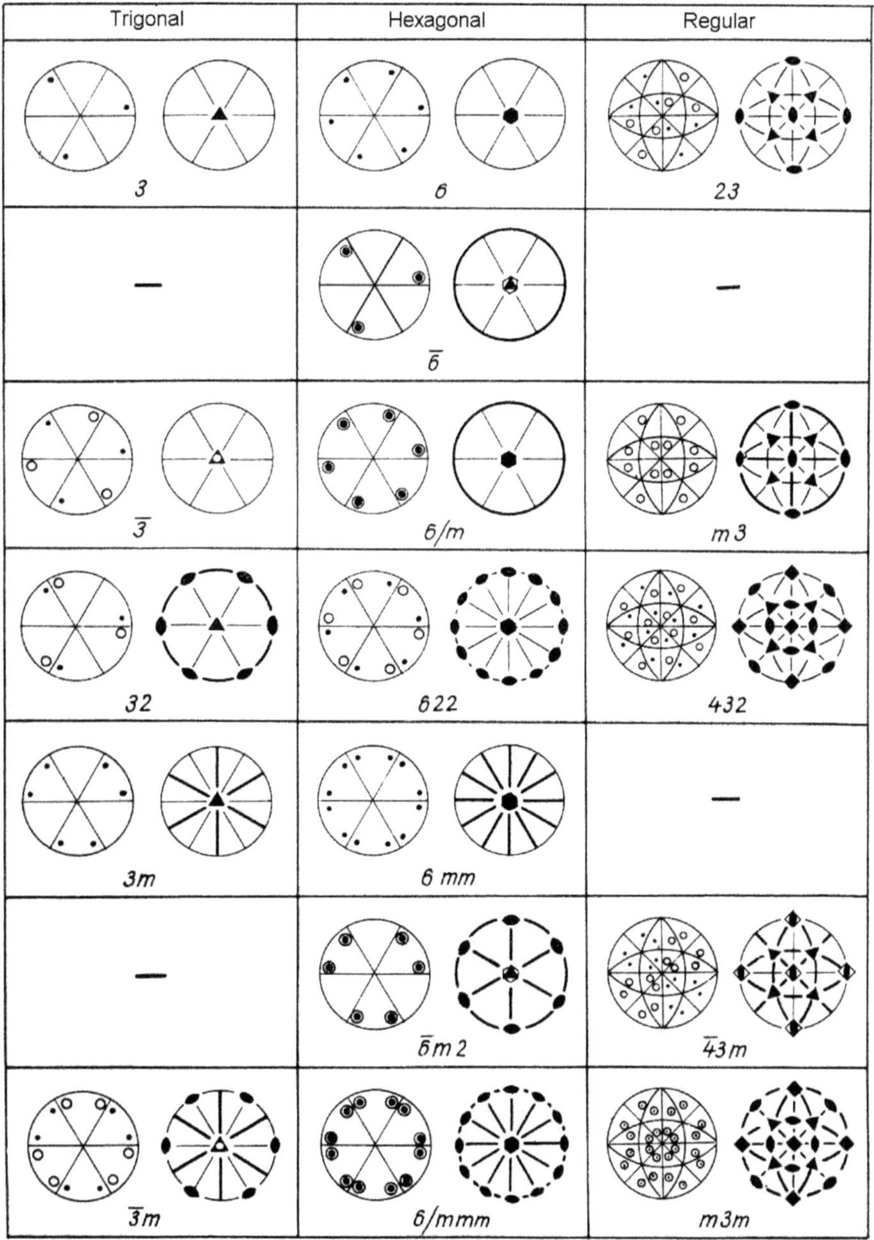

Figure 6.14. (cont.)

In a system with a given fixed point, any set of points can be transferred to another coincidental point set — by maintaining their distance from the origo — with the help of reflections and rotations. It is often suggested that reflection, as an independent operation, could be replaced with a rotation of 180°. To reject this, let us just look at our two hands. There is no rotation than can make one hand overlap with the other. In contrast, by using reflection any point on one hand can be seen to match the relevant point on the other hand. The situation is similar with a left-twisting and a right-twisting screw. Let us try the rotations of the above book to see if it is possible to produce a reflection in one of its edges.

What we have said so far about symmetry operations that can be used in space is true not just for cells, but also for point groups.

With the help of *rotations* and *reflections*, we can characterize all possible point group operations in three dimensions. These can require the following four elements: *axes of rotation, reflection planes, symmetry centre-points* (for inversion), and *inversion axes* (for simultaneous application of rotation and inversion). The *point groups* form the complete set of symmetry operations applied to one point of the cell under which the cell remains invariant. It can be shown that in space there are 32 of these. If we imagine the z axis as perpendicular to the plane of the page, then we present the thirty-two possible 3-D point groups in the stereogram seen in Figure 6.14 (in the figure, the monoclinic ones appear in two positions):

The above considerations were based on the condition that point groups include genuinely point-like elements, by which we mean an idealized state in which the points (or rather the atoms they represent) display complete spherical symmetry, have no internal properties, and that there is no particular spatial direction associated with them.

So far we have examined the symmetries of the cells with a single cell. We can build them up if, in various spatial directions, we place congruent examples of them alongside, behind and above each other. We get a space group from the group of reflections in and rotations around a point if we also add translations in space. To characterize these we have to expand our existing [3 x 3] matrices, in line with the formalism of linear algebra, to $[4 \times 4]$ matrices. This $\begin{bmatrix} \cdot & \cdot & \cdot & \cdot \\ \cdot & \cdot & \cdot & \cdot \\ \cdot & \cdot & \cdot & \cdot \\ \cdot & \cdot & \cdot & 1 \end{bmatrix}$ is how we characterize, in the matrices that describe space groups, the extension of the symmetries of individual cells to include translation as a new property (symmetry operation).

We can arrange the space groups in the points of the space lattice. From experience, we know that space can be filled without gaps with certain cells, for example cubes. Taking the example of the cube, Figure 6.15 presents all its possible symmetry elements.

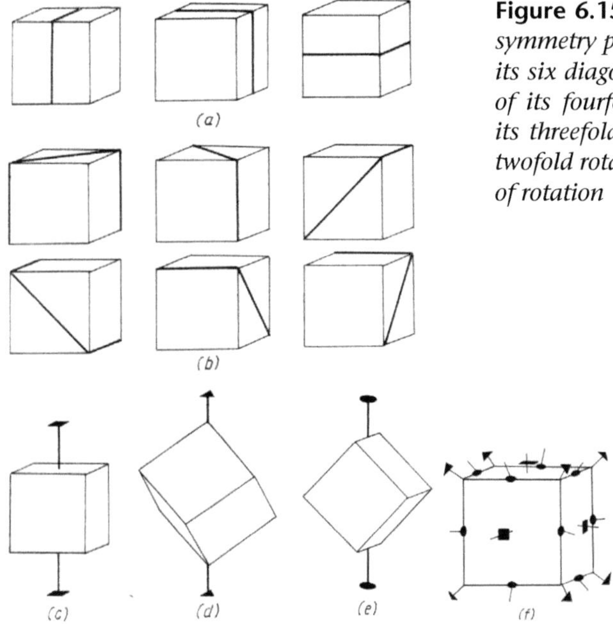

Figure 6.15. *The cube and (a) its three symmetry planes parallel to the faces; (b) its six diagonal symmetry planes; (c) one of its fourfold rotation axes; (d) one of its threefold rotation axes; (e) one of its twofold rotation axes; (f) its thirteen axes of rotation*

The edges and vertices of the cubes form a lattice. As we have seen, we can construct a lattice alongside the cubes out of bricks, out of straight and oblique prisms whose bases fill the plane without gaps, and out of rhomboids. In the case of primitive cells we can place point groups on their vertices, while in a general case we can also place them on the centre-points of their cells, faces and edges, to the extent that their symmetries do not ruin the symmetry operations that can be applied to the given cell. The reverse is also true: we do not always find point groups with suitable symmetries for all space-filling lattices. We use the term space lattice (Bravais lattice) for those lattices (made up of cells placed alongside each other without gaps) demanded by the point group operations applied

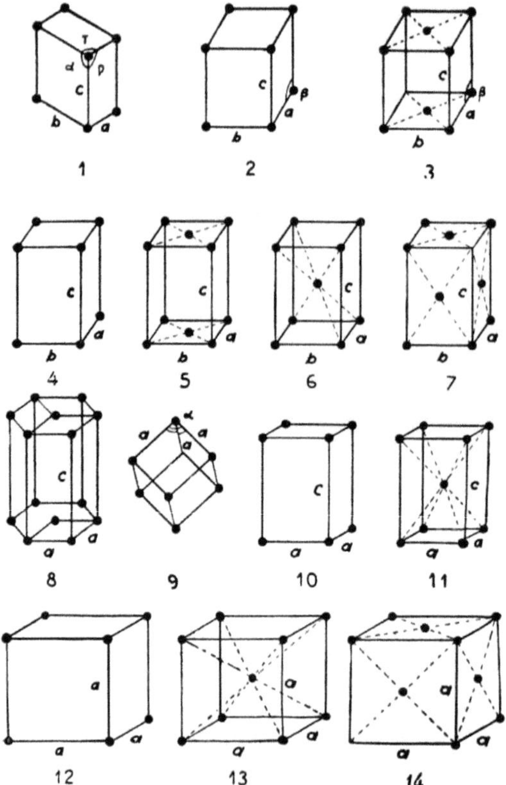

Figure 6.16. *The fourteen Bravais lattices: (1) triclinic; (2, 3) monoclinic; (4, 5, 6, 7) orthorhombic; (8) hexagonal; (9) rhombohedral; (10, 11) tetragonal; (12, 13, 14) cubic*

to the lattice points. In nature, a total of fourteen Bravais or space lattices are possible, and we present these in Figure 6.16.

Together, the point groups and the space lattices contain all the symmetry operations possible in three dimensions. This include the 4 three-dimensional point group operations (the rotation axes, the reflection plane, the symmetry centre-point, in other words the inversion, and the rotoinversion axes), and, in addition, the *screw rotation* (rotation + translation, or helical rotation) and the *glide plane* (reflection plane + translation parallel to it) that can be applied to space lattices.

Point groups are the embodiment of the symmetry elements that can be applied on a point of the space. If we place the point groups at particular points of a space lattice, we get an infinitely repeated series of symmetry elements. This infinitely repeated order we call a space group. Through its symmetry elements, the space group determines the state of equivalent points within the elementary cell. The space group is characterized by the Bravais lattice, the types of point group, and the further symmetry elements placed inside it, as well as by the position of the point group within the elementary cell.

The multiplicity of ways in which the 32 point groups can be placed in the 14 *space lattices*, so as not to break symmetry, makes possible the creation of a total of 230 *space groups*. That there are 230 and no more than 230 was proved, independently, by the Russian Fedorov and the German Schönflies in 1890, followed slightly later by the British W. Barlow (1845–1934) in 1894. Their result was the culmination of the work of many scientists over several decades.

On the basis of the properties of their cells, the fourteen Bravais lattices can be grouped into seven crystal classes: triclinic (number 1 in Figure 6.16), monoclinic (2, 3), (ortho)rhombic (4, 5, 6, 7), tetragonal (10, 11), rhombohedral (9), hexagonal (8) and cubic (12, 13, 14). We note as a matter of interest that these can be related to the seven classical types of kaleidoscope (also identified by Fedorov and Schönflies) — namely the four prismatic kaleidoscopes, and the three types of tetrahedral ones produced by the congruent splitting of a cube — the wall planes of which symmetrically reflect the image of an object placed in-between them in the planes of one another. We can similarly classify the 32 point groups into these seven crystal classes according to the symmetry they demand on the points of the corresponding Bravais lattice. All of the symmetries appearing in the particular crystal classes can be produced by the matrices that generate the 14 symmetry operations. Table 6.4 shows how many symmetry elements the individual point groups belonging to the various crystal classes have, and which of the 14 matrices that generate symmetry operations can be applied to them (after Ervin Hartmann and V. A. Koptsik).

We can obtain *crystal structures* from the point groups and space lattices if we place an elementary set of atoms (with a given point group symmetry) on every lattice point of the space lattice. The space group is not, that is, in itself a crystal structure. What we can say about the relationship between the crystal structure and space group is that if a crystal structure is characterized by a given space group, the structure contains every symmetry element of the space group, but no others.

Crystal System	International symbol	Symbol of the matrix generating the symmetries characteristic of the crystal	Number of symmetry elements
Triclinic	1	M_0	1
	$\bar{1}$	M_1	2
Monoclinic	2	M_2	2
	m	M_3	2
	$2/m$	M_2, $M3$	4
Orthorhombic	222	M_4, M_2	4
	$mm2$	M_5, M_2	4
	mmm	M_5, M_6, M_3	8
Tetragonal	4	M_7	4
	$\bar{4}$	M_8	4
	422	M_7, M_4	8
	$4/m$	M_7, M_3	8
	$4mm$	M_7, M_5	8
	$\bar{4}2m$	M_8, M_4	8
	$4/mmm$	M_7, M_3, M_5	16
Trigonal	3	M_9	3
	$\bar{3}$	M_{10}	6
	32	M_9, M_4	6
	$3m$	M_9, M_5	6
	$\bar{3}m$	M_{10}, M_5	12
Hexagonal	6	M_{11}	6
	$\bar{6}$	M_{12}	6
	$\bar{6}m2$	M_{12}, M_5	12
	622	M_{11}, M_4	12
	$6/m$	M_{11}, M_3	12
	$6/mm$	M_{11}, M_5	12
	$6/mmm$	M_{11}, M_5, M_3	24
Cubic	23	M_{13}, M_2	12
	$m3$	M_{14}, M_2	24
	432	M_{13}, M_7	24
	$\bar{4}3m$	M_{13}, M_8	24
	$m3m$	M_{14}, M_7	48

Table 6.4. The classification of the 32 crystal classes (point groups) on the basis of their symmetries, with the number of the matrices that generate their symmetries and the number of symmetry elements occurring in them

The set of symmetry elements appearing in any of the physical properties of the crystal must include all of the symmetry elements of the point group of the crystal. This was first formulated in this form by Franz Neumann (1798–1859) in 1833. In other words, Neumann's principle states that the point group of a crystal (more precisely, the combined symmetry group of the point groups to be found in the crystal) is a sub-group of the symmetry group of all the (physical and geometrical) properties of the crystal. Expressed in mathematical form: $G_{all\text{-}properties} \supseteq G_{crystal}$.

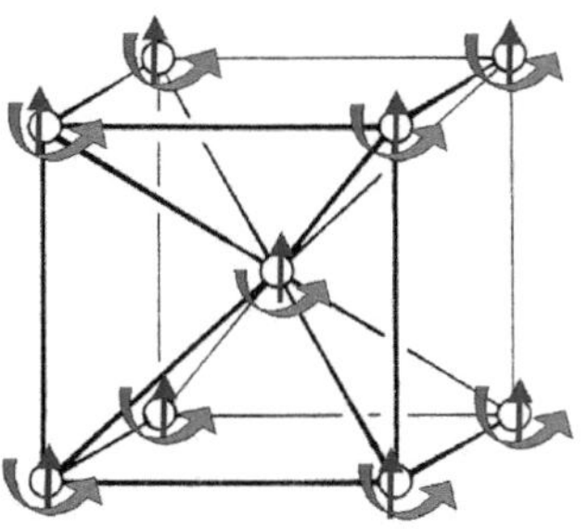

Figure 6.17. *A detail of a crystal lattice, with atoms with internal structures*

This principle seems almost to be a tautology, and would not have caused any problems if crystal were made up of point-like atoms with no internal properties. Atoms are not point-like, however, and have a number of internal characteristics. They have an internal structure, spin, magnetic momentum, etc. At any given moment, the latter assign a spatial direction. Figuratively speaking, these physical properties can be represented as (imaginary) rotations.

Neumann's principle would exclude piezo magnetism from the class of crystals, for example. As we know from experience that there are crystals with piezo magnetic properties, and that these cannot be placed in any of the known classes of crystals, our symmetry concepts had to be generalized according to a suggestion by Bhagavantham. In order to preserve the validity of Neumann's principle, we had to change our symmetry concept of crystals. The achievement of this generalization is attributed to the Russian crystallographer Shubnikov (1887–1970) and the German mathematician Heesch (1906–1995). The essence of their notion was to introduce antisymmetry operations among the symmetries of the crystals: to add to the group a further property that we consider to be a sort of imaginary reflection. (The symmetries thus obtained are combined symmetries.) The antisymmetries can be described with matrices like $\begin{bmatrix} \cdot & \cdot & \cdot & \cdot \\ \cdot & \cdot & \cdot & \cdot \\ \cdot & \cdot & \cdot & \cdot \\ \cdot & \cdot & \cdot & -1 \end{bmatrix}$ (which does not include a translation). This is how we can characterize the spins, magnetic momentums, in short two-state properties that set a direction of the atoms of the point groups placed on the lattice points. By introducing the antisymmetries and taking into consideration all possible antisymmetrical reflections, we extend the thirty-two (hereinafter: grey) point groups with 90 so-called magnetic groups. This gives us a total of 122 point groups. Placing this 122 point groups on the possible Bravais lattices gives us a total of 1651 Shubnikov space groups.

We can characterize two-position states with reflection and antisymmetry operations. But there are also multi-state properties. For example, a given object can have multiple colours. If we characterize double-state properties with an abstract spatial reflection or a 180° rotation ($1 \to -1$), then we can depict triple-state properties with a 120° rotation in an abstract space, and in general we can depict a property with n elements with a rotation through an angle of $2\pi/n$. We are therefore able to extend our concepts to say that properties with multiple elements can be characterized by rotations in an abstract space. We can characterize such a property with multiple elements with an n-fold rotation in an abstract (say complex) space for all n states (say colours) that the given property can have. For example, we can characterize so-called colour symmetries with the matrix $\begin{bmatrix} \cdot & \cdot & \cdot & \cdot \\ \cdot & \cdot & \cdot & \cdot \\ \cdot & \cdot & \cdot & \cdot \\ \cdot & \cdot & \cdot & e^{i\alpha} \end{bmatrix}$, where α is the angle of rotation. (The introduction of colour symmetry groups is associated with Belov.) In this way we can characterize properties that do not have a finite number of elements. (In point of fact, there are not a discrete number of colours, for they can have infinitely many values, that change continuously.) In this fashion we can also describe infinitefold, that is continuous, rotation groups, so-called Curie groups.

At the mention of Pierre Curie — and of piezo electromagnetism — we remember that the exact definition of symmetry concepts we use today (asymmetry, dissymmetry) are thanks to him (1894). Returning to real rotations, we should also mention the important crystallographic principle that he established. The *Curie principle* states that a crystal under an external action, like a deformation characterized by a tensor (let us think of a squeezed cube, for example), possesses those symmetry elements where the symmetry elements of the crystal *per se* and the symmetry elements of the external action *per se* coincide: $G = G_{crystal} \cap G_{effect}$. Curie's principle is in fact a special case of a wider principle that we do not detail here, but have already made use of: the principle of the superposition of symmetry groups. Applying the principle of the superposition of symmetry groups, we can describe further physical properties that come about with the combined realization of a variety of actions. We can see from this definition that the Curie principle reduces the number of symmetry elements that can be applied to any given crystal. Namely to those common symmetry elements which occur separately, in crystals not exposed to the given action, and in the action itself, irrespective of its role on the crystal, that is at the "product" of the symmetry operations of the crystal and the action.

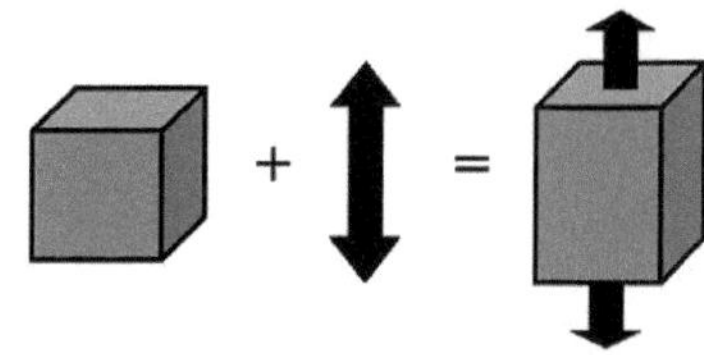

Figure 6.18. *The deformation of a cube (loss of symmetry) as a result of a pulling force*

The significance of the Curie principle goes beyond the physics of crystals. It also applies to biological morphological symmetries. The influence of this principle can be traced, for example, in D'Arcy Thompson's book *On Growth and Form*, when he describes the changes in the morphological symmetries in the living world. In a wider sense, it is the result of this principle, its generalization, more precisely of the external actions affecting living creatures through both their evolution and individual development, that they continuously lose their symmetries during both phylogenesis and ontogenesis.

The symmetry properties of crystals not only determine the structure of matter and its resulting mechanical properties, but have an effect on the magnetic and electric properties of the material, and on its thermodynamic behaviour. Using the extension of the concept of crystal symmetries to interpret the spatial organization of atoms, it is the latter which determine the structure of macromolecules, and which affect the symmetry-asymmetry properties of biological molecules, and through this the phenomena of life.

The effect of an examination of the structure of matter on our world-view

In human thought, the most regular geometric forms discovered in and thanks to crystals are associated with the concept of perfection. They have become symbols, elements of our world-view. Thinking people, becoming aware of their consciousness, saw and sought in the forms presented by crystals the success of a striving for an order governing nature, for perfection. They thought that the order observable in them could be applied in all areas of the description of the natural world and of human creative activity. As perfect forms also became the ideal of beauty, they represented an unbreakable buckle between ration and emotion (and the aesthetic associated with the latter), which would later repeatedly split into two separate worlds.

Let me again remind the reader of Herodotus' already mentioned view of the world: sometimes the belief in symmetry would prove stronger than empirical experience.

The symmetry of the world-view was strengthened by its human-centeredness (anthropomorphism). If we — human beings — are at the centre of the universe, this supposes a world with a discoid or spherical symmetry, at the centre of which we can mark our place. Thousands of years ago, the wise men of antiquity were aware of both the geocentric and heliocentric (sun-centred) views of the world. Both had spherical symmetry. The heliocentric view of the world offered circular planetary orbits, and the Earth-centred one cycloid orbits. The latter gives us a much more complicated description of the world. Yet it was still the more acceptable, because man stood at its centre, and this symmetry proved to be the more powerful. This way of thinking was reflected in ancient philosophy, as well as artistic depiction of nature and of man.

In Far East thinking, the most perfect form was embodied in the yin-yang. It was the symbol of all contrasts, with the world made up of the unity and struggle of opposite pairs. This way of thinking founded on the opposites of the yin-yang determined a universal opinion not just of the nature world and its laws, of the beautiful and the good, but of human nature.

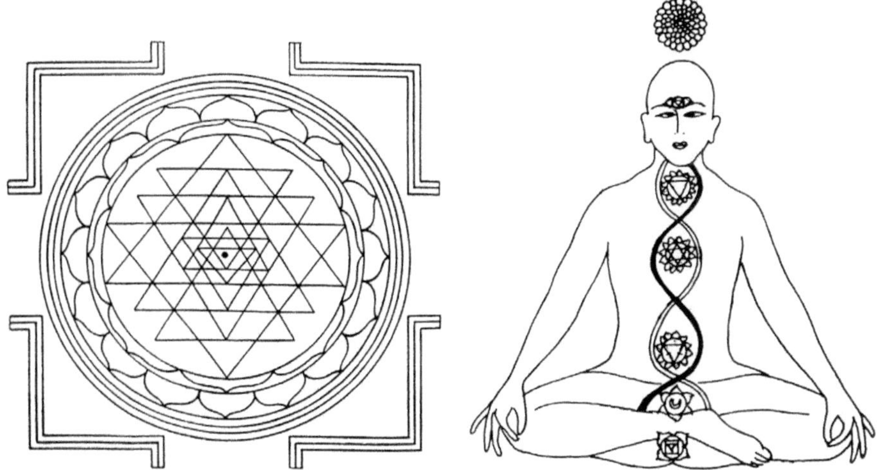

Figure 6.19. *Mandala: Sri Yantra* ('Nava Chakra')

Figure 6.20. *The anthropomorphism of the Hindu cult of Mandala*

In Oriental culture, thinking about nature, man, society, religion and transcendence to this day all form a much more closely connected unit than in Western cultures. The Indian Mandala cult offered natural science and philosophy a geometrical symbol system more complicated than the Chinese one, which at the same time was more diversified, varied and flexible.

In Mandala, essentially all symmetrical figures have some meaning. It exists in planar (painted, drawn) and spatial (e.g. architectural) variants. In the geometric design, a particular meaning is attributed to every element, colour, concentric circle, square, and so on. Figure 6.20 shows us the anthropomorphism of the Hindu Mandala cult, portraying the unity of human thought, the body and the symmetrical cult symbols.

The roots of European thought can be traced back to classical Greek philosophy, which naturally combined elements from a number of earlier cultures in the Middle East. As Heraclitus put it in the 6th century BC, "hidden harmony is stronger than explicit harmony". Explicit harmony was taken to mean the natural reality we perceive and see, that we experience directly. Hidden harmony meant the internal (rational) values we grasp and understand. He gave priority to these. The physical and the mental world (judgments about nature and human thought) that formed unity in the East quickly became separated in Western thought. If we think back to the symmetry concept of the Greeks, namely that for them that which we now term symmetry represented harmony, it is easier for us to understand how the priority on "hidden harmony" can have preferred symmetry even to emotional experience. This could then be projected back onto nature, so much so that the search for harmony, for symmetry, should become the object and even the goal of investigation.

On the other hand, as rationality and aesthetics, science and art began to separate, so symmetry began to mediate between the two, filling the middle ground as a bridge. Socrates — at least as passed on to us by Plato — makes several references to the key role of symmetry in the borderland between beauty and the truth. In Plato's *The Republic*, in the dialogue between Socrates and Adimantus (cited in Chapter 2 above), we find Socrates pointing his student to the relationship between proportionality (measure) and truth (VI: 486d), or, in the second chapter of *Philebus* (65a), in his sentence about the relationship between beauty, symmetry and truth.

The claim that the five regular bodies are the symbols of perfection embodied in nature, mentioned on many occasions above, was first formulated in Plato's work. In no small part influenced by the Pythagoreans, he attached particular significance to mathematical knowledge and geometry. In his opinion, the final elements of all things were the indivisible triangles, the bodiless geometrical atoms. (The "indivisible triangles" and "bodiless geometrical atoms" can appear to be meaningless associations, but do we not do the same in our world-view of today, when we consider so-called elementary particles in one sense to be point-like, and in another sense to have internal geometry, like helicity, spin?) The properties of perfection, which he projected onto all areas of existence — including, for example, the social relations between people — were, for him, of symbolic significance. In his interpretation, symmetry was the harmonious order of the universe, that is the very cosmos itself. He writes of the elements of this generalization in *Gorgias* as follows: "And philosophers tell us, Callicles, that communion and friendship and orderliness and temperance and justice bind together heaven and earth and gods and men, and that this universe is therefore called Cosmos or order, not disorder or misrule, my friend" (trans. B. Jovett). The significance he attributed to forms and their generalization was further developed by Aristotle, his student whose work would remain a decisive influence for centuries.

Aristotle also extended the teaching about forms to cover the rules of human thought in addition to physical bodies. Perfection, or at least the efforts to attain it, appears in both the form and content of our thoughts. The perfection of our logical claims (judgments) is an effort to find the truth. The same objective is served by the conclusions we make, by syllogisms and by deduction.

and	T	F
T	T	F
F	F	F

or	T	F
T	T	T
F	T	F

Figure 6.21. *Logical truth-tables*

In its content, the form of logical statements serves the search for truth; the form of works of art serves the creation of the beauty that they embody. In this regard we are again returned to H. Weyl, who in his book introduces the relationship between the original meaning of the word 'symmetry' and beauty with reference to Polykleitos and his book on proportions, more specifically to his glorification of the perfect harmony to be found in the proportions of ancient statues.

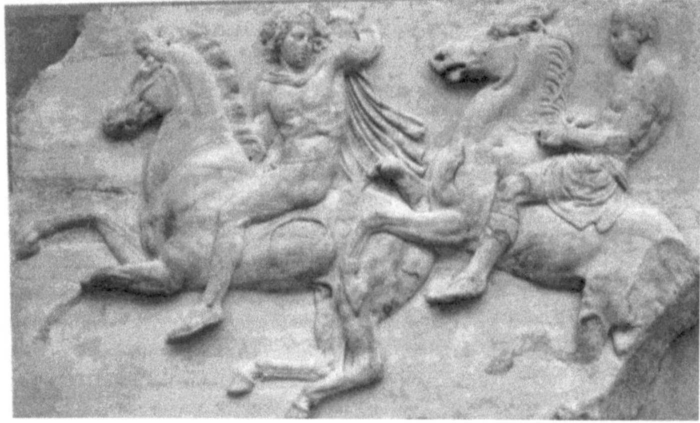

Figure 6.22. *Pheidias: Knights from the western frieze of the Parthenon*

Efforts to achieve perfection in geometrical appearance are characteristic both of the order of our logical statements and the form of our works of art; in terms of content, this contains symmetry, which for Aristotle, in-between the search for the beautiful and the true, means the proportional, the middle way. The middle way, the proportional fitted perfectly into the concept of symmetry, meaning as it did harmony, proportion, measure and moderation. To this was added the form of our actions, which presented itself in the struggle for good.

The trinity of the true, the beautiful and the good, or, put another way, that of science (rationality), art (the aesthetic) and ethics, is later reflected in the construction of the great trilogy of Kant (1724–1804), the *Critique of Pure Reason*, the *Critique of Judgment*, and the *Critique of Practical Reason.* The basic principles behind this later distinction were already there in Aristotle's work. The search for truth and for beauty are connected by a common thought, that of geometrical perfection, and more specifically symmetry (the search for harmony). The search for truth would later lead to the separation of science (the rational, the logical), the search for beauty to the independent category of art (the emotional, the intuitive), while the connecting element, symmetry, represented the bridge between the river's two increasingly distant banks. To all of this, Aristotle added the search for good, which became the basis for ethics.

We mentioned the influence of the Pythagoreans when discussing Plato. Their numerology looked for the same harmony as Plato did in the perfect bodies, and those in later ages in crystals. This Pythagorean "perfection" also included musical harmony in the list of things of artistic beauty. The relationship between the length of strings and the proportions of musical notes was the first law of nature put into mathematical terms. The enumeration of musical sounds and the association of their harmony with the cosmos appear in the music of the spheres, which would later have such an intuitive impact on Kepler's discoveries.

The proportions of musical sounds could be written down in the form of the relationships between the smallest natural numbers. It was out of this success that the numerology of the Pythagoreans grew. In their case, numerology meant tracing the material world back to a limited number (1, 2, 3, 4, 5 or many) of primary elements or substances. The schemes of the world based on substances were an invariant element in many different cultures. In the world of Ancient Greece alone (as in Oriental philosophies) we encounter a few schemes of the world based on different numbers of substances. Of the Pythagoreans it was perhaps Empedocles (c. 495–435 BC), whose theory was based on four primary elements (fire, air, water and earth), who had the strongest influence. To these four substances, Plato added a fifth, the universe (cosmos). This allowed him to associate a substance with each of the five perfect bodies (Figure 6.23).

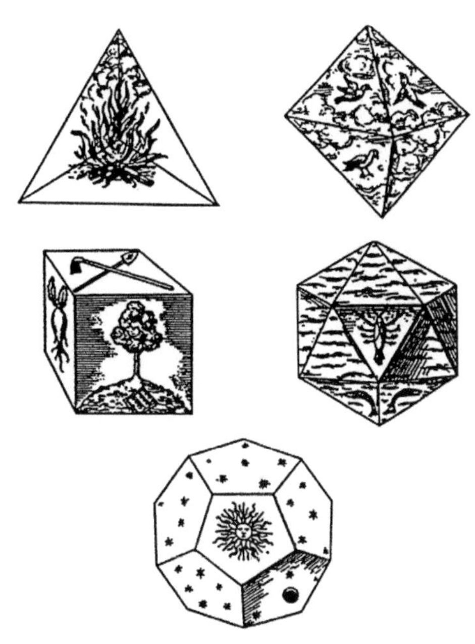

Figure 6.23. *Fire, air, earth, water and the universe in the five Platonic regular bodies*

This was later to become the basis for the harmony in Kepler's picture of the world. In India, at around the same time as Empedocles, a similar role was played by *charvaka* (four substances, comparable to his) and *vaisheshika* (five substances: earth, water, light, air, ether). In China, at around this time, the Confucian Hsu Hsing (c. 300–c. 230 BC) associated five *chi* with the five primary elements (metal, wood, water, fire, earth) that could be deduced from the material dichotomies of the yin-yang. (Here, by *chi* we mean what was originally the only substance, primary element, as elaborated in the Tao by Lao-Tse (c. 6th–5th century BC), though *chi* also means knowledge, wisdom, and intellectual essence).

During the Renaissance in Europe, the Platonic perfect bodies appeared as the embodiment of the divine proportion. This is well illustrated by the drawings which Leonardo prepared for Luca Pacioli's book *Divina Proportione* (Figure 6.24 shows two of these bodies, which are not Platonic, but display symmetry).

In science, the Renaissance brought the triumph of rationality. In contrast to the ancient picture of the world, which, in line with its anthropomorphic attitude, put man at the centre of the universe, and thus preferred

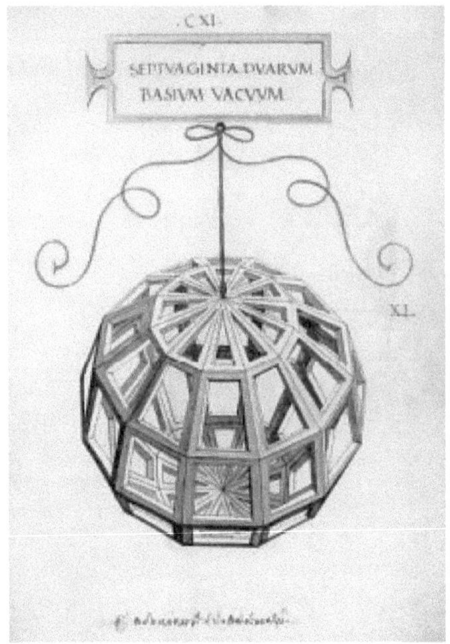

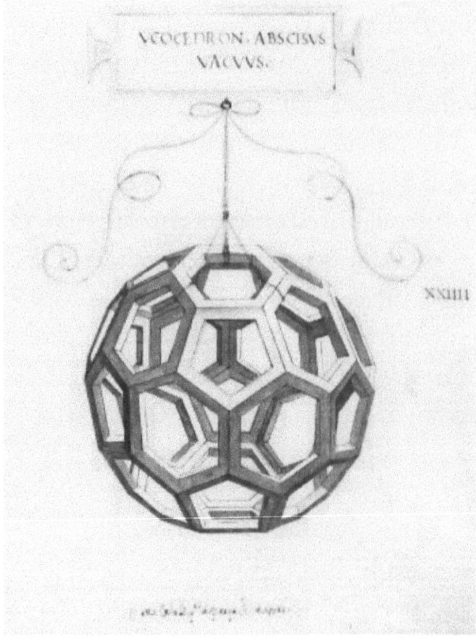

Figure 6.24. *Leonardo's illustrations*

the geocentric world-view with its planets tracing complicated cycloid orbits, *Copernicus' revolution* (as Kant referred to it) brought to the fore the heliocentric world-view which described the orbits of the planets with circles. On the basis of the observations provided by astronomical measurements, both models were suitable for correctly describing the movement of the planets. As empirical experience was not able to decide between the two, the choice fell on the simpler, more symmetrical mode of description. In place of apparent symmetry (if we look up at the starry sky above us, we see it as a hemisphere with us standing at its centre), the victor was the path for the Earth given by the symmetry in the measured data (that put the Earth on the same footing as the other planets, tracing a similar orbit around the Sun).

This also brought the rejuvenation of astronomy. Kepler came to determine the laws of the movement of the planets while searching for harmony in the world. His model was based on nested spheres fitting the five Platonic bodies both internally and externally. He wanted to find the perfect picture of the world, the harmony, symmetry embodied in the world. He was himself the most surprised that the orbits of the planets proved not to be symmetric circles, but rather ellipses where the presence of the Sun favoured one focal point over the other. The destroyed symmetry had to be recovered somehow. Like Plato, Kepler went back to the music of the spheres of the Pythagoreans. He found the proportions of the angular velocity of the various planets when closest (in perihelion) to and farthest (in aphelion) from the Sun to be as follows: for the Earth 16/15, for Mars 3/2, for Saturn 5/4, which can in order be compared with a half-note, quint, and third. It was these that he composed in his musical motifs (see Figure 2.28).

One of the consequences of the problem of drawing regular polyhedra around spheres was Kepler's original formulation of the problem of the densest packing. We know that, on a plane, a circle can be circumscribed by exactly six circles of the same radius, touching it and each other. In space the story is nothing like as simple. We cannot surround a sphere in space with spheres of the same radius in such a way that all the neighbouring spheres touch each other as pairs. The classical Kepler problem is thus as follows: how can we arrange similar spheres in a bowl in the densest fashion (i.e. with the most spheres)? (The planar equivalent of this would be how we could cover the surface of the table in the densest fashion with a given denomination of coin, i.e. with the most coins.) The set of problems concerning the densest tilings and closest packings has, over time, become and increasingly independent branch of mathematics. Perhaps it is surprising that centuries had to pass after Kepler before the problem of the densest filling of space was solved (Figure 6.25). In the nineteenth cen-

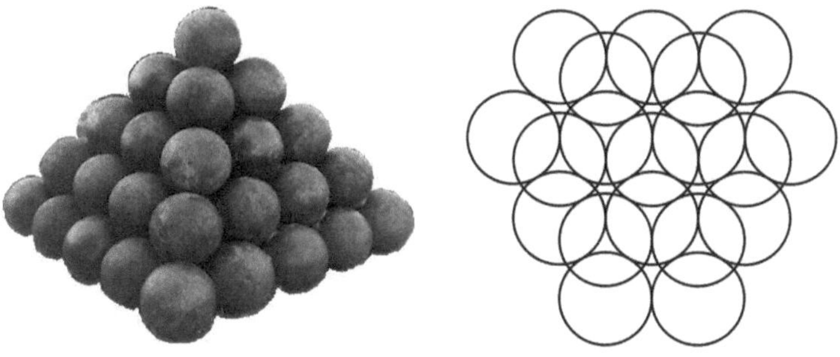

Figure 6.25. *The packing of spheres*

tury, Gauss (1777–1855) provided an estimate that with congruent spheres the space could not be more densely packed than 74.048 % (which coincided with the original Kepler conjecture), but this was only proven in 1998 by Thomas Hales, who reported his proof to the fourth congress of the International Symmetry Society in Haifa in that year.

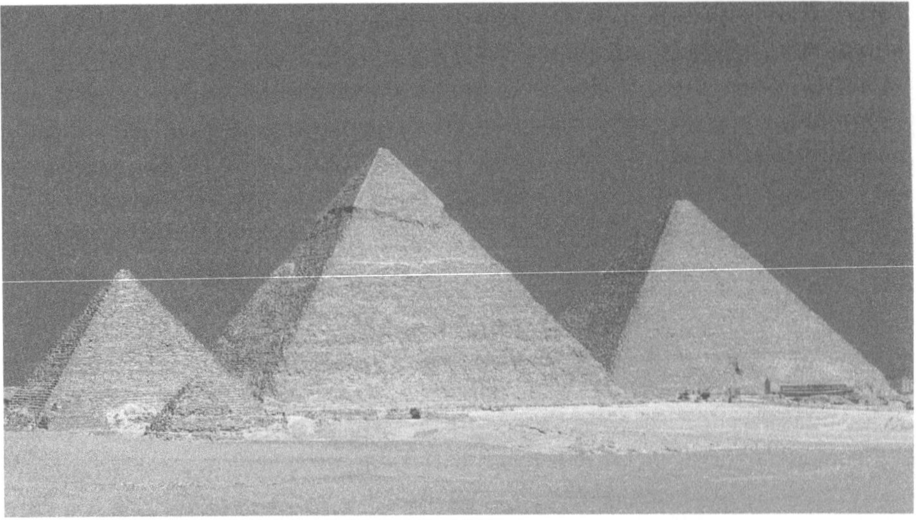

Figure 6.26. *The necropolis of Giza. The pyramids of Mycerinos, Khephren and Kheops, with a smaller pyramid of a queen in the foreground*

This set of problems is not merely an intellectual game. The areas of application of the dense filling of space and close packing of spheres are very wide. Examples of some such areas are: in crystallography, the determination of the optimal position of atoms; in chemistry, determination of optimal bonding directions; in physics, in atomic nuclear models; in biology, in the division of egg cells; in architecture, the selection of the most stable supporting structures.

Architecture has long recognized the significance of the relationship between the application of regular bodies and the requirement of stability. It is in the most symmetrical arrangement that edges (supporting structure elements) meeting at a single point display the most even distribution of load. And the even distribution of load is a statical requirement. The simplest space-covering load-bearing structures, and therefore those used the earliest, were the pyramids, which in geometrical terms can be considered as half octahedra.

Figure 6.27 shows a fine example of modern architecture, the Art Tower Mito, which is constructed of octahedra, each connected to a face of the next.

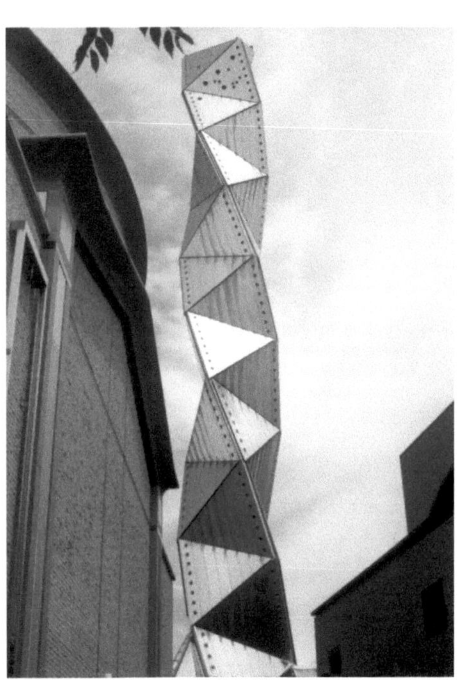
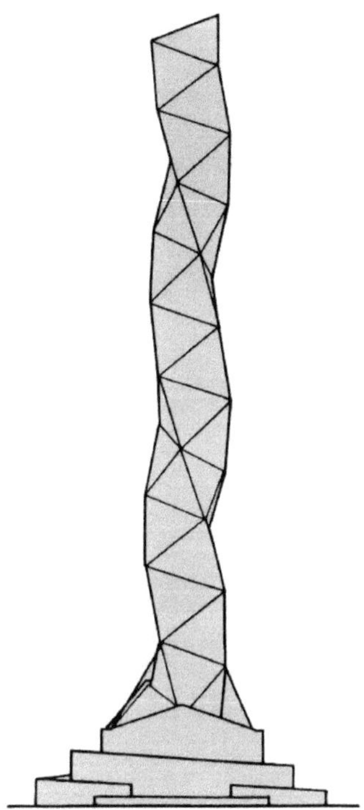

Figure 6.27. *Art Tower Mito (Japan)*

Chapter 7

The mystery of fivefold symmetry

From Dürer to quasi-crystals

Since ancient times, people have thought the number five to have hidden mysteries. What is mystical has symbolic significance. The mystical treatment of the number five belongs to the cults that date back to ancient times. Many cultures treat it as a fairy-tale number, others as a lucky number.

We saw that the number five plays a special role in the algebraic determination of the Fibonacci number (Φ). We cannot fill a plane without gaps using regular pentagons. We can use them to cover the surface of a sphere, but we cannot fill space with dodecahedra, either. We do not encounter fivefold symmetry in inanimate matter (and if we do, as in the case of the recently discovered quasi-crystals, it is very rare), whereas it is very common in living matter.

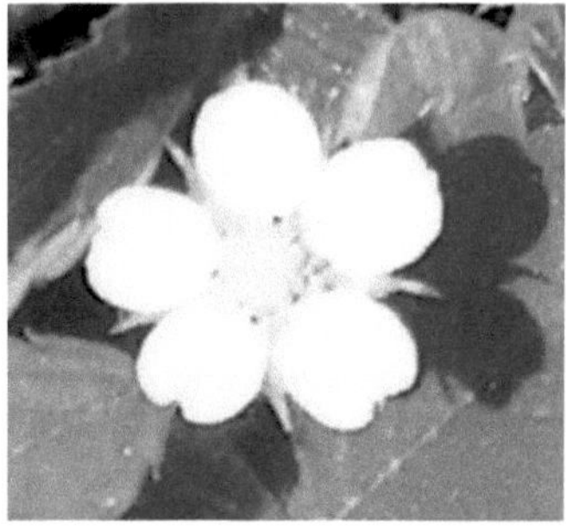

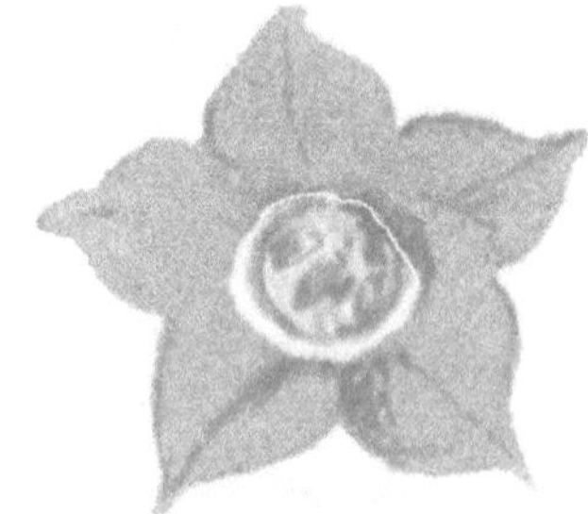

Figure 7.1. *Wild strawberry flower; tobacco flower*

Certain philosophical systems traced the construction of the world back to five principles, or five primary elements. To the four "primary elements" Plato added the universe, with these five corresponding to the five perfect bodies. The five regular bodies had appeared as ancient symbols well before Plato. The complete set of Neolithic "Platonic" bodies seen in Figure 7.2 was found in Scotland, from the period a thousand years before Plato.

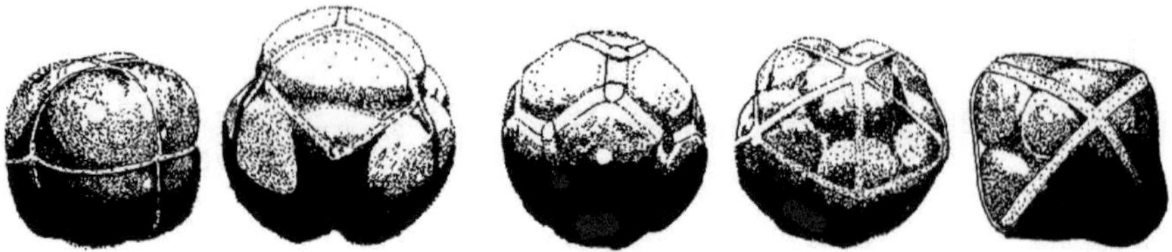

Figure 7.2. *Complete set of regular bodies dating back to the Neolithic age*

The philosophy of *vaisheshika* in India (6th–5th C. BC) distinguished five primary elements, the philosophy of the Hindu school of *samkya* five main principles, then (in the 12th–15th C. AD) *charvaka* also distinguished five substances: earth, water, fire, air and ether. In China, in teachings related to yin and yang (8th–6th C. BC), there were also five primary elements present, and these were later adopted by the Confucian Hsu Hsing as five "*chi*": metal, wood, water, fire, earth (*chi* means wisdom, knowledge, study, principles).

The special role of the number five is based on very simple, elementary experience. We have five fingers on each of our hands. The fingers of our two hands — the combined number of which gives us the basic set for the arithmetic system of our everyday calculations — can be arranged into five pairs. This can clearly be seen in the hands delivering blessings on these details from gravestones (Figure 7.3).

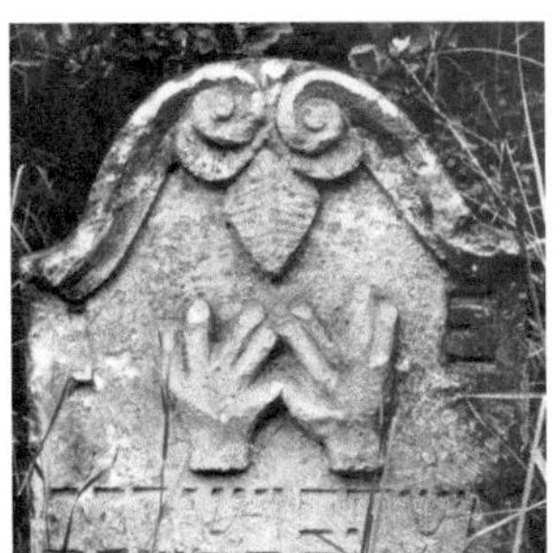

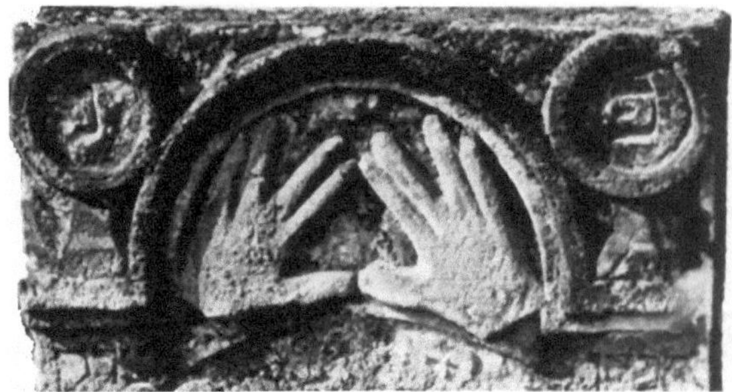

Figure 7.3. *Details of gravestones from the Jewish cemetery in Tokaj, Hungary*

In the geometrical context, the mystery of the number five appears with regard to the construction of the regular pentagon. With classical tools — a compass and a single-edged rule — the construction of a regular pentagon was for a long considered a difficult task. Figure 7.4 shows Dürer's construction, probably on the basis of book I, chapter 9 of the *Almagest* of Ptolemy (2nd C. AD), which he includes without proof.

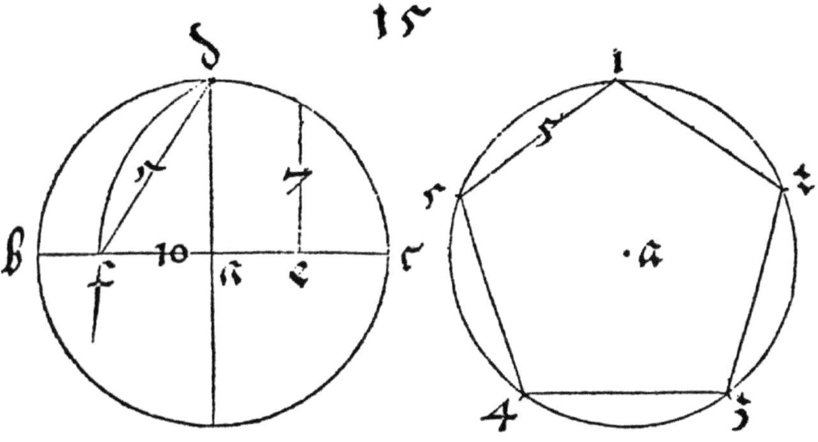

Figure 7.4. *Dürer's construction of a pentagon*

Today, we would approach the construction as follows. We previously saw that for a section of one unit we can construct a section of length Φ. If we draw in one of the diagonals of a regular pentagon, this divides it into a rhombus and a triangle. The two sides and one of the bases of the rhombus are equal in length, and it can be shown that the diagonal (the other base) is Φ times this length. On the basis of this, a rhombus can be constructed and extended into a pentagon.

How much easier it would be to create a regular pentagon with the help of a simple slip of paper! With no need for compass, ruler or pencil, we just fold a loop from the paper, then flatten it. The result is shown in Figure 7.5.

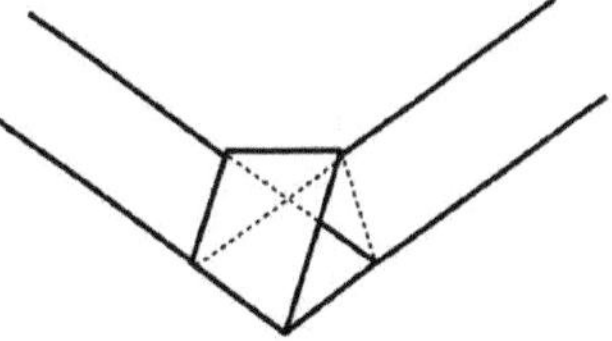

Figure 7.5. *Folding of a regular pentagon*

This technique is used in some cultures to close planar surfaces during wickerwork, and to deal with the curvature of surfaces. The examples in Figure 7.6 are from Mozambique, as published by P. Gerdes (1952–).

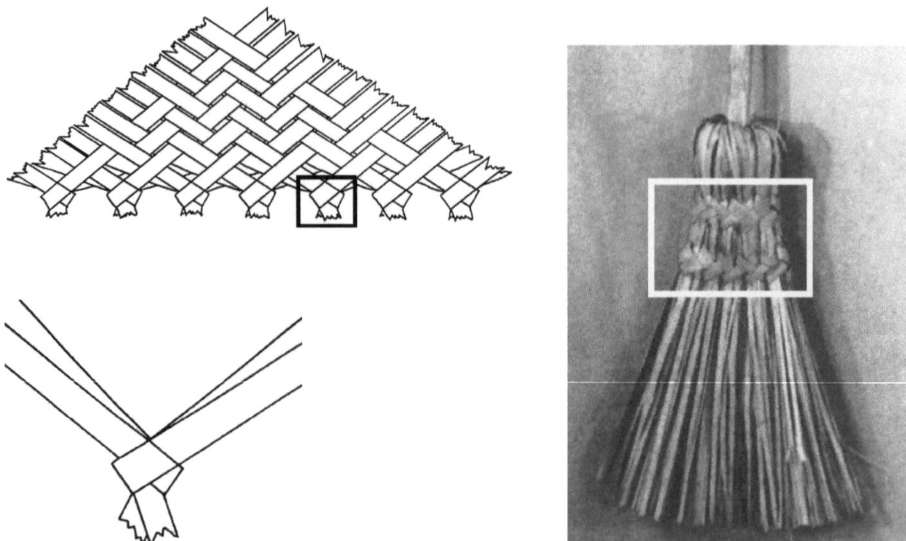

Figure 7.6. *Weaving of a bag in Mozambique (with an enlargement of one knot) and tying together the threads of a broom*

The geometrical properties of the regular pentagon can be seen in the diagram in Figure 7.7. The two diagonals drawn from one vertex divide the 108° inner angle into three equal thirds. The diagonals divide the pentagon into similar triangles and rhombuses. We can observe the division

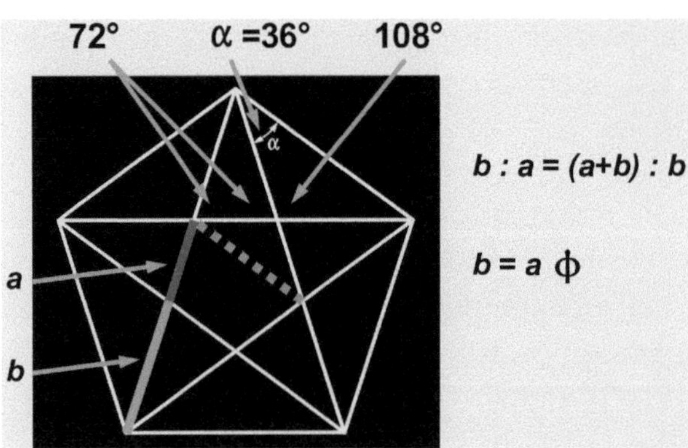

Figure 7.7. *Properties of the regular pentagon*

of the sections of the diagonals according to the golden proportion, as also used in the construction above.

The proportions of the regular pentagon also inspired the artistic thinkers of the age of humanism. There is no end to the studies and schematic and artistic depictions of the proportions of the human body and their relation to the pentagon. The best known is the previously presented drawing of Leonardo da Vinci from the *Codex Urbinax Latinus*. As the study of symmetrical proportions was taken from Vitruvius, diagrams placing the ends of the arms and legs and the head at the five vertices of a pentagon were used to illustrate the translations of Vitruvius that were published in several 'modern' European languages at the time (e.g. the editions of C. Cesariano and F. di G. Martini). Particularly worthy of note are the scheme (Figure 7.8) of Cornelius Agrippa von Nettesheim (1486–1535), and the diagrams of Albrecht Dürer, notated with detailed numerical proportions (*Four Books on Human Proportions*, 1528).

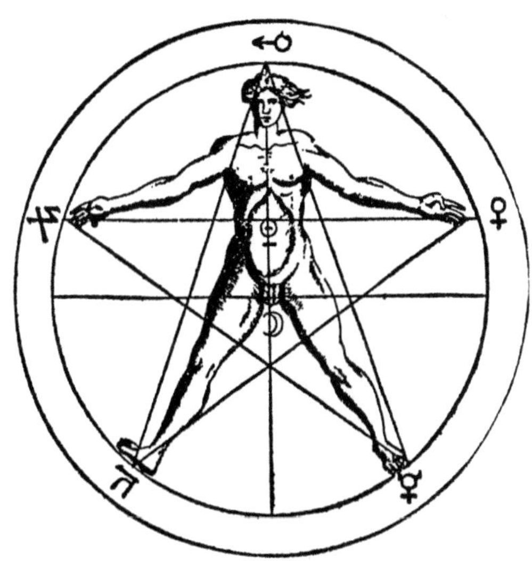

Figure 7.8. *C.A. Von Nettesheim's drawing of the proportions of the human body*

Naturally, neither did Vitruvius take the "mystical" proportions associated with the regular pentagon from his own sources. Ancient architecture provided ample examples. The construction of the internal spaces of the pyramids (semi-octahedra) of Gizeh follows the proportions of the 'golden triangle' (the triangle EFG on Figure 7.9).

A reminder: one of the sides of the right-angled '*golden triangle*' is twice the length of the other, while its hypotenuse is $\sqrt{5}$times the length of the shorter, and this is how it is connected to the issue of the number five. The so-called '*Egyptian triangle*' also appears in other internal spaces. The sides of the 'Egyptian triangle' follow the proportions of

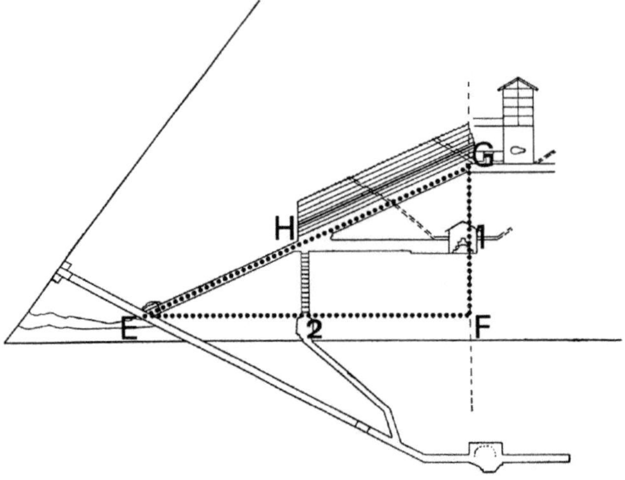

Figure 7.9. *Triangle EFG in the (Great) pyramid of Kheops, marked with a dotted line, is a 'golden triangle'*

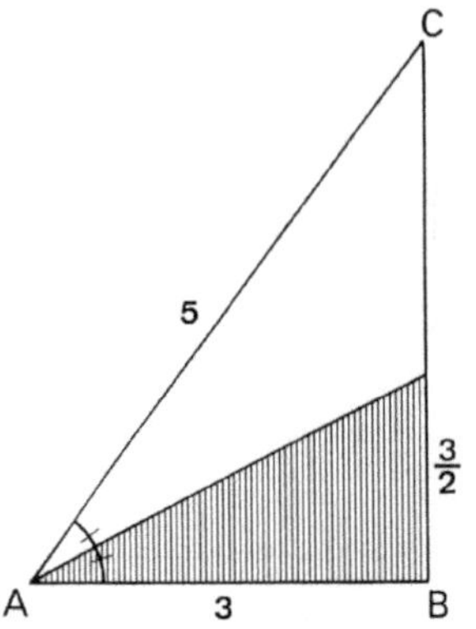

Figure 7.10. *The relationship between the Egyptian triangle and the golden triangle*

the Pythagorean (3, 4, 5) trio of numbers. In increasing order, the lengths corresponded symbolically to Osiris (3), Isis (4) and their child Horus (5). Figure 7.10 presents the relationship between the 'golden' and the 'Egyptian' triangles. If the longer side of the 'golden triangle' corresponds to the shorter (Osiris) side of the 'Egyptian triangle', then the hypotenuse of the former bisects the angle formed by the hypotenuse of the latter.

These two special triangles present themselves in the 'hall of the kings' in the Great Pyramid in Gizeh, along the shaded planes (Figure 7.11).

A comparison of the other measurements of the pyramid bears witness to the use of further special proportions. Continuing to consider areal proportions, the relationship between a quarter of the base and the square of the height is the golden proportion. The external proportions of the pyramid are presented in cross-section (*b*). (P and Q divide the side of the square AB = CD with the golden proportion, from which it follows that the cross-section of the pyramid is half a golden rhombus.) The projection of one face of the pyramid onto the base is in

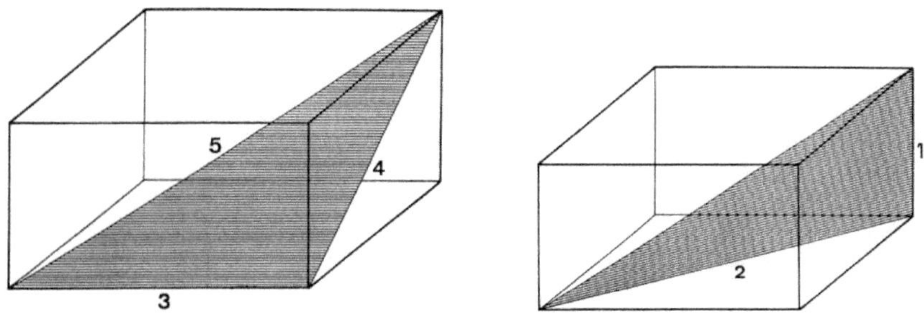

Figure 7.11. *The Egyptian triangle and the golden triangle in the hall of the Kings of the great pyramid*

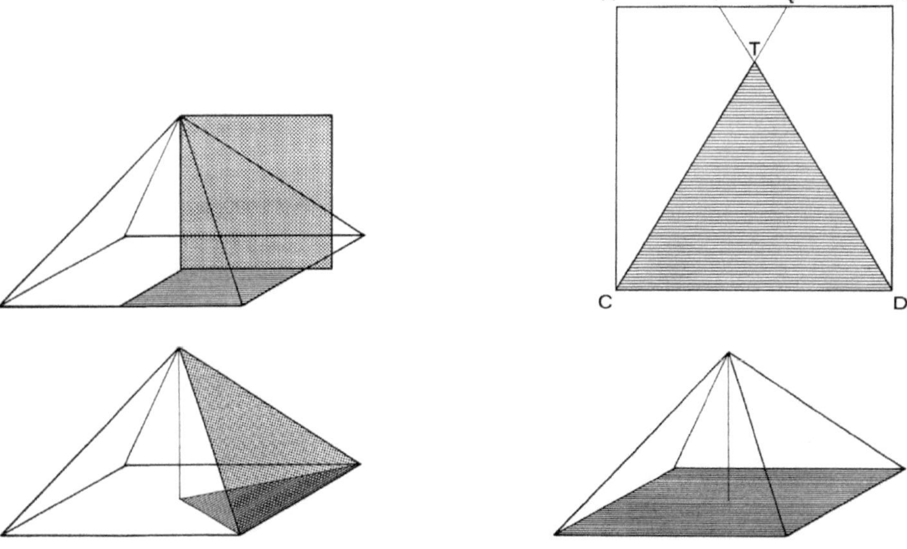

Figure 7.12. *The proportions of the pyramid*

the golden proportion to the area of the face itself. This means that the base of the pyramid as a whole is related by the golden proportion to the total surface area of the pyramid.

In addition to the golden section, the mystique of the regular pentagon also appears in Egyptian culture. In its mythology, Isis, in the form of a swallow, discovers Osiris' grave by flying in the spiral of the forces that emanate from the sarcophagus placed in the centre of a regular pentagon

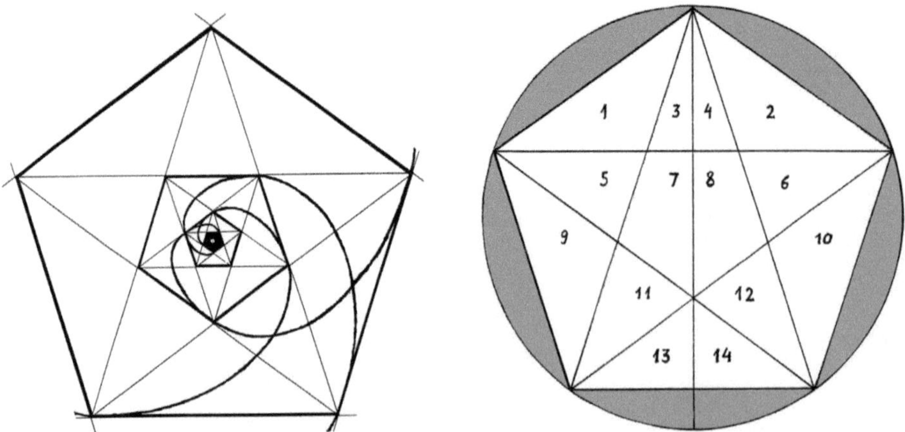

Figure 7.13. *Finding Osiris' sarcophagus (flight of the swallow), and dividing up his corpse*

(as shown in Figure 7.13). Following her path, their brother Seth divides up Osiris' body, which lies in a sarcophagus with a regular pentagonal cross-section, into fourteen pieces, following the mirror symmetric division seen in the right hand figure.

The Egyptians also attributed magical significance to the angles of the regular pentagon. The erection of the Djed column on a relief originating from the temple (Abydos, c. 1300 BC) of Seti I (?–1279 BC) symbolizes the coming to life of the god. The artist depicts the moment (Figure 7.14) at which the column is tilting at an angle of 72° from the side of "death", and at an angle of 108° from the side of "resurrection". The king stands on the latter side.

The relief on the tomb of Amenophis II (Thebes, c. 1450 BC) shows Iwf before being returned to life (Figure 7.15). We see two wings in Iwf's hands, under the eyes of the Sun and the Moon. The angle between the wings and the base is 108°, and the angle between the two wings is precisely 36° (72°/2), which symbolizes the regular pentagon.

The pictorial depiction of both situations suggests that the symbols hidden in them are only visible to those able to measure. The symbol of the return to life is the visualization of the divine principle of the opposition between light and dark.

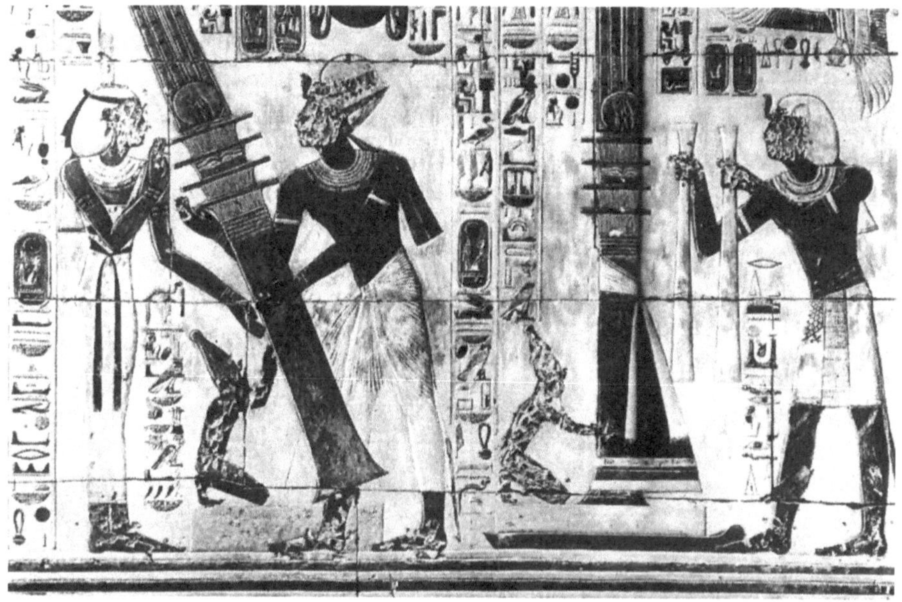

Figure 7.14. *The temple of Seti I, Abydos, c. 1300 BC*

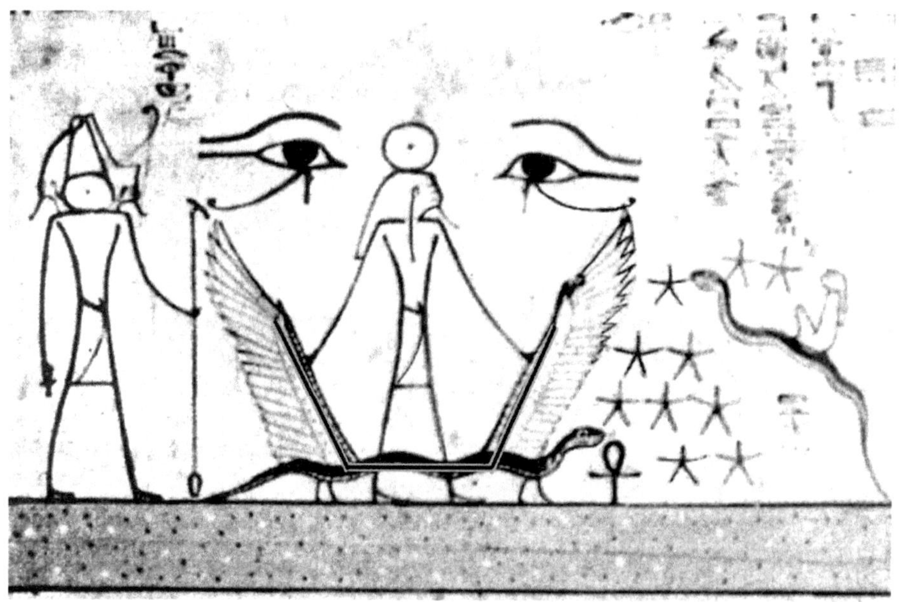

Figure 7.15. *From the tomb of Amenophis II, Thebes, c. 1450 BC*

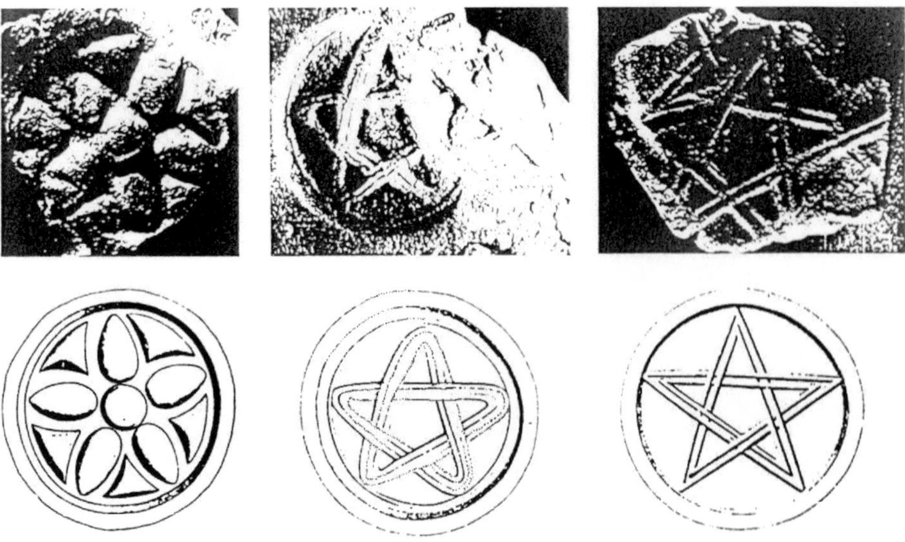

Figure 7.16. *Minoan and Mycenaean stamps*

The ancient myth of the pentagon, the pentagram, also appears on the stamps of Minos and Mycenae (2nd millennium BC) seen in Figure 7.16. Later, in the Classical Greek period, it was the secret sign of the Pythagoreans.

From then on, it was in their footsteps that the study of pentagons became the subject of mathematical enquiry, which for many centuries was developed further at the highest level in the Arab world. Figure 7.17 shows a page from a thirteenth-century Persian translation of the geometry of Abu'l-Wafa al Buzjani (Baghdad, 945–987) on the solution to the so-called *Kunya 5* problem. The pentagons and decagons of the problem studied can clearly be seen in the figure.

It is no accident that it was in Islamic art that regular pentagonal ornamentation flourished the most. Figure 7.18 shows the *Blue Tomb* in Maragha and an enlarged detail of its ornamentation. We can observe a highly interesting early version of pentagonal tiling of the plane, one which includes the motif of the open pentagon.

The open pentagon, or its pentagram variant (concave decagon, five-pointed star) entered Medieval European culture, under the name of

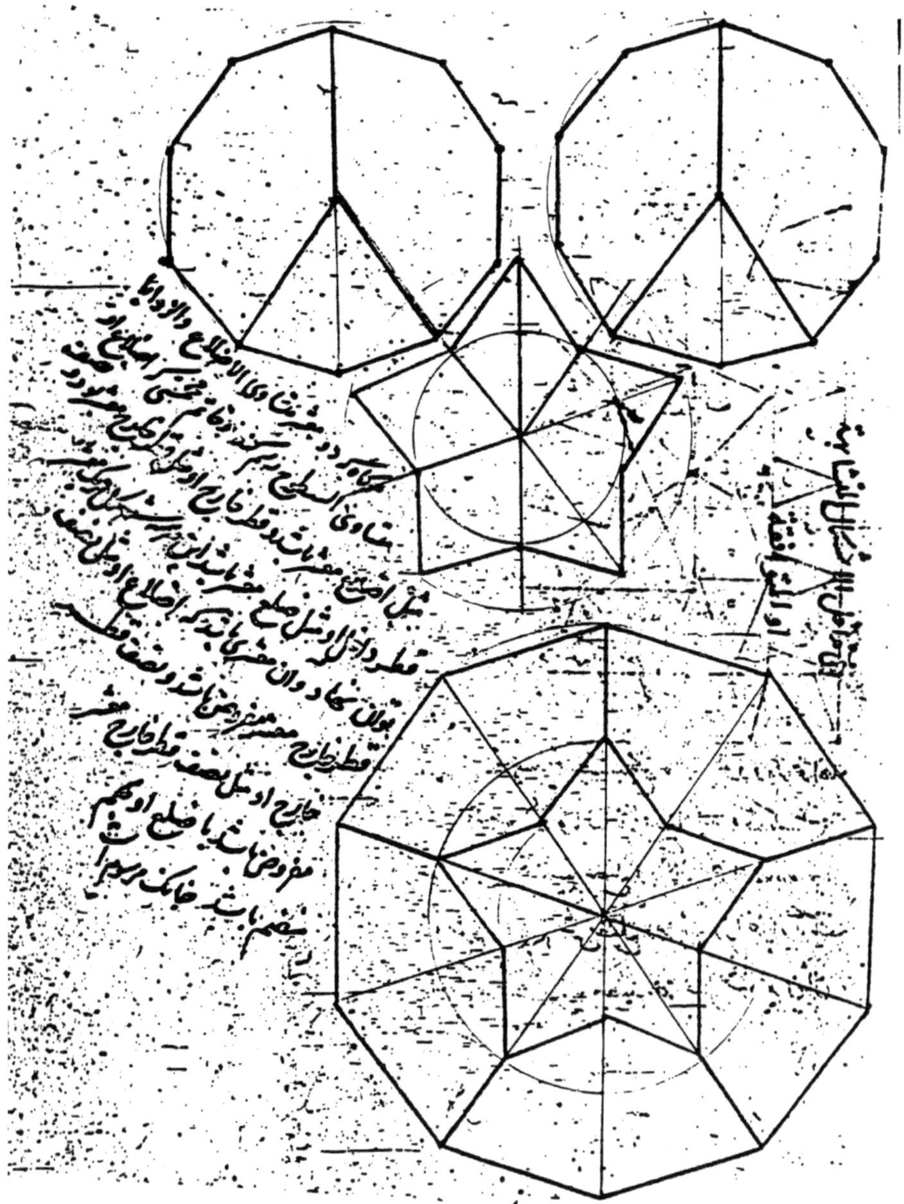

Figure 7.17. *A page from the Persian translation of Abul-Wafa al Buzjani's geometry* (Bibliotheque Nationale Paris, ancient fond Persian Ms. 169, folio 180a)

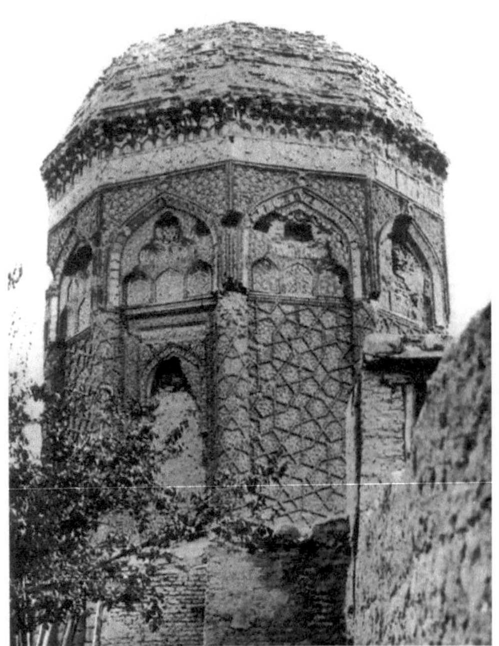

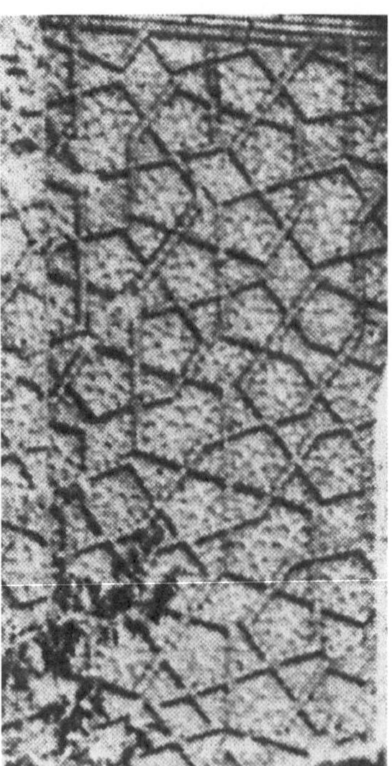

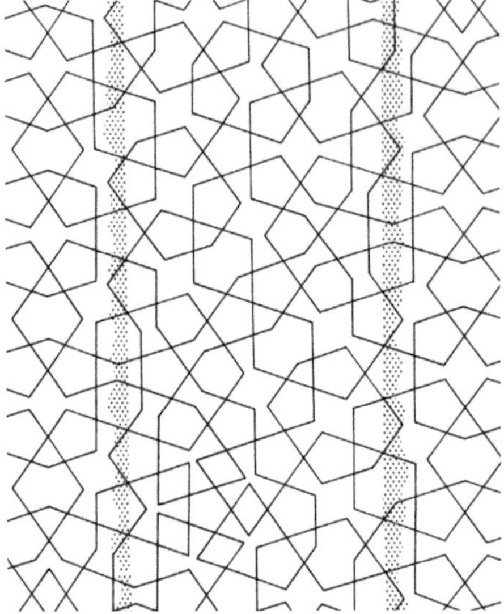

Figure 7.18. *The Blue Shrine and an enlarged detail of its ornamentation, and a reconstructed drawing of it (with the edge-columns shaded in).* (Gunbad i-Qubud, 1196–1197, Maragha, Western Iran)

'witch's foot' (or 'goblin's cross'), as a symbol of bad omens. The pentagram is one of the key elements in the Faust legend passed on in a number of versions in late medieval Europe. There were many adaptations of it, the best known being that of Goethe (1749–1832). Here we quote from its first act, from the scene in which Faust and Mephisto make a pact, and the latter prepares to leave:

"*Mephistopheles:*
[...] But may I for the present go away?

Faust:
Why you should ask, I do not see.
Though we have only met today,
Come as you like and visit me.
Here is a window, here a door, for you,
Besides a certain chimney-flue.

Mephistopheles:
Let me own up! I cannot go away;
A little hindrance bids me stay.
The Witch's foot upon your sill I see.

Faust:
The pentagram? That's in your way?
You son of Hell explain to me,
If that stays you, how came you in today?
And how was such a spirit so betrayed?

Mephistopheles:
Observe it closely! It is not well made;
One angle, on the outer side of it,
Is just a little open, as you see.

Faust:
That was by accident a lucky hit!
And are you then my captive? Can that be?
By happy chance the thing's succeeded!

Mephistopheles:
As he came leaping in, the poodle did not heed it.
The matter now seems turned about;
The Devil's in the house and can't get out.

Faust:
Well, through the window — why not there withdraw?

Mephistopheles:
For devils and for ghosts it is a law:
Where they slipped in, there too must they go out.
The first is free, the second's slaves are we."

The pentagram that is open at one vertex is even capable of keeping the devil at bay!

We can date the nature of the Faust legend as it has developed today back to the first half of the sixteenth century, to the time when a series of translations of Vitruvius started to appear in various European languages. To the same time when a succession of Renaissance artists used the pentagon to frame the proportions of the human body. It was at this same time that the Turkish menace appeared at Europe's border. The construction of castles was booming.

Pentagonal arrangements in architecture

Architecture is, first of all, the shaping of space. In space, pentagons are primarily symbolized by dodecahedra. However, we cannot fill space periodically and without gaps by using dodecahedra (cf. crystal groups). Previously it was believed that pyrites display a continual crystal structure filling space with dodecahedra in a regular fashion, but this was later to contradict the rules of crystallography that had been discovered in the meantime. And it really did transpire that, of the edges of the pentagons circumscribing the dodecahedron of the pyrite crystal, only four are of equal length, with the fifth different, and their angles are accordingly not equal, either. That is, space filling that contains pentagons does exist, but not with regular dodecahedra. It is worth noting that in the most recent era the possibility of construction using dodecahedra has not escaped the imagination of the architectural profession. Figure 7.19 shows the design for a housing estate in an Israeli town by Berlin-based architect Zvi Hecker (1931–).

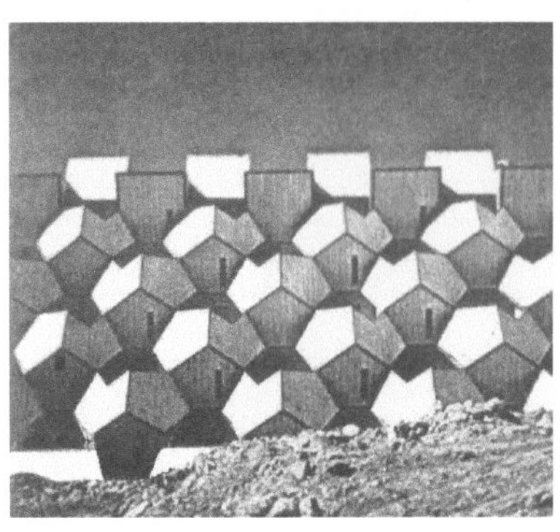

Figure 7.19. *Zvi Hecker's dodecahedral design for a housing estate (Ramot)*

The slanting walls created by the dodecahedron do not offer a single room that is useable living space. Pentagons can be implemented in blueprints, if not for periodic tiling.

Pentagonal architecture has long been known outside the boundaries of Europe. In Japan, where the number five and thereby the pentagon both had greater significance, it has been used for a long time. The excavated ground-plan shown in Figure 7.20 dates back the 2nd–3rd centuries AD.

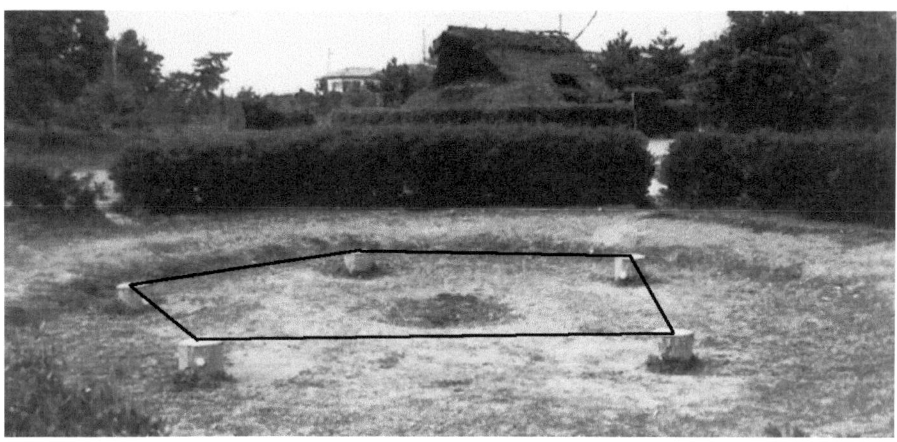

Figure 7.20. *The ground-plan of a Japanese building from the 2nd–3rd c. AD* (Harima, Hyogo prefecture)

In European castle architecture, the so-called *puntone*, the walled gate-protecting bastion (Figure 7.21), appeared in the first half of the cinquecento. This type of fortification system using bastions was originally a pentagon-based fortress which had one side supported by the castle wall, and the other four sides sticking out in wedge-like fashion.

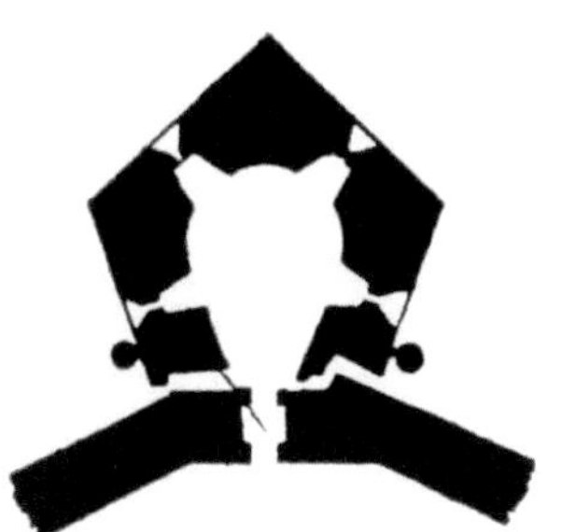

Figure 7.21. *Puntone, 15th C., Italy*

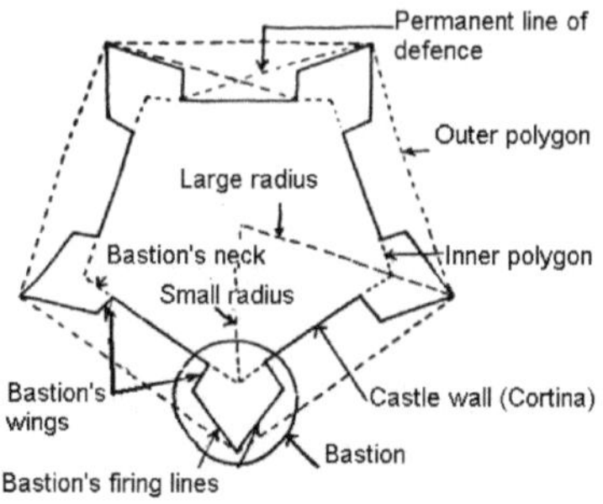

Figure 7.22. *The ground-plan of a system with bastions*

It was soon, however, to serve as the ground-plan for whole castles (Figure 7.22). One of the best-known examples is the *Castel Sant'Angelo* in Rome. The currently visible building with its pentagonal ground-plan was constructed at the beginning of the sixteenth century around Emperor Hadrian's (reigned AD 117–128) square-shaped mausoleum (Figure 7.23).

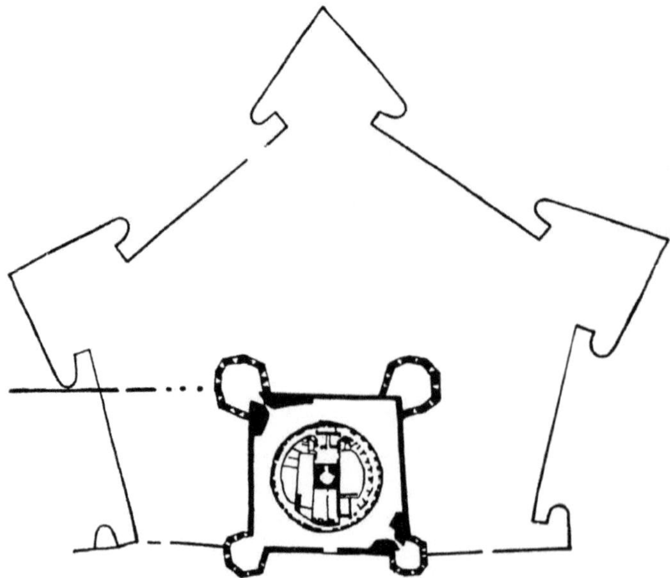

Figure 7.23. *The Castel Sant'Angelo in Rome, early 16th C.*

It was in the sixteenth century that the central, so-called many-angled ideal city plans with rotation symmetry appeared in Europe. Of these, those with an odd number of angles seemed more realistic, bearing in mind, with the risk of over-interpretation, that regular polygons with an even number of sides would, with streets running from wall to wall, have produced overly draughty cities. The puntones are clearly visible on the nine-angled plan seen in Figure 7.24.

The so-called spiral castle schemes — an arrangement developed in stepped fashion alongside the road up a hill — can be dated back to the same period. It combines the symmetry of the spiral with the series of defensive bastions facing the valley at appropriate angles along the spiral.

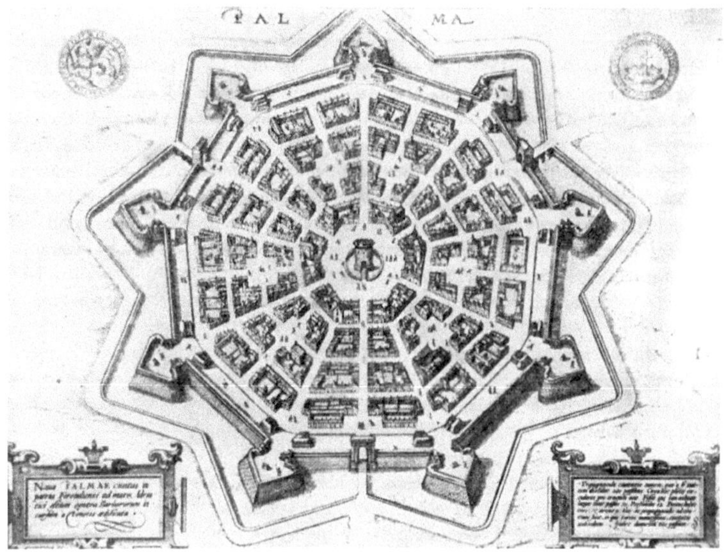

Figure 7.24. *Palma Nuova (1593 plan by Vincenzo Scamozzi* [c. 1552–1616])

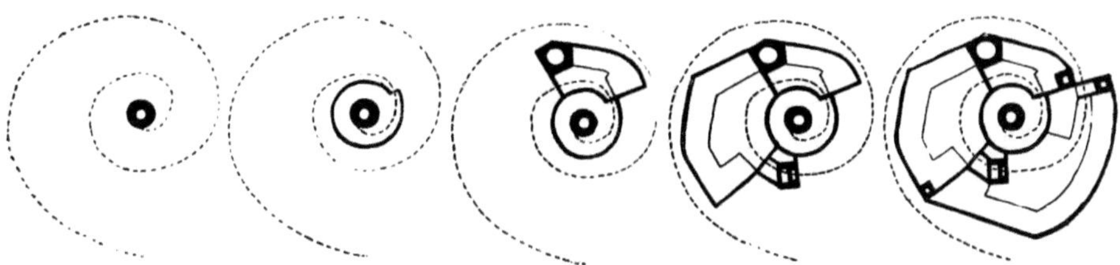

Figure 7.25. *Spiral castle design with a pentagonal gate tower, which in the course of further construction became an inner tower*

We can observe a certain similarity to the phyllotaxis we encountered in flora.

Although it was in Italy that the puntone and pentagonal castle architecture developed and became widespread, we can also find many nice examples of pentagon-based castles in Hungary, too. This is no accident. In the sixteenth century it was primarily under the direction of Italian architects that the system of fortresses was built in Hungary to defend against Turkish expansion. The inner tower of the castle of Hollókő, with a pentagonal ground-plan, is a good example of this (Figure 7.26).

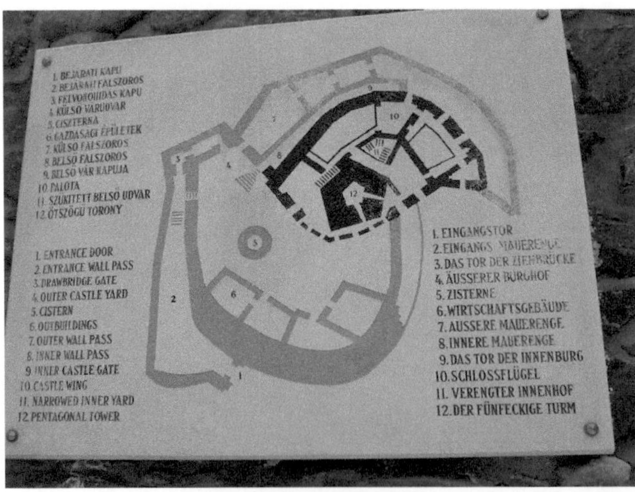

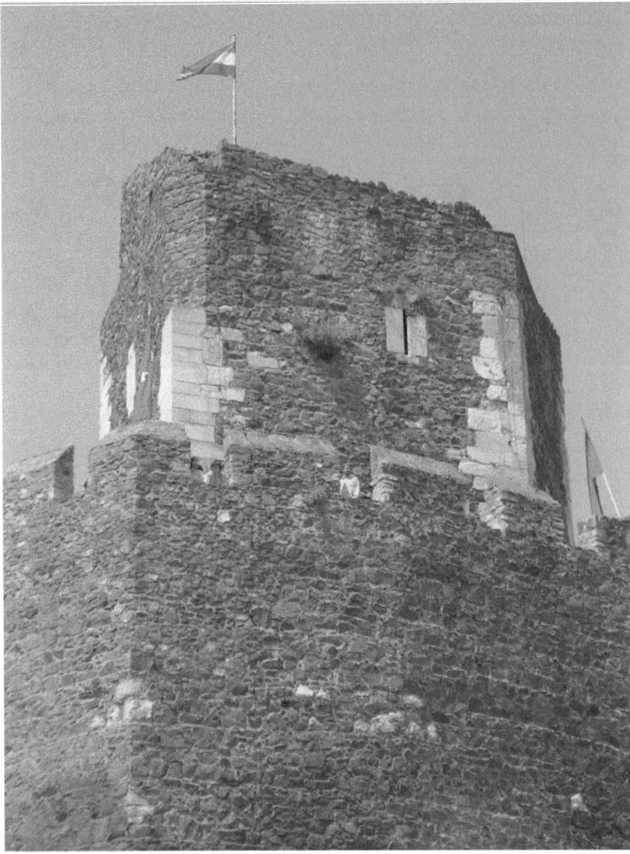

Figure 7.26. *The ground-plan of Hollókő, and its pentagon-based tower*

Castles with much more regular pentagonal designs were constructed in Nagykanizsa, Lenti, Karánsebes and Nagyvárad (Figures 7.27, 7.28, 7.29, 7.30), and we could also mention the example of the castle of Szatmár (Figure 7.31).

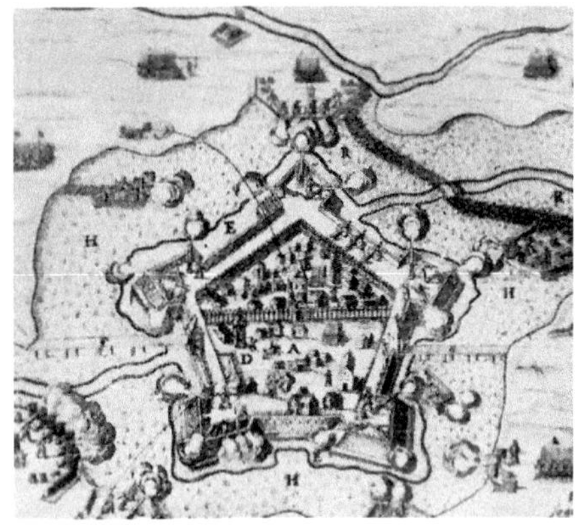

Figure 7.27. *Nagykanizsa castle (1665)*

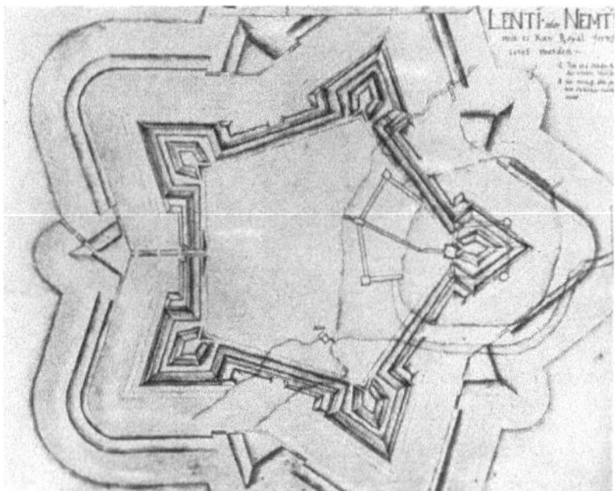

Figure 7.28. *Lenti castle*

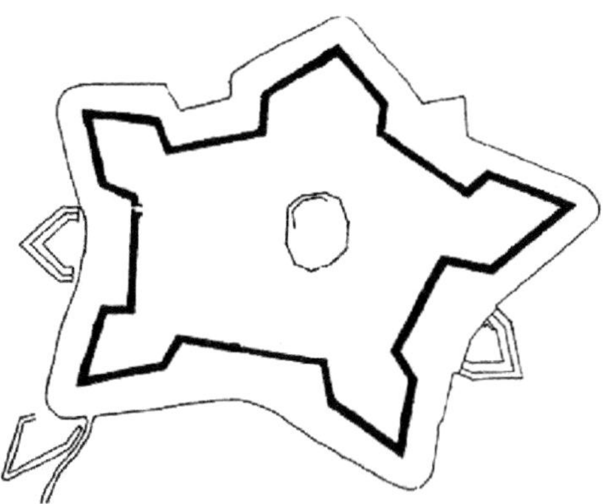

Figure 7.29. *Karánsebes castle (1690)*

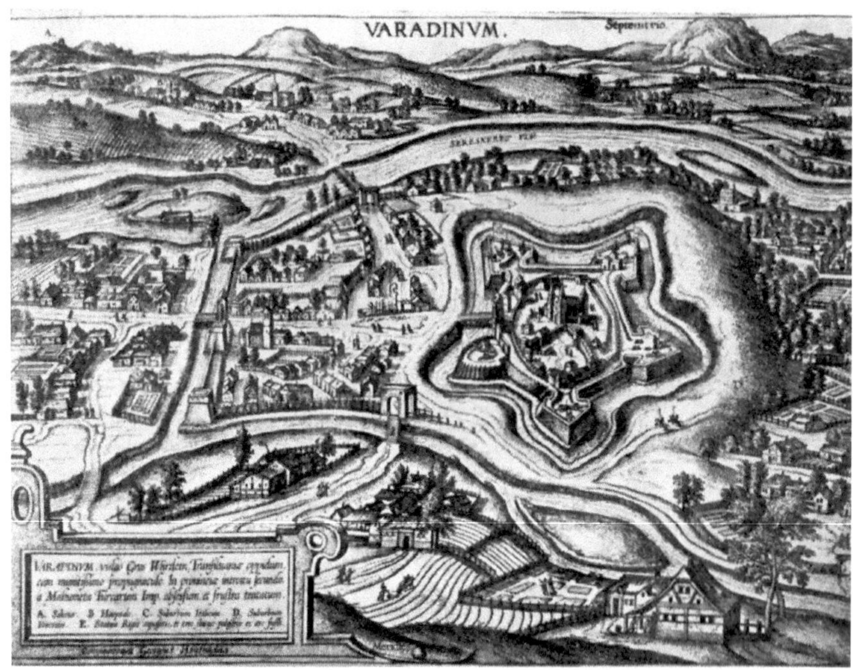

Figure 7.30. *Nagyvárad castle (16th C.)*

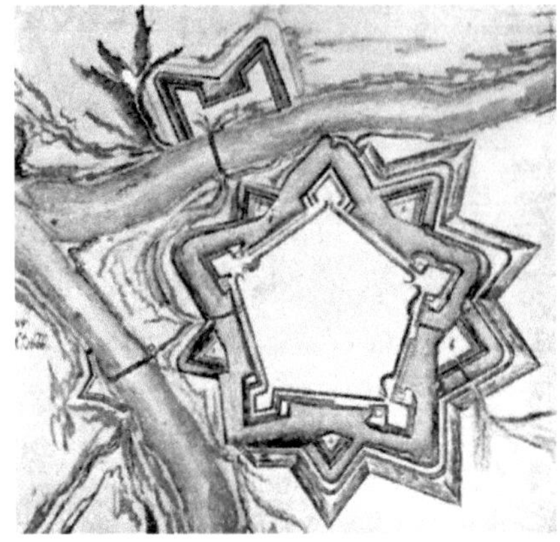

Figure 7.31. *Szatmár castle (1666)*

Figure 7.32 shows Goryokaku Fort (the prefix 'go' means 'five'), constructed in the nineteenth century on the island of Hokkaido, which has a similar layout, and in fact could be associated with the quasi-periodic pentagonal tiling of Penrose.

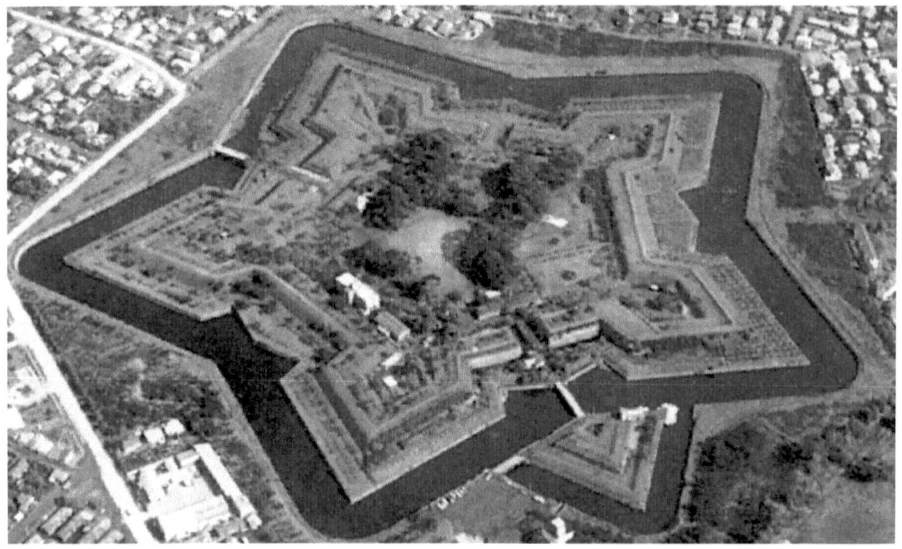

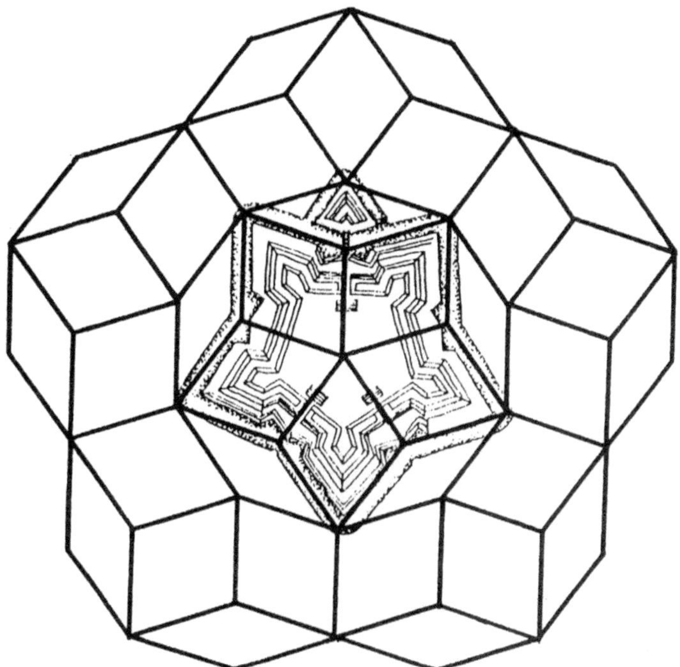

Figure 7.32. *Goryokaku castle* (Hokkaido, Japan, 19th C.)

The *Villa Farnese* in Caprarola, Italy, is an interesting example of a combination of the pentagonal construction introduced to castle architecture and of traditional rectangular arrangements (Figure 7.33, G. Viliola, 1559).

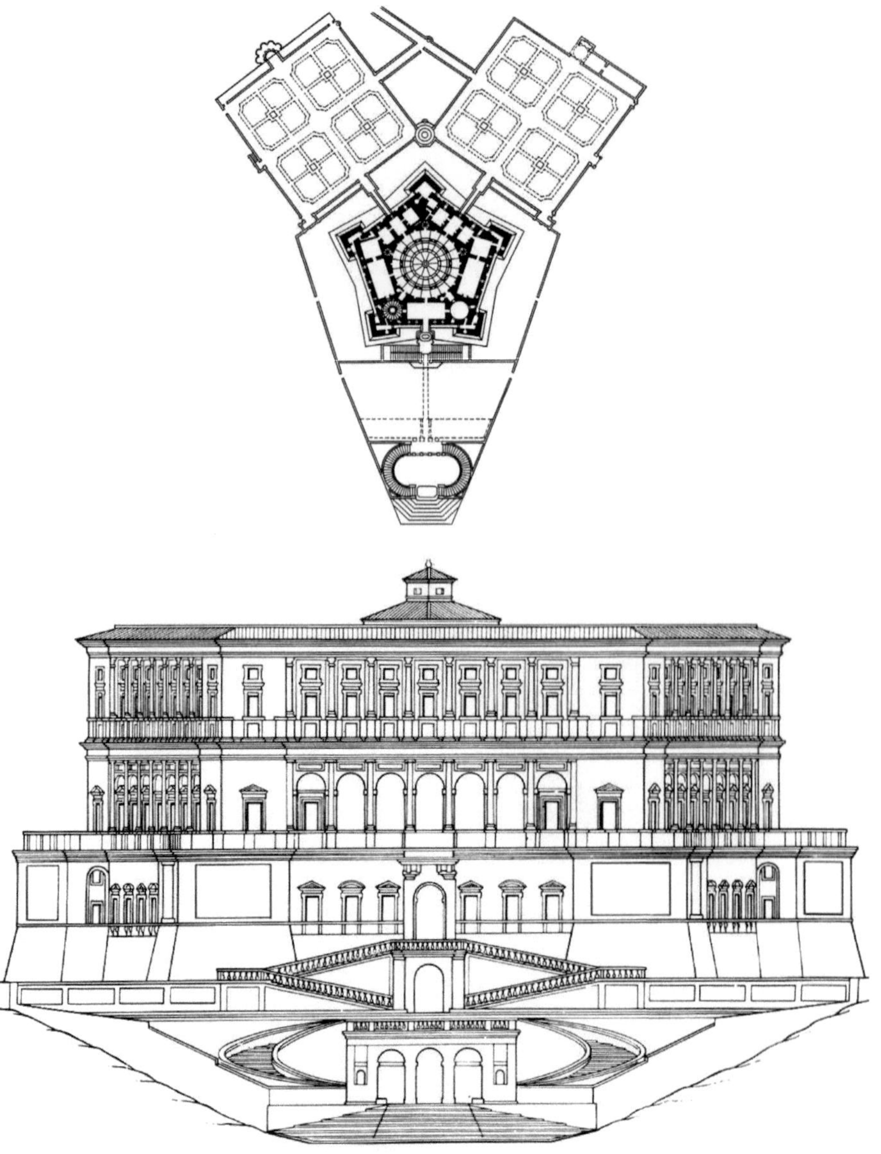

Figure 7.33. *Villa Farnese* (Caprarola)

In the present era, the best-known building around the world with a pentagonal ground-plan is the Pentagon, home to the United States Ministry of Defense, in Arlington (Virginia), on the outskirts of Washington D.C. (Figure 7.34). This enormous complex contains buildings forming five pentagons nested in one another, which are connected by ten wings that cut across them at right angles, while the building as a whole surrounds a place that is a regular pentagon.

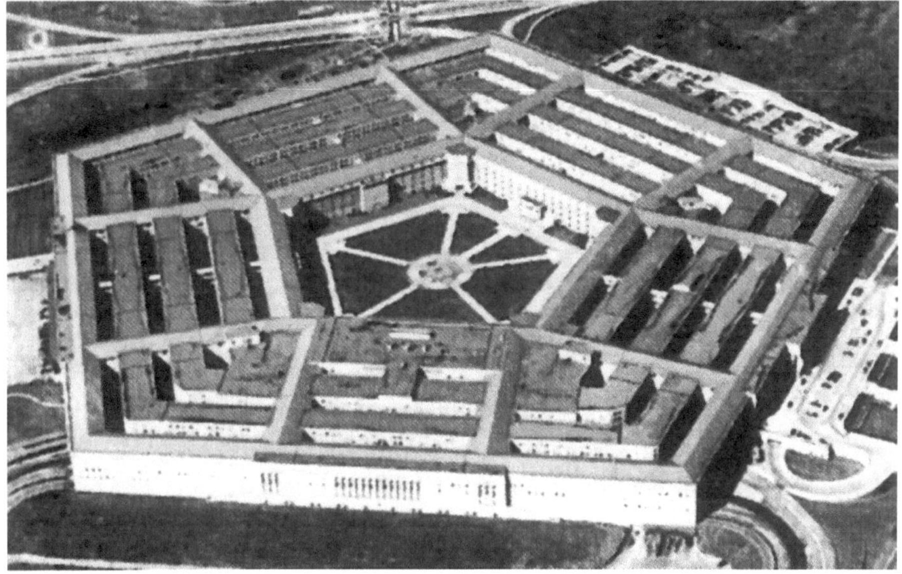

Figure 7.34. *The Ministry of Defense* (Arlington, Virginia, USA)

From Hungarian architecture in the modern age, the Calvinist church on Calvin square in Szeged, built in 1940, with its internal space forming a regular pentagon, is worthy of note (Figure 7.35).

Tiling the plane with pentagons

After this diversion into architecture, let us for the time being return to the planar geometry of ground-plans. We would like to tile the plane without gaps. We have seen that this cannot be achieved simply with congruent regular pentagons. At least one of the conditions has to be abandoned. We have to surrender either the edges being equal, or the angles being

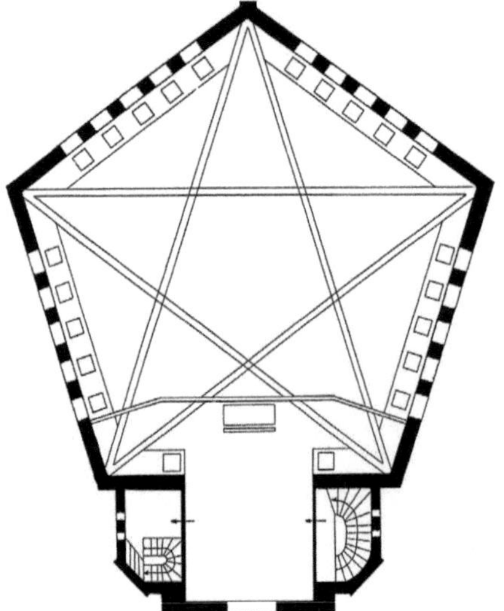

Figure 7.35. *The ground-plan of the Calvinist church on Calvin square, Szeged* (after S. Bérczi)

equal, or the requirement for the construction to involve a single type of element.

Figure 7.36 shows an example of how the plane can be tiled with congruent but not regular pentagons (with varying internal angles). The goal, in any case, is that we give up as few conditions as possible. For example, if we ignore congruence, we should use the fewest and most regular elements. Dürer had already tried to restrict tiling to two types of element. The mirror pages seen in Figure 7.37 are from his book *Underweysung der*

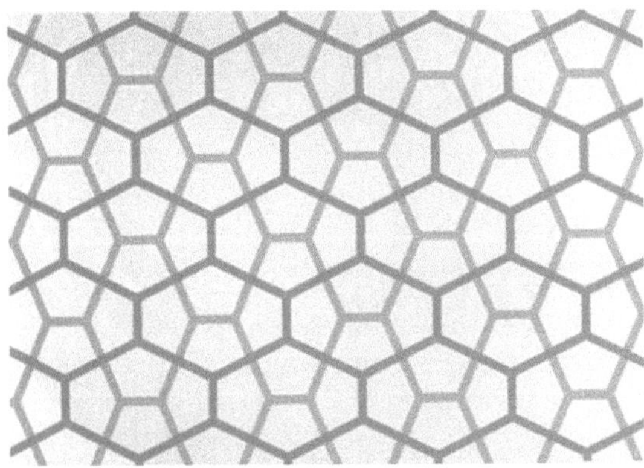

Figure 7.36. *Tiling of the plane without gaps, with congruent but irregular (non-equal-angled) pentagons*

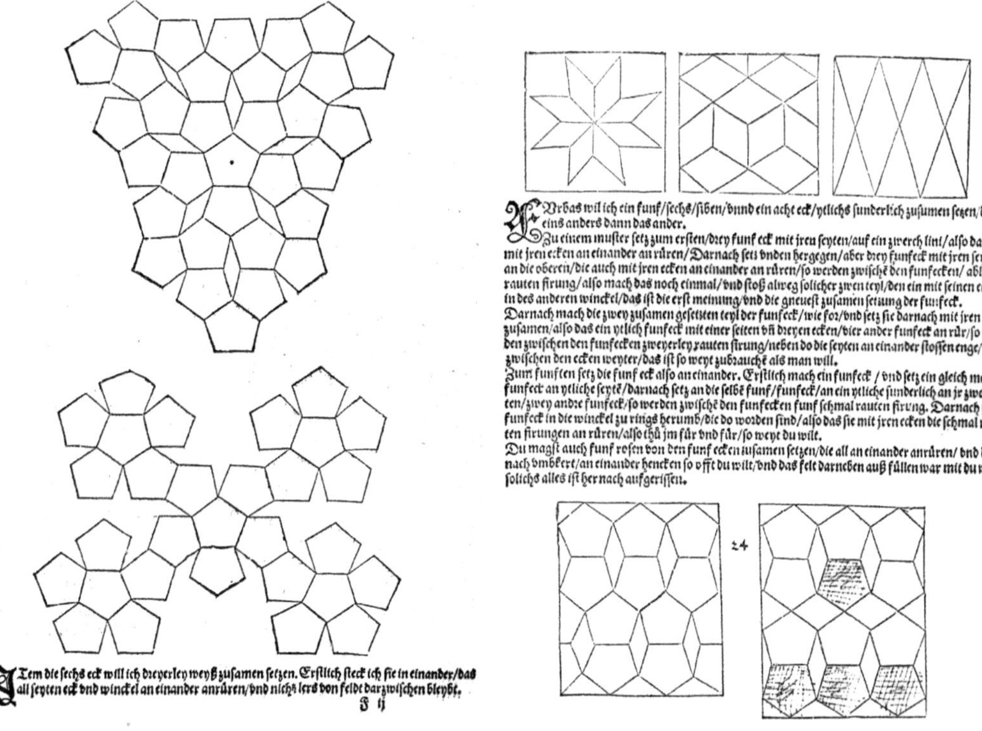

Figure 7.37. *Dürer's attempts at pentagonal tiling of the plane at the beginning of the 16th century*

Messung. The combinations of regular pentagons with rhombuses with the same edge length are clearly visible.

Not for all tilings with two elements can it be proved that the plane can be filled, theoretically, to infinity. The twentieth century saw a number of suggestions that guaranteed periodicity by tiling the plane with three elements, also using rhombuses. New York architect Bernie Kirschenbaum presented a periodic tiling of the plane with two types of rhombus and a regular polygon. (Merely for the sake of interest, I note that it was not in this capacity that he first entered the history books. In 1945, as a soldier in the US Army, he was the first to cross the Elbe, and he met the first soldier of the Red Army arriving from the East. The photograph of the two men waving flags and embracing each other first did the rounds of the world press, than became a standard picture in history textbooks.) He used his periodic tiling of the plane, which applied a combination of two types of rhombus and a regular pentagon and displayed fivefold symmetry, for blueprints of his designs for houses (Figure 7.38). These pentagonal house plans were an almost direct progression from and pentagonal extension of the functional modern house concept of the late Bauhaus, which still followed a rectangular arrangement, as represented at the time in Europe by Max Bill (1908–1994) and in the United States by Luis Kahn (1901–1974).

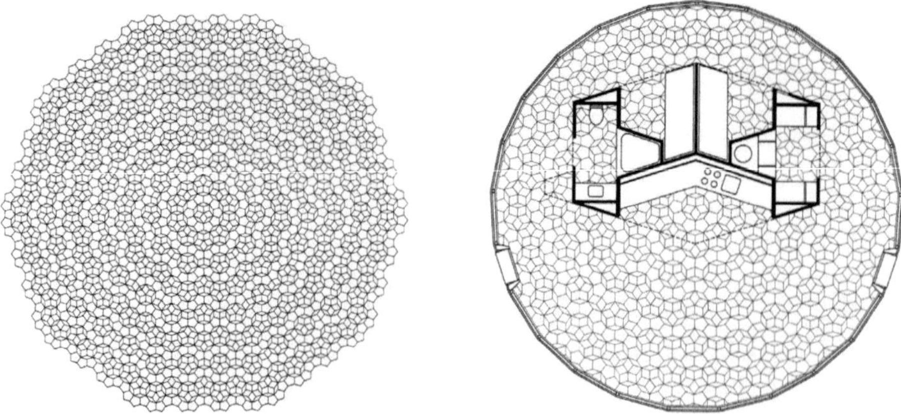

Figure 7.38. *Tiling of the plane with regular pentagons and two types of rhombus, and a design for a house* (B. Kirschenbaum, USA, 1956)

Kirschenbaum further constructed his pentagonal tilings of the plane in the direction of quasi-periodic tilings using the structure of the golden rhombus (Figure 7.39).

At the beginning of the 1970s, Tamás F. Farkas (1951–) used the combination of two types of rhombus and a square to construct designs that displayed local fivefold symmetry while tiling a part of the plane in quasi-periodic fashion (Figure 7.40).

Designs with similar properties were made by Y. Watanabe (Tokyo), by combining, among others, a rhombus, a triangle and a square (Figure 7.41).

Even together, these designs with three elements, and, at least locally, displaying fivefold symmetry, could not be a substitute for a fivefold tiling of the plane using two types of element. The increasing number of three-element solutions suggested that a solution with two elements was not far

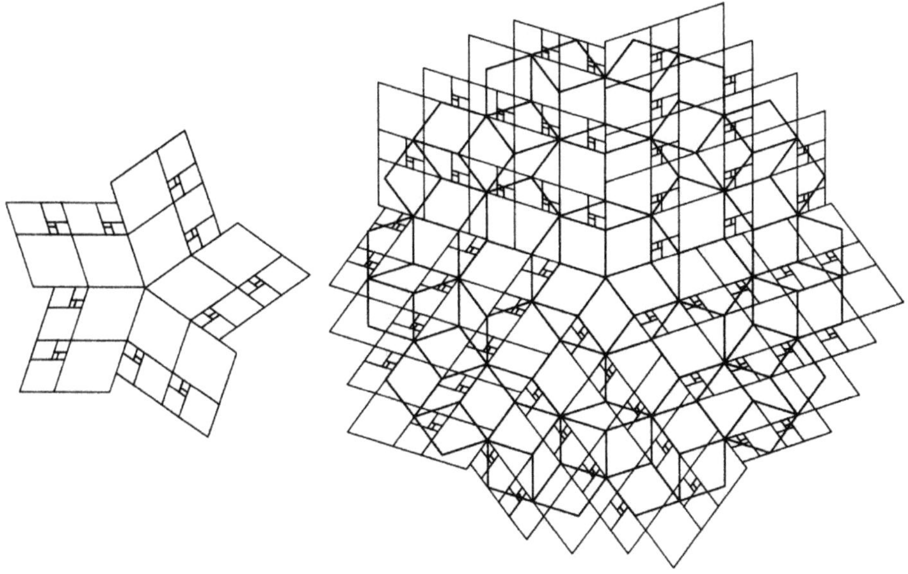

Figure 7.39. *The construction of B. Kirschenbaum's design with golden rhombuses*

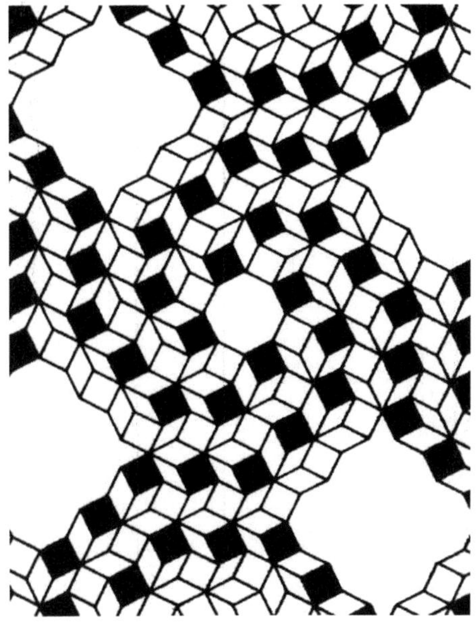

Figure 7.40. *Tamás F. Farkas' design made up of two types of rhombus and a square* (1972)

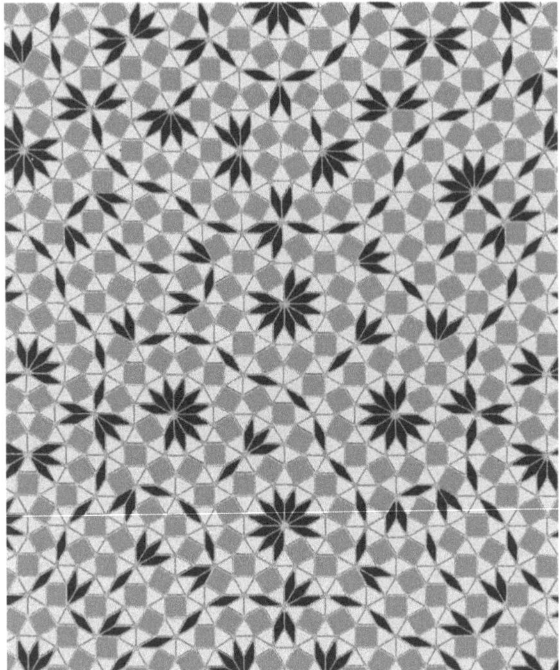

Figure 7.41. *Y. Watanabe: quasi-periodic design using a rhombus, a triangle and a square*

away. For it was clear that if nature did not find a solution to some problem using a single element, then it was more likely to do so by combining two elements than immediately with three separate ones. Furthermore, crystallographers have always been excited by the mystery of the lack of fivefold symmetry in nature. It was to be expected that, even if we excluded from the structures of nature the possibility of fivefold symmetry created from a single element, we might find it put together from a number of elements. The first ray of hope, that in principle a solution might exist, was expected from planar geometry. We just saw, however, that it was not geometers who performed the best in three-element solutions. Finally it was Oxford professor Roger Penrose, one of the mathematicians of our age with the most amazing conception of space — and who, nevertheless, has recently become better known as a cosmologist and theoretical physicist — who was the first to find a solution to tiling with two elements. (Mathematicians think it highly likely that non-periodic tiling of the plane displaying fivefold symmetry at least locally cannot be produced with a single element, but the exact proof of this is yet to be discovered.)

Penrose originally used a convex and a concave rectangle, the so-called dart and kite, to construct the design, which produced a continuous, quasi-periodic tiling of the plane with local fivefold symmetry. In point of fact, he combined the problem of pentagonal tiling with the solution to another problem that was no less of a challenge: how, using as few elements as possible, the plane can be tiled non-periodically. We saw fine examples of periodic but not pentagonal tiling in the works of Escher (e.g. *Birds*, Figure 7.42).

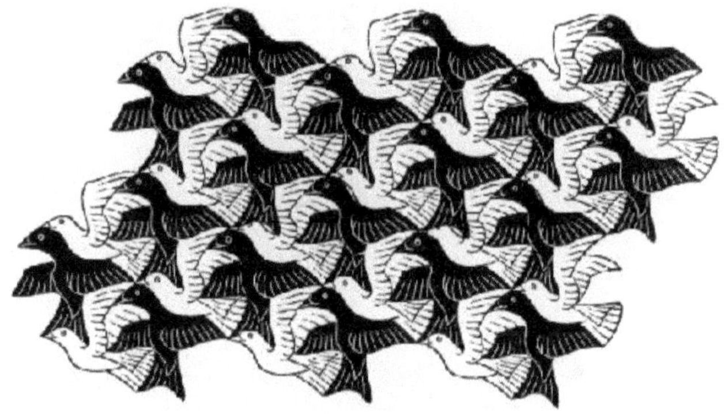

Figure 7.42. *M. C. Escher: Birds*

An example of non-periodic tiling is the three generations of 'sphinxes' seen in Figure 7.43.

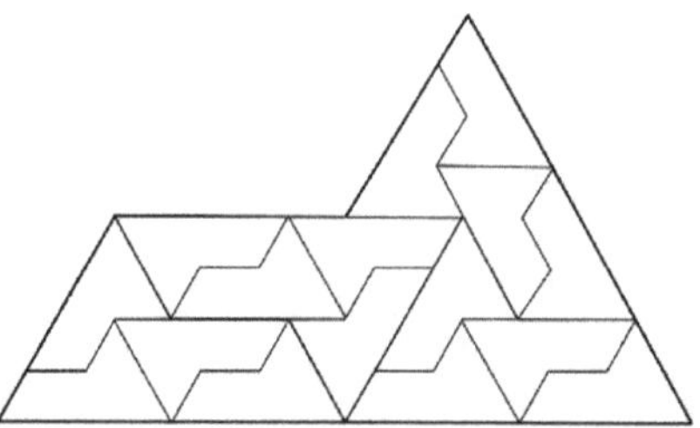

Figure 7.43. *Sphinxes. This type of tiling which leads an enlarged version of itself was given the name "reptiles" by S. W. Golomb, which is a direct reference to Escher's lizards, and also combines the prefix of the word "repetition" with the word "tile" — source: M. Gardner [1914–]. With this type of basic design, however, periodic tiling can also be produced.*

Along the road to finding the solution, Penrose was inspired by the attempts of Kepler. He has stated that at first he was simply playing around with the pentagon. "... I used to doodle just for fun, designing patterns which would repeat-complicated things, different shapes that would make formal repeating patterns. One needed many of them before they would repeat. I was also interested in hierarchical structures, in which the pattern would appear at a larger scale. This was just playing around-there wasn't a feeling that this was science or anything. One thing that must have been influential, although I didn't know it at the time, was a book my father had of Kepler's works. Among these was a picture with many tiling patterns, some of them involving pentagons. I hadn't been thinking about them particularly, but it must have created the feeling that maybe pentagons were things that one could use for interesting designs."

"One can produce pentagons in a kind of hierarchical scheme, but there are always spaces between them. And when these spaces get big, one has to think about what to put in the spaces and make a choice whether to do it one way around or the other way. ... I produced a non-repeating pattern with pentagons, and then somewhat later realized one could force these patterns into a kind of jigsaw puzzle. If one modified the shapes of the pieces a little, then one could assemble them in this way. And this led to six different shapes, which would force one into a non-repeating pattern."

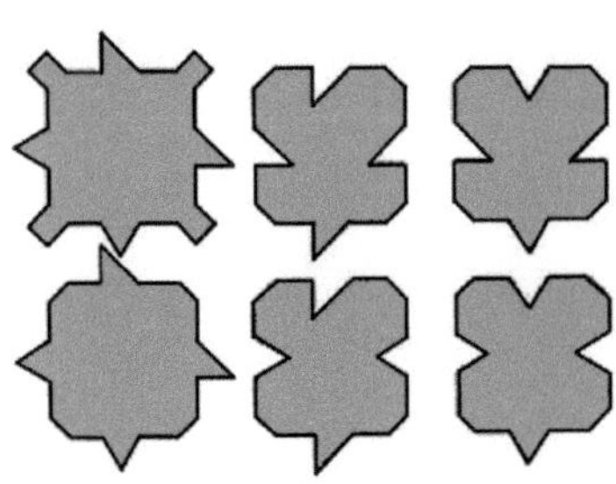

Figure 7.44. *Robinson's six tiles, with which the plane can be tiled exclusively in a non-periodic fashion*

When he discovered a symmetric arrangement made up of pentagons, as inspired by Kepler's picture, he immediately told a friend of the news. It was then that he realized he was able to construct a system consisting of six tiles, where each tile had a different shape, and could be fitted together like pieces of a jigsaw puzzle (Figure 7.44). The friend "reminded me of Raphael Robinson, who had a set of six tiles that would tile a plane in a non-periodic way. He also mentioned that Robinson tried to keep his numbers down to a minimum. He had this non-periodic tiling based on squares — with modifications, but basically squares — and he had six different tiles to force non-periodicity. When I saw this again,

I knew I could do better: my tiles were also six, but there was a redundancy; one could glue two pieces together and get it down to five. This was something of an improvement over what he had, but then I started thinking about it even more. And I realized one could get it down to two" (Dialogue: Roger Penrose and Sato Humitaka).

He reached the lowest number of tiles with which it was possible to construct a non-periodic arrangement (Figure 7.45). Robinson's tiles were characterized by quadratic symmetry, while those of Penrose had the special characteristic that they contained pentagonal symmetry. This pentagonal symmetry was as yet unknown in crystallography. The general theorems stated that only twofold, threefold, fourfold and sixfold symmetries were possible there. If one avoids the theorem, one will find only approximately periodic symmetry, and this is how fivefold symmetry is produced.

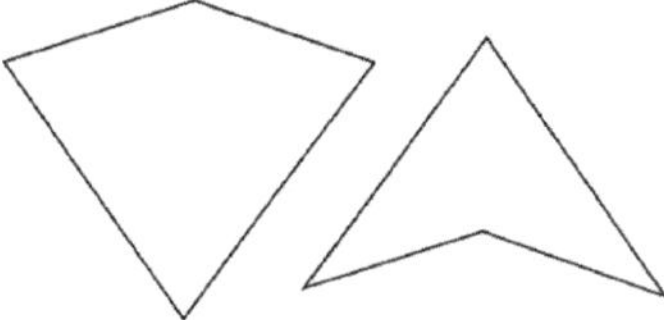

Figure 7.45. *Kite and dart*

Penrose patented his design in 1973. He later succeeded in reducing the darts and kites to two simpler elements, two types of rhombus.

Joined together, the dart and kite form a golden rhombus (Figure 7.46). Placing one on the other, the difference gives us a slimmer rhombus. Both rhombuses display the 36° and 72° angles of the pentagram of the Pythagoreans. The ratio between the various lengths of the edges of the dart and the kite is the golden section. What is more, both are among the shapes produced by dividing a pentagram by drawing in its diagonals. As if we were dividing up Osiris' sarcophagus all over again! It was there all the time, just we did not notice it. Penrose interpreted this conceptual progress by positing that different ideas are born at different times, that there are many levels and stages of thought. Initially his interest was in so-called periodic tiling. The tiles that were later named after him simply came into his head. He was fiddling around, playing with them,

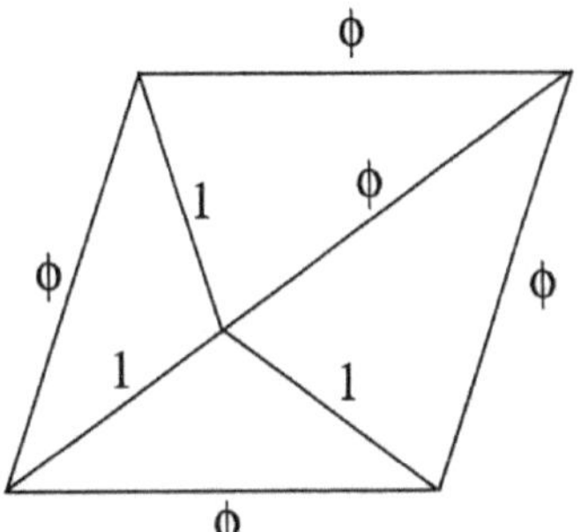

Figure 7.46. *Construction of the golden rhombus from a dart and a kite*

and suddenly the solution came to him. His original intention was not the solution of some particular problem, or the discovery of some hidden principle. He simply started to play with the tile forms because he found this activity interesting. His opinion about the road to discovery was that on almost all occasions we reach a solution in a gradual fashion. Over quite a long period of time, one goes down the road in little steps, with little discoveries. Lightning discoveries are rare. A great number of wonderful ideas can come into being when one brings together two things that previously appeared distant from one another, the relationship between which was at yet unclear. We should add that symmetry considerations have often played just such a heuristic role in the process of discovery.

Generalizing this universal nature of learning, Attila József, one of the best-known Hungarian poets of the twentieth century, put the same phenomenon into the following words with poetic succinctness:

"I am made thus: what for a thousand ages
I've looked upon, now suddenly I see."

(*By the Danube,* trans. Zsuzsanna Ozsváth and Frederick Turner)

Nowadays, Penrose's arrangement, which is not a single motif, and which can be interpreted in a variety of forms, is usually reproduced not with darts and kites but with the two rhombuses containing the respective angles of the regular pentagon (Figure 7.47).

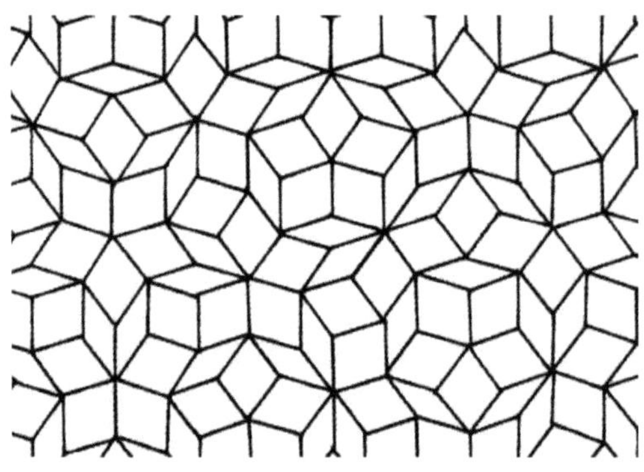

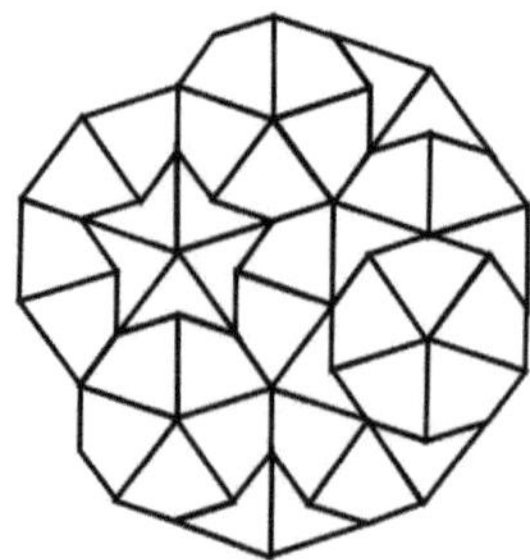

Figure 7.47. *Penrose quasi-periodic tiling of the plane with two types of rhombus, all of which include the angles of a regular pentagon, and with darts and kites*

The advantage of rhombuses over darts and kites is that they present a single edge length rather than two different edge lengths. As far as the notion of perfection is concerned, we have fewer conditions to sacrifice!

John Conway (1936–) proved that the number of Penrose tilings, like the cardinality of the points on a straight, is not countable. And, in addition to the infinite richness of the Penrose tilings, Penrose also succeeded in proving that any finite domain of any of the tilings also appears somewhere in all the other tilings. What is more, it appears an infinite number of times in every design. After all, the fewer elements we use to construct a non-periodic design, the greater the probability that, sooner or later, the design of a few tiles placed alongside one another should be repeated. It is enough for us to consider that there are only seven ways of placing darts and kites around any given vertex.

Conway also proved, inter alia, that we can find a domain tiled similarly to that we happen to have chosen within a very small distance. If the diameter of a chosen (say circular) domain is d, then we reach a domain with similar tiling in some direction from the boundary of the original domain, at a distance of at most $s \leq \frac{\Phi^3}{2}d = 2.11\ldots d$. (As further evidence of the universal connections between symmetries, Φ here is again the golden number we know from the golden section.) This theorem not only gives a certain limit to repetition (even if it is not periodic) — this limit is a surprisingly low, "visually observable" boundary distance. The quasi-periodic arrangement also means that we find certain "local" symmetries in it, which break once past a certain boundary, but which are repeated elsewhere, locally. There are also those that infinitely preserve their symmetry with regard to a certain point, but where the periodic translation of this symmetry centre-point does not make the entire tiling of the plane overlap itself. An example of the latter is the circular symmetrical, infinite arrangement of the "Sun" and "star" designs. No other infinite arrangements displaying regular pentagonal symmetry can be produced (though non-pentagonal can). Let us observe how the two are essentially the same: all of the five kites located in the centre of the "Sun" can be divided up into a dart and two kites. This is also true of the elements to be found surrounding them. The figure thus produced generates the "star" tile design (Figure 7.48).

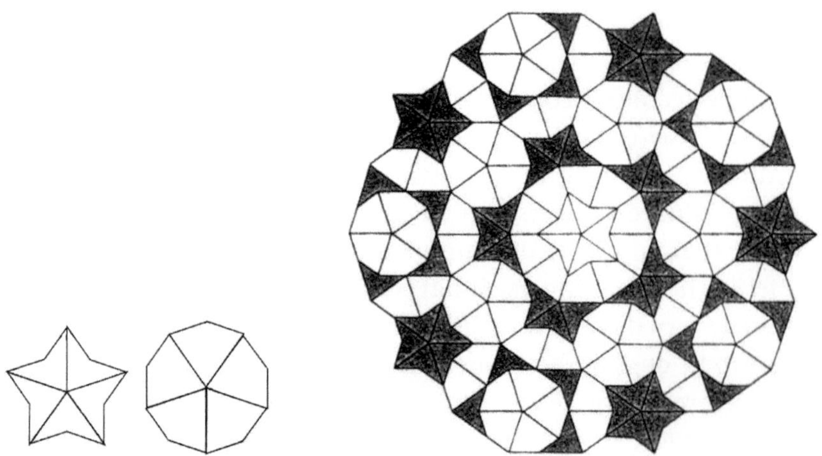

Figure 7.48. *The infinite tile pattern of the star or the Sun*

We have made an interesting observation: the darts and kites can be broken up into smaller darts and kites. The reverse is also true: darts and kites can be use to create larger darts and kites. These shapes, made up as they are of smaller and larger units, result in a Penrose tiling of the plane, that has precisely the same characteristics. All established mathematical theorems are applicable to them.

One of the applicable mathematical theorems with the most interesting symmetry relates to the proportion of the elements used. The proportion of the number of dart and kite shapes is — like the proportion of their area — equal to the golden proportion. 1.618 ... times as many kites are needed as darts. If the tiling is infinite, this number is the precise proportion. The fact that this proportion is not rational is used by Penrose to prove that tiling is not periodic, for if it were periodic, this proportion would have to be a rational number. This theorem — like all those concerning kites and darts — holds true even if the tiling takes place with the Penrose rhombuses, made up of darts and kites, with edges of equal length, and angles of 72° and 108° or 36° and 144° respectively.

Figure 7.49 clearly shows how the mystical pentagram of the Pythagoreans can, just like the Penrose tiles, be indefinitely divided and extended both inwards and outwards with similar elements. Within these, every section is in golden proportion to those one step smaller (this is what we used to construct the regular pentagon). In the figure, the two pairs of Penrose tiles can be seen: the kite in the tetragon *ABCD*, the dart in *AECB*, and the rhombuses (though not in proportion to each other) in the tetragons *AECD* and *ABCF*.

It is worth comparing the open pentagons of the Maragha design with the Penrose designs. In Figure 7.50 we can see the so-called long or short "bow-tie" shapes that can be constructed from darts and kites, which are shown enlarged in Figure 7.51.

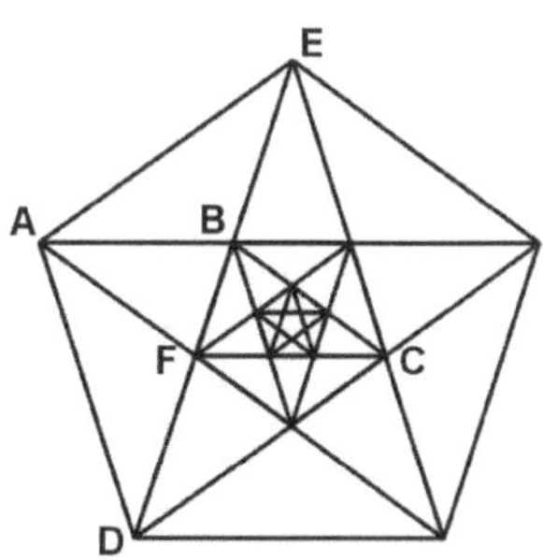

Figure 7.49. *The pentagram of the Pythagoreans*

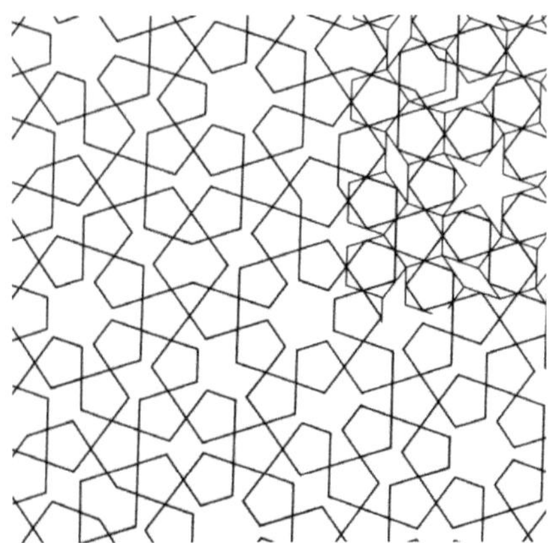

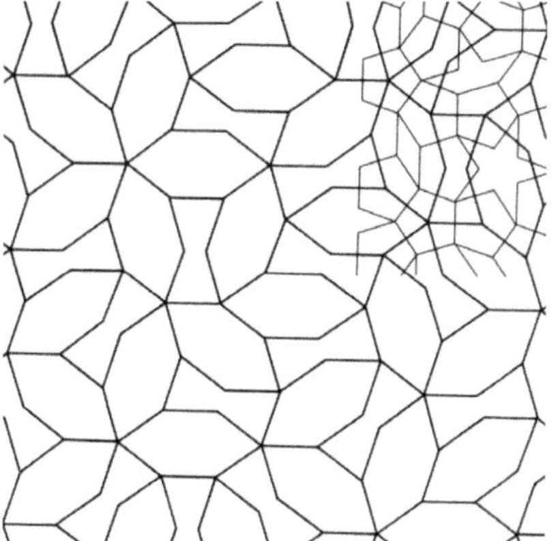

Figure 7.50. *Comparison of the Maragha pattern and the Penrose arrangement. The bow-ties in the right hand figure can be constructed from darts and kites (cf. Figure 7.51), while the hexagons are made up of three rhombuses with the known angles*

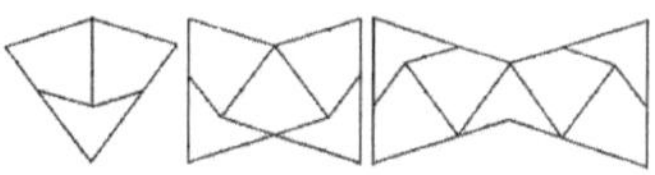

Figure 7.51. *The ace and bow-tie modules made up of darts and kites, which can substitute the darts and kites elements during tiling*

The Maragha pattern is made up of open pentagonal elements, just like the devil's pentagram (witch's foot or goblin's cross) opened at one of its vertices which trapped Mephisto. It produces its own quasi-periodic design from these elements.

In many of his drawings, Escher drew periodic tilings of the plane made up of two or three types of element, in such a way that he replaced the straight edges of the geometrical shapes used for tiling with congruently arced lines (Figure 7.52).

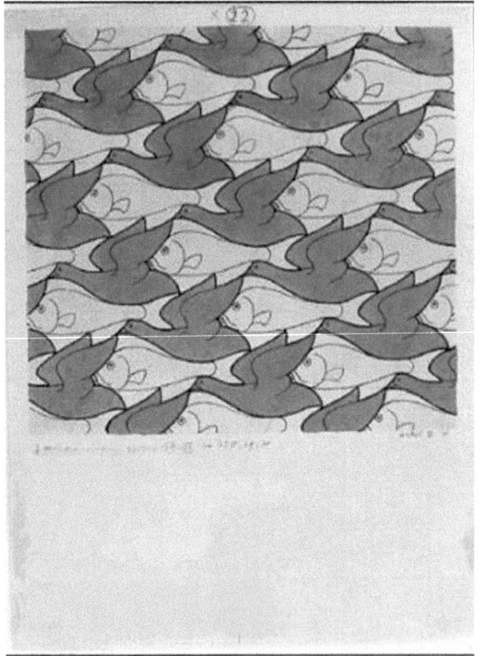

Figure 7.52. *M. C. Escher: Fishes and Birds*

Instead of polygonal shapes, he used "tiles" representing the shapes of animals. Penrose used a similar technique to transform his darts and kites into "chickens" that quasi-periodically filled the plane (Figure 7.53). (He even succeeded in preserving the chicken's direction of orientation.)

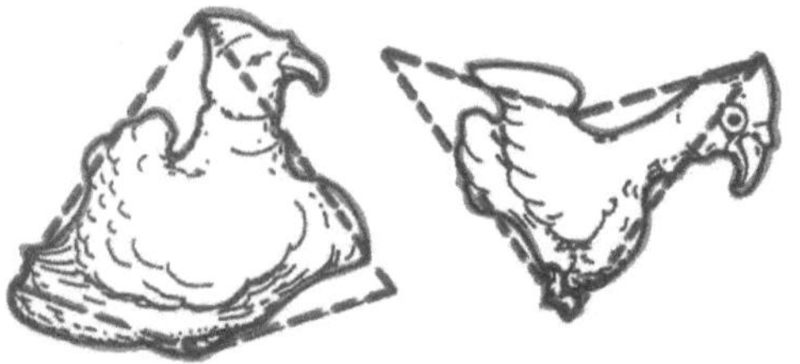

Figure 7.53. *Penrose's non-periodic chicken*

Independently of Penrose, American mathematician Robert Ammann also discovered the possibility of tiling with rhombuses. In 1976, Ammann realized that both pairs of tiles can be used to create patterns that can be laid alongside one another like ribbons, and which are determined by five families of parallel straight lines, namely by having the bars cover the plane such that the straight lines indicating their directions intersect each other at an angle of precisely 72°. If the way in which the lines are laid onto the equivalent points of the design is chosen appropriately, the parallel "Ammann bars" lie at two different distances from each other (Figure 7.54). The ratio between these two distances is the golden proportion. Furthermore, continuing the laying down of the parallel lines throughout the entire plane, within a particular family of parallel lines the proportion of the number of shorter and longer distances between the bars will also tend towards the golden ratio.

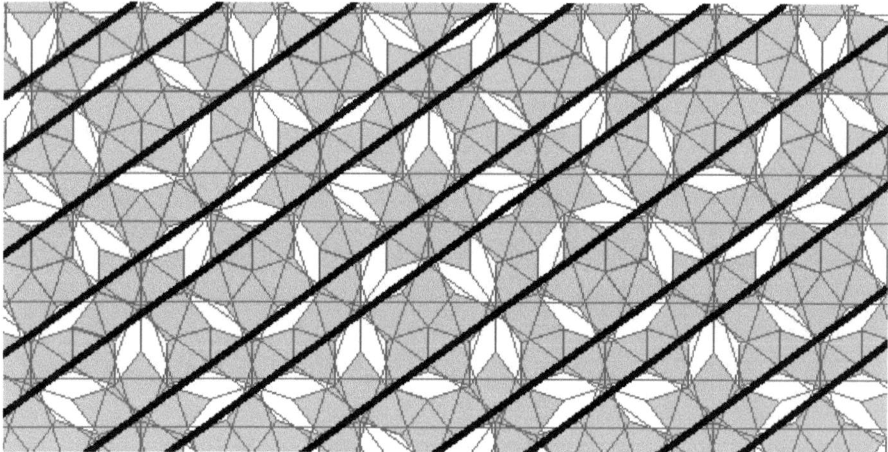

Figure 7.54. *A family of Ammann bars*

Conway produced a number of further results concerning the way in which the Penrose tilings relate to the Fibonacci numbers, which, as we previously saw, display a close similarity to the rules of growth of various plants.

Another discovery of Ammann's (also from 1976) paved the way for the possibility of space filling in non-periodic ways displaying fivefold symmetry. He began by constructing from two rhombohedra which have edges all of the same length, and which are both delimited by a single congruent "golden" rhombus. The proportion between the diagonals of the faces is the golden section. One looks like a cube flattened along one of the diagonals of its body, while the other looks like a cube stretched along one of its diagonals. H. S. M. Coxeter, undoubtedly the modern era's greatest personality in geometry, who died in Toronto only recently, referred to these as "golden rhombohedra". Apart from these two, no other golden rhombohedra exist. Both were already studied by Kepler. Martin Gardner, the great expert in this field, drew Penrose's attention to Ammann's results. Ammann took a set of two rhombohedra, parallelepipeds with six congruent rhombus faces, and showed that, when face-matching rules are applied, they could tile space non-periodically. As the two rhombohedra are golden rhombohedra, the faces have diagonals in the golden ratio.

Penrose came to the conclusion that it might be a model for certain unexplainable molecular formations, such as viruses. Alongside his congratulations to Ammann, Penrose made the following reply to Gardner:

> "... some viruses grow in the shapes of regular dodecahedra and icosahedra. [...] But with Ammann's non-periodic solids as basic units, one would arrive at quasi-periodic 'crystals' involving such seemingly impossible (crystallographically) cleavage directions along dodecahedral or icosahedral planes. Is it possible that the viruses might grow in some such way involving non-periodic basic units, or is the idea too fanciful?"
>
> (R. Penrose to M. Gardner, 4 May 1976)

Now that we have succeeded in tiling the plane with fivefold symmetry using two types of element, albeit "only" quasi-periodically, which in return for their less than perfect periodicity compensate us with other important proportions we have previously encountered in our symmetry studies (not all of which we have had the opportunity to deal with here), the question has already been raised in Penrose's cited letter of whether our new knowledge can be generalized for the space filling without gaps with fivefold symmetry. For we can only draw nearer to answering questions about crystals and the structure of matter if we can answer this one. A number of different people worked on this subject in various parts of the world, independently of one another.

In 1977, Koji Miyazaki at the University of Kyoto reached a discovery similar to but independent of Ammann's. In addition, he found another method of filling space with two golden rhombohedra in a non-periodic way. In this method, five golden rhombohedra with acute angles and five rhombohedra with obtuse angles fit together to form a rhombic triacontahedron. Two such bodies can each be surrounded by a further thirty golden rhombohedra of each type, which results in a larger rhombic triacontahedron, and this extension can be continued infinitely. This gives us a honeycomb-like filling of the space, the centre of which displays icosahedral symmetry.

Naturally — almost the minute Penrose had made his planar discovery — the theoretical work began. The theoretical possibility of the filling of three-dimensional space with atoms with two types of cell stretched by atoms, in quasi-periodic fashion, displaying local fivefold symmetry, was first shown by London crystallographer Alan Mackay in 1978. His solution was rendered more precise by Tohru Ogawa (Tsukuba, Japan, 1936–) in 1981 — in the knowledge of Miyazaki's result. After this, Mackay and Ogawa joined forces to use calculations to confirm the theoretical possibility of creating stable electron bonds in the space directions determined by their models.

The space filling suggested by Ogawa can clearly be seen in Akio Hizume's model (Figure 7.55), where the cells enclosed by bamboo shoots represent pentagonal space filling and rhombohedral cells.

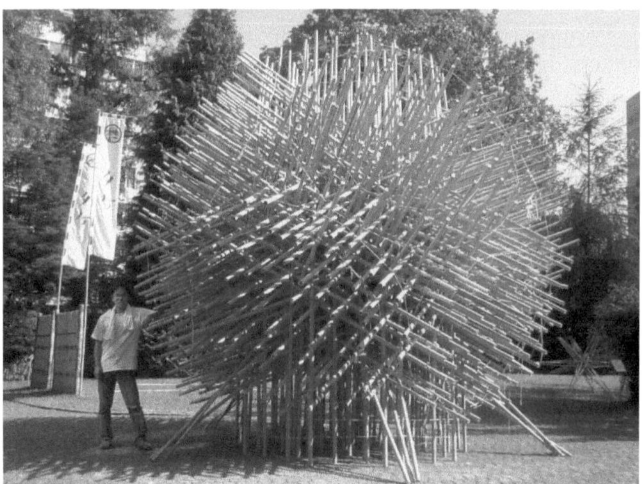

Figure 7.55. *Akio Hizume's model, made of bamboo, showing a pentagonal spatial arrangement and two types of rhombohedral cell*

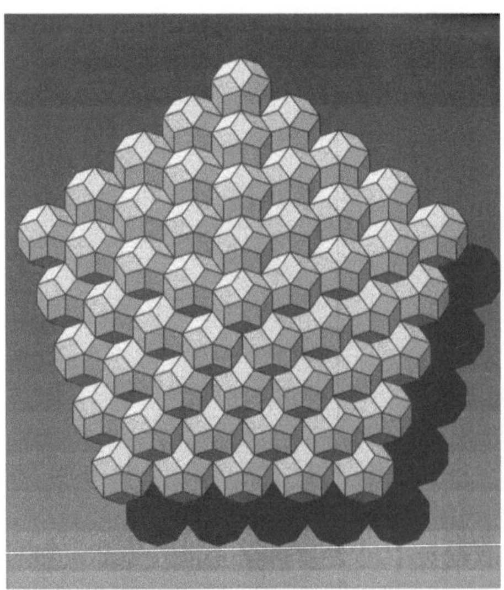

Figure 7.56. *Penrose-type cells, constructed of rhombohedra arranged in space* (R. Dewar, 1989)

Hizume (1960–) is an artist who was inspired by the possibilities for depiction this scientific model offered. He is not the only person to have been interested in the spatial depiction of fivefold symmetry. Ultimately, as soon as we gave up the requirement of filling space in a congruent manner with a single type of cell, in space as on the plane, so the possibility of solving the problem using numerous types of cell presented itself. The juxtaposition of dodecahedra presents itself naturally, with the gaps left when they are joined being considered as other cells. True, these do not display proportions and symmetries like those generated by construction from golden rhombohedra. Californian graphic designer and sculptor Robert Dewar's poster for the founding congress of the International Symmetry Society in 1989 displays an arrangement of rhombohedra displaying dodecahedral symmetry in space (Figure 7.56).

Dewar's poster for the 1995 Washington congress of the Society displays dodeca/icosahedral, non-continuous space filling. It is worth comparing this to the model of the icosahedral structure of the *HTLV-2* virus (better known as the *AIDS* virus) (Figure 7.57).

Staying at the scale that is only visible under an electron microscope, let us in Figure 7.58 wonder at the tenfold symmetry displayed in the cross-section of a DNA molecule.

The last two illustrations show that nature really does produce phenomena with fivefold or tenfold symmetry. Yet, the closed system of crystallography presented above, which canonized the 230 crystallographic groups and the conditions for classification into the world of crystals, ruled out the possibility of accepting fivefold symmetry. The dilemma of crystallography at the beginning of the 1980s was as follows: can quasi-periodic space

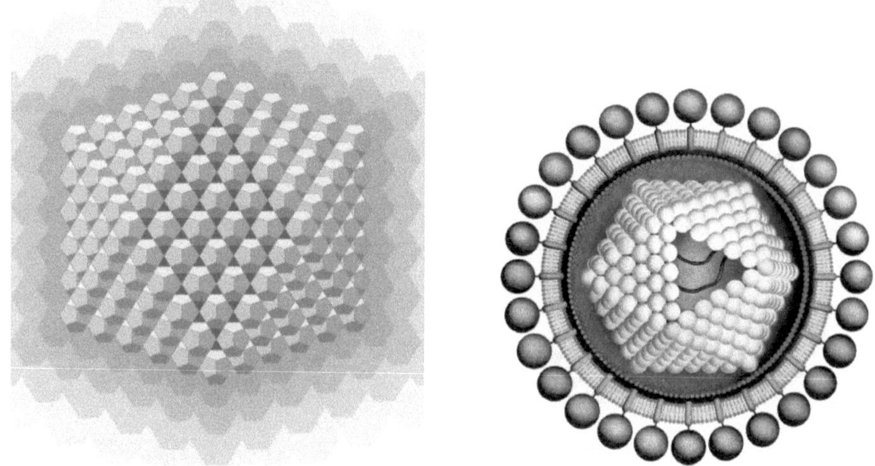

Figure 7.57. *The poster design for the 1995 congress of the International Symmetry Society, and the icosahedral model of the HTLV-2 (AIDS) virus*

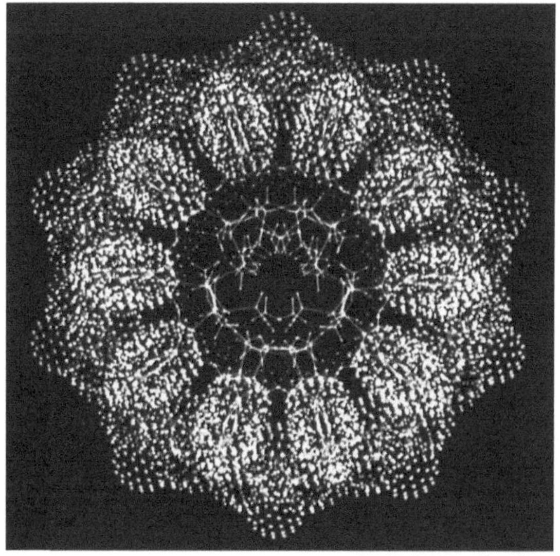

Figure 7.58. *Cross-section of the DNA molecule, displaying tenfold symmetry*

filling be regarded as a crystal? It seemed that, after the theoretical work of A. Mackay and T. Ogawa, even if the concepts of classical crystallography did not allow quasi-periodic space filling into the system, their existence

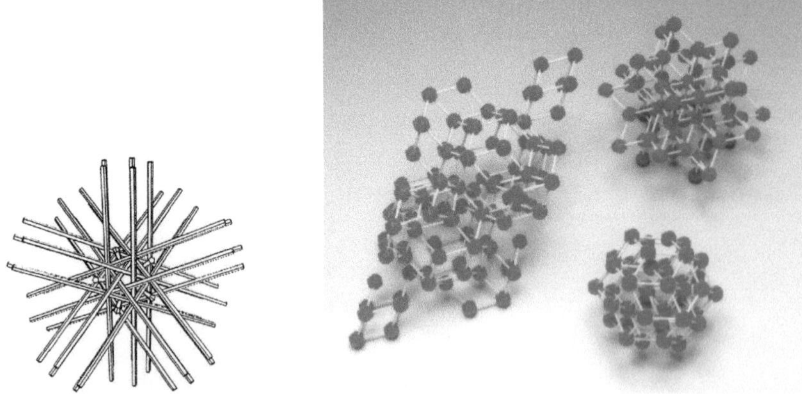

Figure 7.59. Ogawa's bamboo and point lattice models

could no longer be ignored. Figure 7.59 shows T. Ogawa's bamboo and point lattice models.

In Figure 7.60, we see an X-ray diffraction image displaying tenfold symmetry. As it later transpired, a number of crystallographers investigating the structure of matter had already encountered such images. They had always thrown them away as bad exposures.

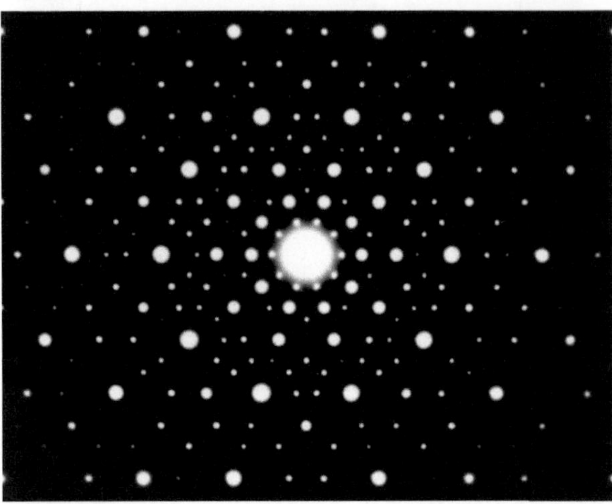

Figure 7.60. *X-ray diffraction image displaying tenfold symmetry* ($Al_{65}Cu_{20}Fe_{15}$ quasicrystal, An Pang Tsai)

This is just what Dan Shechtman always used to do, as he and his colleagues in a laboratory at the American National Institute of Standards and Technology examined the material structure of alloys with the help of X-ray diffraction images. In 1982, when he looked through many hundred exposures in a single night, and rejected at least five seemingly defective images, he was struck by the idea that they were perhaps not defective, after all. It could not be an accident that there were so many of them. Maybe there was some kind of regularity behind their occurrence? With his colleagues, he set about examining the rejected pictures again, together with the sample of the material they had been taken of, a suddenly cooled aluminium-manganese alloy. This was how he discovered the materials that would later be called quasicrystals.

Their result was an encouragement to all those who had previously not dared to believe in the existence of samples of materials displaying five-fold symmetry. From this point on, "faulty" X-ray images were no longer thrown in the rubbish bin, and one alloy after the next was discovered to have similar symmetry. Quasicrystals, with their fivefold symmetry, opened a new chapter in material science. We saw above how the electron microscope helped us become aware of the secrets held deep inside matter. Figure 7.61 shows the electron microscopic photograph of an *Al-Cu-Fe* alloy. The dodecahedral structure created by the connected submicroscopic grains of material is clearly visible.

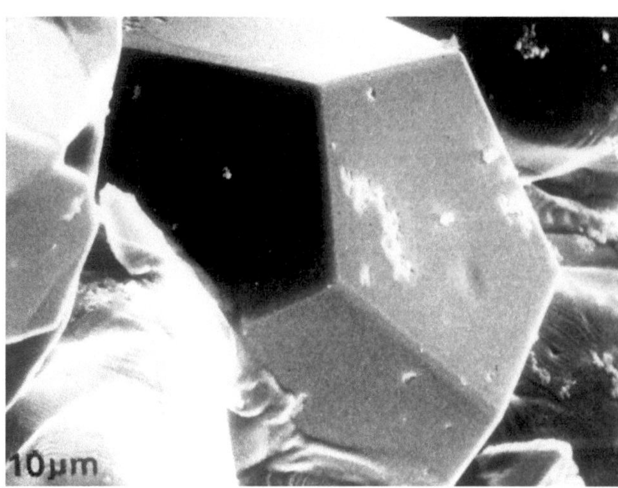

Figure 7.61. *Electron microscope image of the alloy* $Al_{65}Cu_{20}Fe_{15}$ (An Pang Tsai)

Chapter 8

From viruses to fullerene molecules

What do the surface structure of viruses, the morphology of sea radiolaria, the pattern on a golf ball, the weaving of baskets, the geodesic dome structure of Buckminster Fuller, the sewing design on a soccer ball and the C_{60} fullerene molecule all have in common? The feature common to all these and many other phenomena is a family of structures displaying a characteristic symmetry: truncated icosahedra. Truncated icosahedral structures belong to the sizeable set of polyhedra invariant under spatial rotations. It is no accident that a similar structure developed. Their formation involved a common principle of structure construction, that of so-called synergy.

The concept of synergy is difficult to define in a single sentence. German physicist Hermann Haken (1927–) and American architect R. Buckminster Fuller introduced this term, independently of one another, and hence in not entirely corresponding fashion. Haken considers it to be the principle which regulates the interaction of the elements of self-organizing systems and the spatial arrangement, temporal flow and functional structure of the process of self-organization. Fuller's interpretation of synergy is more static, essentially a geometrical arrangement of the structure of matter on the one hand, and, on the other, a construction principle which can be implemented with a minimum of material and energy to give a maximum of stability and endurance. We can encounter the realization of this principle in inanimate and living nature and in the optimal organization of human activity (e.g. our work). The implementation of the principle of synergy is best elucidated by reference to some examples.

Crystals, for example, utilize the optimal filling of space. There is a violation of the spherical symmetry displayed by the charge distribution of the electron shells of the atoms and atom groups that form the crystal; the charges are more likely to arrow to the centre of attraction of the neighbouring atoms. The most stable chemical bonds are created by the optimal spatial directions of charge distributions (which in crystals are symmetrical). Other phenomena also suggest the implementation of the synergy principle: optimal bonding energy corresponds to maximal stability. The implementation of variational principles also guarantees the

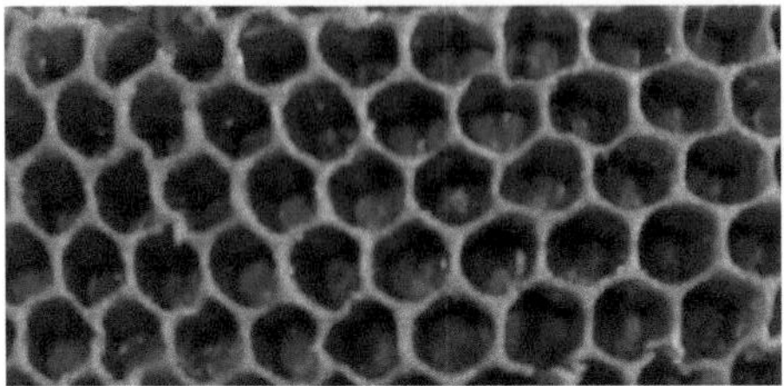

Figure 8.1. *Honeycombs*

realization of these goals. In living nature we can observe the synergetic principle in action in the distribution of cells (e.g. when the cells of morula divide, the blastula cell form develops as a result of this principle), which is followed by the construction of tissue. Neither do bees build their honeycombs as they do out of a particular fondness for regular hexagonal symmetry (Figure 8.1). The optimal pattern for an individual honeycomb — and the one that is easiest for the bees to drip — is one with cylindrical symmetry. From this it follows pragmatically that if circular honeycombs are placed on the plane in the densest formation, each will be surrounded by six other ones. Their dividing walls form the hexagonal lattice structure that our visual memory stores.

We use the same principle for construction in the man-made (artificial) environment. We strive for maximum stability using a minimum amount of material. We plan for maximum bridging of space and endurance with minimum material and expense. We strive for a similarly optimal result in the utilization of human labour. When creating our buildings, machines, tools and accessories, we look for extreme values, are led by variational principles, and accordingly we look for symmetrical solutions in whatever ways the present level of technology allows us.

This can all be clearly traced through the changing architectural styles over history. The polyhedral symmetry of the pyramids is inescapable, and Romanesque architecture saw the realization of the bridging optimums of dense filling of space, while Gothic architecture looked for this in solutions with supporting walls and supporting columns. The architectural technology of the Renaissance made possible the expansion of the horizontal covering of space, of boarded ceiling solutions. The Baroque developed this technique further, applying it to curved surfaces rather than planar ceilings, creating boarded domes that could bridge larger spaces. The architecture of the modern era bridged spaces of unprecedented size with framed supports, which made use of the rich geometric inventory of symmetrical structures.

Before we wonder at the beauty of the results of this lengthy process of development, let us follow the full journey of synergetics from its roots in geometry. The distribution of honeycombs is the instinctive solution to the following geometric problem: how it is possible to fill the plane in the densest way with congruent circles (of the same radius) (Figure 8.2). It is easy to try this out with buttons or coins on the table, or on a piece of paper.

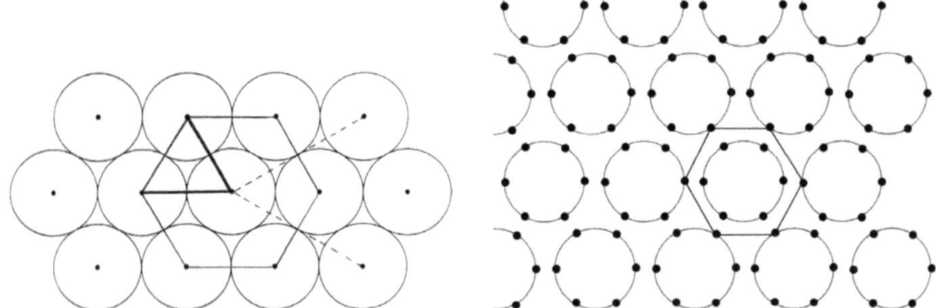

Figure 8.2. *The densest tiling of the plane using circles*

We can observe that each circle is touched by six others in such a way that these also touch each other. Their common tangents form a net of regular hexagons on the plane. The centre-points of the neighbouring circles also form a hexagon. From the centre of a circle (hexagon) six radii can be drawn in the direction of the centres of the neighbouring circles, and another six in that of the tangents that divide these circles. This $6+6$ radii, which in pairs represent the continuation of each other, describe $3+3=6$ axes of symmetry on the plane. This is the most symmetrical tiling possible: the plane cannot be filled in a denser way with congruent circles.

Let us also try tiling curved surfaces! Instead of circles, we can use the regular hexagons that bound them. A cylindrical surface (e.g. a cob of corn) can be tiled with regular hexagons (but not regular pentagons) in the same way as the plane (Figure 8.3).

The same is not true for the sphere. A spherical surface can be tiled without gaps using regular pentagons (dodecahedron), but not with regular hexagons. For this would contradict Euler's theorem. (We encounter a slightly similar problem when faced with the task of close-packing of the space. With spheres laid on a plane, one can be surrounded by six others of equal size in such a way that it touches all six and the others touch each

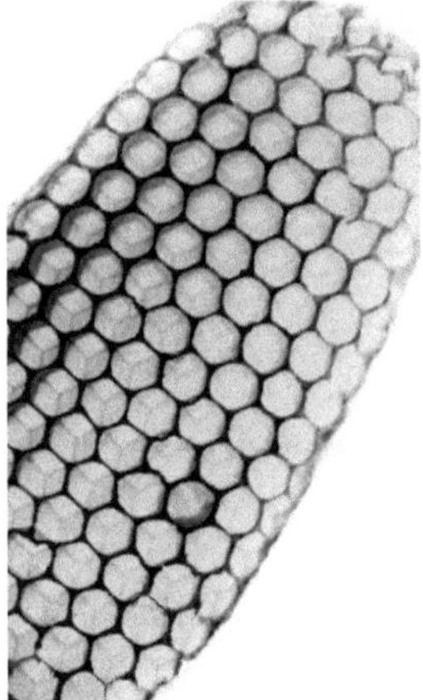

Figure 8.3. *A cob of corn: the densest tiling of a cylindrical surface using circles*

other. The same cannot be achieved in space [cf. the Kepler problem]: we cannot surround the sphere with spheres of the same radius such that each sphere touches all the neighbouring ones. There is a way of using them to fill space in the densest [optimal] way, but this does not display perfect symmetry.)

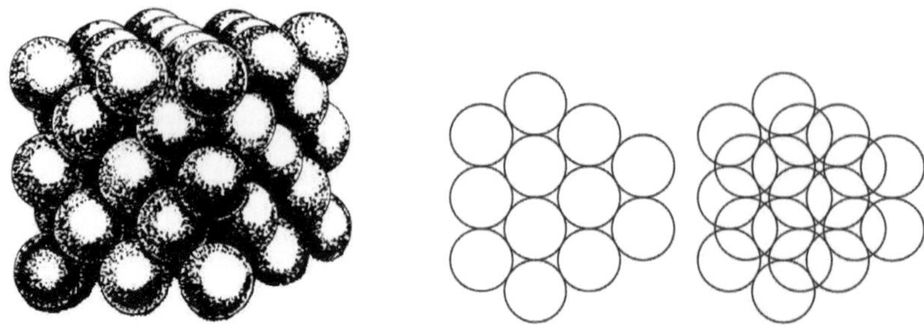

Figure 8.4. *Close-packing of congruent spheres, in space and viewed from above*

Onto the surface of the sphere we can project regular polyhedra that can be inscribed inside it (such that its vertices touch the surface of the sphere). As regular bodies are bound by regular polygons, inside each of these regular polygons we can draw a circle that touches each of their edges at their centre-point, and which thereby touch the circles which can be drawn inside the neighbouring polygons. These circles can also be projected onto the surface of the sphere drawn around the regular body, which they cover when they touch each other. In this way, and only this way, spheres drawn around polyhedra bounded by regular polygons can be covered by the congruent circles projected onto them from the polygonal faces.

If we cover the plane with a plane lattice made up of touching polygons and draw circles inside them that are also touching, by bending the plane the touching edges of the polygons and thereby the touching points of the circles drawn inside them become overlapped. The same occurs with the planar hexagonal lattice if we try to smooth it onto a sphere. Gradually bending them inwards, at first they only partially overlap each other. Bending them further, they would regain their symmetry if two faces were to become completely overlapped. However, if two of the three hexagons which touch at one vertex become completely overlapped, we would return to the plane. In the case of the triangular lattice, in which the vertices of six triangles meet at one point on the plane, the bending of these triangles in space means that at first one, then two, and finally three of them become overlapped with their neighbouring triangles, thus producing a vertex and the touching faces first of an *icosahedron* (when five triangles meet at one vertex), then of an *octahedron* (four triangles), and finally of a *tetrahedron* (three triangles). The same occurs if we bend in space the four squares around a vertex of a quadratic lattice: at a certain point, a square precisely overlaps the neighbouring one, giving us a *cube*. The only regular lattices at our disposal are the hexagonal, quadratic and triangular ones.

In Figure 8.5 we see the covering of a sphere with circular faces stuck to its surface. We can observe that, as in the case of the plane, a circle is

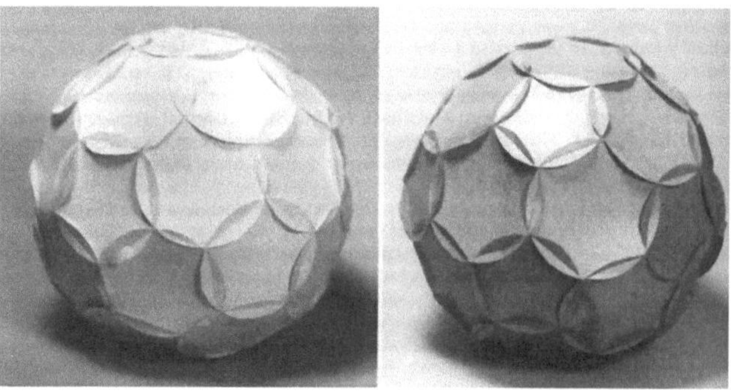

Figure 8.5. *Covering a spherical surface with circles*

surrounded by six neighbouring circles (or displays a congruent overlap). Yet the symmetry is not perfect. However precisely we accomplish the joinings, once a few circular faces have been attached, there will often be joinings where one of them is bounded by five circles. There is no way of avoiding pentagonal joinings.

Why is it not possible to cover the sphere with only regular hexagons? We can find the answer with the help of Euler's theorem. Let f be the number of faces of a body bounded by planar faces, v be the number of its vertices, and e the number of its edges. According to Euler's theorem:

$$f + v = e + 2$$

Let us assume that we can cover it with n triangles! Then:

$$\begin{aligned} f &= n \\ v &= 6n/3 = 2n \\ e &= 6n/2 = 3n \end{aligned}$$

(Three of the vertices and two of the edges are common; we have to divide by this number to avoid counting the same thing twice.)

If our assumption were true, placing these facts into the above equation — according to Euler's theorem — would give us the following:

$$n + 2n \stackrel{?}{=} 3n + 2$$

We have reached an evident contradiction, and so we have proved that our proposition cannot hold.

If we are not satisfied with the opportunities offered by the five regular bodies, and would like to include other polygons (e.g. regular hexagons) in the construction of polyhedra and the covering of the sphere that are as regular as possible, we are left with no choice but to allow the combined use of a number of regular polygons. In the course of this, like in the construction of Archimedean (semi-regular) bodies, we have to use truncation. This technique is also called the cell automata method. We start to cut off the polyhedron at all of its vertices at once, in such a way that we cut thin slices one after another along a plane perpendicular to the axis of symmetry going through the given vertex. This will give us regular polygons with as many sides as the number of edges that came together at a given vertex, while the number of edges of the original faces will double. In the course of the slicing process, the length of the original edges will decrease, while those of the newly-created faces will increase. Sooner or later we will reach the state in which the lengths of the edges are equal. Continuing the truncation, we reach the point where the length of the original edges decreases to zero, while that of the remaining edges is again equal. Further continuing the truncation of this semi-regular body, we again reach a state in which all edges are of equal length, and finally, when the length of one of the family of edges has again decreased in length to zero, this gives us the dual of the original body. In Figure 8.6 we see the simple series of truncations of the cube and the octahedron, indicating the aforementioned intermediate stages. The cube turns into an octahedron, and vice versa.

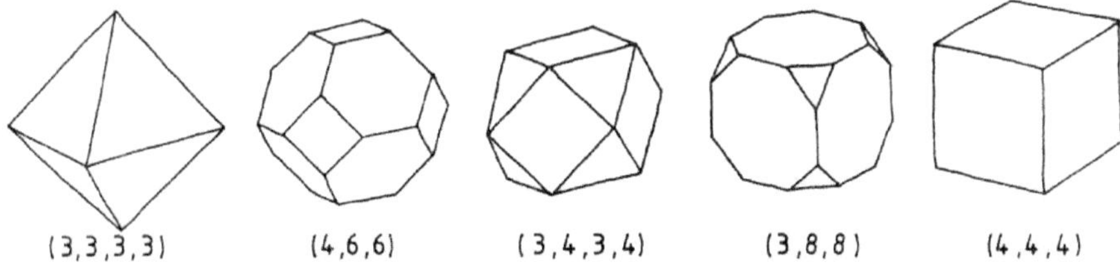

Figure 8.6. *The series of simple truncations of the octahedron or the cube*

In the following figures we show the half-series of the truncation of the icosahedron and that of the dodecahedron, broken down into a number of steps. It can clearly be seen that, from whichever direction we start, the truncation takes us to the same intermediate state: the body bounded by regular pentagons and hexagons with edges of equal length.

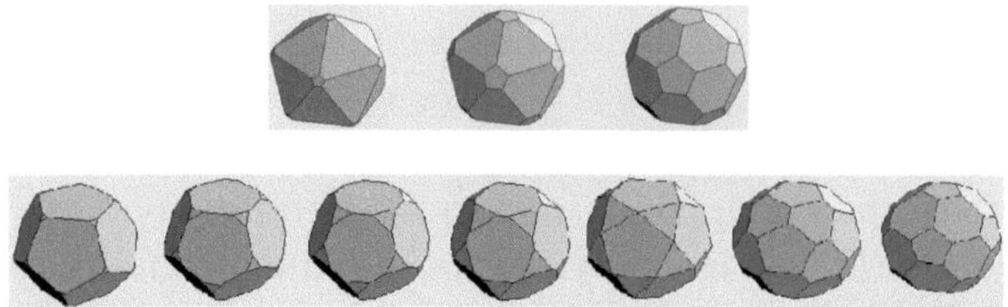

Figure 8.7. *The truncation of the icosahedron or the dodecahedron* (computer graphics by Sándor Kabai)

This truncated icosahedron is a semi-regular body that has long been known. (It was Piero della Francesca who rediscovered the Archimedean bodies for the Renaissance.) Of course — looking at the series from the longer direction — we can equally consider this body to be a truncated dodecahedron. We can find Leonardo's drawing of a truncated icosahedron in Luca Pacioli's book *Divina Proportione* (1509). As can be seen from the figure, Leonardo also derives the truncated body bounded by relatively the most hexagons and preserving the most axes of symmetry from the dodecahedron.

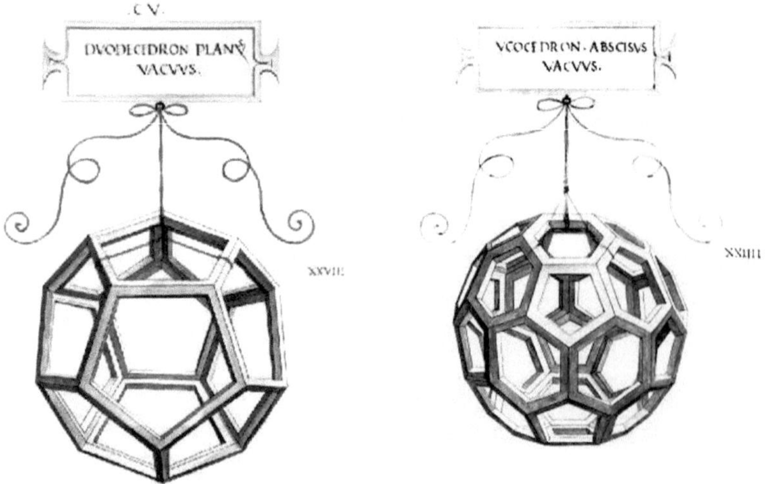

Figure 8.8. *Leonardo's drawings of a dodecahedron and a truncated dodecahedron, from L. Pacioli's book*

Why from the dodecahedron in particular? There are a number of reasons. If we take the dodecahedron as our starting point, we manage to preserve the original shape at least for some of the faces. We can also refer to the Pythagoreans, to the embodiment of harmony in the pentagon and the dodecahedron it covers, and to the secret symbol. We can think of the mystique of pentagonal symmetry, and the similar mystique which surrounded the hexagonal tiling of the surface of the sphere. Finally it was with the truncation of the dodecahedron that at least partial tiling of the surface of the sphere, and thereby the preservation of most of the axes of symmetry, proved possible. (After all, there are twelve pentagonal faces on the truncated dodecahedron just as there are on the original one.)

It is worth examining here how many pentagons we need to place on a body bounded (not exclusively) by hexagons. To determine this we again apply Euler's theorem. Let us assume that the sphere can be covered with n hexagons and m pentagons. Then:

the number of faces: $f = n + m$
the number of vertices: $v = (6n/3) + (5m/3)$
the number of edges: $e = (6n + 5m)/2$
according to Euler's theorem:

$$n + m + 6n/3 + 5m/3 = 6n/2 + 5m/2 + 2$$

This leads us to a surprising result. Rearranging the equations tells us that the number of pentagons to be used is m, irrespective of how many hexagons we use for tiling:

$$m = 12$$

However many hexagons we use to tile the sphere (polyhedron), the number of pentagons we need will always be twelve. This is true irrespective of whether the number of hexagons is zero (this is the "pure" dodecahedron), twenty (this is the truncated dodecahedron), or whether even more hexagons are used to tile the sphere. To summarize, the sphere can be covered (a polyhedron can be constructed) with twelve pentagons and any number of hexagons that satisfies Euler's theorem. (In every case we have assumed that the edges of the pentagons and hexagons are of equal length.) The polyhedra generated in this way we call fullerenes (we will explain the origins of this name later).

The simplest fullerene is the truncated icosahedron/dodecahedron. This is a body bounded by twelve regular pentagons and twenty regular hexagons (by dint of the truncation procedure, this number is the same as the number of vertices of the dodecahedron) and is also the most symmetric fullerene (ignoring the trivial dodecahedron containing $n = 0$ hexagons). It has 60 vertices. Every pentagon is bounded by five hexagons, and every hexagon is bounded by three pentagons and three hexagons. If we imagine carbon atoms with a valence of 4 at its vertices, this gives us

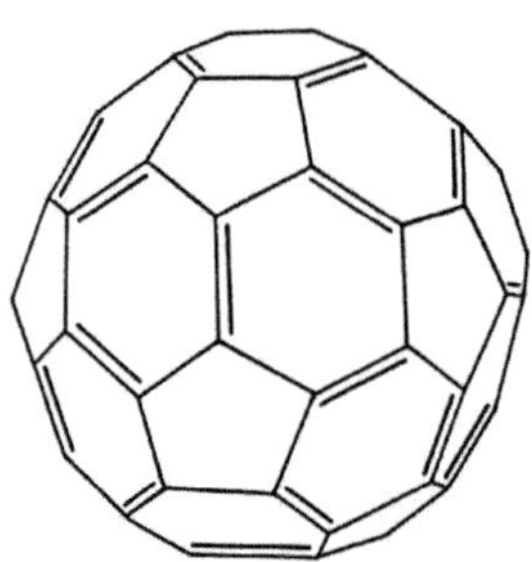

Figure 8.9. *Fullerene molecule of 60 carbon atoms*

a carbon molecule consisting of 60 carbon atoms (Figure 8.9). We assume single bonds along all the edges between the pentagons and hexagons, and double bonds along those where hexagons meet one another.

In Figure 8.10 we compare the structure of fullerenes with 60, 240 and 540 vertices. We can see that, as the number of vertices and faces increases, so symmetry decreases. When increasing the number of hexagonal faces, the number of pentagonal ones does not: in line with the simple calculation above, it remains strictly constant at twelve. More and more hexagons touch each other, indeed more and more hexagons exclusively touch other hexagons, not coming into contact with the pentagons that are increasingly rare by contrast. For this reason — let us think back to the geometrical considerations listed above — the sections covered entirely with hexagons become arranged on surfaces that are less and less curved. (This is demanded not only by their geometry, but also by the synergy of their bonds.) If we greatly increase the number of vertices (carbon atoms), the twelve pentagons will merely mark the vertices of a larger quasi-dodecahedron, while the faces between them will be bound by quasi-planar lattices of hexagons. The beginnings of the quasi-dodecahedral

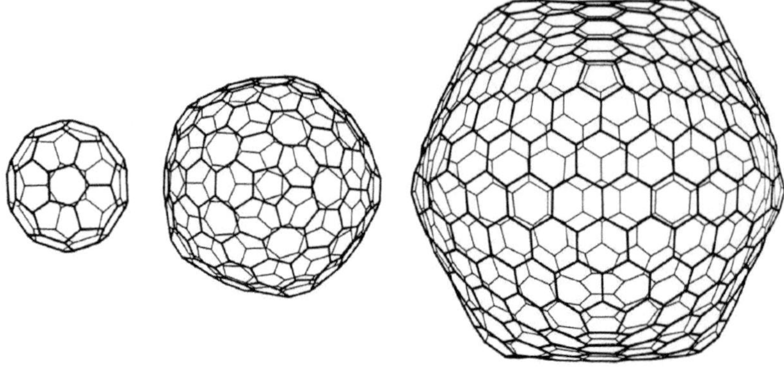

Figure 8.10. *C_{60}, C_{240} and C_{540}: as the number of faces is increased, so the symmetry is reduced, and it becomes more like a quasi-dodecahedron*

structure thus constructed can easily be seen in the figure of fullerene with 540 vertices.

This structure is not entirely new, either. Basket-weavers have long known that a basket surface woven with hexagons needs to be closed at the corners with pentagons (Figure 8.11). This is where ethnomathematics, which examines traditions, meets the results of the most modern research into the structure of matter.

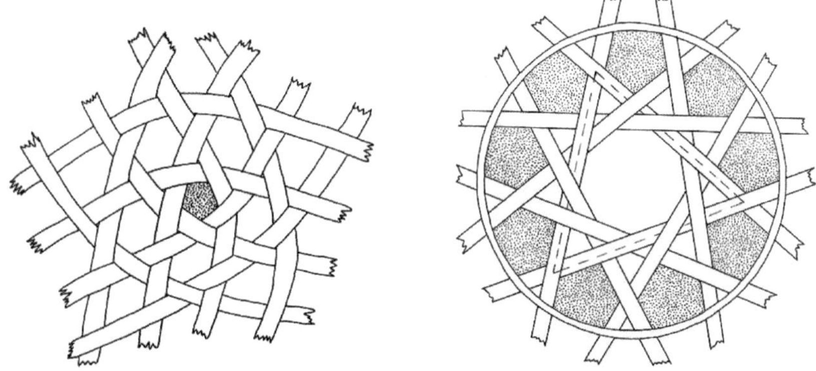

Figure 8.11. *The weaving pattern of a basket from Mozambique; a Japanese basket seen from below, with three triangles bounded by nine pentagons*

It was precisely the experience of Far East basket-weavers that led S. Iijima (researcher at the NEC company, Japan) to the solution of the atomic structure that closed so-called tube fullerene at its ends. Tube fullerene, also referred to as a carbon nanotube, which he discovered in 1991, is a cylindrical carbon molecule covered with a hexagonal lattice, with six pentagonal closures at each of its ends (Figure 8.12).

Truncated icosahedral structures develop not only in geometry, but also through spontaneous self-organization in the living world. The synergetic principle holds true here, too. In nature, the pursuit of the greatest stability is achieved, among other things, through attempts to minimize the ratio between surface area and volume. This is what viruses strive for, as well as multiplying cells (e.g. morula), the cell arrangement of developing tissues, the surface structure of pollen, radiolaria, and so on.

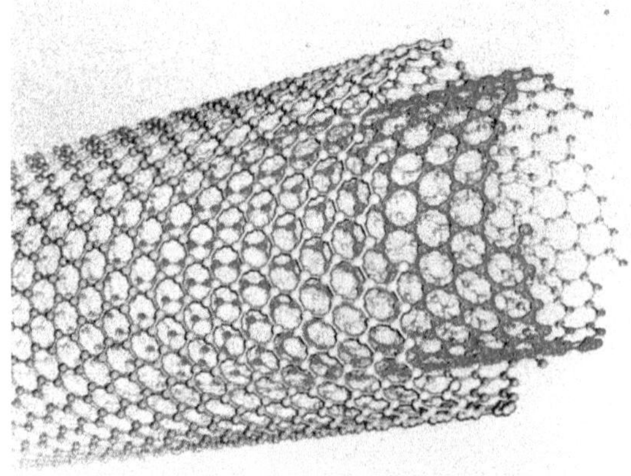

Figure 8.12. *Tube fullerene*

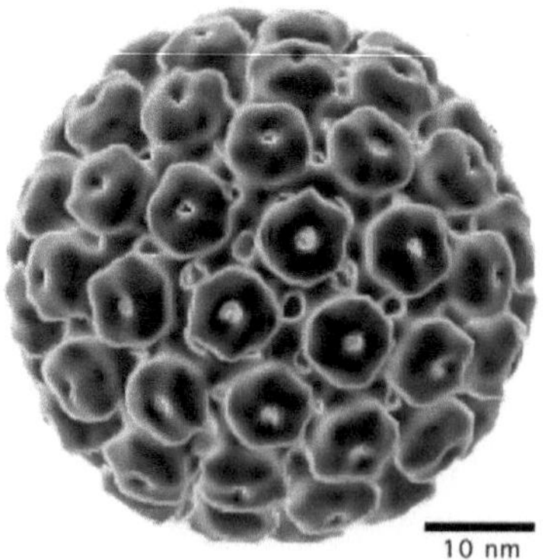

Figure 8.13. *Virus (Polyoma) with an icosahedral surface structure*

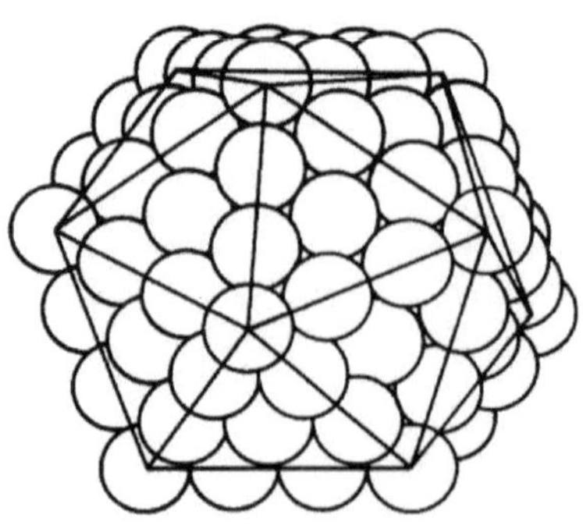

Figure 8.14. *Filling an icosahedron with spheres*

On many electron microscope images of viruses it can be observed that the skeleton of their near-spherical surface displays an icosahedral design (Figure 8.13).

Let us compare this to the filling of an icosahedron with congruent spheres. We recall from the above that the hexagonal arrangement of spheres in space is only possible in a fullerene structure. The spheres located at the vertices of the icosahedron are each surrounded by five neighbours, while all the other spheres on the surface are surrounded by six (Figure 8.14).

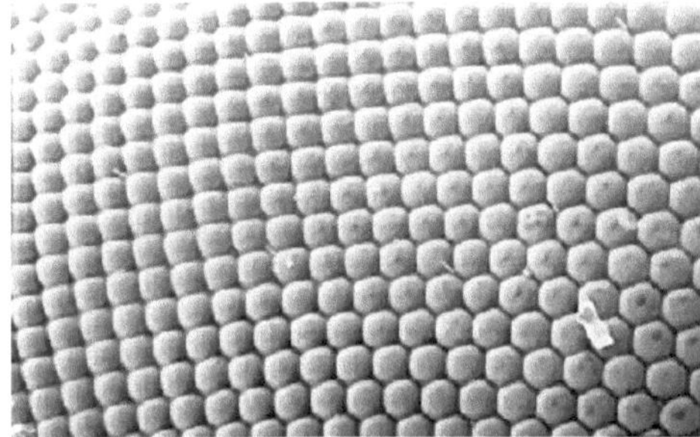

Figure 8.15. *The arrangement of the tissue cells of the eye of* Calliphora erythrocephala

The cells which make up the tissues also strive to fill the space at their disposal in the optimal way. This is well illustrated by the distortion of the surface cells of the eye tissue of the fly *Calliphora erythrocephala* depicted in Figure 8.15. At the left edge of the picture, as a result of inevitable perimeter conditions, the cells still display a quadratic arrangement, but moving towards the bottom right corner of the picture they increasingly transform into a spontaneous hexagonal arrangement.

They apply the same surface structure we saw in the case of honeycombs. The surface of the boundary of the honeycomb is made up of circular arcs, all of which form an angle of 120° with respect to the neighbouring cell walls. The honeycomb aims to form a minimal surface area, in which, in a general sense, it follows the principle of synergy. In this particular instance it is the optimum according to a physical variational principle that represents this minimum. It follows from the realization of the common principle that we encounter hexagonal designs in structures as various as the parenchyma of maize (Figure 8.16), the pigment of the retina of our eye, the cell design of honeycombs, and the shells of diatoms.

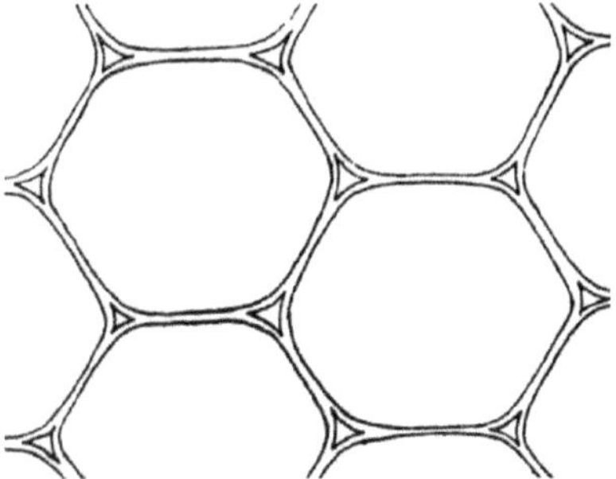

Figure 8.16. *The parenchyma of maize* (after H. Weyl)

Weyl discusses the numerous attempts made over the centuries to uncover the secret of honeycombs, from the tales from the Thousand and One Nights to Maraldi's first measurements in the eighteenth century, and from Réaumur's (1683–1757) conjecture to Koenig's mathematical proof of optimal cell capacity. He quotes Darwin (1809–1882), who considered the construction of honeycombs to be the most wonderful of known instincts, and according to whom "Beyond this stage of perfection in architecture natural selection (which now has replaced divine guidance!) could not lead; for the comb of the hive-bee, as far as we can see, is absolutely perfect in economizing labor and wax."

Analyzing the optimal polyhedral filling of space, Weyl devotes a thorough discussion to the tetrakaidecahedra (Figure 8.17) produced by the truncation of an octahedron. This gives us a polyhedron bounded by six squares and eight hexagons, which is a semi-regular body suitable for the filling of space without gaps. Archimedes (287–323 BC) was already aware of this fact, and Fedorov discovered it anew. Lord Kelvin (1824–1907) displayed how, by bending the sides of this body and curving its edges, the condition of minimizing surface area could be met, thanks to which a more efficient ratio of surface area to volume could be achieved than with the division into rhombododecahedral cells that was previously considered optimal. Weyl was in admiration of Kelvin's shape, which was thought to give us the minimum proportion. Since then there have been newer results (e.g. L. Fejes Tóth [1915–2005], B. Csákány [1932–]) on how to fill the honeycomb with a more efficient use of surface area for a given volume.

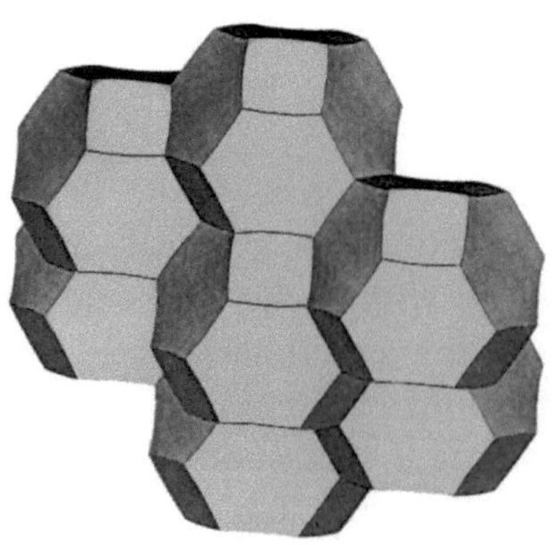

Figure 8.17. *Filling space without gaps using tetrakaidecahedra*

Returning to the subject of how the close-packing appears on the surface, we can observe the arrangement of the icosahedral pattern on the surface of pollen (Figure 8.18), or, for example, in the arrangement of the dandelion flower that has lost its petals.

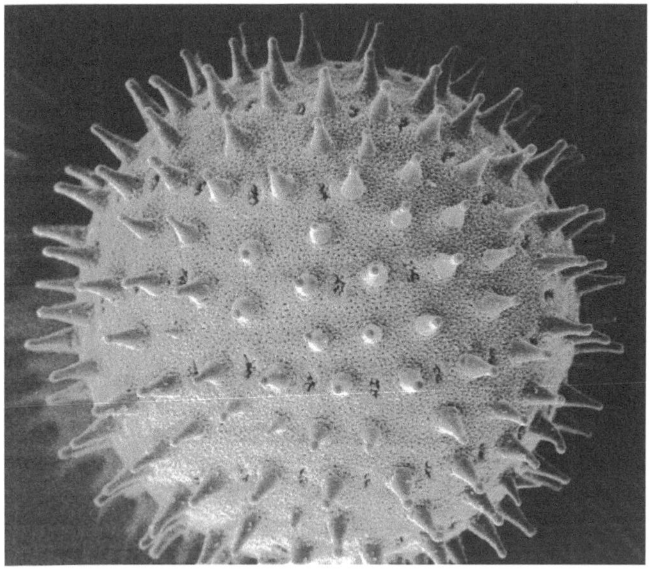

Figure 8.18. *An electron microscope image of the icosahedral surface of the pollen of Sildalcea malviflora*

Most radiolaria display a regular icosahedral structure. Haeckel, who in the last decades of the nineteenth century published a good few volumes on the forms of living creatures and their symmetries, in his book Challenger Monograph (1887) studies the collection from an expedition during which he discovered no less than 3508 species of radiolarian. As a result of their way of life, these little creatures which float in water are exposed only in the least possible measure to the effects of specific space directions (e.g. directions determined by gravity, light, and movement on the surface of the water). For this reason the equivalence of space directions plays a determining role in the development of their bodies — hence their symmetry. For movement and nourishment they require marked points on their bodies. The arrangement of these, however, is characterized by attempts at uniform distribution and a design that follows the placing of vertices in regular (or at least semi-regular) bodies. We find among them examples of body surfaces corresponding to all the regular polyhedra, but the surface's increasing size and the increasing number of cilia located on it demand an icosahedral arrangement (Figure 8.19).

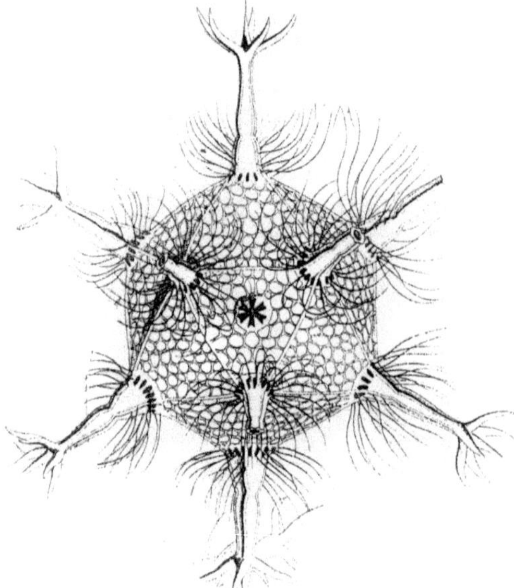

Figure 8.19. *Radiolaria* (Haeckel: *Challenger Monograph*)

In Figure 8.20 we see the diatom framework of the radiolarian *Aulonia hexagona*, also from Haeckel. The hexagonal patterns its name suggested are, however, interspersed on the surface with pentagonal (and heptagonal) ones. The same principle holds true as in the case of the development of fullerene: Euler's theorem holds that the surface of the body cannot be covered purely with hexagons, and in this case they are accompanied

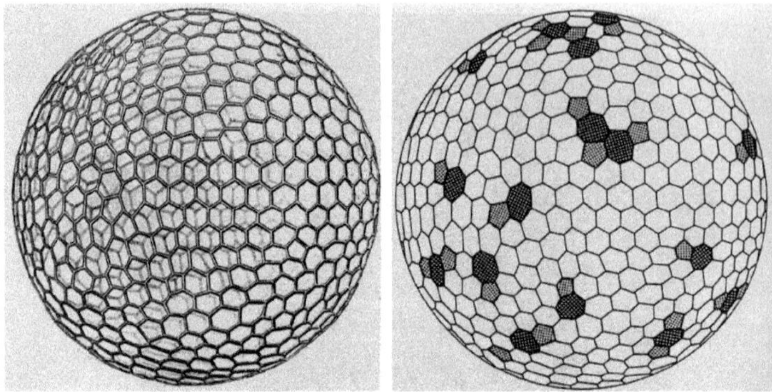

Figure 8.20. *The siliceous skeleton of radiolarian called Aulonia hexagona* (after Weyl) *and the pentagons and heptagons appearing on its surface*

sometimes by pentagons, sometimes by heptagons. The surface structure is very similar to that we saw for pollen, despite the fact that genetically they are very different from one another.

We can observe the same pentagons and heptagons on the surface of another radiolarian, as photographed with a scanning electron microscope (Figure 8.21).

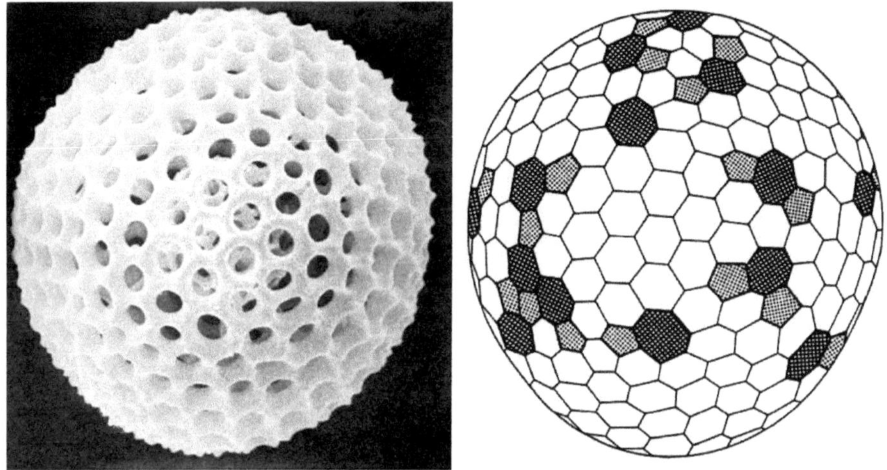

Figure 8.21. *Left, a scanning electron microscope image of a radiolarian; right, the pentagons and heptagons on its surface are emphasized*

Why is uniform surface coverage so important? One reason is that it ensures an optimal distribution of load. We might ask why we do not find the form of the surface of the five regular bodies everywhere. This would not allow the alternation of hexagons with pentagons and, as we have seen, even heptagons. The regular body with the largest number of planar surfaces, the icosahedron, only has twenty faces. In the majority of cases this is not enough. The faces of the icosahedron meet at quite a sharp angle, and surfaces bounded by more faces with less sharp angles between them, ones that are closer to hugging the surface of a sphere, bring advantages which outweigh the symmetry lost from having more types of face. Such surfaces are more like spheres, and better satisfy the requirement of uniform distribution of load.

This last consideration, and hence the synergy requirement, applies in the planning of load-bearing structures like halls, domes and large-scale antennas, which, in addition to bridging large spaces, have to bear the load of their own supporting structure.

The requirement of uniform load-bearing also holds, for example, for the design of the sewing, gluing and surface patterns for certain types of balls. Important considerations in the design of balls are the distribution of response to shock, avoidance of meandering movement to make trajectories predictable, and the symmetry requirement that wherever it is struck by the force, should fly with the same conditions. The optimal stability for the sewing pattern of a soccer ball given its size, for example, is provided by a truncated icosahedral fullerene structure with sixty vertices (Figure 8.22). For this reason, since the middle of the 1970s almost all soccer balls have been produced with this design.

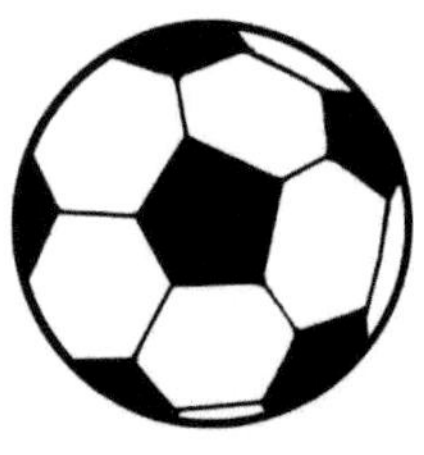

Figure 8.22. *Sewing pattern for a soccer ball*

As load and stability are both a question of measure, it does not necessarily hold that this surface pattern ensures maximum stability for balls for other uses and of other sizes. The tennis ball, which is much smaller but exposed to a much stronger impact, has for decades been stuck together from two rubber shapes in the form of an 8. When we place this shape on a plane, let us observe how similar it is to the design of this wallpaper from ancient Egypt pictured alongside it (Figure 8.23).

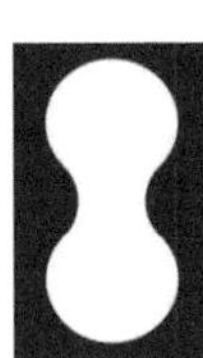

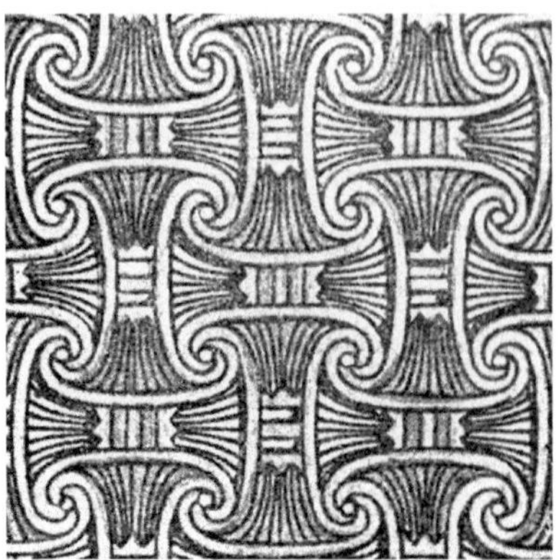

Figure 8.23. *The pattern for a tennis ball in the shape of the number 8 (the ball is made by sticking together two of these), and an ancient Egyptian design tiling a planar surface that displays similar symmetry*

Golf balls are poured from hard material in a dense fashion. In order that the club should not slip on them, their surface is made rustic with small indents. These holes cannot be very large, however, for this would contradict the requirement that the ball can be struck from any direction under the same conditions. Thus circular indentations following the arrangement of a regular body are not suitable on account of their small number. However, as we know, a sphere cannot be covered in uniform fashion with a greater number of circles. To achieve an optimal size and pattern, golf ball producers have brought many different designs to market over the last decades. We see some of these, from the collection of Tibor Tarnai (1943–), in Figure 8.24.

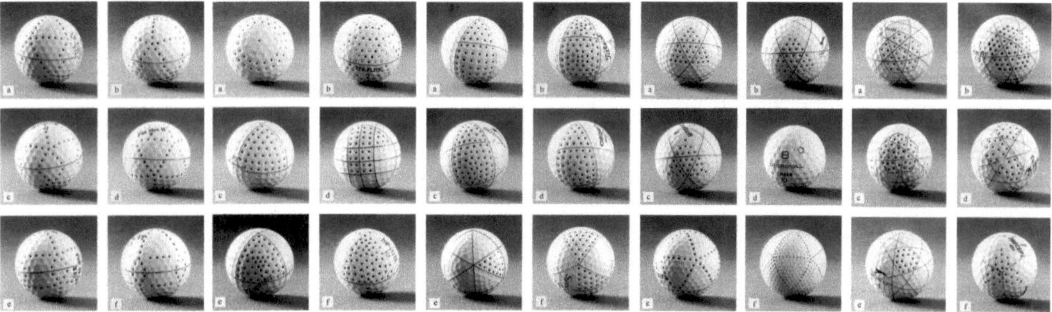

Figure 8.24. *Golf ball designs*

In the figures, it can clearly be seen from the marks on the individual balls that some have more and other fewer so-called 'special' indentations which are not surrounded by exactly six others in a uniform way. It is not hard to understand that if we place a large enough number of indentations in the golf ball (the size and approximate number of which have over the years become standardized), then the twelve 'special' ones, which come to be on the ball in line with the fullerene structure, will be of almost negligible number relative to the 'specially' located indentations on other arrangements. For this reason it is this design that proves the most symmetrical.

The examples mentioned show the many applications of the truncated icosahedral/dodecahedral structure (and in particular that with n = 20 hexagons) to the most diverse areas of existence. This structure got its name from Robert Buckminster Fuller. Fullerene became the 'official' name later used with regard to the C_{60} molecule, but Fuller's friends and students who called him by his nickname refer to it as the Bucky-ball. Fuller was primarily an architect, and looked to this structure as one embodiment of the synergetic principle. He wanted to devise a structure with maximal endurance from minimal elements, following the geometrical and load-bearing optimum. In 1951, on the basis of the drawing shown in Figure 8.25, he patented the structure for the geodesic dome that would later be named after him.

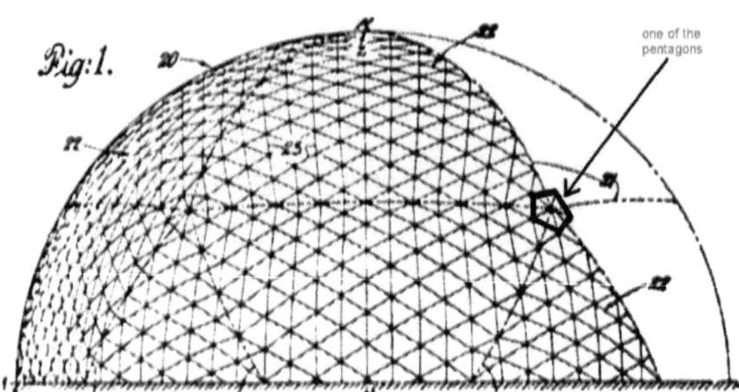

Figure 8.25. *R.B. Fuller's patent design for the geodesic supporting structure* (1951)

For Buckminster Fuller, the term 'geodesic' refers to the shortest distance between two points on a spherical surface. Figures 8.26 and 8.27 show even more clearly the structural details of the geodesic dome, and the alternation between regular hexagonal and pentagonal load-bearing elements, first on a drawing, then in the process of construction of a geodesic structure.

Figure 8.26. *The structural framework of a geodesic dome. An arrow shows one of the pentagons*

We owe it to historical accuracy to point out that the first known geodesic dome was built in 1922 by Walter Bauersfeld (1879–1959) as the supporting frame for the Zeiss planetarium in Jena, Germany. This is considered to be the first lightweight steel-frame lattice structure. The geodesic dome structure is clearly visible on the photograph taken during the construction (Figure 8.28). It cannot be seen in the final building, however, as the supporting structure has a concrete layer on it which hides the framework underneath.

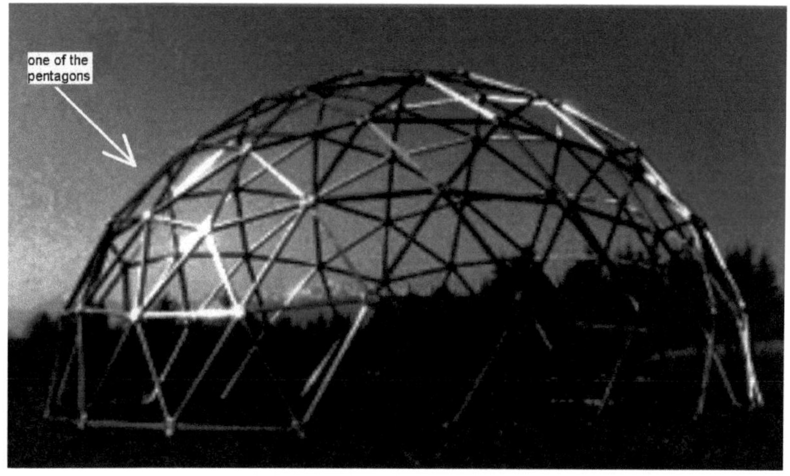

Figure 8.27. *A geodesic structure being built*

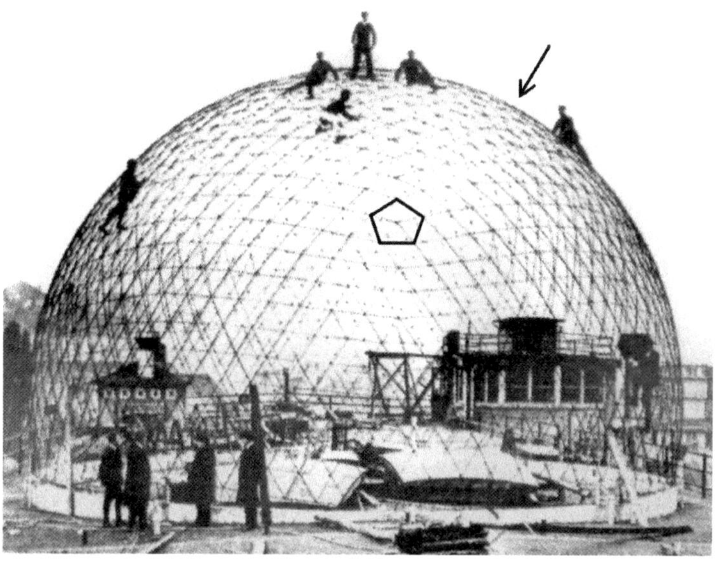

Figure 8.28. *The first geodesic dome* (Walter Bauersfeld, Zeiss planetarium, Jena, 1922)

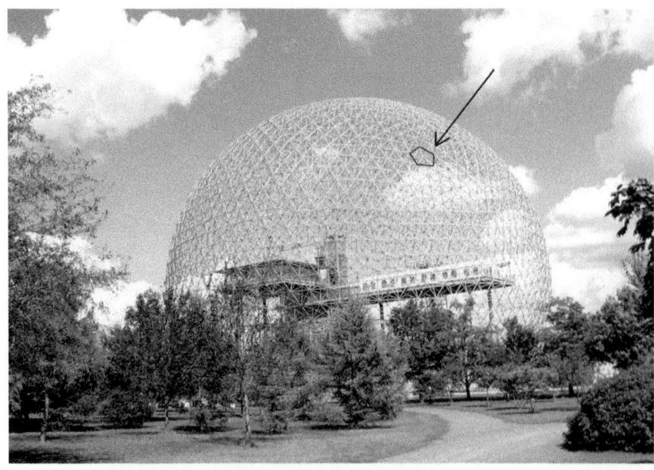

Figure 8.29. *B. Fuller's pavilion for the world expo in Montreal* (1967)

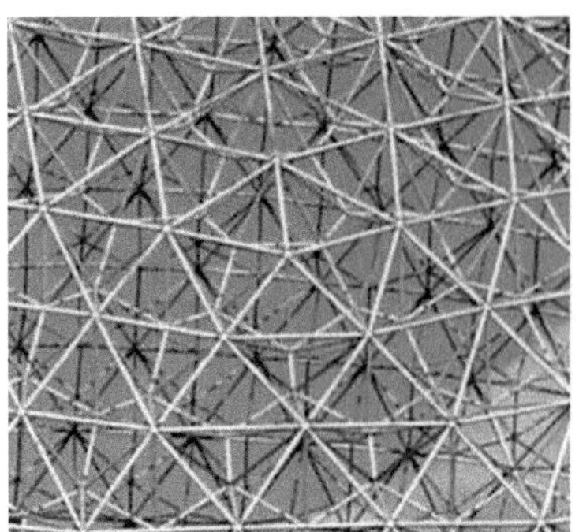

Figure 8.30. *A detail of the Montreal dome*

The best-known example of the geodesic dome structure, which brought worldwide recognition for the structure and its designer, is the hall R. B. Fuller built for the 1967 World Expo in Montreal (Figure 8.29), where the elements of the frame are clearly visible from outside (indeed, in the years that followed, they are all that have remained of the building). The hexagonal lattice of the surface structure and Fuller's two-layer supporting tetrahedron solution are clearly visible in the close-up photograph in Figure 8.30.

With the spread of the construction of lightweight buildings, the fullerene framework became a relatively common solution (Figure 8.31) in the construction of domes (in Mexico City alone, for example, there are around half a dozen such roof structures).

The Montreal dome has a special place in the history of fullerenes because it was this monumental building (or at least the many photographs taken of it) that made the structure known to the public at large around the world. It was this dome — and, indirectly, the calculations which were the

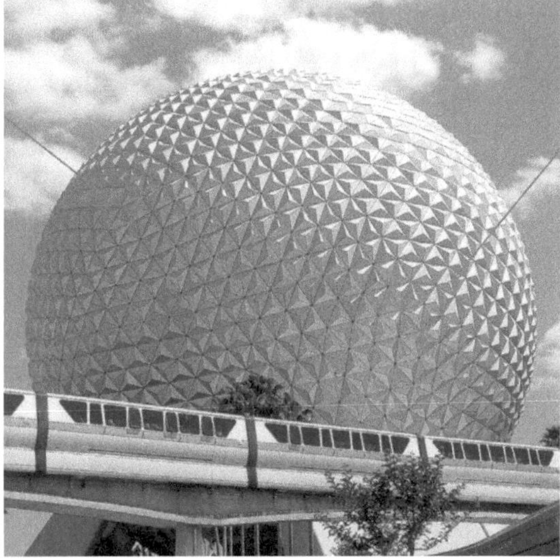

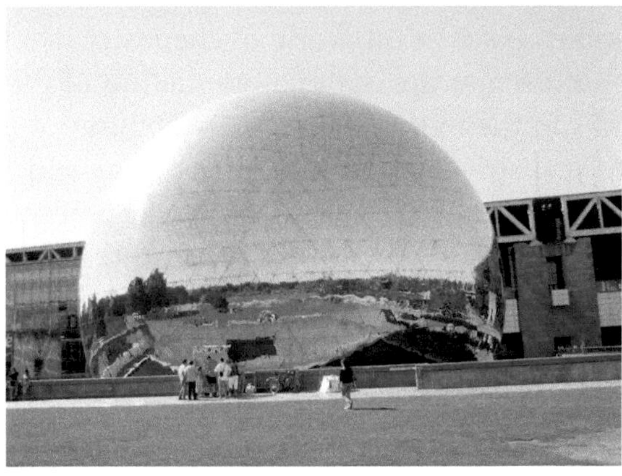

Figure 8.31. *Geodesic dome in Disneyland, Florida, and the Dome Villette in Paris*

basis for its stability — which started to generate awareness of the fact that it was the same principle of synergy as common force for forming space that lay behind all the examples we have just listed. It is this same principle which is behind the structures of viruses, the morphology of marine radiolaria, the design of a golf ball, basket-weaving, the sewing pattern of a soccer ball, Buckminster Fuller's dome framework, and, as we will see, the stability of the bonds of the C_{60} molecule.

A principle of this kind acts as a heuristic force on the path to other scientific discoveries. In general, we can say that symmetry considerations have often played a role in scientific discoveries. An example of this was the dream of Kekulé (1829–1896) about little devils holding hands and dancing, which led to the discovery of the ring structure of benzole. While in retrospect many debate the authenticity of this heuristic dream — Kekulé only mentioned it decades after the event, and the story exists in public knowledge in a number of versions — it is certainly worthy of mention, to support the claim that heuristic symmetry observations can play a role in scientific discovery. It was an insight inspired by symmetry — albeit a rather less romantic one — which led J. D. Watson and F. C. Crick to the final identification of the structure of the DNA molecule, and we will see the heuristic role of the eightfold way of Buddha (c. 560–480 BC) in the discovery of the symmetries in physical microparticles and their classification.

Synergy found its place in the world of chemistry in a natural manner. In the bonds of molecules the optimal orientation of the polarization of charges is (thanks to the optimal energy distribution of electrons) in the direction of the real spatial arrangement of atoms and the direction of the lattice edges of the crystals. Thus if the Bucky-ball's structure displays optimal load-bearing (e.g. in the case of a supporting frame, the supporting bars withstand the greatest load by pointing in the direction of the least tension), the same can be true for a molecular structure made up of atoms. We knew of a good few examples from organic chemistry of the large-scale connection of identical atoms. But, as far as we knew at the time, these did not produce closed spatial surfaces, only chains and rings. There was, therefore, the theoretical possibility that a fullerene structure might come about from carbon atoms, as we saw above, if a carbon atom were to be located in each of its vertices, with a single bond along the edge of every pentagon and a double bond along the common edges of the hexagons.

Yet the discovery did not happen as the result of a conscious investigation based on this logic. The background to the discovery also involved researchers who were ahead of their time, but to whom attention was not paid. The Japanese scientist Osawa published an article in 1970, as did Soviet researchers Bochvar and Galpern in 1973, independently of him, which discussed the theoretical possibility that such molecules might come into being. The former was published in Japanese in the journal *Kagaku,* the latter in Russian in the chemistry section of *Dokladi Akademii Nauk SSSR.* For researchers exclusively educated in Latin script, not even the captions were comprehensible. Retrospectively, after the discovery of fullerene in 1985, notice was taken that there had been these two articles on the subject, but in their own day neither were there any references that drew any attention to them, nor did their theoretical predictions inspire any experiments.

The discovery itself took place at Rice University, seemingly by accident. Together with British guest researcher H. Kroto, R. Smalley and his colleagues found carbon molecules of large mass in the course of the laser-bunched vaporization of graphite. The mass spectrum of carbon molecules is displayed in Figure 8.32.

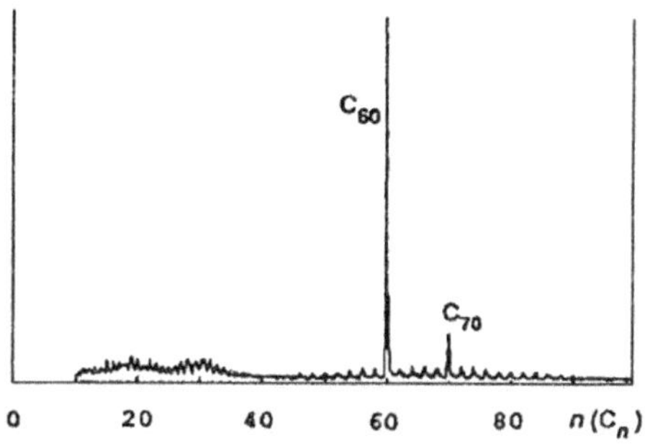

Figure 8.32. *Mass spectrum generated by the laser scopiform vaporization of graphite at Rice University, on 6 September 1985*

Of these, the carbon molecules with the most common mass were C_{60}, C_{70}, C_{84} and C_{540}. At first they did not themselves know what they were up against. The events of the first hours and days are remembered by Tibor Braun (1932–) in his book *The Glorious C_{60} Molecule,* as follows:

"[...] They were evidently thinking of the hexagonal structure of graphite when they attempted to put together a polyhedron with sixty vertices exclusively from hexagonal faces, but at the time they did not know that Euler had already proved in the eighteenth century that it was not possible to construct a regular, closed polyhedral shape merely out of hexagons. Kroto had a vague memory that the geodesic dome he had seen at the Montreal World Expo in 1967 was made up of hexagons and pentagons, and even that he had once put together a shape for his son from hexagons and pentagons, and he seemed to remember that the shape had had exactly sixty vertices. If the time difference had not meant that in England it was late at night, he would have rung his wife to ask her to look around for this old toy. In the early hours of Tuesday morning, Smalley attempted to construct this shape at home, first with computerized molecule modelling, then, when this did not succeed, with hexagons and pentagons cut out of paper. His efforts were rewarded, and the next morning Kroto saw that the almost spherical model Smalley had built was the same as that which he had made for his son many years earlier.

"It seemed unlikely to Smalley that such a regular shape could be unknown in geometry. He asked the head of the mathematical institute at Rice University, who soon gave him the answer: of course it is long well-known, and, what is more, the (modern) soccer ball is precisely this shape. Now the biggest task was to find a name for C_{60}. The following ideas were brought up: footballene, soccerene, soccerballene, etc., until finally they agreed on Kroto's suggestion, buckminsterfullerene. True, it was a little long, but it was pretty conspicuous and paid tribute to an eminent architect and scholar [Fuller died two years previously – G.D.], whose work had, however indirectly, had an influence on the discovery of the structure" There was no mention of Leonardo or Piero della Francesca.

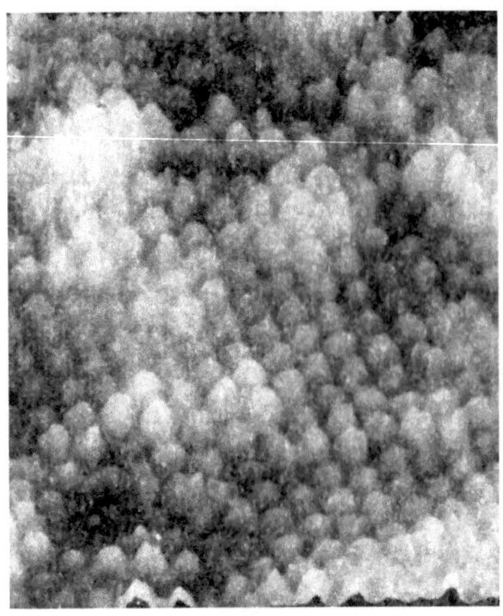

Figure 8.33. *An image of carbon globules taken with a scanning electron microscope*

Figure 8.33 shows a photograph of carbon globules made with a scanning tunnel electron microscope. In 1991 fullerene became the molecule of the year, while in 1996 its discovery was rewarded with the Nobel Prize.

These were individual, solo molecules. Their mass production for industrial use (1988) is associated with the name of W. Krätschmer.

The future of fullerenes is determined by the possibility that they could be used to produce a greater number of molecules and compounds than exist in organic chemistry. The large number of dual bonds makes it possible to connect atoms and groups of atoms to any of the atoms of the fullerene, whether from inside or outside. If only two atoms are connected to two carbon atoms of C_{60}, this results, the symmetrical arrangements notwithstanding, in a very high number of molecules that differ stereochemically. Increasing the number of connected radicals, meanwhile, causes exponential growth in the number of molecules that can be created. Fullerene can also be used to lock molecules inside it, rather like inside a cage, so they can be transported securely before being released. This primarily presents opportunities for pharmacology.

Even today, fullerene research is the branch of chemistry developing the most quickly. After quasi-crystals, it is the second good example from the last two decades of a landmark discovery born out of symmetry considerations, and of the influence of an architectural design on a scientific idea.

Symmetry in Inanimate Nature

Chapter 9

Cosmological symmetries

The symmetries of polyhedra were primarily rotational. Turned around one of their axes of symmetry, they would overlap themselves. A body displaying n-fold symmetry around a given axis will, after n rotations, not just overlap itself in terms of the space taken up by its shape, but really return to exactly the same position: if we mark a particular point on it (e.g. the one we use to rotate it), it will return to the same place. If we continue to repeat the rotations of angle $2\pi/n$, on each nth occasion we will return to the same position (though, unless we place a mark on the polyhedron, we will not be able to distinguish the intermediate stages from one another). During the rotation, that is, the shape will periodically repeat itself.

It is not only when rotating regular bodies that we encounter *periodically repeating* phenomena. The majority of natural processes belong to the set of periodic phenomena. As time passes — as it completes a uniform translation, that is — we can depict the phenomenon along a closed curve as a series of states that revolve again and again. In this way we can determine that periodic phenomena embody the combined presence of rotational and translation symmetry.

Let us place a little light bulb on the edge of the wheel of an upside-down bicycle, and spin the wheel. If we look at it from the side, the light of the bulb describes a circle. If we watch the movement of the light from the plane of the wheel (for example from behind) from a reasonable distance, we see that it moves up and down between the highest and lowest points of the wheel. If we depict the movement in time of this point of light in a coordinate system, with distance from the centre on the vertical axis, and the passage of time on the horizontal axis, we get a sine curve. If we spin the wheel with uniform angular velocity, then one unit of angular rotation is associated with one unit of time, and the horizontal axis also measures the angular rotation of the given point. The curve returns to its

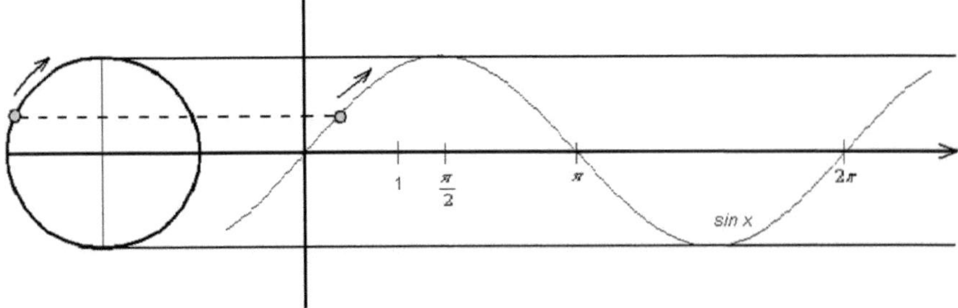

Figure 9.1. *Rotation and translation symmetry. Sinusoidal movement*

original state after precisely 2π rotations, i.e. after completing a full circle (Figure 9.1).

We can observe many phenomena in nature that return to their original state at regular intervals. This is true of the alternation of days, months, seasons, years, and the return of comets. For the peoples of Middle Eastern cultures, the alternation of floods and harvests behaved likewise. We can also add to the list changes in the biological cycles of plants, animals and human beings, and reproductiveness.

With regard to cyclically repeating natural phenomena, the question arises of whether they are really periodic in a matching way. The answer might seem very simple at first glance, but we have to add a reservation: taken one at a time, they match, but not compared to one another. Let us consider, for example, how the changes in the light of the moon are incommensurable with the length of the year, or days with the length of the year, from which it follows that while the return of solar and lunar eclipses can be calculated, they do not display uniform periods, especially not within one generation. The fractional units of months and days compared to the length of a year have not, however, deterred human beings from producing calendars. Their world-view — the faith in symmetry — was yet again stronger than their empirical experience.

Cosmological symmetries primarily emerged during the preparation of calendars. It is precisely as a result of these discrepancies, of the necessity of approximation, that various different ages and civilizations have left behind so many different calendars. Their richness and diversity is a significant treasure of human civilization. They show how many different ways the same periods can be approximated and depicted with the same natural phenomena.

As well as measurements and calculations, the preparation of calendars required making, recording and organizing observations over long periods, over many human generations. Neither the measurement of time nor that of distance were easy tasks, not to mention the other units of measurement that can be traced back to these, like angles. The preparation of calendars meant the harmonized ordering of days, seasons, years and the periodicity of lunar and solar cycles. As an example, Figure 9.2 shows (as interpreted by A. Marschack) bone engravings of cavemen of the phase changes of the moon from about 30,000 years ago (after K. Simonyi).

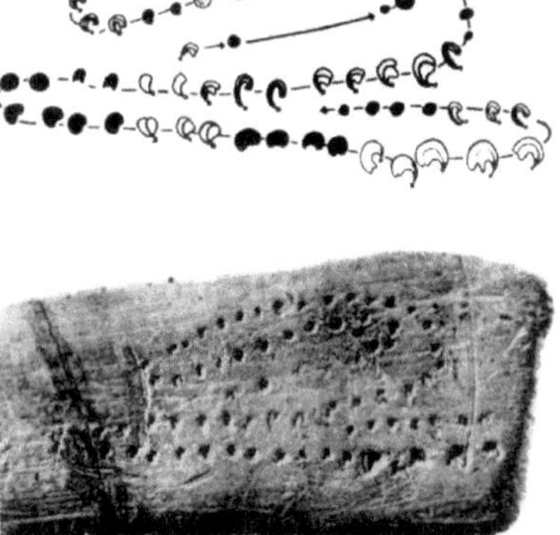

Figure 9.2. *The phase changes of the Moon as depicted around 30,000 years ago*

Phenomenologically, it is relatively easy to observe and note the movements of the sun and the stars in the sky, but marking the changing position on the horizon each day of sunrise and sunset, for example, is a rather more complicated task. Similarly: it is relatively simple to determine the location of the sun in individual constellations, but harder to register the height of the sun's culmination point or fixed stars in the sky. Let us try to think how it would have been possible to determine the moment of midnight with the tools of our ancestors 3,000–5,000 years ago. Yet we can still say that the most advanced science (in the modern sense) of the peoples of early antiquity was astronomy. This sophistication was a result precisely of the periodicity (i.e. calculability) of this set of phenomena.

The peoples of ancient cultures had a developed view of the geocentric solar system model, just like they did of the heliocentric one, of the orbits of the planets, and even, albeit to a limited degree, of galaxy models. To understand their models, we have to establish some notion of the precision of their measurements. The angle-width of the sun in the sky is around half a degree. In principle, with their instruments they could measure even a quarter of this. The question remains: how? Exact astronomical observation requires measurements to be taken at the same time every day. What time would this have been? Furthermore, in a given place, the measurement for one day had to be compared to that of the previous day. How were they able to store the previous day's result? How were they able to compare measurements taken in different locations? Árpád Szabó (1913–2001) dealt with these questions in detail. Let us here illustrate the problem with a single example: how was it possible, more than 2200 years ago, to determine the size of the earth?

Eratosthenes (275–194 BC ?) made his measurement, which today we would rather call an estimate, as follows (Figure 9.3). At the summer solstice, the sun had reached its culminating point above his head in Syene (today Assuan). He measured the angle of the sun at its highest point on the same day in Alexandria, which lies almost exactly to the north of Syene. He found that this value was a fiftieth of the entire circle. Thus the circumference of the earth is fifty times the distance between Syene and Alexandria. What was harder was estimating this distance. He knew that a caravan of camels could travel this distance in fifty days, and that the caravan would on average cover a hundred stadia in one day. With this, he made an estimate of the earth's circumference which is not too far from the value accepted today. Of course, we do not know the precise value of the distance of 250,000 stadia, as the Greeks as well as the early and late Egyptians all measured the length of a stadion (stadium) as a different number of feet, but even the extreme values are not a world away from the data we have at our disposal today (K. Simonyi).

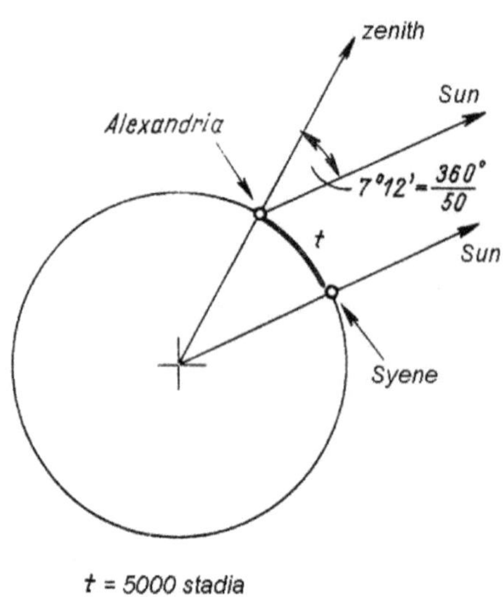

Figure 9.3. *Eratosthenes' measurement*

The precision of these measurements give us a reference point for understanding the astronomical symmetries of the ancients. The changes in

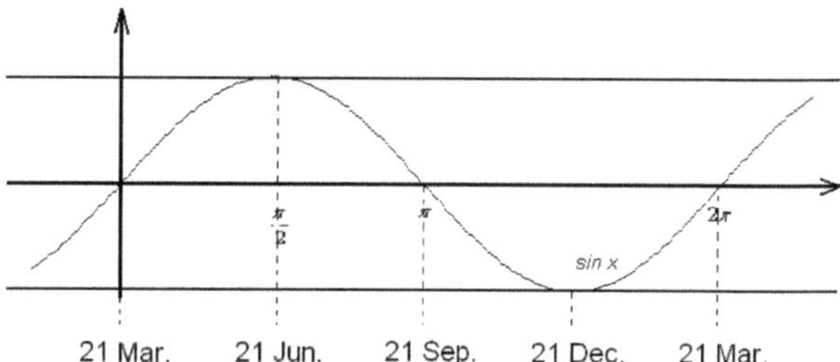

Figure 9.4. *The sun's culmination point during the year*

the culmination point of the sun over the year can be characterized with the sine curve seen in Figure 9.4.

In the roughly 90 days from the spring equinox to the summer solstice the sun's highest point rises by nearly 23.5°. This is just one degree away from a quarter of a right angle. We could say that they were able to measure such a difference in angle exactly, but in order to draw the curve, they had to measure the difference not every three months, but every day. 23.5° works out at an average of around ¼° a day: 23.5°/90 = 0.26° (ignoring the fact that the change in highest point takes place not in linear but in sinusoidal terms). The change in angle corresponding to a quarter of a right angle would be 22.5°/90 = 0.25°. As they were not able to measure with an accuracy of 1/100 of a degree, they could reasonably have assumed that the change in the sun's height in the sky from December to June, between the two tropics, was precisely half a right angle (180 times a quarter degree). Figure 9.5 shows the angle of the sun above the horizon in both extreme instances, projected onto the sine curve.

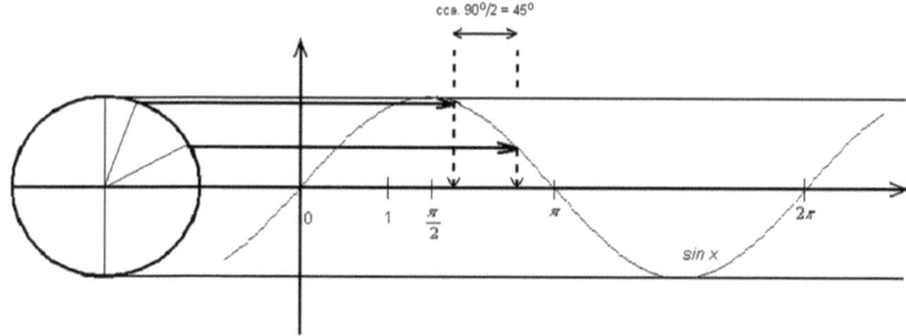

Figure 9.5. *The angle of the sun above the horizon at the summer and winter solstices. The changing height of the sun in the sky between the tropics of Cancer and Capricorn*

According to this depiction, the apparent movement of the sun relative to fixed stars also displays a symmetric shape. The tropics cut the distance between the equator and the poles almost into four. Making use of our current knowledge, the polar circles also lie at a quarter of this distance. The approximate symmetries of the earth can be seen in Figure 9.6.

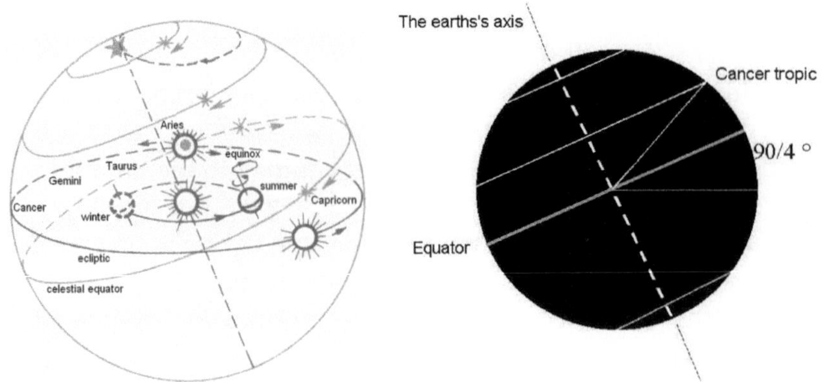

Figure 9.6. *The sun's apparent yearly movement across the sky, and the approximate symmetries of the earth*

All this makes it easy to understand the selection of the angle scale most widespread to this day. The sun goes through an entire cycle in about 360 days. Its daily movement relative to the constellations is about 1°. Thus in the space of a year it covers an angle of 360° (Figure 9.7).

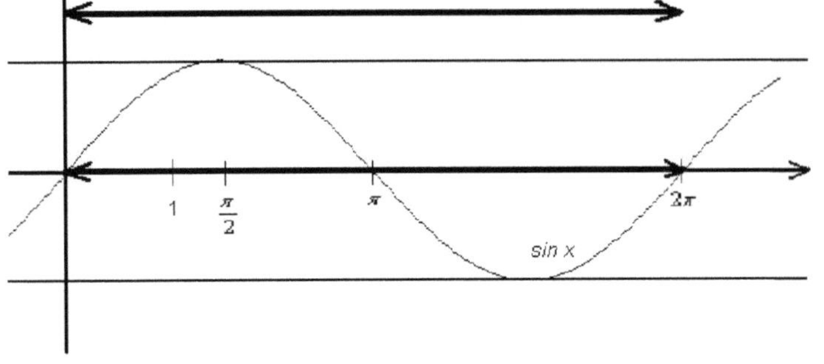

Figure 9.7. *Choice of degree scale (1)*

Its apparent vertical movement is about a quarter of this. At its culmination, the sun on average rises or drops 1° every four days. Selecting an angle scale along both the vertical and horizontal axes, the movement of the sun's culmination point over the year can be traced as in Figure 9.8.

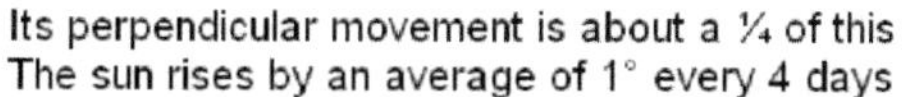

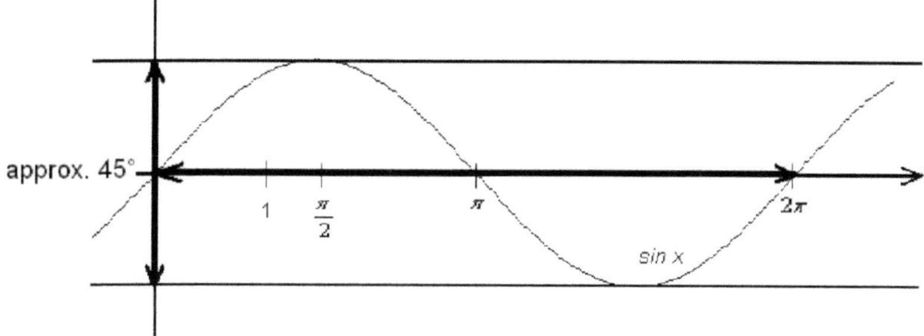

Figure 9.8. *Choice of degree scale (2)*

In order to appreciate the extent of this, let us note that in Szeged, which lies about a degree and a half to the south of Budapest, on average over many years, and ignoring other climatic factors and deviations, spring will start six days earlier and autumn six days later. Such deviations can be seen, for example, in the time at which plants flower, or the date chosen for harvests to begin.

It was not only in the preparation of calendars that world-view and experience clashed. There have been other respects in which symmetry considerations have had an influence on the development of our cosmological world-view. Our concepts of the earth and of the world have all generated models with continuous rotational symmetry (e.g. discs or spheres). This is no accident, as it was these that were in harmony with the observed phenomena of periodic nature which displayed continuous change. We have seen how geocentric and heliocentric world-views were both well known, but different ages placed the emphasis on one or the other. In the historical chapter we referred to the fact that Copernicus was able to choose between two modes of description that were essentially equal in rank. Empirical observation was on the side of the more complicated geocentric world-view displaying cycloid planetary orbits. Reason, a world-view consciously constructed, pointed towards the heliocentric world-view serving approximately circular planetary orbits. Copernicus' choice was of simplicity, perfection, of symmetry. As we have also seen previously, it was this struggle for perfection, the absolute priority of the guiding principle of symmetry which led Kepler to his results. The heuristic significance of the principles was greater than its scientific precision: despite his more or less inaccurate starting points (the circular orbits of the planets, the assumed relationship between the radii of their paths and the radii of the spheres inscribed in and circumscribed around perfect bodies, the relationship between of spheres and the harmonies of music) it proved to be a productive working hypothesis for finding the right laws.

Kepler regarded the number of known planets and their known moons as absolute and irrevocable. The six planets were perfectly suited to the number of spheres which could be inscribed inside or around the five regular bodies. Kepler compared the proportions of the radii of the spheres drawn in and around the regular bodies nested in one another with the proportions of the radii of the neighbouring planetary orbits (Figure 9.9).

Kepler's proportions	The ratio of the spheres inscribed inside and circumscribed outside the given regular body	The ratio of the radii of the orbits of two neighbouring planets	
		1	Saturn
cube	0.577	0.672	Jupiter/Saturn
tetrahedron	0.333	0.290	Mars/Jupiter
dodecahedron	0.795	0.658	Earth/Mars
icosahedron	0.795	0.719	Venus/Earth
octahedron	0.577	0.500	Mercury/Venus

Figure 9.9. *Table of the proportions of the planetary orbits*

On the basis of this approximate conformity, he determined the orbits of which planets could be identified with the spheres drawn between which bodies, i.e. the order in which the regular bodies had to be nested inside one another (Figure 9.10).

Although all these deductions seem unscientific by today's standards, the approximately correct proportions provided him with a heuristic model. The symmetry of the model did not guarantee its correctness, but it helped to find mathematically precise laws. It is a sign of Kepler's intellec-

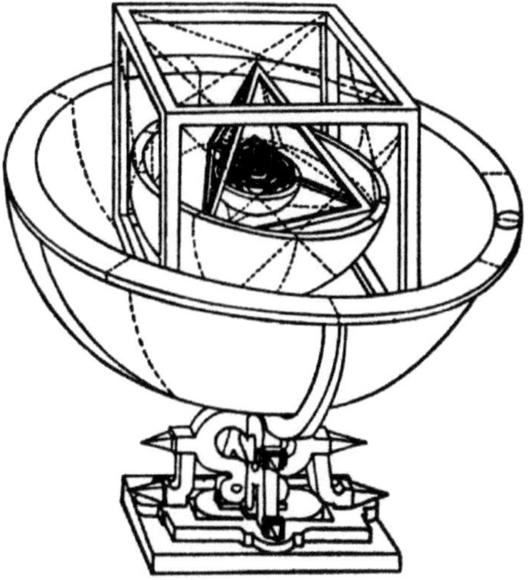

Figure 9.10. *Kepler's drawing of the nesting of planetary orbits in regular bodies and the spheres that can be drawn around them*

tual greatness that he rose above his former hypotheses and accepted the results dictated by reason: elliptical planetary orbits. It was in this way that the scientist of the modern age was distinct from that of the ancient one: the result of calculations made on the basis of experimental measurements became a more powerful argument than intuitive symmetry.

The elliptical planetary orbits, on the other hand, gave science a new kind of symmetry. The discovery of the laws of gravity in principle gave meaning, under given initial conditions, to movements along all conic sections.

We succeeded in locating the earth in the solar system. Later we located the solar system in the hierarchy of the galaxies in the universe. With telescopes we discovered order in the star clouds. The observation of distant star systems gave us the beauty of spiral symmetry. We know that we, too, are found on the arm of such a spiral. The swirling movement of the stars is regulated by the same laws as the ones we can observe on the surface of the earth in the cloud systems of meteorological cyclones (Figure 9.11), the eddies in the water of a river, or the sediment of stirred sugar at the bottom of a cup.

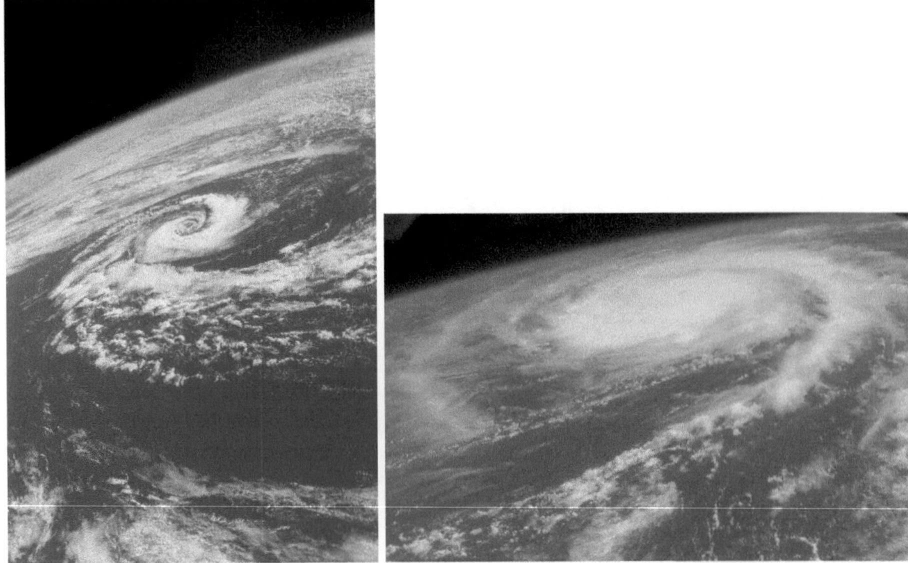

Figure 9.11a,b. *The spiral arms of atmospheric cyclones*

Images taken by NASA astronauts; publication courtesy of NASA

(a) STS109-E-6003 (March 10, 2002) — The astronauts on board the Space Shuttle Columbia took this digital picture featuring a well-defined subtropical cyclone. The view looks southwestward over the Tasman Sea (between Australia and New Zealand). According to meteorologists studying the STS-109 photo collection, such circulations are recognized as hybrids, lacking the tight banding and convection of tropical cyclones, and the strong temperature contrast and frontal boundaries of polar storms. The image was recorded with a digital still camera.

(b) ISS009E20440 (August 27, 2004) — This photo of Hurricane Frances was taken by Astronaut Mike Fincke aboard the International Space Station as he flew 230 statute miles above the storm at about 9 a.m. CDT Friday, August 27, 2004. At the time, Frances was located 820 miles east of the Lesser Antilles in the Atlantic Ocean, moving west-northwest at 10 miles per hour, with maximum sustained winds of 105 miles per hour. Fincke, the NASA ISS Science Officer and Flight Engineer, and Expedition 9 Commander Gennady Padalka are in the fifth month of a six-month flight aboard the Station.

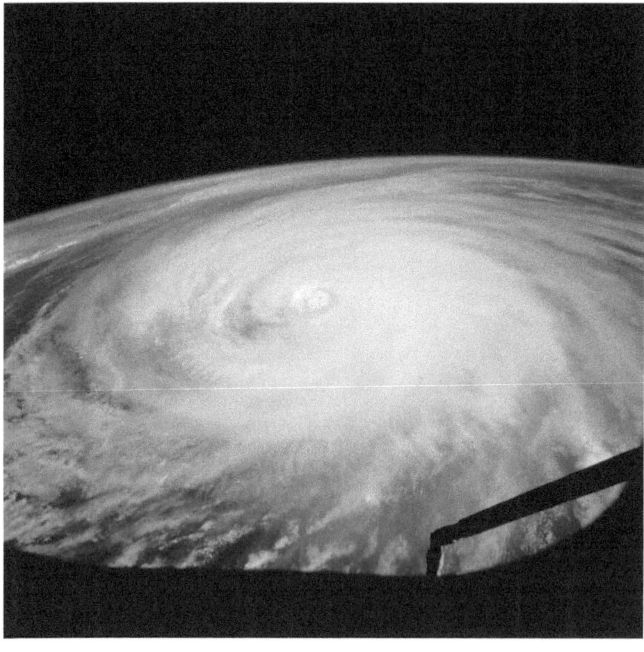

Figure 9.11c. *The spiral arms of atmospheric cyclones* (cont.)

(c) STS069-731-031 (September 9, 1995) — Hurricane Luis is captured on film in its latter days in the Caribbean in this 70 mm frame. During the 11-plus day mission, the astronauts aboard the Space Shuttle Endeavour caught with their cameras at least two large oceanic storms. A second one later in the flight, named Marilyn, followed a similar path, leaving havoc in its wake on several islands. Endeavour with a five-member crew, launched on September 7, 1995, from the Kennedy Space Center (KSC) and ended its mission there September 18, 1995, with a successful landing on Runway 33. The multifaceted mission carried astronauts David M. Walker, mission commander; Kenneth D. Cockrell, pilot; and James S. Voss (payload commander), James H. Newman, Michael L. Gernhardt, all mission specialists.

Chapter 10

Sight and Hearing

The harmony of the world of colours and sounds

Periodically repeating processes form a very broad class of natural phenomena. We encountered the symmetries of periodically repeated patterns during the mathematical and crystallographical description of friezes and wallpaper motifs. They are to be found as decorative elements almost everywhere in our environment. When we were at elementary school, we decorated our notebooks with serial designs. It is periodically repeating patterns that decorate the façades of most houses, our fences, railings, and this is how lampposts and rows of trees are laid out. Periodic decorative designs could be found in all periods in all folk decorative arts. They were used as much for decorating personal items and textiles as for decorative products made specifically to be looked at. Their fashion changed from age to age and culture to culture, with their characteristic features making it possible to identify and distinguish ancient Persian designs from Greek ones, or decorative elements of the European classicism of the seventeenth and eighteenth centuries from the *art deco* of the twentieth.

The set of periodically repeating phenomena includes light, sound, and rhythm, to name but a few. Among phenomena displaying translation symmetry, it is those that change continuously that are in the majority, and these can be related to some wave motion. This is shown by a number of experiments (some of which can be conducted at home), and in cases where we do not have direct empirical evidence for this, physicists can nevertheless describe them as wave phenomena and treat them mathematically. The fact that these descriptions lead to a result that squares with experience allows us to deduce that the wave description was justified.

In the case of cosmological phenomena, we looked at the periodically repeating movements of mechanical bodies revolving along a closed curve. Another set of periodic phenomena is associated with oscillations. In terms of their physical description, the two phenomena are very similar. We saw with the example of the movement of the bicycle wheel that in a certain projection the circular motion appears like oscillation, and also that, projected in time, circular motion and oscillation can be described with a sine curve. We can feel the periodicity of our heartbeat, and we know that this is the result of the continuous repeated contraction and extension of our heart muscles. This is the movement that the cardiologist projects on our electrocardiogram on a distorted sine curve, modulated by the operation of the heart. What we hear is a few distinct moments from this continuous motion, those when the cardiac valves open and close; the motion of our muscles is in fact continuous. If we observe one of Mars' moons continuously with a telescope, we see this as an oscillation, too: it moves first in one direction relative to Mars, then back in the opposite direction, towards the other extremity of its orbit. We can deduce that this apparent oscillation is in fact the projection of a circular motion from the fact that in one direction the moon appears in front of the planet, while in the other it is behind it.

We can also consider waves as oscillations that move in a direction in space in addition to their temporal periodicity. They display combined symmetry, superposing spatial periodicity and temporal translation. We distinguish two essential forms of wave.

One of these groups is made up of *transverse* waves. As their name suggests, the two movements they are composed of are perpendicular to one another. The direction of vibration is at right angles to the direction of the wave's translation. A well-known example is the wave travelling along the surface of water, the vertical amplitudes of which travel horizontally across the water's surface.

The other group of waves are *longitudinal* waves. Here the direction of oscillation is the same as that of the progress of the waves. The best-known example of a longitudinal wave is the sound wave. If we use our mouths to make the particles in air vibrate, denser and thinner stretches of air alternate, depending on whether the air coming out of our mouths compresses the molecules to a lesser or greater extent. The direction of the pressure is in the direction away from our mouths. This is the direction in which the air "vibrates", condensing and thinning. This direction is also that in which the sound spreads.

This classification is based on one particular characteristic. We could group waves according to other ones. According to what it is that vibrates, for example. The movement of mechanical bodies or particles with periodic changing of place or thinning or thickening, as we have seen in the above examples, produces mechanical waves. If the vibrating or wave movement is not produced by the mass of the body, but for example by the periodic changing of (the vectors of) an electromagnetic field, we call it an electromagnetic wave. In this instance the wave involves no visible mechanical alteration or movement. It is by measuring its effect that we can deduce the presence of the wave. Examples of electromagnetic waves are light or radio waves.

Electromagnetic waves

We group electromagnetic waves according to their wavelength (Figure 10.1). Present knowledge suggests that they cover about 14–15 scales from 10^{-10}m to around 10^4m.

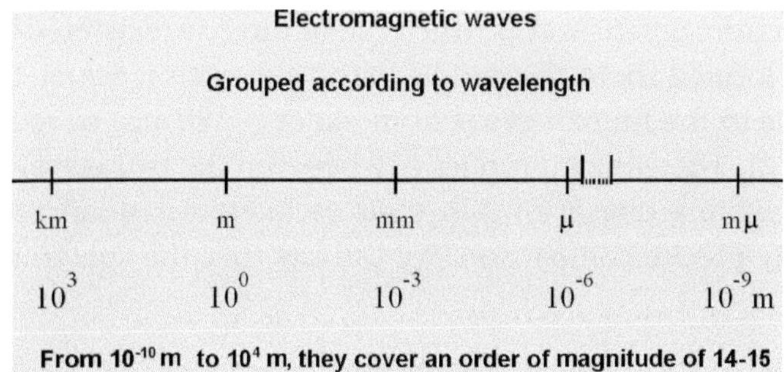

Figure 10.1. *The wavelength of electromagnetic waves on a logarithmic scale, marking the wavelength range of visible light*

The electromagnetic waves that fall into the various different wavelength domains have different physical effects, and are thus given different names. Table 10.1 shows the spectrum of electromagnetic waves and the names of the waves that fall into the various categories. We can also see from the table how there is no sharp boundary between the different electromagnetic waves: their spectra flow into one another in continuous fashion. We distinguish them on the basis of their physical effects, the phenomena they generate. The boundary values marked in the table are not exact or absolute boundaries; indeed, different sources quote boundary value data that vary slightly.

~km – ~ m	long, medium and short radio waves
~cm – ~ mm	microwaves
~mm – ~ 3μ	thermal radiation
~mm – 25 μ	far infrared
25 μ – 0,76 μ	near infrared
0,76 μ – 0,38 μ	**range of visible colours**
0,38 μ – 0,3 μ	near ultraviolet
0,3 μ – 0,2 μ	far ultraviolet
0,2 μ – 2 mμ	extreme ultraviolet
2 mμ – 0,1 mμ	soft X-ray radiation
0,1 mμ –	hard X-ray radiation and *gamma* rays

Table 10.1. *The names given to the various parts of the electromagnetic spectrum*

The electromagnetic waves which can be directly registered by our sense organs are located about the middle of the logarithmic scale. The range of light visible to the human eye is from violet (~380 mμ) to red (~760 mμ) (Figure 10.2). There is a multiplier of 2 between the two wavelengths (this is equivalent to a divisor of 2 in their oscillation frequency). To take a not entirely precise comparison, we can say that the spectrum of visible

The spectrum of visible light

from violet (380mμ) to red (760mμ)

multiplying wavelength by a factor of two,
and dividing frequency by a factor of two:

1 octave

with 7 basic colours in-between:

red, orange, yellow, green, cyan, blue, violet

the transition is contiuous

Figure 10.2. *The spectrum of visible light* (see colour version on page 491)

light covers about an octave. Although there are no clear boundaries between colours on the scale in-between the two extremes, and the colours flow into one another, our eyes nevertheless distinguish sharp and mixed colours. By convention, we usually highlight seven basic colours from the spectrum. These are red, orange, yellow, green, cyan, blue and violet, which we also refer to as the colours of the rainbow.

The spectrum of the rainbow covers the entire range of visible colours. Figure 10.3 shows the spectrum of the sun's light stretched along a strip according to the wavelength of the colours found in it. At the two edges a part of the ultraviolet and infrared scale appears, which we cannot "see" with our eyes, and so perceive as black (colourless). The clear, dark lines in the spectrum are the absorption bands of the material in the solar corona and the Earth's atmosphere. The light at the wavelength this material absorbs does not reach our photographic plate, and so a black line appears.

The spectrum of the rainbow

The spectrum of the sun. At the two edges, part of the ultraviolet and part of the infrared scales are visible. The black lines in the spectrum are the absorption bands of the solar corona and the material in the earth's atmosphere.

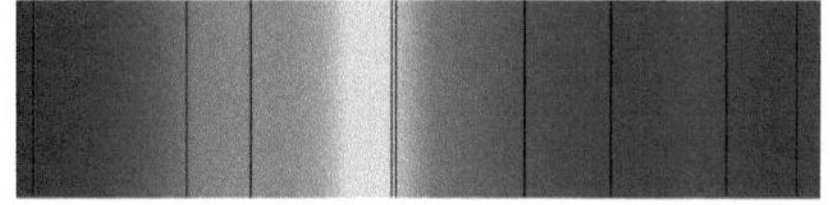

Figure 10.3. *The spectrum of the rainbow*

We see an object as being of the "colour" of the wavelength of the rays (electromagnetic waves) reflected from the electron shells of the atoms on their surface. There is great variation in individual sensitivities to colour vision. Firstly, the boundaries of the spectrum are not the same for all. It varies from person to person which of the different shades of red or violet they see and which they do not. Secondly, we do not all see the strength of individual colours, and thereby the composition of certain mixed colours, in the same way. If we look at the connected, uniform colour spots of particular objects first with one eye, then with the other, we can observe slight differences: the colour vision of the two eyes of the same person is not identical. Outside the edge of the spectrum we can sense further frequency ranges, but not with our eyes. We feel them indirectly thanks to their biological effect, for example that of the heat rays of the infrared range, or the tanning effect on our skin of ultraviolet rays. The range of the vision of certain living creatures differs from that of humans. Those which sense infrared waves are better at orientating themselves in the dark, and, for example, it is because of a different range of colour vision that the bull is more responsive to the colour red.

Our brain is capable of distinguishing many tens of thousands of shades of colour. In the light of this it is surprising to find that our eye in fact only senses three colours, from which our brain "mixes together" and reconstructs the colours in question. There are three types of little socket in the retina of our eye, which sense the red, blue and green from the incoming light. Our eye breaks the incoming light into these three colours, and our consciousness blends the approximate original wavelength of the light falling on a given point, and the sensation of colour associated with it. These sensations of colour are not an absolute physical property (like a wavelength, which can be given a value), and subjective experiences, memories and impressions can influence the feelings that become associated with them, whether we consider them beautiful or uncomfortable, vivid or vague. We can present and sense the various colours by breaking them down into the proportions between the three fundamental basic colours (analysis) or reconstructing this (synthesis). This is what we term biological colour addition.

A lack of all three colours generates the sensation of the colour called black. The equal presence of all three colours generates the sensation of the colour called white. It is well known that any colour can be broken down into its components, and that the spectrum of the components can be presented independently of each other if a light of that colour is shone through a refracting prism and captured on a screen. The basic colours can be chosen from such a spectrum, and we can use them to experiment with mixing colours of different wavelengths. This phenomenon is based on a particular feature of waves, namely that the amplitudes of waves of differing wavelengths (frequencies), when combined, sometimes complement one another, and sometimes cancel one another out. This is how they can be used to produce waves of other frequencies. This type of solution is suitable for experimentation, but the generation of particular colours is tricky.

To win colours from light sources, we normally use colour filters, which let through wavelengths of certain colours, but not others. It is more common, however, to make use of light reflected from the surface of objects. We cover the objects with a material whose atoms have electron shells that can reflect light of a given frequency or combination of frequencies. We call this kind of material paint. Its preparation was once a secret kept by shamans; today a whole industry is involved in it. Multiatomic molecules or alloys of these can generate a wide spectrum of the various mixed frequencies. This is what gives our natural environment its richness of colour.

The reconstruction of colours with paints is not always a ready solution, either. If we have to transmit the information about the reconstruction of a particular colour to a distant location, it is not enough to describe its colour and shade (even if we have enough expressions in our language to express the richness of the mixed colours) — we would have to include a recipe book for each point of the image, describing what is needed to produce the paint suitable for blending the colour required. We do not even have words to denominate all seven basic colours of the rainbow (while we have words for other colours, e.g. brown). The word cyan refers to a compound substance that reflects mixed colours as light of this colour. The words violet or orange refer respectively to the plant or fruit which reflect light of this colour to our eyes. In point of fact, these are items of chemical information: the composition of the materials found on their surface has atoms with electron shells that, when lit, are capable of emitting electromagnetic rays of this wavelength. We encounter just such a "recipe transfer" problem when we transmit visual information via the medium of television, and we are faced with a similar task when we want to save, transfer, retrieve and revisualize colour information on a computer.

In these instances it is best if we turn for help to the wave properties of colours and the designs of nature — in this instance the construction of our eyes. We break each individual colour down into three basic colours, code the saturation of each according to an agreed scale, then transmit these codes. Where necessary, we can synthesize the coded colour anew from the components of the three basic colours. This is what happens on a television screen, where the molecules of three layers sensitive to three different colours appear just as they do on the three different types of socket in the retina of our eye. The sensation of colour for each point of the im-

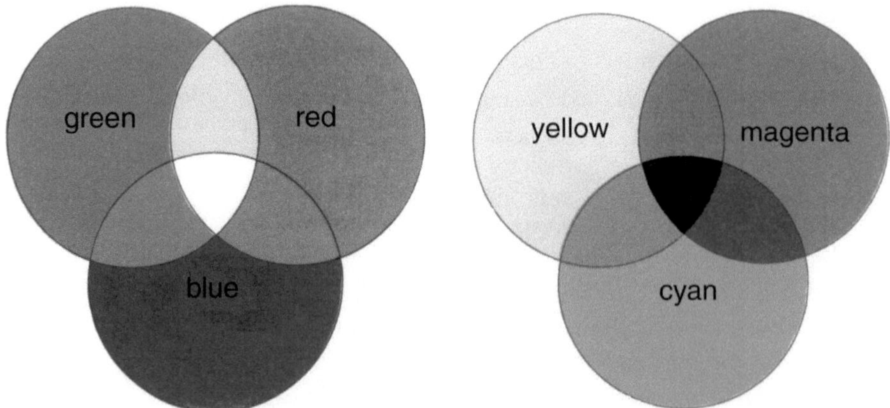

Figure 10.4. *Colour synthesis with addition*

Figure 10.5. *Colour subtraction*

age is made up of the effects of the three different light waves amplifying and cancelling each other. The three light waves each display a nice, regular, periodically repeated, i.e. symmetrical, sine wave — the sine waves of the oscillations of the electromagnetic field. Even the (multicolour) light wave considered the most asymmetrical can be broken down into three symmetrical waves (Figure 10.4).

It is not only in the way described above, with so-called colour addition, that we can produce colours from symmetrical waves (i.e. ones that are regular, sinusoidally periodic electromagnetic), but also with so-called colour subtraction. Just as yellow can be made with the combination of green and red, or magenta from that of blue and red, and white is the result of combining all three basic colours, we can also perform subtractions. By subtracting yellow and cyan we can produce green, by subtracting magenta and cyan we get blue, by subtracting yellow and magenta we get red, and by subtracting all three we are left with a colourless spot, i.e. black (Figure 10.5). This procedure is based on the same wave properties, and as capable of generating any colour as the previous one.

Thinking about all this, we can again wonder at the astonishing aesthetic richness that nature can generate from phenomena that display a few simple regularities, symmetries.

Mechanical waves

In contrast to electromagnetic waves, mechanical waves represent spatial oscillations of an entirely different type of material, namely mechanical matter. Their effect is also different. We can, however, discover some similarities, due to their symmetry properties and wave character.

In general, appliances that produce mechanical vibrations sooner or later cause the mechanical material surrounding them to vibrate, including the air. We sense the air particles made to oscillate as sounds, at least within a certain frequency range. For this reason we can regard the majority of devices that generate mechanical oscillations as musical instruments.

One way we can classify these instruments is according to how many dimensions the vibrating body oscillates in. We can further classify three-dimensional bodies in terms of the consistency of the material that is vibrated in the first instance. We list some examples in the right hand column of Table 10.2.

(1D)	strings	string instruments, piano, cimbalom
(2D)	plates	drum, cymbal
(3D)	– solid states	triangle
	– liquid columns	water organ
	– air columns	wind instruments, harmonica, organ, pipes

Table 10.2.

The common feature of each is that they transmit sound waves to our ears by vibrating the air that surrounds them, which transfers information to us on the one hand, and produces an aesthetic experience on the other.

The spectrum of sounds that can be heard by the human ear ranges from a frequency of around 40 Hz to one of around 20,000 Hz — with very considerable individual variations. The difference in the value of these extreme frequencies also changes with age. Vibrations of a lower frequency are sensed not by our ears but by our skin; vibrations of higher frequency than this range we call ultrasound. We can produce devices to generate such high frequency mechanical oscillations, just as we can use appliances to detect them. Ultrasound oscillations, transmitted from a single point and reflecting off various surfaces, can, if detected with devices, then processed and transformed with computers, be used to produce images. The medical uses of this are well known, but there are also many industrial applications.

The sounds within the 40–20,000 Hz range span 8–9 octaves (this is how many times the 40 vibrations per second frequency doubles within this range). Within this, if we vibrate a string with a given frequency typical of it, we can also sense so-called upper harmonics. (Upper harmonics are two, three, four, etc., times the given frequency, or half, a third, a quarter, etc., of it.) So we cannot really speak of pure sounds. Depending on the strength and composition of the upper harmonics, we can speak of the tone of the sound-producing device in question. The same is true of the human vocal chords. This is what gives our speech its unique characteristics. As mechanical waves display all the mathematical properties of waves that we encountered during the discussion of electromagnetic waves, they can also be combined, whereby their amplitudes reinforce and cancel one another, thereby creating rich possibilities for the generation of tones. It is no accident that we speak of 'tones' as referring both to sounds and to colours.

Let us add to the list of similarities the fact that within one octave we distinguish seven musical whole notes. Although the frequencies of the notes (as in the case of colours) represent a continuous spectrum, it is by selecting the sensation of these seven musical notes that we produce musical harmony. The "distance" between the notes and the proportion of their frequencies is the subject of mathematical and musicological examination. The Pythagoreans already defined harmony as the state in which the lengths of the strings relate to one another like the small integers. And the basic vibration count of a string is relative to its length. We consider this the first quantitative law of nature (Figure 10.6) to have expressed a physical fact in mathematical form. It was already known by the Egyptians, and its first appearance was in the illustration of the strings of the harp on the tomb of Rekhmire the Vizier of the 18th dynasty (Figure 10.7).

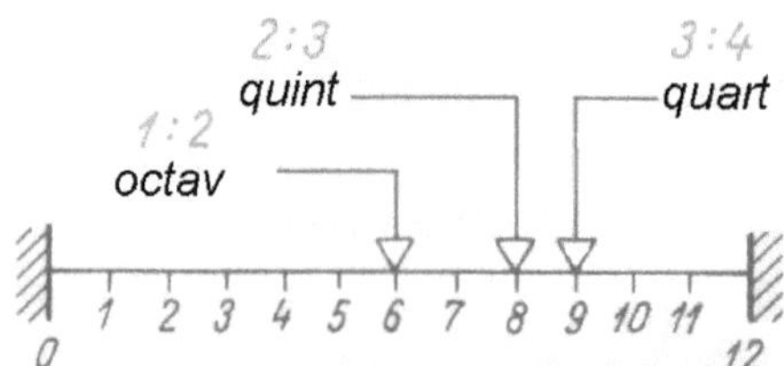

Figure 10.6. *The first quantitative law of nature*

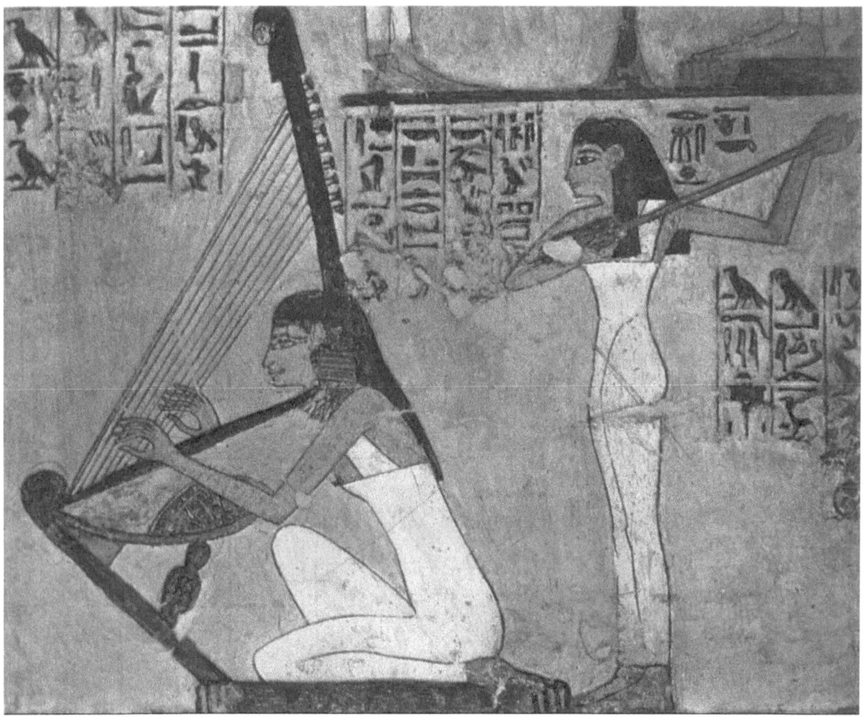

Figure 10.7. *A harp from the tomb of Rekhmire the Vizier* (18th dynasty)

The Pythagoreans defined the proportions of the quint (2: 3) and the quart (3: 4), which were consistent with the proportions of the lengths of the strings producing harmonic sounds, to the division of the octave, which also included half-tones.

It was the Pythagoreans who first referred to the world as the cosmos, which means beautiful, harmonic order. The cosmos, this harmonic, beautiful order, can, however, like the length of taught strings, be expressed in integers. The cosmos, the world is constructed according to their order. In research into numbers, it was they who succeeded in discovering the various proportions summarized in Table 10.3.

Although the various means are related to one another, it is the harmonic mean to which the musical proportions of taught strings are ultimately related (the proportions listed in the lower half of the table are not independent of those above). The world constructed according to

The search for the perfect proportion
(Pythagoreans)

Cosmos = beautiful order	arithmetic mean	$A = \frac{p+q}{2}$
	geometric mean	$G = \sqrt{pq}$
Harmonic beautiful order = the proportions of the length of taught strings	harmonic mean	$H = \frac{2p\,q}{p+q}$
		$(G = \sqrt{AH})$
	perfect proportion	$A : G = G : H$
	musical proportion	$p : A = H : q$

Table 10.3. *The Pythagoreans' search for perfect proportions*

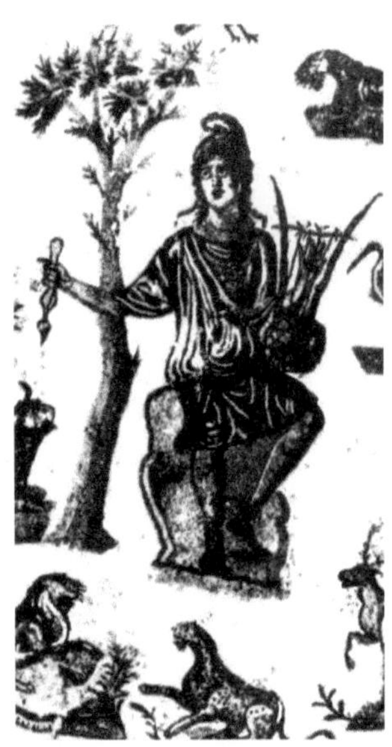

Figure 10.8. *Orpheus taming wild animals. Even wild animals resonate with harmony*

the harmonic order expressed by the lowest integers displays a close connection to the cult of Orpheus. The Pythagoreans operated in Southern Italy, and it was this region which witnessed the spread of the Orphean cult, which held that the reason that Orpheus was able to have an effect on both the living and the dead was that the whole world was constructed in harmonic fashion and thereby resonated to the sounds of Orpheus' music. Figure 10.8 shows Orpheus as he tames wild animals, because they too resonate with the harmony of the music. And harmony is numbers.

The study of the harmony (cosmos) constructed by the Pythagoreans from the proportions of the low integers not only had an influence on those spreading the cult of Orpheus. The music of the spheres that they had dreamed up also had a heuristic influence on the creation of Kepler's models at the turn of the seventeenth century. Lo and behold, we have again returned to cosmological models, in connection with a different class of periodic phenomena. The idea survived, however, and (on more than one occasion) went on to inspire Hindemith in the middle of the twentieth century.

The effect of the proportions of musical harmony also left its mark on twentieth-century science. The more than decade-long road to the elaboration of the theory of quantum mechanics, which began with the first models of the atom but particularly with the introduction of the quantum condition of Bohr (1885–1962), was paved with the study and systematization of the spectrum of electromagnetic waves reflected by atoms. We will no longer be surprised to hear that the proportions of the distance of the spectrum lines, i.e. their frequency, and thereby indirectly their energy, were the proportions of the small integers. It is this that led to the defining theory of twentieth-century physics, in which the solution of physical problems was served by so-called eigenvalue equations. These equations establish a relationship between functions describing continuous waves and the so-called eigenvalues, which are discrete numbers (usually small integers, that is natural numbers). This is how they became suitable for the simultaneous explanation and description firstly of the wave nature of matter, and secondly of the interpretation of discrete values measured in the structure of atoms.

In 1925, immediately after the birth of this new theory, one of the eminent physicists of the period, A. Sommerfeld (1868–1951), wrote the following: "The spectral lines governed by integers are essentially generalizations of the ancient triad of the lyre from which the Pythagoreans inferred the 'integer' harmony of natural phenomena 2500 years ago; our quanta remind us of the role that the Pythagorean doctrine seems to have ascribed to the integers, not merely as attributes, but as the real essence of physical phenomena."

There is no mention of any mysticism or esoterica, nor that the Pythagoreans might have anticipated something of the modern laws of physics. The suggestion was simply that there are universal regularities in nature which have an effect in very remote fields of knowledge, independently of one another, and sooner or later scientists become aware of them. Symmetry, the harmony of the cosmos, is one such means of intermediation which helps the discovery of similar regularities in different phenomena.

And the periodicity characteristic of wave phenomena is an example of such a symmetry.

Music

The development of musical culture went way beyond the followers of Orpheus, and, of course, has had an influence not just on physicists. The regularities of musical harmony can be displayed at various different levels of a given work. In addition to the *distance between notes,* they can be recognized in the work's *rhythm, melody,* and *composition.*

With regard to *distance between notes,* we saw above that the position of the frequencies of musical notes along the scale of sounds follows the harmonic proportion.

Musical *rhythm,* the harmony of repetition (i.e. translation symmetry) is found in well-known rhythm schemes. The alternation of shorter and longer notes and pauses within short units displays the clearly visible order of proportions between the low integers in the same way as the difference between the level of the notes. Rhythm is a physiological part of our lives: we hardly notice the rhythm of our heartbeat or our breathing. When we enjoy the harmony of musical rhythm, we in fact achieve harmony with the rhythm of our lives. Heightened rhythm, on the other hand, can bring us to a state of excitement, even ecstasy.

The harmony of repetition also presents itself in poetry. The rhythm of poems lies in their *rhyme,* their *metre,* their *rhyming schemes,* and their *alliteration.* In a well-known poem by Edgar Allen Poe (1809–1849), "The Raven", we find rhyme, repetition and even reversion (reflection) in a single oft-quoted line:

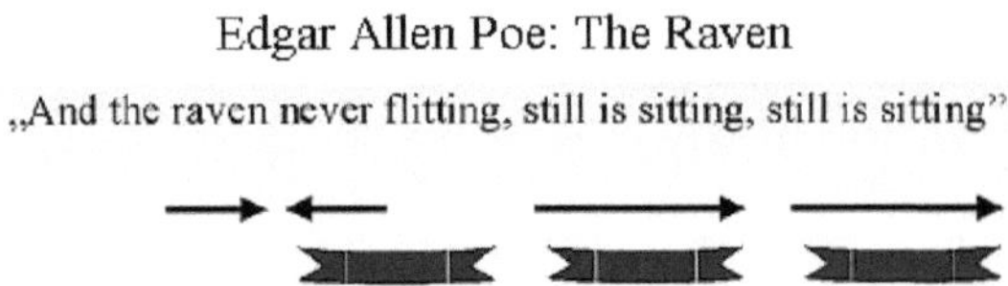

Figure 10.9. *Edgar Allen Poe: The Raven*

We can recognize familiar geometrical symmetry transformations in the way *melodies* are notated in a musical score. One example is the way

Béla Bartók (1881–1945) varies the first three notes in a line of his 24th Concerto using reflection, reversion, and gliding (translation).

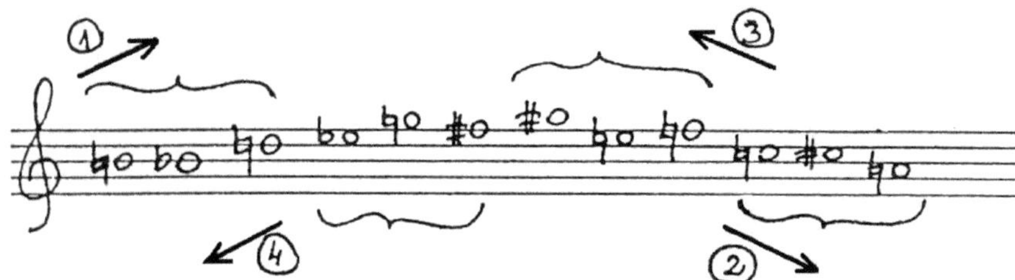

Figure 10.10. *Reflection, reversion and gliding in a line of Bartók's 24th Concerto* (after E. Lendvai and K. Fittler)

As shown in Figure 10.11 (after K. Fittler), we find two axes of reflection in two lines of the second cantata of Weber (1786–1826). We see the temporal reversion of the basic motif, then the inversion of the basic melody — the transposition of the rising and falling parts of the melody — and finally the reversion of the inverted motif.

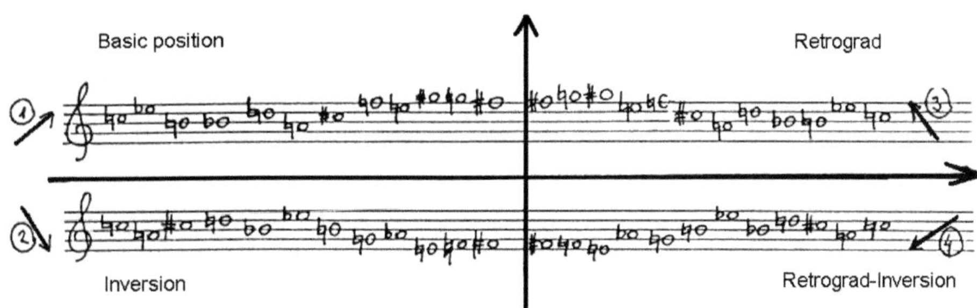

Figure 10.11. *The reversion and inversion of the basic melody, then the reversion of the inverted melody in Weber's Second Cantata*

We must mention an exceptional appearance of a musical application of the symmetry principle. The fugue is well known as a musical genre based on symmetry. The manuscript for J. S. Bach's *Kunst der Fuge* was left to posterity in mixed-up pages. It was not performed for some two centuries. In the 1920s, Wolfgang Graeser, a Swiss mathematics student (and A. Speiser's pupil), reconstructed the manuscript on the basis of geometrical symmetry considerations. It is in the form in which he put it together that we listen to the *Kunst der Fuge* to this day.

Finally, musical *composition* is the proportion of the various parts of the work to one another. The results which Ernő Lendvai (1925–1993) achieved in the study of this enjoyed world acclaim. I would like to take one example from his highly-respected life's work to illustrate the results that his research led to with regard to the role played by the Fibonacci sequence and the golden section in musical composition. Figure 10.12 analyzes and presents the way in which both the details and the entirety of the structure of Bartók's *sonata for two pianos and percussion* follows the golden section.

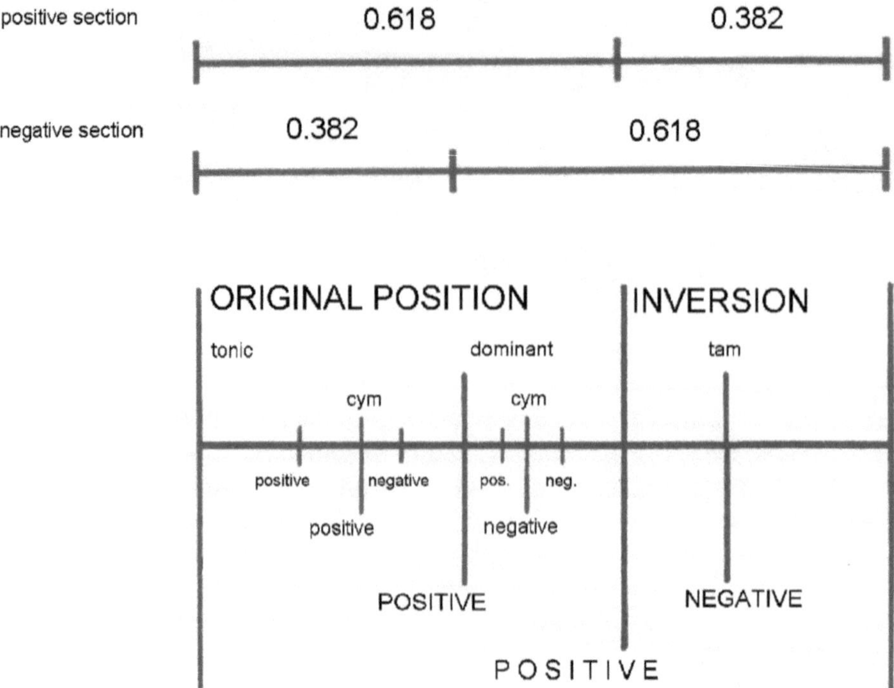

Figure 10.12. *The construction of Bartók's sonata for two pianos and percussion*

Chapter 11

Symmetries and symmetry breakings in inanimate nature

The generalization of the concept of symmetry in physics

If we still have any illusions left about the symmetric arrangement of the world, the world-view of physics will be enough to demolish them. The investigation of periodic phenomena briefly raised our hopes that the world might be arranged in a harmonic fashion. A deeper awareness of the physical structure of matter and its laws of motion questions the faith we have placed in the supremacy of symmetry. Is it possible that our belief in the beautiful, harmonious order of the cosmos was just an illusion, after all?

The entire history of physics is accompanied by symmetries. The whole edifice that is physics, and all of its individual theories, are built directly or indirectly on symmetry principles, and yet again and again we encounter symmetry breakings. The *physics saga* is a series of discoveries of symmetries, followed by their breakings; at deeper or more general levels we find new symmetries, only to be faced with their breakings at the next stage. Like the rhythm of a well-composed work of music, so a gleam of hope appears in physicists' eyes from time to time that they have happened upon a law containing the universal symmetry regulating all phenomena in the material world, and then they confront more symmetry breakings. There is no last beat to this symphony, however, after which the conductor can lower his baton with satisfaction and take a bow.

The study of the laws of nature coincided with the birth of modern science at the beginning of the modern age. While in Italy Galileo (1564–1642) studied natural phenomena in practice and determined the laws that governed them, in England Francis Bacon outlined the conditions for modern science on a theoretical basis. The foundations for a modern science based on experiments was laid down by Galileo's relativity principle. According to this, the result of our experiments does not depend on where and when we performed them (assuming, of course, that the conditions for them are unchanged), even if we switch to a system of reference moving at a constant speed relative to us. In Prague, Kepler may have encountered the same regularities that he had met in Italy, and his successors today may reach the same conclusions, with the greatest difference being the precision of their measurements. Galileo established the first symmetry principle: it is worth experimenting, because experiments can be reconstructed, repeated, and the regularities observed will be valid irrespective of the place, time and system of reference. This principle became the basis for experimental science. This same principle was outlined in a more general form by his contemporary Francis Bacon in England in his *Novum Organum*, in which he set out the philosophical foundations of scientific methodology, announcing a programme of science based on conclusions derived from experiments. His body of work was a triumph of the generalization of the principle of symmetry, and to this day our research activity is built on its subconscious application. For our research projects are justified by the fact that an experiment conducted in our laboratory, and its result, will be valid and applicable for other colleagues in every part of the globe, either today or in the future. This statement is true despite the fact that we only see this with today's eyes, and with today's concepts: at that time the concept of symmetry was not extended to philosophical principles, and it did not even find a place in the specialist terminology of physics.

The success of the principle continued into the seventeenth century with the framing of Newton's laws. His laws work within a homogeneous and isotropic world — which, as we know, expresses the symmetries of space — and they also comprise the first conservation laws, which represent invariance with regard to the passage of time. We have every reason to be glad about the successes of the early modern period in discovering the symmetries of the world.

The first tell-tale signs were already present at the very beginning of this era, however. We saw in our examples taken from the history of cosmology how passionately Kepler, Galileo's contemporary, pursued research under the influence of a belief in a perfectly symmetrical world. We saw his first cosmic model, which arranged the orbits of the planets around the Platonic perfect bodies. For him, the search for symmetry had a heuristic force: it helped him along the road to his scientific discoveries. However inspirational symmetry considerations may have been, his research convinced him that the world was not so perfectly symmetrical as he had at first supposed. Firstly, he had to come to terms with the fact that the planets revolve on elliptical rather than circular orbits, which in geometrical terms represent a less symmetrical shape. Secondly, he had to accept that while the elliptical orbit has two symmetrically placed foci, one of these is marked by the sun, while the other is empty. Thirdly — and this was the biggest shock to the symmetry of a world believed to be isotropic — the planets of the solar system all travel in the same direction around the sun, on planes that are close to one another, resulting in the angular momentum of each one's orbit pointing in almost the same direction, marking a spatial direction in the cosmos. In the spirit of the modern age, reason triumphed over the feelings that favoured symmetry: Kepler accepted the laws deduced from the experimental results even though they did not corroborate his preconceptions.

The next significant dissonance to trouble the world-view of physics came a good century and half later. This was the appearance of electric charges. The fact that there are two types of electric charge (positive and negative) was just the beginning. What fitted into the world-view with much greater difficulty was the fact that, while all previously known (mechanical) phenomena could be described with the help of the geometry of the world, electric charges assumed new kinds of fields. The realization of this was made up of the observation of many small phenomena over many decades, before finally taking general shape in the Maxwell equations, then in the discovery of the Lorentz invariance of these equations, before finally becoming a unified doctrine in the form of the special theory of relativity.

We previously mentioned the young Mach's reflections on the symmetries of electromagnetic fields and geometric space: how can it be possible that a physical phenomenon (electricity passing through the conductor) taking place along a plane (determined by the conductor and magnetic needle that are parallel to one another) can result in an effect that leaves this plane (the magnetic needle turns to the side), as seen in Figure 11.1 This can only be possible if electric charge and its effects cannot be described with our geometric concepts, or at least not in the same way as in mechanics.

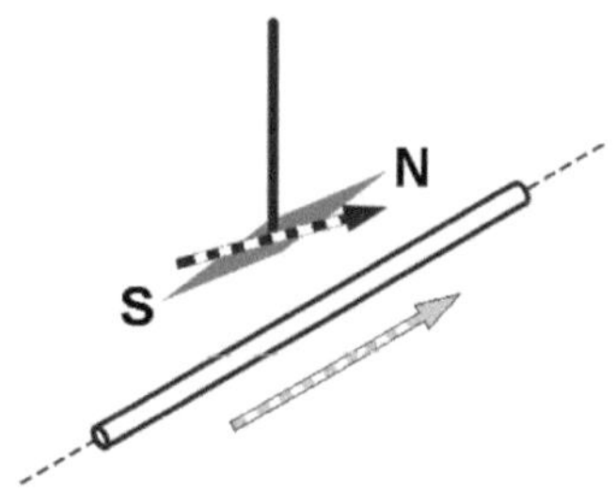

Figure 11.1.

Let us place a magnetic needle used as a compass on a conductor with electricity running through it, then view its image in a mirror. The plane of the mirror will reflect the geometrical properties of the experimental arrangement, but will reveal nothing of the electric and magnetic processes taking place inside them (Figure 1.12 in Chapter 1, left hand image). How can we decide which of the two images is the real experimental state and which is the reflection? Help is given by the letters drawn on the compass on the respective diagram, but we must be able to answer the question without them, too. We know from experience that the northern pole of the magnetic needle always turns to the side relative to the direction of the current, according to the so-called right hand rule. As the direction of the current in the mirror parallel to it has not changed, in the mirror it looks as if the northern pole of the needle has turned to the left. This contradicts our experience. Is it possible that in the mirrored world it is the laws of a different physics that prevail? We feel like Alice before she steps into the Looking-glass House: perhaps Looking-glass milk isn't good to drink?

Let us in our imagination draw inside the needle the direction of the circular currents that determine the direction of its magnetization (Figure 11.2, right hand diagram below). Seeing the deviation, and knowing which is the needle's northern pole, we can draw in the direction of these circular currents on the basis of the right hand rule. Let us look at the image of these directed circles in the mirror. The mirror likewise reverses their direction. If in the mirrored world we insist on the right hand rule that has in our experience proved so effectively valid, then in the mirrored needle the directions of an elementary magnet are commuted, and the northern pole is to be found at the opposite end. The needles we see in the two mirrors are one and the same, and point in the same direction, just the interpretation of their magnetic poles has changed. In the first instance, in the course of reflection we lost the parity of space — right and left were swapped. In the second case we tried to preserve spatial parity, at the price of having to reinterpret the direction of magnetization. It is possible that Looking-glass milk really is not the same as real milk.

Under certain transformations, the symmetries of things characterized by a number of properties are only conserved together, and can be broken individually. James W. Cronin and Val L. Fitch were awarded the Nobel

Prize in 1980 for the discovery of combined symmetry conservations, in particular that space reflection only displays exactly conserved symmetry in combination with charge conjugation and temporal reflection (*CPT*).

A vector is a quantity containing a number of properties from the outset. Over and above the fact that it is characterized by its value and direction, its direction in geometric space marks a right hand twisting direction. By convention, two vectors are always multiplied together in such a way that the product vector perpendicular to their plane — which from the point of view of space reflection is not even a vector, but a pseudo-vector or axial vector — points in a direction in space according to the right hand rule. (The convention could equally stipulate the opposite direction, which would not cause great discomfort, but then we would have consistently to stick to that direction in the same way instead.) The angular momentum vector in mechanics displays such properties, for example. Figure 11.2 shows two vectors together with their associated right hand twist direction. Let us reflect both of them in a mirror! There is no doubt that the direction of the depicted vector is reversed in both instances. But what happens to their associated twist direction?

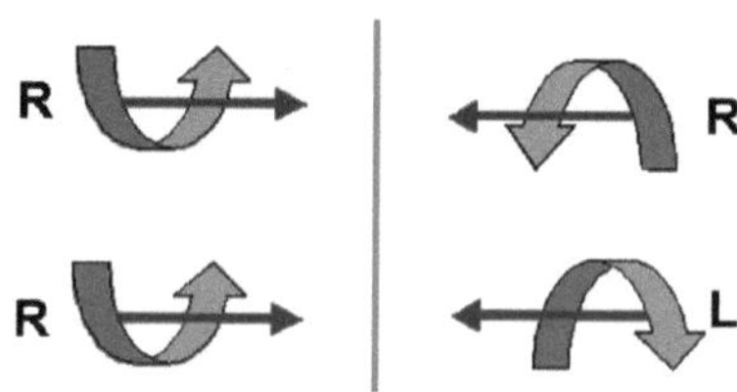

Figure 11.2. *Reflection of a vector*

If we insist on the right hand rule, the direction of rotation parallel to the mirror plane is commuted, which geometrically speaking should not be mirrored. If we insist on preserving the direction of rotation, we lose the right-handedness. Which reflection is the real one? (We are reminded of the comparison in the first chapter of our likeness in a photograph and in the mirror.) Let us take a broom-handle in our right hand and grab it such that our extended thumb points in the direction of the broom-head, then hold the broom in front of the mirror. We can see that in the mirror the direction of the broom is the opposite of that in our hand, as is the image of our thumb to the real one, while our four fingers are clasped around the broom-handle in the same direction in the mirror as in real life. In the mirror it is as if a left hand were grasping the broom-handle. The mirror shows a left-handed world.

If we hold the broom-handle vertically, with the head at the top, and our thumb also points upwards, neither the broom nor our thumb changes direction on reflection. The direction in which our fingers are clasped around the broom is reversed in the mirror, however. Again, in the mirror it appears that a left hand is holding the broom-handle. This also shows us a left-handed world, despite the fact that the broom is now held parallel to the mirror. This arrangement corresponds to the reflection of the magnetic needle in the previous example. The vector is itself a one-dimensional object, which we reflect in the two-dimensional plane. Yet, together with its direction of rotation, our vector is three-dimensional, and it is this which is reflected in the two-dimensional mirror plane. We encountered matrices that described reflections like this in Chapter 6. We must take care when interpreting the result of the reflection of a number of properties, because one or more symmetry characteristics can be lost or be reversed.

So where then is the beautiful, perfectly symmetrical world that we always thought we could find in the nature all around us, from which we thought it an accident that a few particular phenomena deviate, while our laws of nature preserve symmetry? The above phenomena are equally governed by regularities, just as rotations are.

Let us take out a book, holding it with its spine downward and its front cover facing us. Let us rotate the book, first around the spine and away from us, in such a way that its front cover ends up on top. Let us then rotate it around a vertical axis through the corner of the spine closest to our left hand, in such a way that the book's inscription be readable for someone standing opposite us. Let us make a note of the book's position. Let us take the book in our hand in its original position and perform the two rotations in reverse order. First let us turn it around its left hand edge and away from us. At this point the book stands with its spine downward and along a plane perpendicular to our body. Then let us rotate it away from us around an axis perpendicular to the plane of the book. The book is now in a standing position, along a plane perpendicular to our body, with its front cover pointing to our right. The book ends up in a different position from before (Figure 11.3). Despite the fact that we have performed the same two rotations on it, around the same axes and with the same directions of rotation, the result does not prove to be the same. The only difference is that we changed the order in which we performed the rotations. Put

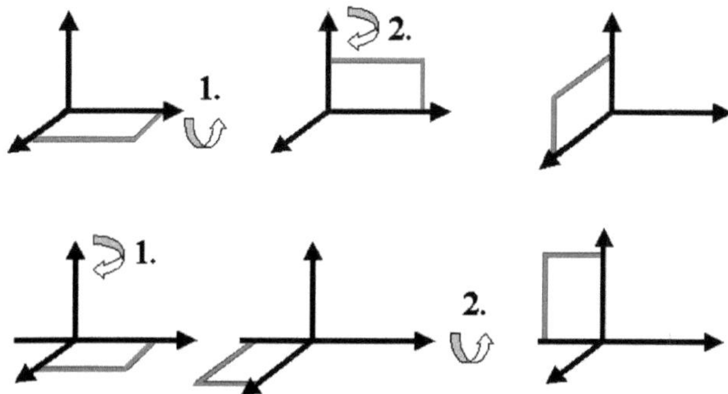

Figure 11.3. *Changing the order of simple spatial rotations around an axis*

in mathematical terms, we perform a right-angled rotation in a positive direction around the x axis, then in a negative direction around the y axis. Let us repeat this in such a way that we first make a right-angled rotation in a negative direction around the y axis, then a right-angled rotation in a positive direction around the x axis. The result is that the order of the two rotations cannot be commuted. Let us remind ourselves of the fact that spatial rotations form a group. And it was not a criterion of groups that their elements display commutativity (that their order can be commuted). If we perform the group operations on them in different order, this does not bring a result outside the group, but we get different elements of it.

Let us write this down in symbols rather than words. Let us use the following symbols:

rotation around the X axis	$\boldsymbol{P}$
rotation around the Y axis	$\boldsymbol{Q}$
the book's initial state	φ
the book's state after operation P	$\boldsymbol{P}\varphi$
the book's state after operation Q	$\boldsymbol{Q}\varphi$
the book's state after operation P, then Q	$\boldsymbol{QP}\varphi$
the book's state after operation Q, then P	$\boldsymbol{PQ}\varphi$
This gives us the final result that	$\boldsymbol{PQ}\varphi \neq \boldsymbol{QP}\varphi$.

Physics uses a similar system of description, particularly since the introduction of quantum-mechanic description, to note physical events. Physical phenomena occur in some kind of — usually abstract — field. We characterize the state of a physical object at each moment with a state function, marking the event that occurs to it, which changes some of its physical attributes (its state), by using an operator that has an effect on this state function and transfers it from one state to another. Put very roughly (concentrating here only the symbols, ignoring the likelihood of the states in question), we can, for example, describe the change of the linear momentum of an object by saying that the momentum operator effects it, describe a change in its energy as the effect of the energy operator, and so on. This mode of description is made possible — in a way analogous to spatial rotations — by interpreting the effect of individual operators as if they were rotations of the state functions in a particular abstract field.

Certain operations on individual physical quantities, that is, correspond to so-called *operators*. The states of physical objects are described by state functions. An operator $\boldsymbol{Q}$ has an effect on state function φ such that it moves it to another state, φ'.

$$\boldsymbol{Q}\varphi = \varphi'$$

In general, if two operators are applied to a state function one after the other, the order of the two operators cannot be commuted: $\boldsymbol{PQ}\varphi \neq \boldsymbol{QP}\varphi$. We could also say that two rotations applied in different order in an (abstract) state field do not result in the same final state. We can symbolize this with the introduction of the so-called {Poisson brackets} as follows:

$$\boldsymbol{PQ} - \boldsymbol{QP} = \{\boldsymbol{P}, \boldsymbol{Q}\} \neq 0$$

In general, a physical state is not left unchanged under two physical operations performed in a different order. There are, however, instances in which it is unchanged. In a physical sense, it counts as an operation if for example the momentum or energy operator affects a physical state, and changes the momentum or energy of the object characterized with the given state function. In the sense of the narrower quantum-mechanical definition of symmetry, if a unitary transformation operator $\boldsymbol{T}$ can be commuted with the Hamiltonian operator of energy $\boldsymbol{H}$, then we call $\boldsymbol{T}$ a symmetry. We mark commutativity in the following way:

$$\boldsymbol{TH} = \boldsymbol{HT} \text{ or } \{\boldsymbol{T}, \boldsymbol{H}\} = 0$$
$$\boldsymbol{THT}^{-1} = \boldsymbol{H}$$

In a more general sense, we use the term symmetry to describe the situation in which the resulting physical (or other type of) state remains unchanged, i.e. invariant under the physical (or other) operations performed in two different orders. Put another way, we speak of symmetry if the pairs of operations $\boldsymbol{TH}$ and $\boldsymbol{HT}$ transform the same state φ into the same final state (in a more precise quantum-mechanical interpretation, the same probability states). In this way we have succeeded in generalizing the symmetry properties of rotations interpreted in Euclidean space to rotations in abstract fields, and afterwards in applying these to the effects on physical states of the physical operators used in quantum mechanics.

This mode of description of the phenomena of inanimate nature did not grow out of the Newtonian physics of the seventeenth century based on the concept of force and the quest for the laws of nature, which had been so successful at the time. From the eighteenth century onwards, another branch of physics was developing in parallel, based on the variational principles. In fact it was in the light of the results of the twentieth century that we were able to decide which mode of description draws a clear caesura between classical physics and modern physics, and which is the one where new theories can continuously be constructed on the basis of earlier approaches.

The key problem addressed by the variational principles is finding the shortest of the many possible paths between two points. Here paths do not necessarily signify distance, and can represent time, completion of work, anything. We are looking for the minimal amount of a physical quantity (e.g. distance, time, work) needed to achieve a particular objective. We vary the possible curves that can be drawn between two points, and look for what is, from some particular perspective, the extreme, in most cases the minimum, value. The minimum means an extreme value, with the associated variation being 0. We encountered similar tasks in our elementary mathematical studies. The inequality between the arithmetic and the geometric mean $\sqrt{ab} \leq \frac{(a+b)}{2}$ turns into an equality if $a = b$, that is if the distribution is symmetrical. Of rectangles with a given area, it is the most symmetrical, the square, that has the minimal boundary. Intuitively, we have known for centuries that variational principles are expressions of invariance principles. It was only the Noether theorems (1918), however, which revealed to us the real content of the relationship between variational principles and symmetries. These two theorems proved (indirectly) the relationship between the *conservation laws* — which play such an important role in physics — and the *variational problems* invariant under the transformations of the two categories of symmetry groups. With the help of integration, we can use action functions with nil variation to give us conservation laws — quantities that do not change with time.

One of the first variational principles in physics was *Fermat's* (1601–1665) *principle*, which stated that the optical distance of the ray of light passing between the two points — quite possibly through media displaying reflections and refractions and with different indices of refraction — is the minimal path among the possible neighbouring curves. The next important (this time mechanical) principle was the *D'Alembert principle* (1743), according to which, as a result of the vanishing of the virtual work of the constraint forces, the material point moves in such a way that the virtual work of the exertion of force $(F - m\ddot{r})$ — the difference between free forces and inertia forces — is zero at all moments in time. The principle of D'Alembert (1717–1783) was later extended to cover point systems, with the so-called principle of virtual work. According to the *principle of virtual work*, if no inertia forces affect a system, it is at rest if and only if the total virtual work of the free forces affecting the system is zero. By deducting the inertia forces from the free forces, D'Alembert formally traced the dynamic problem back to a static one. This brought him to an equilibrium problem. The so-called *principle of least action* (1747) is associated with the name of Maupertuis (1698–1759), and holds that the integral of the linear momentum over the path element ds is a minimum for real motions. In the same century further mechanical variational principles were elaborated, all more or less independent of one another. These were associated with Euler, then later with Jacobi (1804–1851), and comprise the combination of the above variational principles and their expression in various mathematical forms.

The variational principles reached their largest-scale generalization in the *Hamilton principle* (1834). Hamilton (1805–1865) can have taken the mathematically well-elaborated general variational problem as his starting point, according to which he was looking for the function $x(t)$ for which the definite integral $I \equiv \int_{t_0}^{t_1} F(t, x, \dot{x})dt$ has an extreme value. The task can be traced back to the Euler–Lagrange differential equations $\frac{d}{dt}\frac{\partial F}{\partial \dot{x}_i} - \frac{\partial F}{\partial x_i} = 0$, the universality of which is shown by their continued role in modern physics. Hamilton traced the problem of the motion of the conservative point system back to a variational problem in the most general form possible, regarding the Lagrangian [Lagrange (1736–1813)] as the argument F of the integral. The mathematical expression of the Hamilton principle is as follows: $\delta \int Ldt = 0$ (the Lagrangian $L = T - V$ is the difference between the kinetic and potential energy of the system). Thus, according to the Hamilton principle, the variation of the time integral of the Lagrangian, in other words of the action integral, is zero. The Hamilton principle is also called the principle of least action. One great virtue of the Hamilton principle is that the parameters of the function acting in it can be chosen freely, and thus it can be used not only with the customary time-space coordinates and their derivatives, but also with arbitrary generalized (abstract) coordinates, which also helped it to outlive classical physics. With the help of the Hamilton principle, the Euler-Lagrange equations of systems, not necessarily conservative ones, could be put in an enduring form that is evidently symmetrical, even to the layman's eye, and which later survived the test of the criteria of relativity theory and quantum physics. According to this, the Lagrangian L can be described as the function of the generalized momenta and coordinates p_i and $q_i \cdot L = L(q_1, \ldots, q_f, \dot{q}_1, \ldots., \dot{q}_f, t)$. The generalized momenta p_i can be interpreted with the equation $p_i = \frac{\partial L}{\partial \dot{q}_i}$, from which, as a result of the Euler–Lagrange equations, $\dot{p}_i = \frac{\partial L}{\partial q_i}$. With the interpretation of p and q, and with the Lagrangian, the Hamiltonian describing the total energy of the system can be written in the following form: $H \equiv \sum p_i \dot{q}_i - L$. After this, the motion of a material system can be described with the following beautifully symmetrical $2f$ number of first-order ordinary differential equations. We call these Hamilton's canonical equations; we call the functions $p_i(t)$ and $q_i(t)$ canonical variables, and we call p_i a momentum canonically conjugated to q_i.

$$\dot{q}_i = \frac{\partial H}{\partial p_i} \quad \text{and} \quad \dot{p}_i = -\frac{\partial H}{\partial q_i} \quad (i = 1, 2, \ldots, f).$$

These impressively symmetrical equations hold their ground in modern physics, in multidimensional abstract fields, for example in field theory.

The mechanical principles were later complemented by some other ones. This include the *principle of least constraint* in the formulation of Gauss (1829), which in essence is a generalization of the D'Alembert principle. The *principle of Hertz* (1857–1894) in the second half of the nineteenth century established the principle of the straightest path, in other words of the least curvature. It is significant for two reasons: firstly, because it aimed at the elimination of the concept of force; secondly, because it prepared the description of motion along geodesic lines for the general theory of relativity.

In addition to the various variational principles of physics, the general formulation of the principle of least action, which is still valid today, and the recognition of the action integral, many chapters of mathematics have contributed to the discovery of the relationship between conservation laws and symmetries.

In the historical chapter we became acquainted with the development of group theory. Here I would like to emphasize two of its results. One is the Erlangen programme of Felix Klein from 1872. One of its descriptions is as follows: "Geometry is the science which studies the properties of figures preserved under the transformations of a certain group of transformations". In another description Klein abstracts from figures. "Geometry is the science which studies the invariants of a group of transformations." Almost half a century later, Klein was present at the eminent mathematical workshop in Göttingen which was reinvigorated by the spirit of the general theory of relativity, and where David Hilbert's pupil Emmy Noether proved the theorems that would welcome a new era in physics. The other result on the journey here that deserves a mention is a generalization of finite groups by Sophus Lie: the theory of continuous groups.

Noether's two theorems, the importance of which cannot be overstated, are not easily digested, not even for physicists well-versed in abstract mathematics. In their original form, moreover, they can appear to be rather secessionist in style. Their significance obliges us to give them a mention, albeit in a slightly simplified form. According to *Noether's first theorem*, if the integral of a variational problem is invariant (up to a divergence group) under the (finite, continuous) transformation group G_ρ, then of Lagrangian expressions $L_i(f)$ there exist ρ linear connections, each of which is a divergence. The reverse of the theorem concludes the existence of the group G_ρ from ρ number of relations. *Noether's second theorem* states that if integral I of the variational problem is invariant under the (infinite, continuous) transformation group $G_{\infty\rho}$, in whose infinitesimal transformations the arbitrary functions $p_s(x)$ and their differential quotients appear up to the order l, then ρ number of identical relations exists between the Lagrangian expressions and their differential quotients (up to order l).

For everyday physics, the point of the two theorems is that every single symmetry (transformation) corresponds to a conserved quantity, and all conserved quantities correspond to a symmetry.

After the variational principles, we have to deal with the relativity principles. In connection with Galileo we saw that the relativity principles are also symmetry principles. In its contemporary formulation, the symmetry property of the Galileo principle appears self-evident: systems of reference moving at constant speed relative to each other are equivalent.

$$x' = x - vt$$
$$t' = t$$

In other words, in the event of a transition to a different coordinate system moving at a constant speed relative to ours, the form of the equations of motion does not change. The same laws govern the motion of bodies in all inertial systems.

By the end of the nineteenth century, once the laws of electromagnetism and electromagnetic waves had come to form a complete system, and as a consequence of certain experimental results, there was no avoiding confrontation with the concept of the speed of light as a limit speed. Indeed, this meant two new discoveries. One was the symmetry that the speed at which light travels is invariant to the choice of inertial system in which we measure it. The second discovery states that this velocity coincides with the greatest limit speed at which an electromagnetic or mechanical action of inertia can travel. The two are inseparable.

The discovery of this symmetry represented a decisive step in the process during which physics generalized our concept of symmetry. In their day, no correspondence was stipulated between earlier known invariances and geometrical symmetries and their associated crystallographic symmetries. In the 1890s, however, when the crystallographic world was just enjoying the triumph of making the symmetries of crystal groups complete, P. Curie established the definitions of symmetry, asymmetry and dissymmetry that are still used and suitable for generalization to this day. It was also at this time that he made his famous claim, that has given scientists something to do ever since, that "dissymmetry makes the phenomenon". The invariance of the speed of light with regard to the choice of inertial system was one of the first conscious generalizations of the geometrical concept of symmetry along the road presented in the introductory chapter, and certainly the most influential. From then on, it was physics that took over the initiative in the step-by-step generalization of the concept of symmetry.

The Galilean transformation had also to be redefined in the light of the newer, principally wave theoretical results in the study of electromagnetism, and of the new discoveries. For assuming that there exists a limit speed that cannot be exceeded, the rule of velocity addition had to be reassessed. Simple addition was no longer adequate. The suitable form of the new coordinate transformation, taking finite limit speed into account, proved to be the *Lorentz* (1853–1928) *transformation.* It showed how time and space coordinates are transformed in a right-angled coordinate system if we change to a system moving at a constant speed v relative to it. Assuming a single spatial dimension and a velocity pointing in this direction, the coordinates are transformed in the following way:

$$x' = \alpha(x - vt)$$
$$t' = \alpha[t - (v/c^2)x] \qquad \text{where} \quad \alpha = 1/\sqrt{1 - (v/c)^2}$$

At speeds that are very low relative to the speed of light c, the Lorentz transformation turns into the Galilean transformation. The Lorentz transformation expresses the same property, only — as the result of including newer discoveries — more precisely: our laws of nature are invariant under this transformation, i.e. their form does not change if we move from one system to another inertial system moving at a constant speed relative to it.

As early as 1887, W. Voigt examined and found approximating formulas for the fact that the description of such a transformation is necessary for the wave equation of light to remain invariant in coordinate systems of various speeds. Almost two decades had to pass before Lorentz found the suitable formula. Its success was to make it possible to ignore the assumption of the existence of a distinguished (absolute) coordinate system and absolute time, though Lorentz himself did not recognize this at the time. In 1899 he reached the correct transformational formula for space coordinates, before spending another five years looking for the equation of the right transformation of time, which he finally published in 1904. In the following year, Poincaré discovered that the Lorentz transformation leaves the quantity $x^2 + y^2 + z^2 - c^2t^2$, the so-called infinitesimal line segment square, intact, and therefore it can be considered as a rotation in the four-dimensional "space" x, y, z, ict. That is, Lorentz transformations with different speeds form a group. It is this transformation and the transformation group it determines that embody *Lorentz's relativity principle*, which after three hundred years replaced Galileo's similar principle.

Although Lorentz's relativity principle expresses a mechanical transformation — which is based on being aware of the constancy of the speed of light — we must remember that it was born of the need, as Lorentz' article discusses, to ensure the invariance of the equations of the laws of electromagnetism and electromagnetic waves. Put more precisely: the order of events was in some respects the opposite, for the Maxwell equations were from the outset invariant under the Lorentz transformation, and the corresponding transformational formula had to be found for this. We could also say that Maxwell, without being aware of it, created a theory which proved to be invariant with regard to a symmetry that was unknown at the time, and only discovered decades later.

The Lorentz transformation expresses rotations in space-time. Taking the independent parameters into account, the Lorentz group describing the rotations is a six-parameter group. It does not include translations. If we add the translations in space and time to the rotations of the Lorentz group, this gives us a group with ten parameters, which includes all the spatial and temporal transformations. We call this the Poincaré group.

The ten-parameter Poincaré group describes all geometrical invariances. All transformations occurring in the group generate a conserved property, and all of the independent parameters belonging to the transformations generate a different conserved quantity. In mechanics, on account of their generation, we call these conserved quantities the integrals of Lagrangian equations of motion. Their existence can be demonstrated using the Hamilton principle. It was later built into the much more general Noether theorems.

Geometrical invariances:
The ten integrals of the Lagrange equations

Transformation	**Number of parameters**	**Conserved property**	**Number of quantities conserved**
Spatial translation	3	Linear momentum	3
Temporal translation	1	Energy	1
Rotation	3	Angular momentum	3
Constant speed translation	3	Centre of mass	3
Total	10		10

Table 11.1.

Einstein (1879–1955) claimed that he was not aware of Lorentz's latest results concerning the transformation and Poincaré's interpretation when in 1905 he published his article *On the Electrodynamics of Moving Bodies.* In this, he made a definitive break with the concepts of absolute time and absolute space, with the distinguished role of any system, with the concept of the ether, and regarded all inertial systems as equivalent, from which it follows that the speed of light must be equal in all directions and in all inertial systems. In so doing, he applied the velocity transformation based on the constancy of the speed of light to electromagnetic field quantities in a more adequate fashion than Lorentz had done. This is what we consider to be *special relativity theory* and the *Einstein relativity principle.* The novelty of Einstein's much-quoted relativity theory lay in the latter fact, that he applied the Lorentz transformation to the transformation of the components of electromagnetic field quantities in such a way that he did not assume a distinguished coordinate system.

Thus special relativity theory was in fact the expression and extension of a symmetry principle to the laws of electromagnetism, which remain unchanged while we pass from one inertial system to another.

Einstein successfully connected the laws of mechanics and of electricity with the help of an invariance principle. The theory dotted the 'i' of an issue that had dragged on for decades (if we can refer to the birth of the theory as dotting the 'i' of an issue, for a good part of the scientific world debated it for decades and refused to accept it, and, as we know, relativity theory was never honoured with the Nobel Prize). This issue — which we emphasize again, on the basis of our current knowledge, in today's terminology — was as follows. Newton's laws, considered the foundation of mechanics, and the Maxwell equations, considered the foundation of electromagnetism, were each invariant under different transformations. (As we saw, these were the Galileo and Lorentz transformations, respectively.) Are Newton's laws the absolute ones, or the Maxwell equations? Well, the Lorentz transformation includes the Galilean transformation as an extreme case. The reverse is not true. Einstein's choice of accepting Lorentz's transformations as a general invariance principle reflected the discovery that the Lorentz transformation leaves the structure of the Maxwell equations unchanged. In a low velocity approximation, however, it does not question the correctness of Newtonian mechanics, and thus as an extreme case of the former results of classical physics — which are in accordance with our experimental experiences — can be built into relativistic physics.

Einstein later (1916) extended the connection of the laws of classical mechanics and the laws of electromagnetism formulated in the special relativity theory to the general theory of relativity, which included the laws of gravitational fields. The basis for this was the equivalence principle, which stated that the inertial and gravitational force fields are equivalent to one another, and to which the so-called *general covariance principle* of the *laws of nature* was added. Another invariance principle that governs the world. While Galileo's relativity principle stated the equivalence of systems at rest and systems moving at a constant speed, and the special relativity principle extended these to the inertial systems moving at various constant speeds and to the constancy of the forms of our electromagnetic laws (over the mechanical ones) within these systems, so the latter in its turn declared the equivalence of accelerating systems.

By the middle of the 1910s, a whole family of invariance principles were at our disposal, which took all physical interactions known at the time under their jurisdiction. The results got mathematicians as well as physicists excited. It was around this time that a significant brains trust came together in Göttingen, the bastion of mathematics of the day, and formerly the university of Gauss and of Bolyai (1802–1860). Those working there included D. Hilbert, who laid down the modern mathematical foundations for the analytical description of physical fields, and F. Klein, who played a pioneering role in the systematic development of group theory (which became the basis for the description of symmetries), and also contributed with significant results to the theory of non-Euclidean geometries (which represented the mathematical basis for the general theory of relativity). Klein's interest throughout his life was characterized by the way in which he kept an eye on the physical applications of his theories. In addition, a whole list of their protégés also worked there alongside them, including E. Noether and H. Weyl. The eminent physicists of the age would make regular appearances there. It is no accident that they would soon inspire two mathematical theorems, which we learned of above as the Noether theorems, that would collect symmetries, invariance principles and conservation laws into one single unified theory.

Noether's theorems were not limited to the description of invariances known at the time. They laid the foundation for the mathematical basis of all the further symmetry principles and physical conservation laws that have been discovered to this day. Mathematically they have still not be exhausted, i.e. they still contain the possibility of invariances for which, with our present knowledge, there does not appear to be any associated physical content.

The conserved quantities of mechanics (energy or mass, linear momentum, angular momentum, centre of gravity) exist in Euclidean or at least in metric space, and can be described using the parameters of these. In line with the Lorentz transformation, the conservation laws announced for them — that is, that they are constant in time, in other words invariant to changes to the time coordinate — are universally applicable. This means that their extent — which has proven to be constant with respect to time — is the same not only whenever, but wherever we measure it in space. This follows from the fact their space coordinates transform together with their time coordinates. Their symmetry we call global symmetry.

The *universality* of Noether's theorems also showed itself in the following. We get every symmetry from a variational problem. A variational problem leads us to a symmetry if it (its integral) is invariant with respect to a group. Thus the problem must involve a (transformation) group, the problem's symmetry group. The elements of this group depend on a certain number of parameters (in theory, an arbitrary finite number of parameters is allowed). The *parameters* meanwhile *can themselves be the functions of certain space-time coordinates* (even of an arbitrarily large number of generalized abstract spatial coordinates). Put more simply: in general, symmetry groups can be not only the functions of the three space and time coordinates as parameters. In special cases, the parameters appearing in the laws can be identical to the real (metric) space-time coordinates, but when arbitrary parameters are chosen, it is allowed for the elements of transformation groups only to be indirect functions of the coordinates of real space-time.

In the case of geometrical symmetries (which produced the conserved quantities of mechanics), the parameters are themselves the metric space-time coordinates.

There are physical quantities that we do not directly interpret in real space. One example is that of charges (and all other physical properties that behave like charges). We interpret and describe these in *abstract fields*, for example in the electromagnetic field. The behaviour of electric charges cannot be described using the parameters of Euclidean or any metric space, nor with mechanical force fields. The electromagnetic field — in which they exist, and in which their properties and dynamics can be interpreted — is an abstract field by comparison to the Euclidean (or any metric) space.

When we discuss the conservation of (e.g. electrical) charge, we define a group in the electromagnetic field which describes "rotations" in this field. (I use inverted commas because these are not real rotations in real space, just behave like rotations in an abstract field.) The parameters of this group are the quantities of the electromagnetic field. We construct the Lagrangians of the electromagnetic field from the derivatives of these parameters, of the potentials of the field. The value of these quantities, however, varies from place to place and from one moment in time to another, i.e. they are the functions of metric space-time coordinates.

If our variational problem (more precisely, its integral) *is invariant under rotations of this group (in an abstract field) then,* according to Noether's theorems, *the Lagrangians appearing in the integral of the variation* are not independent of one another, i.e. *there are connections between them.* (In the

Second law, the connections are differential quotients of the parameter functions. In the First law, they take the form of divergence. For readers less briefed in mathematics: divergence is an expression of the differentiation of vectors.) *The connections represent limitations both on them and on the parameters they depend on.* Passing over the mathematical details of this claim, these limitations embody the consequence that *the same fields can belong to different values of the parameters.* (To draw a crude comparison, a function determines its derivatives in explicit fashion. The opposite is not true, however: a function is associated with a veritable multitude of integrals. The same is true for abstract vector fields defined in a completely general fashion.) This means that *various potentials* appearing in the variational problem *can result in the same* (e.g. electromagnetic) *field.* That is, *the fields appearing in the variational problem are invariant with respect to certain changes to the potentials which* (in their role as parameters of the field) *produce them.* In other words, *we are allowed to change the scale of the parameters arbitrarily.* The field will be invariant to the choice and the translation of the starting-point for the scale. After H. Weyl, we refer to *the invariance under scale translation* as invariance with respect to a change in gauge, or simply *gauge invariance.*

Let us briefly look over the above train of thought again. The rotations of Euclidean space (and time) that we described with the Lorentz transformation are characterized by the fact that the parameters of their group are the space and time coordinates themselves which are transformed together. Therefore those (conserved) quantities whose value does not depend on the time coordinate are independent of the space coordinates, too.

Not all transformation groups display these properties. As a result, the physical quantities that cannot exclusively be described using the (x_i, t) parameters $(i = 1, 2, 3)$ of metric space do not necessarily display these transformation characteristics. If a physical quantity $\boldsymbol{F}$ depends on a few parameters p, and these parameters depend on the space-time coordinates (x_i, t), i.e. $\boldsymbol{F} = \boldsymbol{F}\ (p_r(x_i, t))$, this can transform in the gauge field in such a way that while it results in a conserved quantity, the dependence of the parameters on the space coordinate $p_r(x_i)$ is preserved. Thus we cannot exclude the possibility of transformation groups for which the time integral of the variational problem gives a constant value, i.e. leads to a conserved quantity, while the value of the given quantity, albeit independent of the time coordinate, is dependent on the space coordinates. We refer to such conservation laws and their associated symmetries as *local* (place-dependent) *symmetries* (as opposed to global symmetries). Gauge invariances are generally local symmetries.

Gauge invariances display the property that a transformation performed merely in a gauge field does not lead to a conservation law. This is easy to see, for a transformation performed in a gauge field changes quantity $\boldsymbol{F}(p)$ to quantity $\boldsymbol{F}'(p)$ within a gauge field stretched by parameters p. Conservation, that is invariance with regard to temporal translation, on the other hand, expresses constancy with a parameter of metric space, namely invariance under the passage of time. The requirement of conservation, that is, means invariance under a transformation that we have to perform in metric space: $(p(x_i, t) \rightarrow (p(x_i', t')$. As a result, in order to derive conservation laws, the transformations of the gauge field have by necessity to be connected to the transformations of metric space.

H. Weyl immediately recognized the significance of local symmetries for potential physical applications, and almost as soon as the theorems had been published, he tried to use the first Noether theorem to prove the conservation of electric charge. There had long been intuitive conceptions of charge conservation, as no experimental evidence contradicted it, but no one had yet succeeded in providing exact mathematical proof for it. Although Weyl's first attempt was not a complete success, eleven years later (in 1929) it was he who was the first to provide an exact proof of the conservation of electric charge.

In a free electromagnetic field, the wave equations can be written with a vector potential A_i $(i = 1, 2, 3)$ and a scalar potential Φ. We can consider these quantities as the parameters of the electromagnetic field. On the basis of the Maxwell equations, $\mathbf{A}$ and Φ can be written down with the help of the magnetic and electric field strength:

$$\begin{aligned} \mathbf{B} &= \mathrm{rot}\mathbf{A} \\ \mathbf{E} &= -\mathrm{grad}\Phi - (1/c)(\partial \mathbf{A}/\partial t) \end{aligned}$$

It is clear that $\mathbf{E}$ and $\mathbf{B}$ are unambiguously determined by $\mathbf{A}$ and Φ. The previously noted surprise, however, is that the reverse is not true: the potentials are not unambiguously determined by the field. Potentials **A** and Φ, different by a gradient, produce the same fields. In order to fix the scale of **A** and Φ, we use the so-called Lorentz condition. (Of course we could use other scale transformations, any scale transformation that does not contradict the Maxwell equations and the above definitions of **A** and Φ). Let

$$\operatorname{div} \mathbf{A} + (1/c)(\partial\Phi/\partial t) = 0$$

The advantage of choosing the Lorentz condition is that it is a covariant, continuity equation. If we consider **A** and Φ as the components of a four-vector, then it precisely expresses a divergence, and with its help the first two Maxwell equations (in a simple state, in a vacuum) can be written down in the form of the following two inhomogeneous wave equations:

$$\Box\mathbf{A} = \Delta\mathbf{A} - \frac{1}{c^2}\frac{\partial^2 A}{\partial t^2} = -\frac{4\pi}{c}\rho\mathbf{v}$$

and

$$\Box\varphi = \Delta\varphi - \frac{1}{c^2}\frac{\partial^2\varphi}{\partial t^2} = -4\pi\rho$$

where ρ signifies charge density and **v** the speed of the charge currents. We did not, however, denote that potentials **A** and Φ, as the parameters characterizing the electromagnetic field, depend on the space-time coordinates (x_i, t) of metric space. Yet electromagnetic phenomena take place in space and time. Considering the components of **A** together with Φ as the elements of a four-potential A_μ ($\mu = 1, 2, 3, 4$), we can also write the scale transformation described by the Lorentz condition by saying that fields **E** and **B** are invariant under a

$$A_\mu \rightarrow A_\mu + \delta_\mu\varepsilon(x_\mu)$$

transformation. We regard this transformation as a gauge transformation performed in the gauge field (electromagnetic field) stretched by the A_μ-s. Any change to **A** and Φ which leaves **E** and **B** unchanged is a gauge transformation, while the conservation of field strengths is a gauge invariance. With the dependence of $\varepsilon(x_\mu)$ on space-time coordinates x_μ, we have expressed that the scale transformation (in other words the gauge invariance) of the potentials of the electromagnetic field, in other words its gauge invariance, depends on local coordinates, which means it is a local invariance.

In order more clearly to see the analogy with the function describing the parameter-dependent physical quantity $\boldsymbol{F} = \boldsymbol{F}\ (p_r(x_i, t))$ just generally defined, let us look at the $[4 \times 4]$ tensor of the electromagnetic field composed by the electric and magnetic field strengths **E** and **B**, which we also designate $\boldsymbol{F}$. In this case, $\boldsymbol{F}(E_i, B_i) = \boldsymbol{F}\ (A_\mu(x_i, t))$.

After this we derive charge conservation by noting the interaction between the electromagnetic field and the metric space. As a result, we have to connect the gauge transformation $A_\mu \rightarrow A'_\mu$ performed in the electromagnetic field with the transformation $x_\mu \rightarrow x'_\mu$ performed in the metric space. The Lagrangian which describes the interaction conserves its form under the combined effect of the two transformations performed, which complement one another. It remains invariant under gauge transformations. This invariance of the Lagrangian results in the conservation of electric charge. Because of the coordinate dependence of gauge transformations, the conservation of electric charge is a local invariance.

Gauge invariances are not the result of the geometry of the space. The source of the gauge-invariant field is electric charge. Other physical quantities also generate gauge-invariant fields. We regard these as charge-like quantities. To distinguish them from geometrical invariances, we refer to them, using Eugene Wigner's term, as dynamic invariances.

Geometrical invariances express the invariances of *natural phenomena*. Geometrical invariance principles refer to the events themselves. Time-translation invariance, for example, expresses the fact that the correlations between events only depend on the length of the time periods that separate them, but are independent of the moment of time at which the first event occurs.

Dynamical invariances express the invariances of *laws of nature*. Dynamical invariance principles refer to laws of nature. The gauge invariance of electromagnetic interaction, for example, refers to a special law of nature which determines the electromagnetic field generated by the charges, and the effect of the electromagnetic field on motion of the charges. When deriving the charge conservation, we demanded that the form of the Lagrangian remain invariant under the gauge transformations of the potentials describing the motion of the charges.

In 1964 Wigner wrote the following on this subject: "The geometrical principles of invariance [...] are formulated in terms of the events themselves. [...] the dynamical principles of invariance are formulated in terms of the laws of nature. They apply to specific types of interaction, rather than to any correlation between events." [...] "there is a great similarity between the relation of the laws of nature to the events on the one hand, and the relation of symmetry principles to the laws of nature on the other. [...] The role of the invariance principles is to give the structure of the laws of nature and the relations between them, like the laws of nature give the structure of the series of events and the relations between the events."

That is, Wigner constructed the hierarchy presented in the following diagram (the direction of the arrows denotes what they refer to):

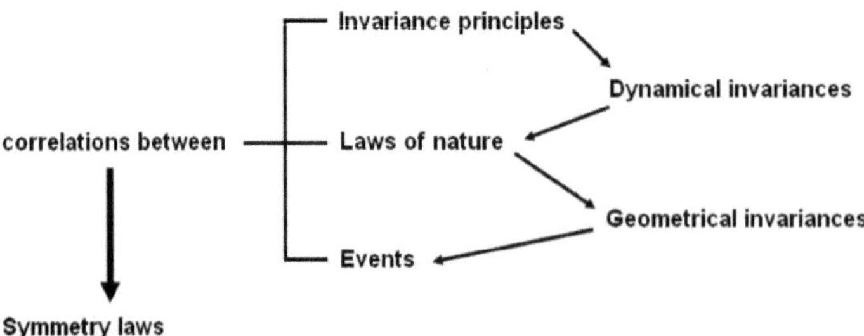

Figure 11.4.

From the proof of the conservation of charge onwards, the history of physics has to this day essentially been made up of a series of researches into symmetries as conservation laws and symmetry breakings. New particles and new properties associated with them have appeared on the scene one after another. In a certain sense, the new properties all behaved like "charges". By this we mean that they are accompanied by a gauge field, transformation properties, and conservation laws that hold within certain conditions. The latter restriction refers to the fact that certain interactions can only come into effect if their effect is not squeezed out by another, more powerful interaction. In such instances, the symmetry of the former only proves to be approximate. (For example, within a nucleus of an atom, where strong interaction prevails, we cannot regard as significant the effect of the electromagnetic forces which are several magnitudes weaker.)

The newly appeared (abstract) charges have new physical characteristics, which display gauge invariance under certain transformations. It is most common to give these transformations as the representations of the groups that characterize them, in the form of providing the matrices of the transformations belonging to the group. Discovered between the 1930s and 1950s, such abstract charges included: spin, isotopic spin, baryon number (also known as baryon charge), lepton number (lepton charge), strangeness, hypercharge (a combined quantity). These were later joined by others, and to this day we cannot consider the list finalized. Their symmetries correspond to invariances under rotations of an abstract vector in a particular abstract field. While rotations are continuous transformations, the rules of quantum field theory dictate that these vectors can only occupy certain discrete positions in their field.

There is, however, a symmetry which is discrete from the outset and which for a long time appeared to be the most unbreakable. And this is the invariance with regard to space reflection, the specialist term for which is parity. Parity signifies that something points to the right or to the left in real space, while its symmetry expresses the equivalence of the space directions. Parity as a particle property also behaves as a charge-like quantity. In 1956 the two American Chinese physicists Ch. N. Yang and T. D. Lee predicted in theoretical terms the potential violation of parity. Furthermore, of the known interactions, parity is only violated in one, namely so-called weak interaction. This discovery shook the world. It seemed that a philosophical taboo had been broken. It was not long, moreover, before Ch. S. Wu succeeded in displaying the predicted phenomenon in an experiment.

We imagine atoms, and their nuclei, as displaying spherical symmetry. If the space directions in the world are equivalent, then — when decay occurs in a nucleus — the decay products must be emitted in any direction with equal probability. If we set the atoms in a certain space direction (for example with an external magnetic field), we should expect the emission of the same number of particles to the right as to the left, or 'upwards' as 'downwards'.

The most everyday weak interaction is the β decay of atomic nuclei. Wu and her colleagues examined the β decay of deep-cooled Co^{60} nuclei in an outer magnetic field, which put the spin of almost all the nuclei into

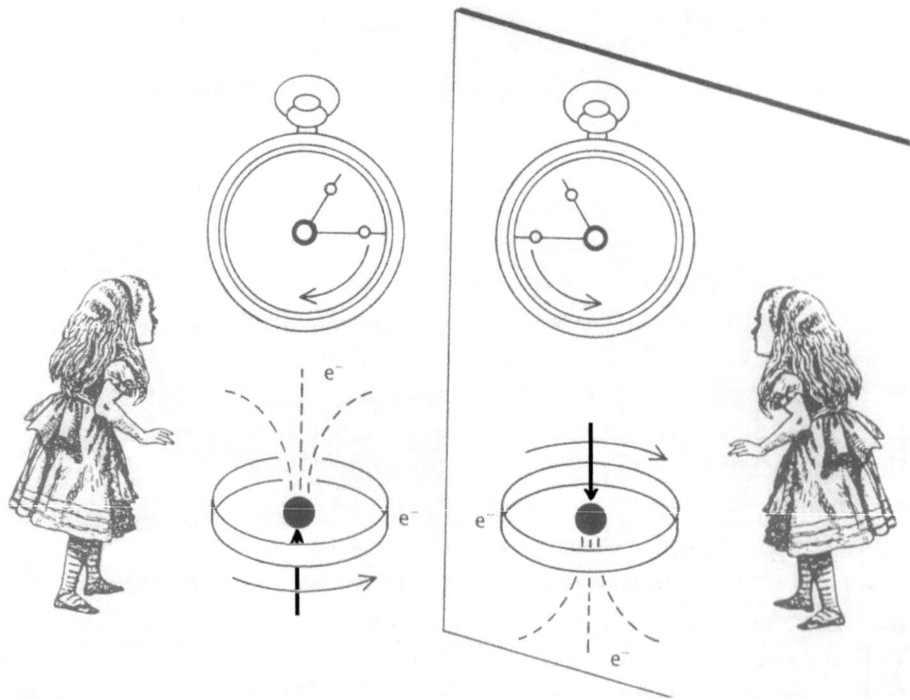

Figure 11.5. *In the parity mirror, Alice sees that the space directions are mirrored, but that electric charge is not (the negative charge remains negative in the mirror, too). The direction of the magnetic field generated by electric charge, however, is mirrored in a similar way to the hand of the clock, as a result of which, to Alice's surprise, during the beta decay of the atom placed in the middle of the wires generating the magnetic field, the direction of the emitted electrons has to change. In the real world, the electrons are emitted from the downwards-pointed magnetic field in the opposite direction, upwards. In the mirror, the electrons revolve in the wires in the opposite direction, so the magnetic field points upwards, and the electrons should be emitted downwards. This does not, however, correspond to the reflected image Alice was expecting. The symmetry of the parity mirror has been broken.*

a single direction — parallel to itself. The figure shows the experimental arrangement and its assumed mirror image. In the mirror the direction of the outside magnetic field parallel to the plane of the mirror is unchanged, while the direction of rotation of the mirror spin is reversed. As a result of this, in the mirror it is the nuclei with downward spin that are visible. Assuming the original *Co* nuclei were right-handed, the mirror nuclei will be left-handed. The real experiment showed that more electrons were

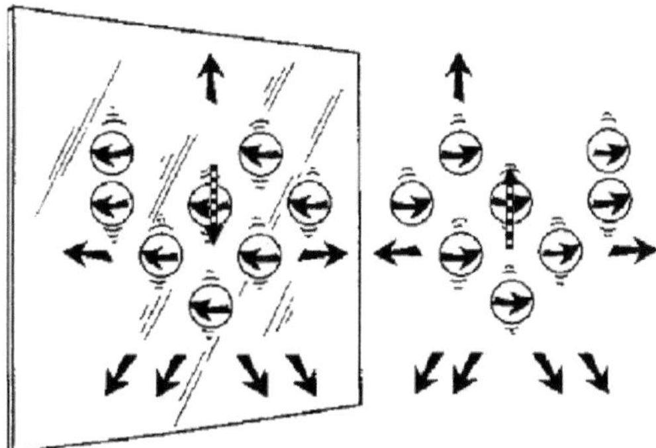

Figure 11.6. *The mirror image of atoms with spin in the same direction (the upward direction is marked with a dotted arrow). The upward spin corresponds to a rightward direction of rotation. The mirror image of this will be rotation to the left, which — assuming the mirror image preserves right-handedness, i.e. parity — will correspond to a downward spin. If in an experiment a higher number of particles are emitted in a direction opposite to the spin, the mirror image of this will demonstrate that there are more particles emitted in the direction coinciding with that of the spin (after Ne'eman and Kirsh)*

emitted in the opposite direction to that of the nuclei's spin, i.e. downwards, as did in parallel with the nucleic spin, i.e. upwards. In Figure 11.6, the greater number of electrons leaving downwards is shown by the arrows pointing downwards. The mirror image of the arrows representing a greater number of electrons pointing downwards is more downward electrons. True, but in this instance the mirror electrons display an opposing correlation to the spin of the mirror nuclei, for the greater number of electrons are emitted in the direction coinciding with the downward-point nucleus spins. In this instance the parity of the space has been violated. If the parity of the space were not violated, then we would get the result that the mirrored electrons give preference to the direction opposite to that of the nucleic spins, as in the real world, i.e. we would see more of them being emitted upwards. This, however, would contradict our expectation the mirror-image of downwards should be downwards. It is this that the experiment has to decide. Only one result would have proved the conservation of parity: if the β rays (electrons) had been emitted upwards and downwards with the same likelihood. The inspection of the result had to be strengthened during the mirror experiment. Wu and her

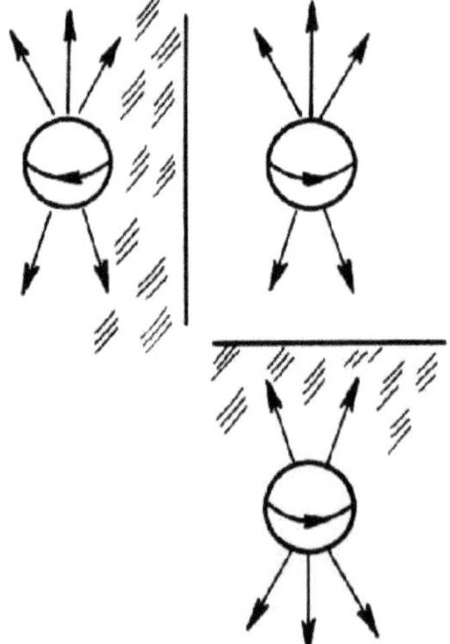

Figure 11.7. *The two options for the location of the mirror*

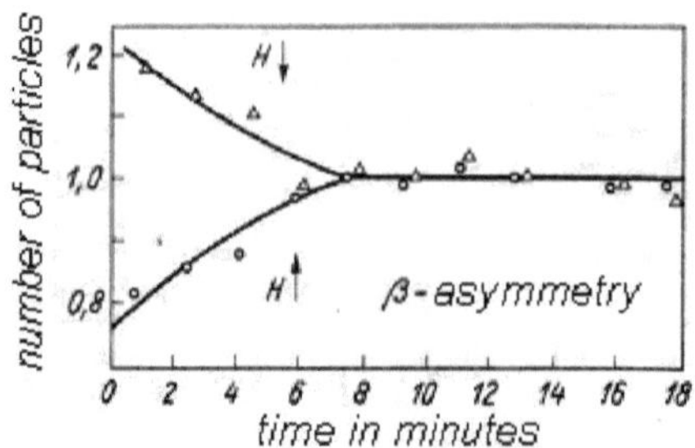

Figure 11.8. *The results of Wu's experiments*

colleagues modelled the mirrored world by reversing the outside magnetic field. In point of fact, this corresponds to the situation in which the mirror is placed perpendicularly rather than parallel to the direction of the magnetic field, though this does not influence the former deductions made from the asymmetrical distribution (Figure 11.7).

Their experiment performed in a mirrored situation confirmed the results. The results they measured before and after the reflection of the magnetic field are shown in Figure 11.8.

Previous to Wu's experiments, we had had the following picture of the world: parity is conserved in all interactions; all interactions are invariant under reflection; all processes appear to be realizable in the form in which we see them in a mirror; nature does not distinguish between left and right.

In contrast, the Wu experiment proved that parity is not conserved in weak interactions; that weak interactions are not invariant under reflection; that the mirror image of a weak interaction might describe an impossible process; that nature does distinguish between left and right.

Previously, the world of physics had been universally convinced that any physical process could be realized in a mirrored world just as in the real one. That is to say, Alice would not have had any trouble satisfying her daily need for milk in the Looking-glass world any more than in the real one. This belief was laid to rest in a short period of time. In 1957 Wu's experiment was followed by a few others in the space of a few months, which verified that parity was violated in weak interactions. The result shocked the world of physics. Never had a Nobel Prize been awarded for a discovery in so short a length of time! For without the experimental verification conducted by Wu and her colleagues on 4 January 1957 and published toward the end of that month, Ch. N. Yang and T. D. Lee would never have been awarded the Nobel Prize within that same year. The press and the public at large also picked up on the news.

Now we could exclaim: but we have already encountered just such a phenomenon! The situation was indeed very similar in the case of the experiment that gave the young Mach such intellectual trauma. Only, *firstly*, the source of Mach's problem was slightly different. He was pondering on the way in which an entirely symmetrical arrangement in a single plane that the conductor and the magnetic needle represented could lead to an asymmetrical result turning out of that plane. The cause, the current passing through the conductor, appeared to be symmetrical. According to P. Curie's principle, a symmetrical cause cannot generate an asymmetrical effect. Yet there the effect was an asymmetrical one. We were able to explain the outcome with a process occurring at a deeper level, namely that in the magnetic needle the magnetism is generated by atomic-scale circular currents that *ab ovo* turned away from the plane of the experimental arrangement. *Secondly*, as far as the experiment's claims that relate to reflection are concerned, they were explained by the visual paradox that while in the mirror situation the magnetic poles at the two ends of the needle were commuted, the new north pole of the mirrored magnet was not suddenly painted black, just as the south pole was not suddenly painted white.

This is not the case in Wu's experiment. Here we genuinely experience things in the mirror world that could not take place in the real one. And yet! Nature has treated the left hand world differently from the right hand world, after all. It has favoured the left over the right, or vice versa. The latter is a matter of point of view, but does not affect the fact that the difference exists.

As a first step to understanding this, let us go back to geometry. Let us place a right-angled coordinate system in front of the mirror, whose *yz* plane is parallel to the mirror, and whose *x* axis is perpendicular to this, say in a direction away from the mirror. Let the three axes be right-handed in alphabetic order. The mirror image of this coordinate system will a left-handed coordinate system in which the direction of the *y* and *z* axes does not change, while the *x* axis points in the opposite direction, similarly pointing away from the mirror, but from its other side, in the other direction. As two of the coordinates of any point in the system have preserved their sign, and the third has its sign changed, so the parity of the system has switched its sign (Figure 11.9).

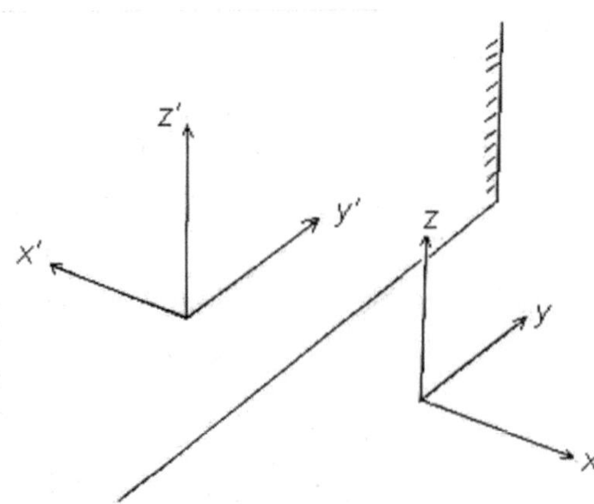

Figure 11.9. *The mirror image of a right-handed coordinate system is a left-handed one*

While this is not what we mean by real space reflection, this is a good model for the creation of a mirrored world, because essentially the same takes place. In precise terms, what we mean by space reflection is when all three coordinates of a given point change their sign (central reflection). In this instance, the direction of all three axes is swapped, and as the parity changes are multiplied together, $(-1) * (-1) * (-1) = (-1)$, the parity also changes sign after an odd number of reflections.

In and of itself, the change of the sign of the parity does not upset the symmetry of nature. We can speak of the breaking of symmetry when we have a physical experiment at our disposal which we can use to determine whether we are in the real world or the mirrored one. Previously we believed that our laws of physics do not make this possible. This belief has been laid to rest. Alice is able to decide what kind of milk she is drinking.

What happens when we reflect two coordinate axes? This is the situation that comes about upon reflection in an axis. For example, if the axis of reflection points in direction *z*, upon reflection the *z* coordinates of a point will remain unchanged, while its *x* and *y* coordinates will change sign, while the direction of rotation of the three axes relative to one another remains unchanged. Thus neither does the parity change: $(-1) * (-1) = 1$. In general it is true that in the case of the reflection of an even number of axes, parity does not change sign. Thus in a four-dimensional space, for example in four-dimensional space-time, if we perform spatial and time reflection at the same time, the parity will remain unchanged. We can interpret time reflection as physical processes taking place "backwards" compared to those in real space-time. It does not follow from all this, however, that in the course of spatial reflection nature should not treat right-handed physical systems in the same way it treats left-handed ones.

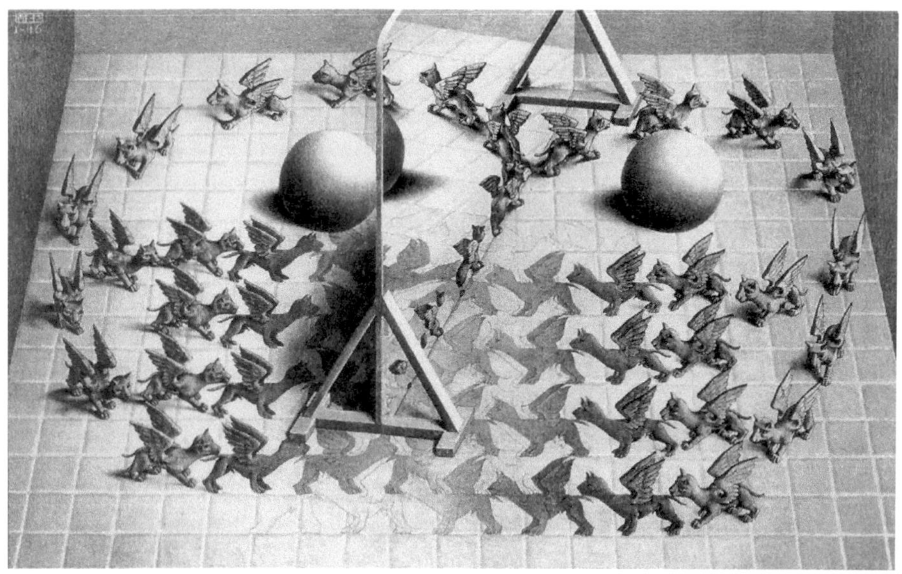

Figure 11.10. *M. C. Escher's magic mirror* (1946)

Yet this is exactly what nature does. It must have its reasons. In Smurf Village every smurf knows that if they experience some extraordinary phenomenon, they should suspect that a satanic wizard they call Gargamel lies behind it. Gargamel can ruin everything in the smurfs' world, and every bit of Brainy Smurf's cunning and time is required to uncover Gargamel's pranks. In the world of physical particles, Gargamel is called a *neutrino*, and possesses the imperfect quality that it only travels in a left-handed form. Its spin always displays a left-handed direction of rotation relative to the direction of its motion. To this day, no opposite spin neutrino has been found. We now also know the brothers of neutrinos and antineutrinos (μ- and τ-neutrinos), but we do not know a chiral pair for either of them. In contrast to neutrinos, antineutrinos are all right-handed; in their family there are no left-handed examples to be found. Apart from neutrinos, all other particles exist in forms with both types of spin. In the case of those possessing mass this is not so surprising, but even the photon, with zero rest mass, can have spin in either direction.

Parity violation only occurs in the course of weak interaction. Neutrinos only participate in weak interactions (ignoring gravitational interactions). For example, negative β-decay produces an electron and an antineutrino. Half of the integral spin lost by the transforming nucleus is taken by the electron, the other half by the antineutrino, which also carries energy and momentum. As the half-spin it takes points in the direction in which the nucleus' spin pointed, it also has to leave in the same direction. (By necessity, the electron has to leave in the opposite direction.) The neutrino is therefore Gargamel, who is responsible for the asymmetrical spatial emission and the parity violation of weak interaction. The devilish law which governed Mephisto's departure also ties Gargamel's hands.

At present, not only are we unaware of any right-handed version of neutrinos, neither do we know of any law of nature which might conclusively rule out their existence. It would no doubt warm many hearts if one fine day, near a high-powered accelerator or somewhere in space, it proved possible to catch one or two examples. But attention should be drawn to the fact that this would not in and of itself make a significant change to the asymmetry of nature. Fortunately, in their everyday lives, the inhabitants of Smurf Village will still only encounter a Gargamel with two left hands. His right-handed version, if it exists at all, probably only occurs in nature in very small numbers, or in circumstances so extreme as unlikely to be faced by an ordinary smurf. It is possible that both these conditions exist. In such an instance we can start to consider what the cause of this symmetry breaking might be.

The existence of neutrinos was predicted by W. Pauli in 1933, and their first experimental demonstration was in 1955. John Updike (1932–) even wrote a poem about them, entitled *Cosmic Gall*: "Neutrinos, they are very small./They have no charge and have no mass/And do not interact at all./ The earth is just a silly ball/To them, through which they simply pass,/Like dustmaids dawn a drafty hall/Or photons through a sheet of glass ... " We should note that for the comprehension and acceptance of certain physical facts, and the experimental verification of theoretical predictions, the right conditions have to emerge. Assumptions must find themselves in harmony with the other observations we have made of the world in order for us to take them seriously. H. Weyl, whose name we have repeatedly mentioned in connection with symmetries and their physical applications, wrote an article as early as 1929 which included an equation which we now know was capable of describing a particle (the as yet unknown neutrino) that is responsible for parity violation. This article was rejected by the physicists of the time, however, because it evidently violated parity. The Herodotean reflex that people had acclimatized to over millennia was still at work. The belief in the symmetry of the world pushed rational arguments into the background. The significance of Weyl's equation was acknowledged in 1956, one year after his death, following the discovery of Lee and Yang.

Let us remember the even later example of quasicrystals. The spatial arrangement of molecules displaying fivefold symmetry was for a long time rejected by the world of crystallography. When D. Shechtman and his colleagues proved that they had found such material and that this was not just the result of an erroneous image, but an existing structure, it did not take long to identify dozens of quasicrystals. The discovery of parity violation dissolved similar mental obstacles. Aside from the fact that parity violation was displayed in a number of weak interaction processes, it transpired in the same year that weak interaction violates charge conjugation, too. Gargamel's reach stretches further than we first thought! No only do natural processes distinguish between left-handed and right-handed worlds, they also violate the invariance of the commutativity of positive and negative charges. Space reflection and charge conjugation originally appeared to be independent phenomena. And yet, as soon as we build the violation of space reflection into a theory that reflects the experimental facts, the theory leads us to the observation that the charge conjugation is also, by necessity, violated.

The simultaneous violation of parity and charge conjugation raised hopes that, if reflected together, the two might be conserved. If we were able to prepare a magic mirror that not only changes the left to the right, but replaces particles with their antiparticles while so doing, this mirror world would be governed by the same laws of physics as the real world. That is, while nature violates parity (*P*) and charge conjugation (*C*), it remains invariant under the simultaneous execution of the two operations (*CP*). Symmetry would not be broken if in Wu's experiment the Co^{60} atomic nuclei were replaced by antiatomic nuclei.

We can depict the symmetry of charge conjugation if we substitute the charge as a property with a property that can be represented graphically, for example a colour. Such complex geometrical + colour reflection invariances appear, for example, in a family of drawings by M. C. Escher (Figure 11.11).

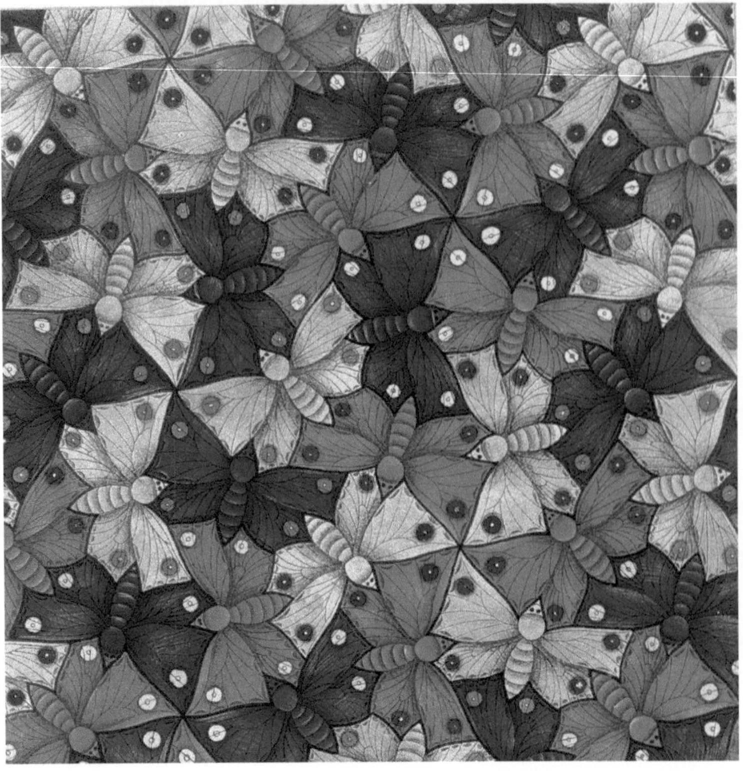

Figure 11.11. *M. C. Escher: Symmetrical drawing 70* (1948)

Ever since Descartes, doubt has been an accepted methodological principle of scientific thought. No sooner was the seed of the doubt sown, as indeed parity violation sowed it, that it was to sprout again and again. Not many years had to pass before the absolute truth of *CP* invariance would be thrown into doubt. As we mentioned, in 1964 V. Fitch and J. Cronin discovered a *CP*-violating experimental phenomenon when examining the decay of *K* mesons. What is more, this was during an experiment with which they wanted to verify *CP* invariance. Their experiment showed that the invariance with respect to time reflection (*T*) was also broken. The symmetry breaking of time reflection means that a given process cannot take place in nature in reverse. This all occurred in precisely the small number of instances in which *CP* invariance was violated. Nature still appeared to be invariant to the combined execution of space reflection, time reflection and charge conjugation. We call this combined symmetry *CPT* invariance, and so far we have not encountered a process in nature which would violate this combined symmetry.

Table 11.2 displays the conservation or breaking of various physical properties in individual basic physical interactions.

Physical quantity	Is it conserved in *strong*	*electromagnetic*	*weak* interactions?
linear momentum	yes	yes	yes
energy/mass	yes	yes	yes
angular momentum	yes	yes	yes
electric charge (*Q*)	yes	yes	yes
baryon number (*B*)	yes	yes	yes
electronic lepton number L_e	yes	yes	yes
muonic lepton number L_μ	yes	yes	yes
tau lepton number L_τ	yes	yes	yes
strangeness (*S*)	yes	yes	no
3rd component of isotopic spin (I_3)	yes	yes	no
isotopic spin (*I*)	yes	no	no
parity (*P*)	yes	yes	no
charge conjugation (*C*)	yes	yes	almost always yes
time reflection (*T*) or (*CP*)	yes	yes	no
(*CPT*)	yes	yes	yes

Table 11.2.

It can be seen that in strong interactions all symmetries hold. In weak interactions a number of symmetries are violated. Of electromagnetic interactions, it is only isotopic spin which is violated, but its third component (projection) is conserved. The question of the invariance of isotopic spin and its projection deserves attention because it represented a step forward towards the understanding of the structure of a deeper level of matter.

We are only able to sense electromagnetic interaction in a part of space where it is not suppressed by another, stronger interaction. Inside the atomic nucleus the strong interaction suppresses the electromagnetic forces. If this were not the case, the repelling forces of the protons would make the nucleus explode. It is the task of the force holding the nucleus together, the strong interaction, to counterbalance this. Where, that is, the strong interaction suppresses the electromagnetic interaction, we are not able to distinguish the particles with the same or differing charge, because there is no measurable effect with which to do so. This is rather like the way that cats at night are all black. Inside the atomic nucleus the positively-charged proton and the neutrally-charged neutron behave like cats at night. We have to regard them as if they were different states of the two identical particles. We call this double-value state isotopic spin, or isospin for short. Isotopic spin states can be interpreted as charge-like quantities (Yang and Mills 1954) that exist in an abstract charge field, the isospin field. They can rotate in this field, but there are only two stable states in which they can position themselves. Their projections made on these directed positions (I_3) prove to be conserved quantities. Exposed to strong interaction, the proton and the neutron cannot be distinguished, but once they are freed, they can.

Isotopic spin joined the list in which double-valued quantities satisfying the condition of gauge invariance became the subjects of a particular conservation law. As the first quantity with this property was electric charge, we consider all the quantities belonging to this group as charge-like quantities. They are all invariant under a two-parameter special (volume-conserving, unimodular) and so-called unitary (conserving its length of one unit) symmetry transformation, and so we refer to their conservation law as $SU(2)$ symmetry.

The conservation of isotopic spin, however, is thanks to the transformation of a different kind of gauge field from that of electric charge (in similar fashion to the other charge-like quantities). In the second half of the 1950s a few models were developed, which combined the charge-like quantities of particles, primarily strongly interacting ones that satisfied most symmetries. The reason for this was that, thanks to improving experimental opportunities, the number of particles increased quickly, and it was time to make order among them. Around this time it was considered evident that the organizing principle could only be their symmetries and their violation. This approach was to considerable degree thanks to the work of Eugene Wigner, who was awarded the Nobel Prize in 1963 "for his contributions to the theory of the atomic nucleus and the elementary particles, particularly through the discovery and application of fundamental symmetry principles".

The need to establish order was accompanied by the need for prediction. Let us think back to Mendeleyev's (1834–1907) periodic table of the chemical elements, which not only put the elements known at the time in logical order, but made it possible to predict the existence, and even the properties, of the as yet undiscovered elements vying for the empty places on it. In the world of so-called elementary particles, too, the time had come for them to be put in an order from which it would be possible to predict the properties of the particles still expected. Mendeleyev could not know in advance which symmetries his periodic table would involve (in the structure of the electron shells of the atoms, as it would later transpire); now the world of physics put more emphasis than ever before on research into symmetries and their breakings.

The cathartic experience of parity violation, extended with the ambivalent behaviour of isotopic spin under interactions of various strengths, drew physicists' attention to the fact that the effectiveness or breaking of a symmetry is not an absolute phenomenon, and can depend on the circumstances in which it is examined. Western-style thinking tended to favour dichotomous or black and white solutions. This is what was dictated by the classical traditions of European philosophy and approach to nature, from Spinoza (1632–1677) to the antinomies of Kant, from the logical structures based on the theses-antitheses of Hegel (1770–1831) to Russell (1872–1970), Wittgenstein (1889–1951) and the logical positivist view of science of the Vienna Circle. Phenomena described with two-parameter symmetry fitted well into these philosophical systems, at least until empirical experience began to stretch the boundaries of $SU(2)$.

We can characterize a pair of opposites with the properties of its members that are alike or that are dissimilar. Western thought puts the emphasis on the former, Eastern on the latter. Vienna-born Erwin Schrödinger (1887–1961), one of the fathers of the wave description of quantum mechanics, would have thought as follows: the cats locked in the room are in fact *identical*; at best I can distinguish that one is black and the other is white in the particular event in which I go in and turn on the light. An Eastern thinker would think of the same situation as follows: the two cats are *different*; at best, in the event that they are locked in a dark room — where their colour, as a property, no longer plays a role — I am unable to distinguish them.

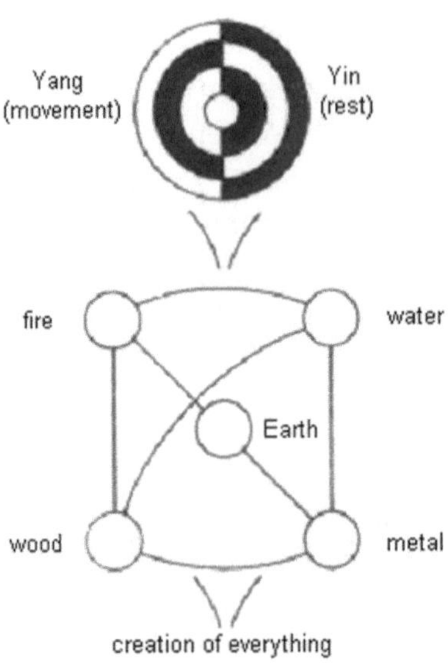

Figure 11.12. *This is how the Confucian Chao-Tse (11th C.) established a connection between teaching on the antithetical pairs of the yin-yang and the interaction of cosmic powers, using them to trace all existing things back to five primary elements*

In the physical world, the violation of left-right symmetry unquestionably brought the Eastern approach to the fore. In this way of thinking, dichotomy is represented not by symmetry but by antisymmetry. The classical symbol of antisymmetry is the yin-yang. The yin-yang is the embodiment of every kind of antithetical pair — whether material or spiritual (Figure 11.12).

The symmetrical equivalent of the antisymmetrical Chinese yin-yang is the Japanese tomoe. We have seen, however, that the tomoe exists in variants that display not only twofold, but manifold, e.g. threefold symmetry. We occasionally encounter similar ones from the earliest period in China, in the form of the triplet of fishes (Figure 11.13), the double version of which, with the fish biting each other's tails, led to the abstraction of the yin-yang.

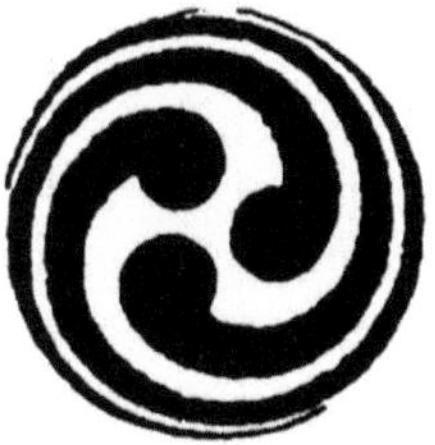
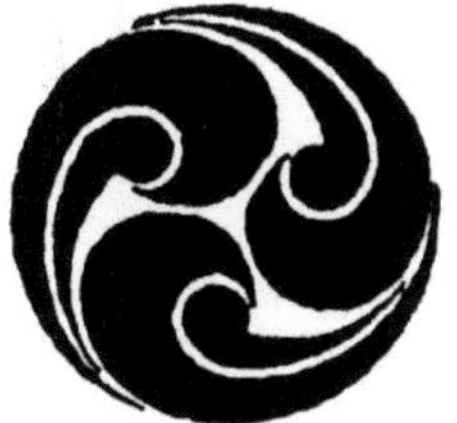
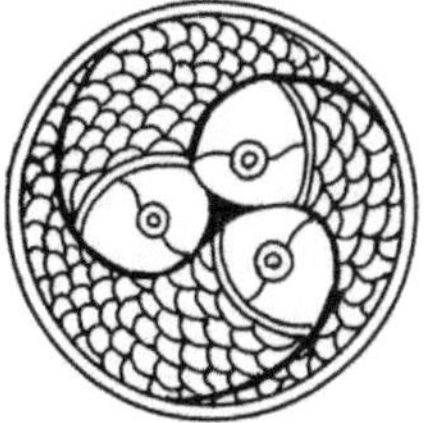

Figure 11.13. *Triple tomoes from Japan and the fish motif with threefold symmetry from China*

The duality of the yin-yang can be used to model the five ancient Chinese primary elements. Trichotomous systems are equally suitable for both human interpretations and those describing nature, like dichotomous ones. Triplicity also appears in the yin-yang, that symbolizes antisymmetry. It is with the triplicity of antithetical pairs that it is possible to generate another interpretation of the Chinese primary elements, the symbol of the eight primary elements (Figure 11.14).

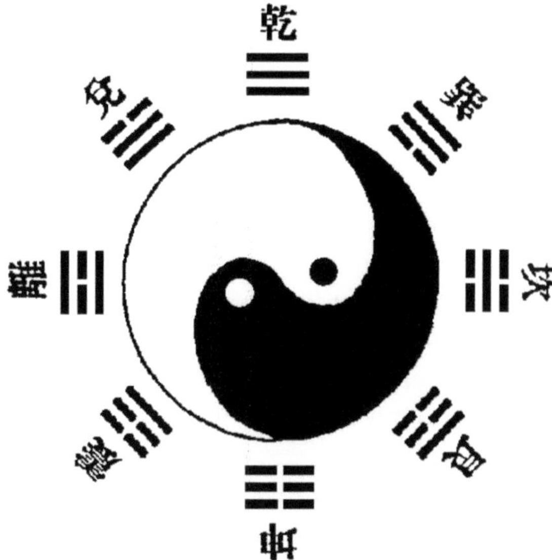

Figure 11.14. *The eight primary symbols of The Book of Changes and their relationship to the yin-yang*

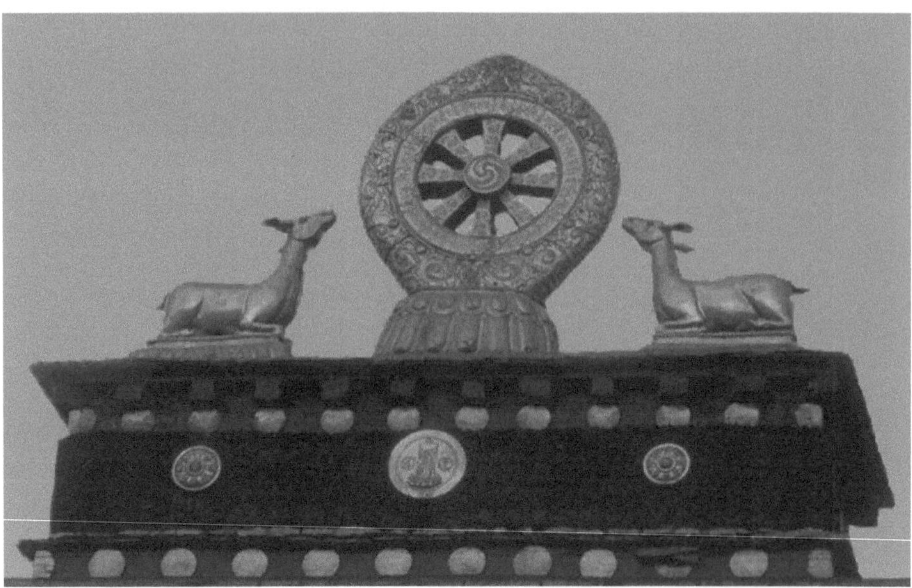

Figure 11.15. *The combination of the eightfold way and threefold rotational symmetry: holy symbols made of gold on the façade of the* Gtsug-lag-khang, *the main church of Lhasa*

With two we get five, and with triplicity we get eight! Yet again the well-known Fibonacci numbers are back to haunt us. The Tibetan cult stone diptych seen in Figure 11.15 is from the façade of the main Buddhist church in Lhasa. It displays the harmonious coexistence of duality, triplicity and eightfold symmetry. For those who grew up in the world of these symbols, these do not appear mystical, but natural. They can be used to symbolize the order that is embodied in the world. For them, the staggering consequences of 1950s particle physics were easier to accept; their imagination could more freely deal with new models. For the most part it was the wise men and women of the East who nurtured the birth of the new symmetries.

After the discoveries of Yang, Lee and Wu, it was Nambu (1921–) who established the first theory of spontaneous symmetry breaking (1959). Sakata produced the first model based on three characteristics (1956) in which he tried to organize heavy particles in multiplets, specifically octets. He took three characteristics into account as organizing principles (the baryon number, the electric charge and the third component of isotopic spin), and built upon three hadrons (proton, neutron, Λ^0) and their antiparticles. By combining their properties, he succeeded in fitting eight semi-stable mesons into a table which represented bound states of the aforementioned hadrons. It appeared that it was possible to construct further particles from a triplet made up of the three hadrons. Instead of $SU(2)$, Ohnuki recommended the use of the three-dimensional continuous Lie group $SU(3)$ to describe the heavy particles regarded as being made up of three baryons. Gell-Mann and Nishijima (1926–) recommended the introduction (1955) of a new (conserved) quantum number, that of strangeness (S), with the help of which, by replacing the baryon number with the sum of the baryon number (B) and strangeness (S), it was possible to establish a simple relationship between these and the projection of isotopic spin (I_3) and electric charge (Q): $Q = I_3 + ½(B + S)$.

If we take a look at the list of the authors and co-authors of the publications which paved the way to the great breakthrough, Eastern physicists are present throughout. In 1961, the American-born Murray Gell-Mann — who can boast of a very healthy number of Eastern co-authors — and Israeli-born Yuval Ne'eman independently recommended the replacement of the Sakata model with a new model of the classification of heavy particles. The new model preserved the $SU(3)$ symmetry recommended by Ohnuki, but rejected the hypothesis that all hadrons were made up of the triplet of proton, neutron and lambda. Their starting point was that the group $SU(3)$ represents a triplet. The operations that transform the elements of these into each other can be characterized by eight mutually independent parameters. They each take the form of a [3 x 3] matrix which transforms the respective states of a given particle. It seemed obvious that eight of the particles to be transformed according to the $SU(3)$ group would each form a multiplet. According to this, p, n, and Λ^0 are parts of an octet (which form a sub-group $SU(2)$ within $SU(3)$). If, however, the hadrons are not built up of these, then they must be built up something else. Gell-Mann referred to the group of heavy particles described by the group $SU(3)$, and to the mode of description as a whole, as the *eightfold way*. The eightfold way that led Buddha to enlightenment played a role in the choice of name. Depicting the octet of the semi-stable baryons as a function of

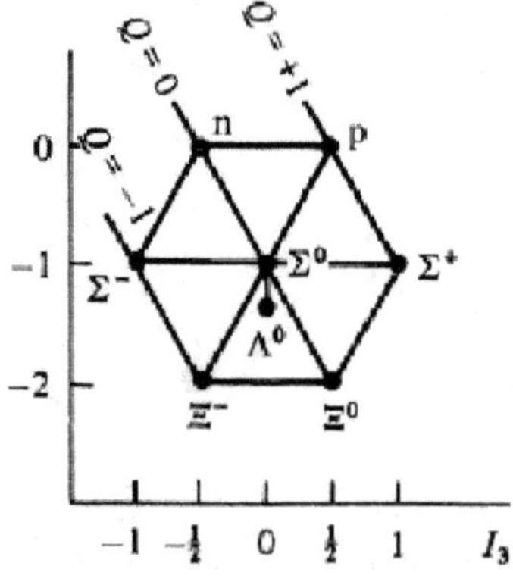

Figure 11.16.

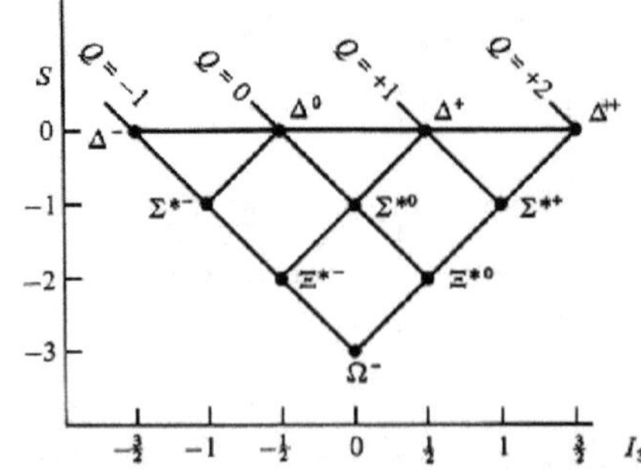

Figure 11.17. *The decuplet of heavy baryons, with the Ω^- (seen below) predicted by the theory. The theory also provided a good estimate of the expected mass of Ω^-.*

charge, strangeness and isospin projection generated the regular picture shown in Figure 11.16.

The model also proved to be suitable for the description of integer spin and less stable particles. We noted that the $SU(2)$ symmetry was approximate in nature; that of $SU(3)$ is even less exact. It could be assumed that it is another (as yet unknown) medium-strength interaction which breaks the symmetry. Its effect depends on the circumstances, and is particularly dependent on energy. This affects the mass of the particles that can be classified in the same multiplet. Gell-Mann and Okubo established a relationship between the different masses of the members of the multiplets, which corresponded well with experimental data. The $SU(3)$ symmetry proved to be effective and a correct description of nature.

In principle, there are four multiplet representations that can correspond to $SU(3)$ symmetry: 1 (singlet), 8 (octet), 10 (decuplet), and 27. In the case of particle multiplets allowing absolute I_3 and S values greater than for those in the octets found first, the particles no longer fit into the octet framework. $SU(3)$ symmetry also worked in the case of decuplets (groups of ten), however. It was with the help of this that it proved possible to predict, then in 1964 experimentally to demonstrate, the existence of the Ω^- particle (Figure 11.17).

Previously, theoretical descriptions had been found for the available experimental evidence which verified its symmetry. Now the situation was reversed. It was on the basis of the symmetry properties predicted by the theory that the first particle was discovered (as, of course, were others afterwards). This success dismissed any doubts about *SU*(3), if they still remained, for good. This was not merely the success of a theory. The history of physics has known many such successes. Here it was the predictive function of symmetry which triumphed. In 1969, the discovery was honoured with the Nobel Prize (for Gall-Mann) and the Einstein medal (for Ne'eman).

The triumph of the eightfold way was not over. Indeed, it opened a new chapter in the history of physics that is still being written to this day.

In 1962, as the extension to the eightfold way, Ne'eman recommended a model according to which all members of the semi-stable baryon octet would be built up from three "more elementary" particles, each with a ⅓ baryon charge. This did not generate interest at the time, but in 1964 Gell-Mann and Zweig (1937–) also presented a model of their own. This model was also based on three particles, and by this time it was extended to include all hadrons. Gall-Mann called these quarks. As at the time quarks were such hypothetical particles, and there was little hope of their being real, the obvious thing seemed to be to give them a name that had no meaning at all. The word 'quark' is a reference to the protagonist of *Finnegan's Wake* by James Joyce (1882–1941), who, on entering the pub, orders "three quarks", and though the word had no meaning even there, the landlord knew what to pour for the gentleman. (Certain gourmets claim that it was a reference to the German equivalent of a type of cottage cheese.)

Quarks not only split the baryon numbers in three, but also the electric charge of the particles. Most of all, they flavoured our world with a new asymmetry. The electric charge of a quark is either ⅔ or –⅓ of the previously known elementary charge, but never –⅔ or ⅓ of it. With antiquarks, precisely the opposite is true. Their charge asymmetry is similar to the spin of the neutrino, with the difference that while the neutrino's spin is asymmetric in real space, the quark's spin is asymmetric in a gauge field. Quarks only establish bonds with one another in combinations in which the sum of their charges is an integer. Furthermore, as far as our present knowledge shows, they only exist in bonds, and never independently.

The number of quarks later grew: we know of essentially six types of quark (and six types of antiquark). Physicists gave the nickname "flavour" to the property that distinguishes the six quarks from each other. It is more precise to say, therefore, that we know of quarks of six different flavours.

The world of the hadrons constructed from three quarks concealed new mysteries for physicists. Among the hadrons, there are particles that are made up of three identical quarks. One example was the famous Ω^-, which is made up of three *s* quarks. Its existence could not be called into doubt, for it had been proved experimentally, yet this contradicted the Pauli principle that in a quantum mechanical system there cannot be two particles in the same state. The theory predicted that the spin of Ω^- would be $3/2$, and in experiments it proved to be so. Yet this could only occur in the event that the spin of all three *s* quarks stands in parallel, in the same direction. Either the Pauli principle so often verified by natural experience had to be abandoned, or the assumption had to be made that quark flavours until then considered identical did in fact differ with regard to another property. Physicists called this further quality "colour", though it has as much in common with the real colours that our eyes sense as electromagnetic waves as their flavour has with curried chicken. The name was rather an expression of the effort that after flavours we should preserve the elegance of "quantum-aestheto-dynamics" (*QAD*) and stay in the domain of the attractive and the pleasant.

There was need for at least three colours to ensure that no two quarks could be in the same quantum state. According to the new, expanded theory, namely that of quantum chromodynamics (*QCD*), there could be quarks of any flavour within a hadron, but at any one time all three had to be of different colours. As quarks cannot be separated, they are held together by an interaction so strong that they each exist "confined" inside a single hadron, with the forces between them transmitted by the members of an eight-member particle family called gluon (from the word 'glue'), which prevent them from separating from each other. In contrast, they keep them in continuous motion, and the individual quarks are continually changing their colour, which is why we "see" them from outside as "white", i.e. as completely mixed in colour, and we are not able to detect quarks of one particular colour or another.

The triplet model seems to be in harmony with a growing body of experimental evidence. It proved possible to extend the $SU(3)$ symmetry to the flavours of quarks, and, further, to their colours, too. This is more evidence of the significance of the discovery of the eightfold way.

Meanwhile the world of light particles, the leptons, also expanded. This was later to help the development of a more symmetrical picture. In addition to the electron and the accompanying neutrino, the muons and the accompanying mu-neutrinos also appeared, as, later, did the tau particles and the accompanying tau-neutrinos. We consider them as three generations of pairs.

Though the details are beyond the scope of this volume, the end of the 1960s saw the birth of the unified theory of electromagnetic and weak interaction. This connected the (e, μ, τ) trio and their corresponding three types of neutrino. Quarks, the hadrons made up of them, as well as leptons are all half-integer spin particles (fermions). The interactions between them are transmitted by medium-mass particles, the so-called mesons. The mesons are all made up of two quarks, more precisely a quark and an antiquark. This makes them able, at appropriate energies, to exchange quarks between hadrons. The mesons are all integer spin (bosons). Four bosons transmit in electroweak interactions: the photon, and three so-called weak vector bosons (W^+, W^-, Z^0). Eight types of gluon transmit between the quarks and change their colours.

To simplify things greatly, this was the development of the picture of the microstructure of matter that is at the disposal of physics to this day and that displays such a great amount of symmetry. Because of its level of acceptance, and despite the multitude of questions still open, physicists call it the Standard Model.

The Standard Model arranges the world of leptons and quarks into three generations. Most of the mass of the everyday matter that surrounds us is made up of elements of the first generation (Table 11.3). This generation comprises the electron and its neutrino, and quarks u and d in all three of their colour varieties. Protons and neutrons are made up of quarks u and d, while atoms are made up of protons, neutrons, and electrons.

$$\begin{pmatrix} \nu_e & u_r & u_g & u_b \\ e^- & d_r & d_g & d_b \end{pmatrix}$$

Table 11.3.

In Table 11.4, which contains all three generations, the subscript indices alongside the particles signify their colour (with the first letter of the basic colours red, green and blue), and a separate column marks their electric charge. Their antiparticles can be arranged into a similar table.

	Leptons		Quarks					
Third generation	ν_τ	0	t_r	2/3	t_g	2/3	t_b	2/3
	τ^-	-1	b_r	-1/3	b_g	-1/3	b_b	-1/3
Second generation	ν_μ	0	c_r	2/3	c_g	2/3	c_b	2/3
	μ^-	-1	s_r	-1/3	s_g	-1/3	s_b	-1/3
First generation	ν_e	0	u_r	2/3	u_g	2/3	u_b	2/3
	e^-	-1	d_r	-1/3	d_g	-1/3	d_b	-1/3

Table 11.4. *The generations of the most basic particles*

The Standard Model offers an opportunity to find a way to describe electroweak and strong interactions at higher energies with a common symmetry. To this end, there have for a long time been attempts to find multiparameter, combined versions of the *SU* symmetry. Attempts with models were trying to reach out towards a supersymmetry. The creation of a unifying theory that also involves the theory of gravity appears much more distant and problematic. Physicists are agreed on one thing, however: a new, general unified theory (or theories, and GUT for short) can be obtained with the help of new *symmetries* that correspond with nature.

The road from nature to man

Chapter 12

Chirality

Morphological and functional symmetry breakings

Alice's worries about the gastronomic value of Looking-glass milk were not so unfounded, after all! Taking a look into the microworld of the physical *structure of matter*, we can have been convinced that the world is fundamentally not mirror symmetric. Before we begin to judge the consequences of this, let us also take a look at the physical *processes*. Events in the material world tend towards lessening differences. A hot and a cold body that touch each other are not going to interact in such a way that the colder gives heat to the hotter one and cools down even further, while the hotter one takes on heat and heats up even more. The majority of physical processes have a direction, and this direction cannot be reversed.

If we split a box in two along the middle, and we fill the two parts with gases of different densities (one will contain more molecules, the other fewer), then open a door in the middle, the less dense gas is not going to flow through the gap and increase the density of the other one. Previously it was thought that if a little demon were placed alongside the door, which observes when and from which direction a gas molecule happens to approach the door, and opens or closes it accordingly, it can pass the gas into one of the halves of the space. The statistical laws of the microworld do not make the operation of such a so-called Maxwell demon possible, however. Even seemingly spontaneous processes are in fact governed by internal regularities. Their direction, that is, is set by a law. The direction of processes signifies the direction or anisotropy, i.e. the asymmetry, of time.

The beta decay of atomic nuclei is also a spontaneous process, but on average, over a longer period of time, we can say what proportion of the atoms of the material will decay during this time. What we are not able

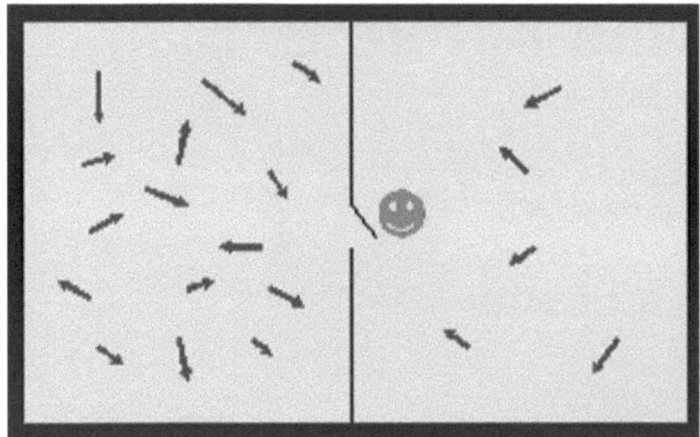

Figure 12.1. *The Maxwell demon*

to say is when each individual one will decay, but statistically speaking we can say how many will do so after a given time, and also that the process of decay will not be reversed.

It is not only amongst physical phenomena that we encounter spontaneous processes. There is the well-known tale of the three dwarves who could not decide which of them was the cleverest. The property of "cleverness" was distributed amongst them with equal probability, so the story goes, and yet one of them was still the "cleverest", i.e. the symmetrical distribution could only be realized in such a way that all three thirds of the probability distributed between the three of them was in fact concentrated in only one of them. In order to find out which of them this was, they all went to the wise elder to ask him to decide which of them was the cleverest. The wise elder acted as follows. He sat the dwarves in a circle, facing one another, and told them: "Here are five hats, three red and two white. I will put one on each of your heads. The first one of you to tell me what colour hat he has on his head is the cleverest." And so it was: he put three red hats on their heads, and hid the two white ones. After a brief silence, one of the dwarves spoke up: "The hat on my head is red!". The question is: how did he work this out?

The cleverest dwarf deserved this honour by assuming symmetry. He considered that if the hat on his head were white, his fellow dwarf would see a red and a white hat, and his fellow dwarf would know that the hat on his head could not be white, because the third dwarf could immediately have exclaimed that he saw two white hats, so his must be red. But as

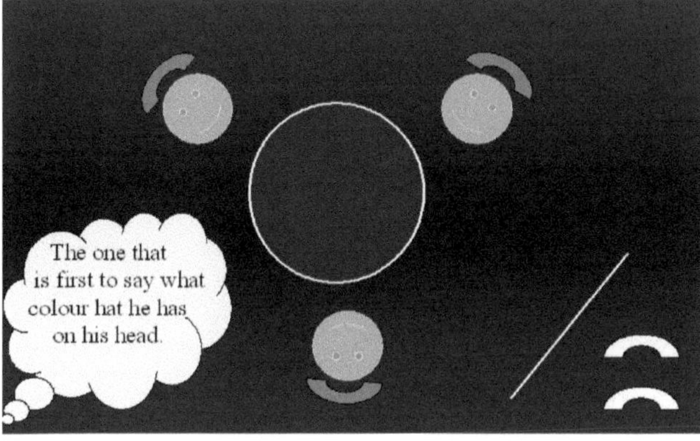

Figure 12.2. *Which is the cleverest dwarf?*

neither of them exclaimed anything, our dwarf deduced that the other two dwarves also saw two red hats.

Naturally, each of the dwarves could have reached this same conclusion with equal probability; the fact that the cleverest of the three exclaimed first is the result of a spontaneous symmetry breaking. Atoms decay with a similar spontaneous symmetry breaking. A single spontaneous symmetry breaking can affect the behaviour of all of the rest of the system. One of the greatest personalities in physics of our age, S. Weinberg, who was honoured in the most enduring fashion for his theory of weak interactions

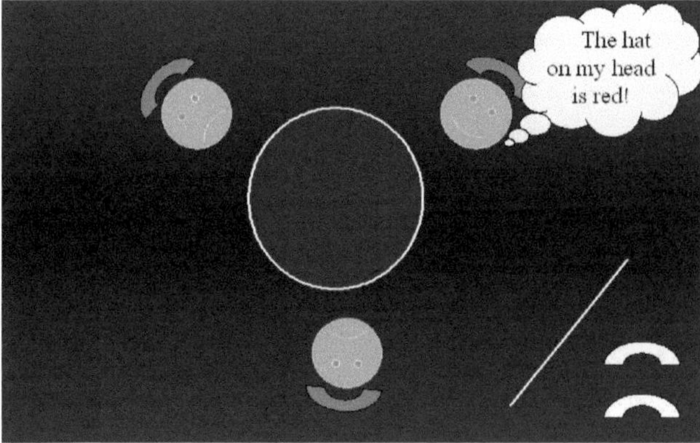

Figure 12.3. *The cleverest dwarf*

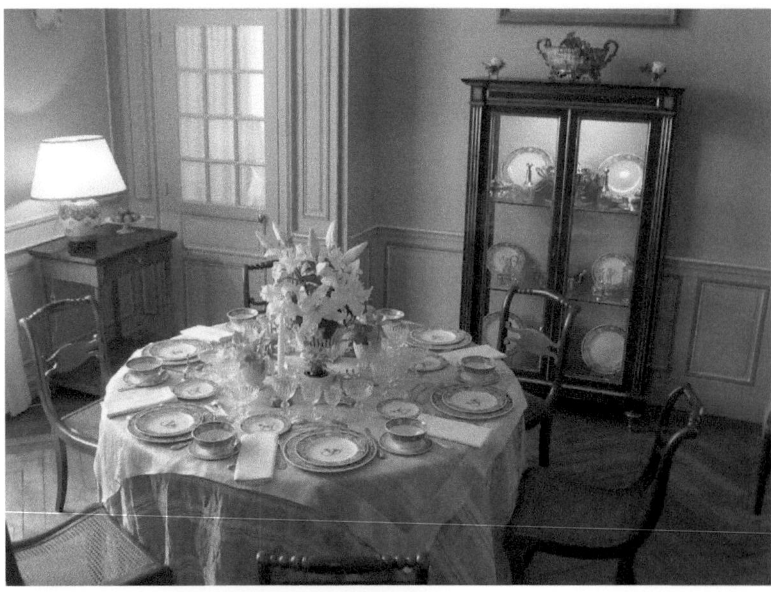

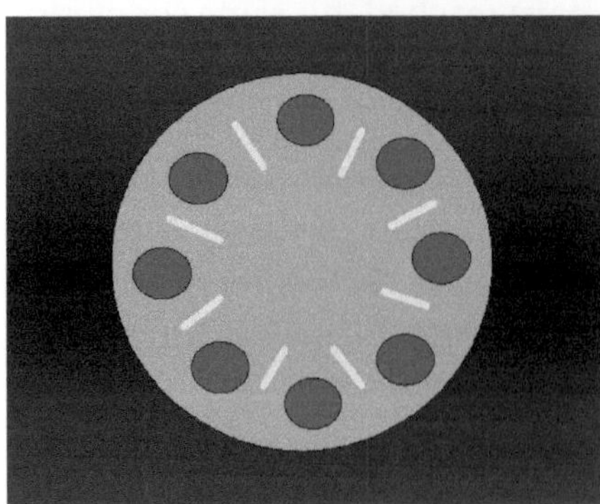

Figure 12.4. *S. Weinberg's example: symmetrical setting of the table*

(he received the Nobel Prize), explained this with the following everyday example (Figure 12.4).

Let us imagine a round table which has been laid for the guests in rotational symmetric form. A napkin has been placed in-between each pair of places that have been laid. The guests take their places, and everyone looks at their neighbour awkwardly, wondering which napkin belongs to their place at the table. Until, that is, one guest calmly picks one up off the table and places it in his or her lap. From that point on, everyone knows that they must reach out for their napkin with the same hand as the

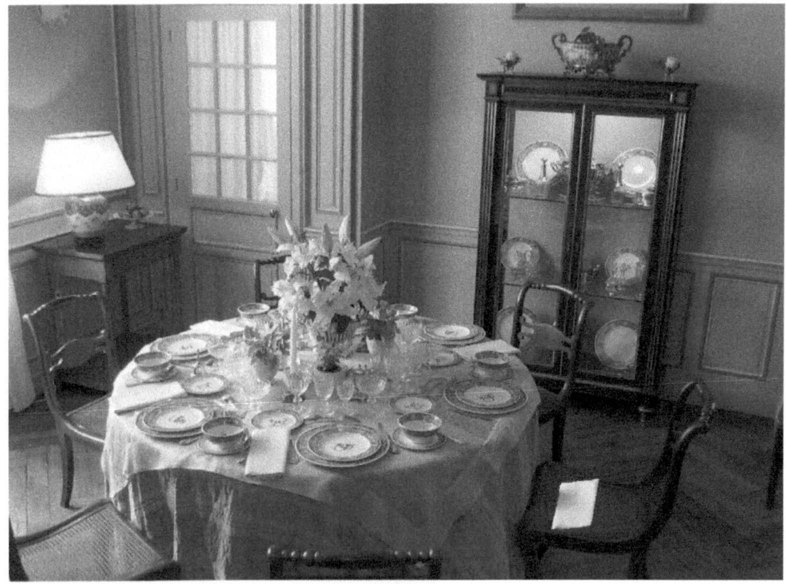

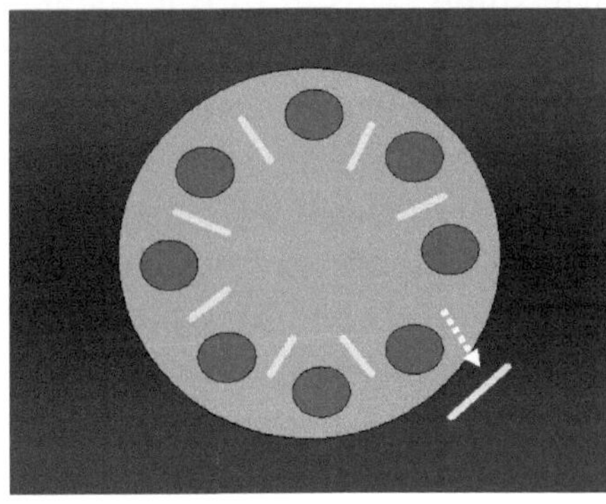

Figure 12.5. *Spontaneous symmetry breaking*

first person did. The first spontaneous symmetry breaking determined the order around the whole table in the blink of an eye. A moment's miniscule symmetry breaking can have an effect on the processes occurring in the whole of a system that is magnitudes larger.

This phenomenon became known as the butterfly effect. As originally used, this term refers to the hypothetical situation in which a butterfly above the Andes flaps its wings, thereby causing a drop of rain falling just at that point to drip down the other side of the crest of the Andes. This trivial change upsets the daily equilibrium between the cloud systems

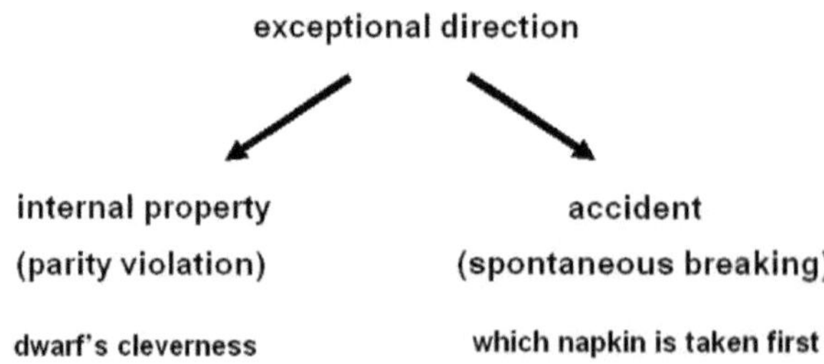

Figure 12.6.

above the catchment areas for the Atlantic and Pacific oceans, the flow of which can determine whether the day after tomorrow it will rain in Western Australia or whether the sun will shine. This in turn influences the expected harvest there, which affects next week's price for grain on the Chicago stock exchange. This in turn decides whether the interest rate for Japanese yen is raised or not, as a consequence of which it will transpire whether the little daughter of employee N. N. at the Japanese car assembly plant in Esztergom, Hungary, will get a Barbie doll for her birthday.

The question of the isotropy or anisotropy of the world concerns the problem of whether or not our world has determined directions. With regard to physical processes, we saw that the spatial preference of direction is related to the temporal one, i.e. the asymmetry of the structure of matter with respect to space reflection brings with it the irreversibility of phenomena with respect to the passage of time, and vice versa. The question for philosophers is whether there is a direction set for the world at all, or whether these observations are only applicable to well-defined isolated phenomena observed under particular circumstances. Put another way: is the existence of a determined direction an *internal* (immanent) *property*, the realization of which is inevitable (parity violation), like the cleverness of the dwarf, or is it rather the result of an *accident* (spontaneous breaking), like the taking away of the first napkin?

The answers given to this question that look back to the historical past signify very much an anthropomorphic approach.

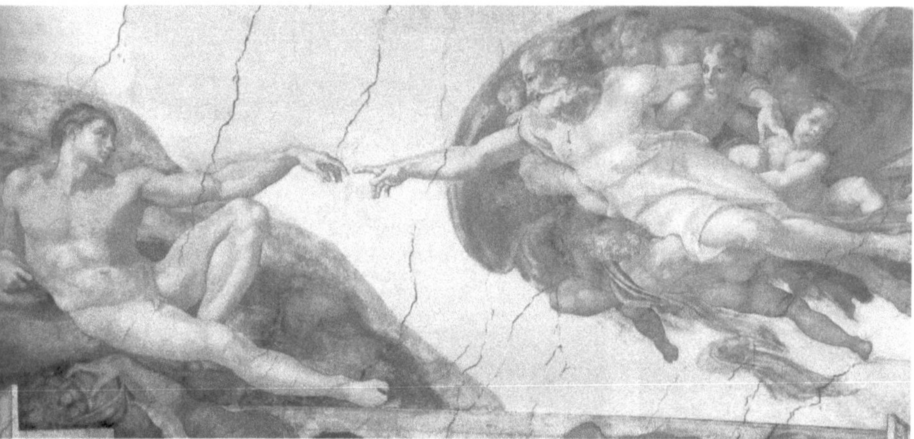

Figure 12.7. *The creation of man. Detail of the fresco on the ceiling of the Sistine Chapel in the Vatican (Michelangelo, 1508–1512)*

The meaning of left and right already display interesting antisymmetries. One meaning of *left* is the left hand direction. It also means bad, as in the multiple meanings of the Latin word *sinister, sinistra, sinistrum*, which in addition to the direction refer to ill fortune and even clumsiness (cf. 'two left feet'). In the approach of the Greeks, the West, if looking North, was to the left, and thereby meant unfortunate, dangerous, ominous. (True, this can be contrasted with the approach of the Romans, according to which, if the prophet looked South, then East was to the left, which had a fortunate, good meaning.) One meaning of right is the right hand direction. As in English, in many languages, for example German, Russian, or Hungarian, it means proper, true, or legitimate. Let us observe the Creation of Adam on the fresco of Michelangelo (1475–1564) on the ceiling of the Sistine Chapel: God creates with his right hand (Figure 12.7).

The thief crucified at Christ's right hand joins him in paradise: in the description of the Last Judgment in the gospel according to Saint Matthew, those to the right rise to heaven, while those to the left are condemned to hell. For most of us, our right hand is our more capable hand. When shaking hands, we extend our right hand. The symmetry of left-right is the symbol of equilibrium (the balance). The disruption of this symmetry means the disruption of equilibrium.

The only question is: is our anthropomorphic asymmetry connected to the butterfly wings of our neutrinos?

We can say that the world around us is *potentially* isotropic. This means that whatever direction we set off in it, and however far we travel, we will encounter similar circumstances. By similar circumstances we mean that processes will take place in the same way. Leaving all external conditions (that do not play a role in the given phenomenon) the same, our experiments will lead to the same result, and we will arrive at the same laws of nature. At the same time our world is *locally* anisotropic. Certain experiments will show that the given phenomenon distinguishes left and right space directions. This is the kind of phenomenon we encounter when studying weak interactions in physics.

By the distinction between potential and local isotropy or anisotropy, we mean that wherever we go in our potentially isotropic world, and whenever we conduct an experiment, we reach the same result. For example, whichever direction we move in and whatever the distance we travelled in a given experiment, we have the same chance of experiencing the same local violation of space directions.

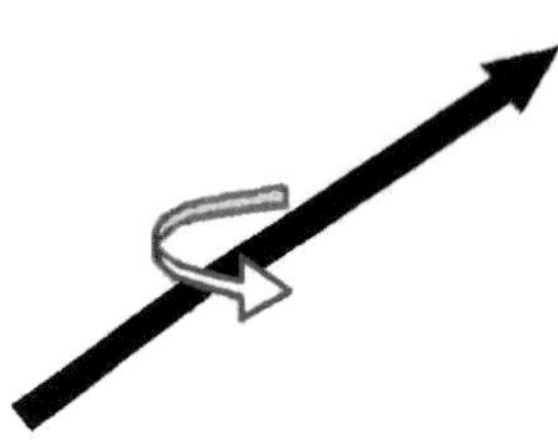

Figure 12.8. *The anisotropy of the world. Direction determined by a moving charged particle*

Local anisotropy is not only a property of weak interactions. This is also true of the direction determined by the moving charged particle (Figure 12.8). It is on the basis of this that we can speak of right- or left-handed systems. (The name is a question of convention, but the fact of the determination of direction is not.) In addition to the electromagnetic example, the angular momentum of rotating mechanical systems sets a space direction for which a conservation law also holds. In the microworld, the own momentum of particles, the spin, and all spin-like properties represent the local determination of space directions. Quantities of this nature behave like a sort of axial vector (pseudo-vector), with the applicable reflection properties encountered previously (if we commute space directions, the direction of the vectors in space is reversed, while axial vectors do not change direction, as they are insensitive to reflection).

To all this we could say 'No problem!': in the microworld we have asymmetries like these, but in the event of a large number of particles, they balance each other out, and at the macro level we do not even notice them. The situation is not so simple, however, if we think of the effects of spontaneous symmetry breakings, for example the case of the napkins. This has aroused in us the suspicion that it is worth being cautious in all instances. Take permanent magnets, for example. The magnetic momentum of a handful of atoms becomes set in a particular direction, and the whole of the crystal adjusts likewise. Their effects are added together and become detectable at the macroscopic level. Chirality, or handedness, is here in our everyday lives, in our appliances, in the Earth's own magnetic field, which we make good use of with compasses, in the radiation originating from the magnetic field of the Sun, and in their effects on Earth.

In line with the well-known Pauli principle, in a single system there cannot be two fermions (half-spin particles, like electrons, protons, neutrons, etc., that which everyday matter is made up of) that are identical in all of their characteristics. Without going into detail, we note that this is the consequence of the commutation relations, thus ultimately of the symmetry properties of the elementary operators that make up the energy (Hamilton) operator of the system. This means that there cannot be two particles — with spin in two possible tracking directions each — in the same energy state. This determines the structure of the electron shells of atoms. The Pauli principle is present in the bonding of atoms, too. If two or more atoms want to form a molecule, it is equally valid for this new system. If, in the course of the formation of the molecule, filling in the states prohibits the implementation of this principle, this circumstance has the same effect as if there were a strong repulsion force in operation between the molecules. The asymmetries created at atomic level are inherited in more complicated formations of material structures. And asymmetries of varying magnitude are indeed present in our atoms.

Let us begin by saying that the world we are made up of is itself constituted of atomic nuclei with positive charge and the clouds of electrons with negative charge revolving and swirling around them. Central forces act between them, and the momentums associated with these central forces are added to the own momentums carried by the particles from the outset. The nuclei of the atoms are made up of baryons (in this instance from

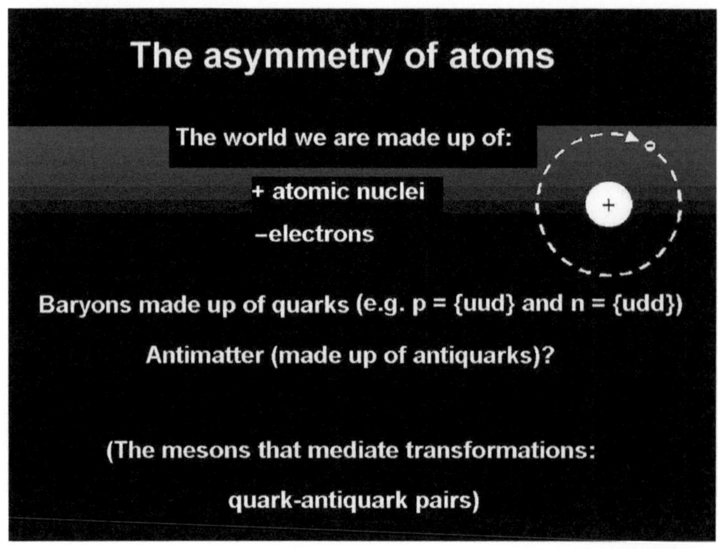

Figure 12.9.

nucleons, that is from protons and neutrons). The individual baryons are made up of quarks. The proton, for example, is made up of two *u* (up) quarks and a *d* (down) quark: $p = \{udu\}$. The neutron, on the other hand, is made up of two *d* quarks and one *u* quark: $n = \{dud\}$.

Antimatter can also be imagined, made up of positrons revolving around negatively charged nuclei, but these can only be produced amidst extreme laboratory conditions, and we have not yet encountered them under natural conditions. However, antimatter is nothing like as rare a phenomenon lost in a distant world of mystery as we might think. It is here in our everyday lives. Antiquarks, or the antimatter of quarks *u* and *d* that make up protons and neutrons, are considered to be simply those quarks which correspond to *u* and *d* in every respect, but with electric charge of the opposite sign. The properties of quarks and antiquarks are shown in Table 12.1.

We saw that in the course of weak interaction protons and neutrons can transform into one another. This transformation does not only occur in weak interaction, however. In the case of baryon-baryon collisions, their transformation is transmitted by mesons. For example, a proton can transform into a neutron in such a way that the process is transmitted by

	electric charge	baryon number	isospin
u	2/3	1/3	1/2
$\bar{u}$	–2/3	1/3	1/2
d	–1/3	1/3	1/2
$\bar{d}$	1/3	1/3	1/2

Table 12.1.

a meson with positive charge π^+. One possible way this can occur is as follows:

$$p + p \to p + n + \pi^+$$

Figure 12.10 shows the nature of the process in four stage, on the basis of the so-called string theory.

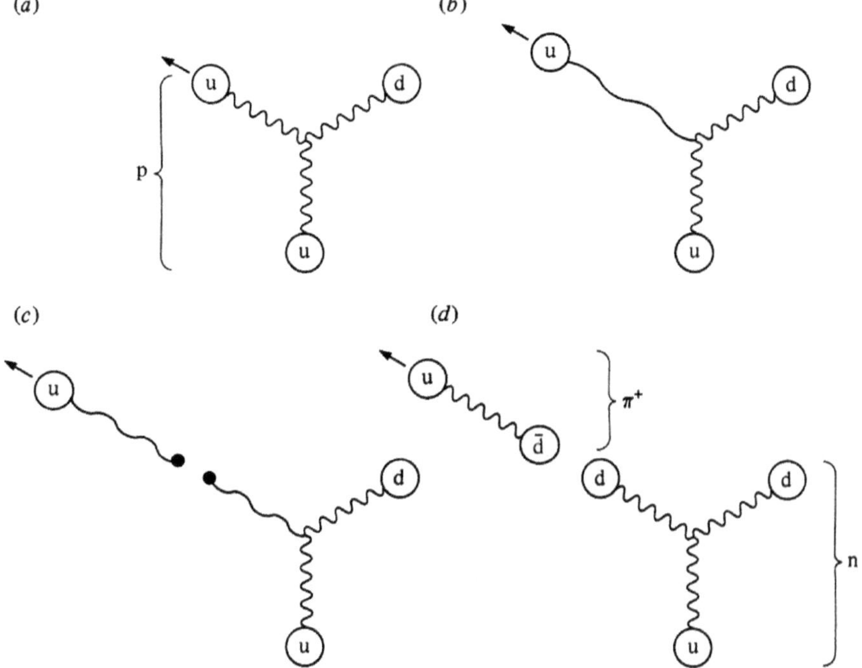

Figure 12.10. *The creation of a quark-antiquark pair (meson) (after Ne'eman and Kirsh)*

One of the strings holding together the quarks of proton {*udu*} (*a*) is stretched (*b*), then snaps (*c*). A quark-antiquark pair ($d - \bar{d}$) is formed at the broken ends. As soon as the string has snapped, the *d* quark becomes connected to the original baryon, creating a {*dud*} formation, i.e. a neutron. The quark-antiquark pair that becomes detached, meanwhile, will create a ($u\bar{d}$) meson we refer to as π^+-(image *d*). All mesons are made up of quark-antiquark pairs. As intermediaries they are present in all interactions that involve the transformation of baryons.

Mesons are materials that are permanently present in nature, and thus through them antimatter is present everywhere in our environment in the form of antiquarks. The fact that the prefix *anti* merely refers to their charge having an opposite sign can perhaps help disperse some of the mystique we attribute to antimatter. For we have long been aware that electrons have their positive equivalents. We know of positive β-decay just as we know of negative β-decay, and the former is just as natural a phenomenon as the latter. Why should quarks not have brothers with opposite charge among the family of particles, just as the electron has a brother with (+) charge?

Where protons collide, the π^+ mesons thereby produced will in most instances decay. Most often they decay into a muon and a muon-neutrino; more rarely, a photon also appears amongst the decay product. Even more rarely — on average once in every 8000 instances — a positron and an electron-neutrino are created ($\pi^+ \rightarrow e^+ + \nu_e$). There is a high probability that the positive muons further decay into positrons and neutrinos. We detect the latter two events as positive β-decay. Neutrons also have their decay processes, with similar decay products. Wherever we encounter protons and neutrons, (and this is what our atoms are like, after all), Gargamel has stuck his finger in things: neutrinos are present in the everyday lives of our atoms. As neutrinos and antineutrinos only appear in nature in one possible chiral form, asymmetry is present everywhere in atomic processes.

The asymmetry of the spin of neutrinos cannot be dismissed with a single gesture, that in great numbers they will counter each other's effects. However many atoms we cram into a body, each will only produce neutrinos with one type of chirality. This asymmetrical property, which is present in the atoms, is further inherited by molecules and macromolecules, and by any larger material structure made up of them. All of them carry this tiny asymmetry inside, and we cannot know in advance when it will lead to a (butterfly) effect that involves the whole material system.

The asymmetry hidden in atoms is very small indeed. In most interactions it can be ignored. If we ignore it, then symmetrical atoms generate symmetrical pairs of molecules in nature with equal probability.

Symmetrical pairs of molecules are easy to separate from one another. Let us take our two hands as an example. If we hold our palms in front of our eyes, we see them as symmetrical. And yet, however we rotate them, we cannot make them overlap each other. On a planar projection, our hands display mirror symmetry, but as they also have spatial thickness, the reflection cannot in practice be performed. However we rotate a left hand glove, we cannot pull it over our right hand. This kind of symmetry, where left-handed and right-handed pairs cannot in practice be overlapped, we call *chiral symmetry* (from the Greek word for hand and handedness); we refer to the phenomenon as a whole as *chirality*. We encounter many such phenomena (for example two columns twisted in opposing directions, as seen in Figure 12.11).

In chemistry — similarly using a term of Greek origin — we refer to groups of atoms that are mirror symmetric but cannot be overlapped as *enantiomorph* pairs, or enantiomers. Let us imagine a regular tetrahedron, say. Let us imagine an identical atom at each of its four vertices. The molecule has mirror and rotational symmetry. If, however, we exchange one of the atoms for one from an atom group that does not display three-fold symmetry, then this has an effect on the binding directions stretching from the vertices of the tetrahedron and its mirror image (which can 'ruin' the regularity of the tetrahedron). The two mirror-symmetric molecules cannot be overlapped with their own chiral mirror images (Figure 12.12). Both molecules can be realized, but they are distinct from one another. (In general, we refer to molecules that display different spatial structures but the same chemical composition as stereoisomers of one another. Enantiomery is a special case of this.)

The more complicated the molecule we create, the greater the chance of finding structurally different molecules among the chemically identical molecules created. Of these, the mirror image pairs that cannot be

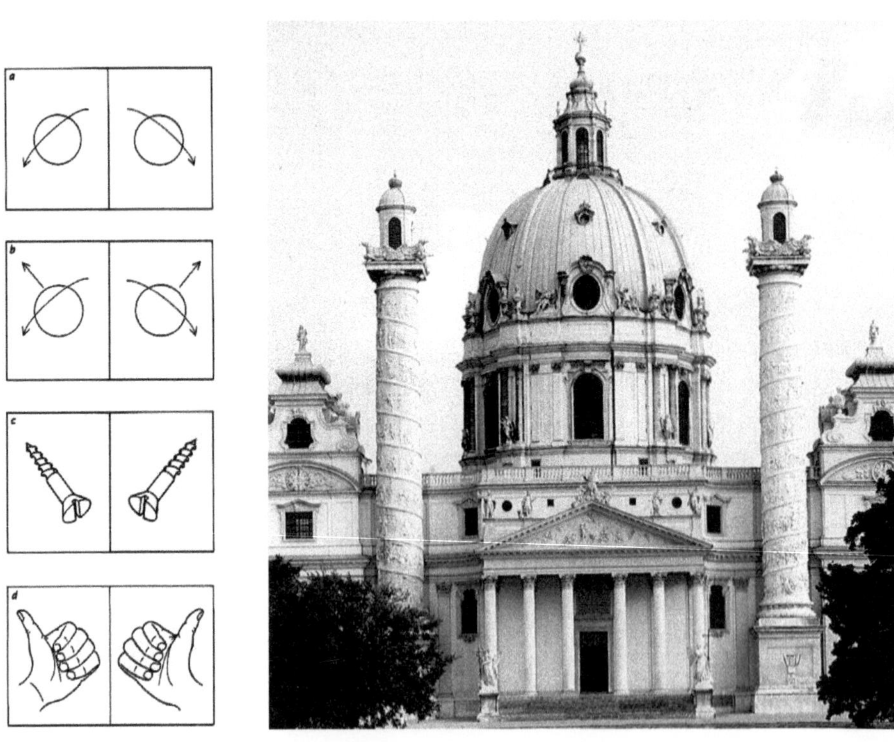

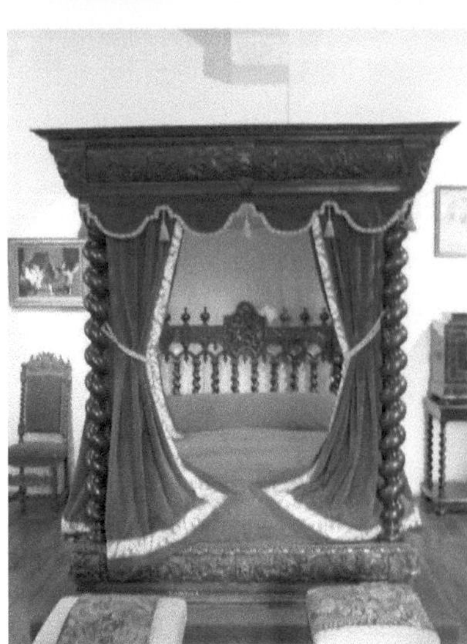

Figure 12.11. *Chirality schematically and in contrarily twisting pairs of columns. The columns of the Karlskirche in Vienna, and Leonardo da Vinci's bed in Clos Luce castle by the Loire, in which he died*

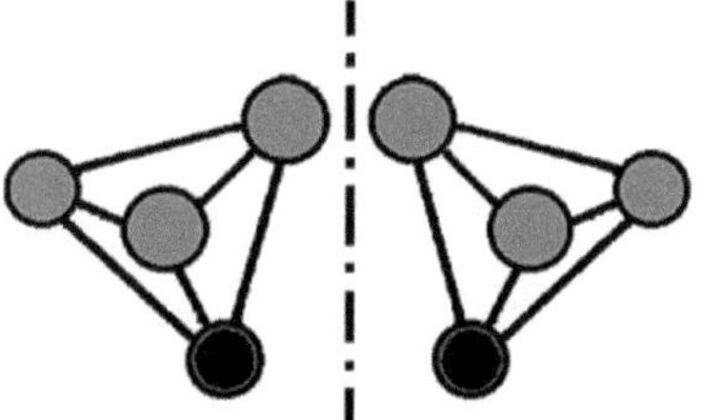

Figure 12.12. *Mirror symmetric group of atoms. Entaniomorph pair*

overlapped are enantiomers. Figure 12.13 shows a right-handed and a left-handed amino acid molecule.

We must mention the chirality of crystals, for this was not discussed previously. The majority of the 32 three-dimensional point groups are mirror symmetric. For the rest, we can find enantiomorph crystals. And in the case of enantiomorph crystals, we speak of optical chirality when they rotate the plane of the polarized light to the left or to the right. Even a long time ago, it was possible to observe this by using a simple experiment.

If we assume of individual atoms that they are symmetrical, and that, internally as well as morphologically, they do not contain asymmetries, then we could expect nature to create both members of the various enantiomer molecule pairs in equal proportion. In laboratory conditions this

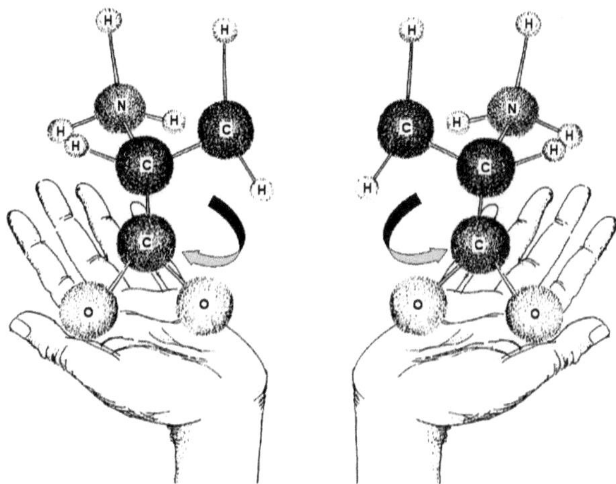

Figure 12.13. *Right-handed and left-handed amino acid molecules (after R. N. Bracewell)*

is generally what happens. This is not surprising, for we know that the internal asymmetry of our atoms is trivial in size.

More interesting is the examination of particular organic enantiomers. The differences are clearly visible on Figure 12.14 in the *H–C–OH* and *HO–C–H* atom groups, which are reversed in the two instances with respect to the others. We refer to the different variations as *D*- (right-handed, *dextro*-form) or *L*- (left-handed, *laevo*-form) molecules:

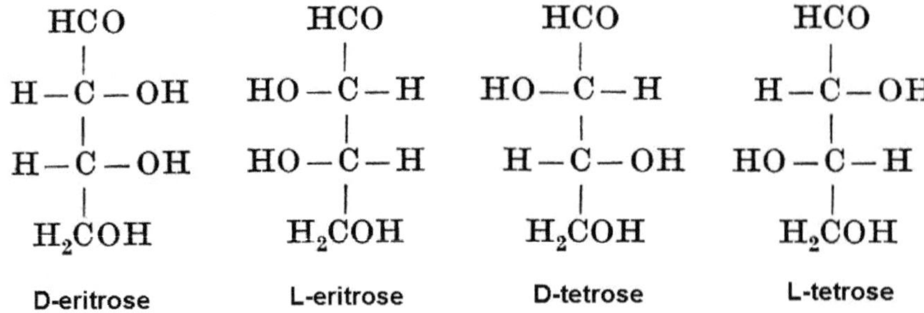

Figure 12.14. *Variations of aldo-tetroses that rotate optical light to the right and to the left*

The *D*- and *L*-modifications are mirror images of one another. Carbohydrates are important elements of our diet, for our bodies produce further organic materials through their fermentation. The well-known simple sugar molecules $C_6H_{12}O_6$ play an important role in this process. Both types of sugar molecule have an enantiomer pair. The structural formula of their optically right-rotating variations can be seen in Figure 12.15, and their mirror images are the *L*-modifications.

It is an interesting property of sugar molecules that human and animal bodies can only process dextro-active variations of grape sugar (glucose) and laevo-active variations of fruit sugar (fructose). So Alice's worries were not unfounded. True, Looking-glass milk will taste just as good as real milk, but has no nutritional value, as her body will not be able to metabolize it.

One big question of science is why living material distinguishes the members of enantiomer molecular pairs depending on the direction in which they are rotated. Macroscopically, we have witnessed such phenomena: the grape tendril or the tendrils of a bean are only wound round the stem in one direction, we only rarely find snails whose shells twist in the opposite direction, and for 99.98 % of us our bowels twist in the

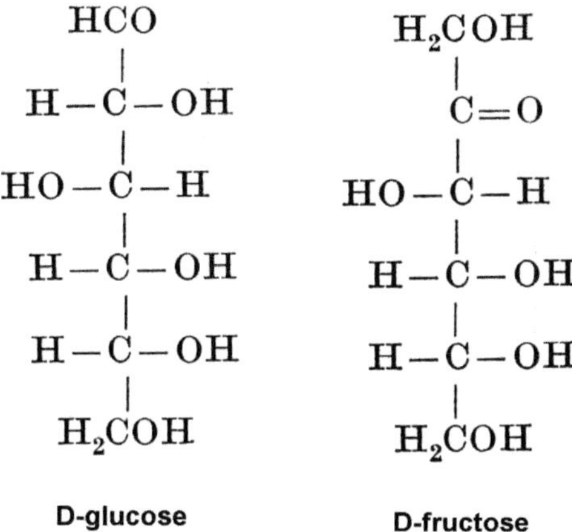

Figure 12.15. *D-glucose (grape sugar) and D-fructose (fruit sugar)*

same direction, to mention but some of the many examples. More difficult is finding relations with molecular examples, correlations between these, and discovering the causes of all these phenomena.

Further questions arise, such as why our bodies prefer only those glucose molecules that are rotated to the right, while in the case of fructose ones they prefer the opposite. The mysteries only deepened when it transpired that the fundamental ingredients of living matter, amino acids, are only present in living creatures in left-handed form. Moreover, the nucleotids, the building blocks of nucleic acids (RNA, DNA), only appear in right-handed form.

To summarize these experiences, we can say that the living organism only includes the laevo-rotatory forms of some compounds, and only the dextro-rotatory forms of others. The opposite form is either unprocessable for the organism (e.g. *L*-glucose), or it is even poisonous. The so-called Contergan scandal that broke out in the 1960s made it clear that the stereoisomers of molecules produced synthetically for pharmaceutical use must be tested separately for their potential effects and side-effects, and that they must be separated during production.

In any event, the phenomenon can be used to separate left-handed and right-handed molecules using living organisms. This is what lay behind Louis Pasteur's classic experiment in 1848. Pasteur investigated the optical modifications of tartaric acid (more precisely its potassium salt). Three optical modifications of tartaric acid (tartarite) are known (Figure 12.16): one rotating the plane of polarized light clockwise (dextro-rotatory), another rotating the plane of polarized light counter-clockwise (laevo-rotatory), and an optically inactive (meso) modification.

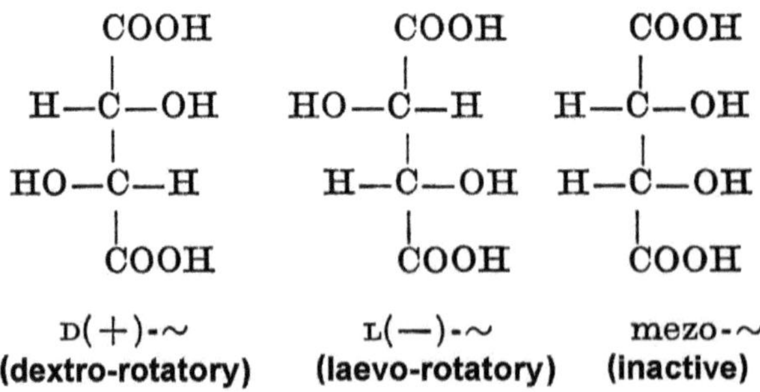

Figure 12.16. *Tartaric acid*

A mix of the two optically active forms with 1-1 proportion (which we normally refer to as racemate or raceme) is *DL* tartaric acid. Produced under laboratory conditions, these symmetrical *D*- and *L*-modifications appear with the same probability, and in equal proportions. From a solution, Pasteur recrystallized (the salt of) optically neutral (raceme) tartaric acid. Molecules of one modification are only capable of crystallization with others of the same modification, so he attained two types of crystal which were mirror images of each other. After separation he found them to be chemically equivalent, but they rotated the polarizational plane of the polarized light in the opposite direction. He identified the dextro-rotatory with the tartaric acid to be found in nature. The laevo-rotatory one was not found in nature.

Pasteur then fed the synthetically produced racemate with microbes that fed on tartaric acid. He found that as the originally optically inactive solution gradually became laevo-rotatory, the proportion of the two forms of molecules shifted in favour of the left-handed ones. The microbes were only capable of metabolizing the natural (*D*-) tartaric acid. Pasteur later concluded (1860) that the substance of life is the creation of optically active compounds: "This is perhaps the only well-marked line of demarcation that can at present be drawn between the chemistry of dead and living matter."

With today's knowledge we see the problem in a slightly different way. Today we would put Pasteur's problem as why, of the enantiomorph pairs, nature often chose only the modification which almost certainly originated from living organisms. To answer this we would even turn the logical order of the question around: why are there asymmetries in living matter?

There are two types of possible answer to the question. The first possible explanation is that the asymmetries in the structure of matter from before life began are what affect the development of the structure of the molecules that play the most important role in living matter. The other possible explanation holds that it can have been caused by spontaneous symmetry breakings brought about during the appearance of the first biologically active molecules.

Beginning with the latter, it must have been such spontaneous symmetry breakings that can have been caused by small force differences, for example Coriolis force (Coriolis, G.G. [1792–1843]), resulting from the rotation of the Earth, or the force generated by the direction that the Earth's magnetic field happened to take at the time, which favoured the appearance of molecules rotating in one direction rather than the other, which then became more widespread, with the asymmetry of further, more complex molecules being the consequence of this. As far as the effect of the rotation of the Earth or the separate or combined effect of the magnetic fields of the Earth and the Sun are concerned, H. Weyl, in his 1952 book, interestingly made a categorical rejection of P. Jordan's (1902–1980) suggestion that these could represent a valid explanation. L. Keszthelyi and K. Kovács undertook partial examinations in this regard at the Szeged Biological Centre, but to the best of my knowledge the decisive experiments that could explicitly have ruled out this opportunity have not been carried out to this day, even though the experimental conditions would now be at our disposal. In principle, it is also possible to surmise that both members of the enantiomer pairs came into being, but one proved to be less stable than the other, and the smallest disturbance upset the metastable balance. In this case our situation is the same as if we assume than an asymmetry is present in the structure of matter from the outset.

As we have shown, the small asymmetries in the structure of matter are the consequence of the role of neutrinos. Some authors have explained this as the result of *CP* symmetry violation, but this explanation is equally rooted in weak interaction. Very small asymmetries can generate similarly small discrepancies in bonding energies and the stability of bonds. We can estimate the size of these if we know that the spin of electrons makes a minor contribution to their orbital momentum, and a much smaller additional contribution can be made by any effect of neutrinos. Thus even if this cause exists, its significance in the strength of a bond can only be very minor. On the basis of mathematical model experiments, however, we know calculations demonstrating the existence of metastable systems in which, causing slight disturbance, the system develops in a direction which does not compensate deviations, but rather reinforces them, which increases rather than decreases the number of differing elements. We do not yet know the precise answer, but this possibility cannot be ruled out, either.

However they might have come into being, it is a fact that one member of the enantiomer pairs has become widespread or rather entirely dominant among molecules which play a key biological role, and this member has become built into the mechanisms of generation, spread and inheritance for biologically active molecules. In the living organisms that are created, the processes are completed under the effect of specific enzymes, and from the outset only one of the two enantiomers has a chance to form, because the enzyme only catalyzes the transformation of one of them. This is not an explanation for the development of asymmetry, however; it merely transfers the question to the development of the asymmetry of the enzymes. The living matter we know of today carries with it an asymmetry that developed systematically and is inherited genetically. This asymmetry accompanies the entire process of the phylogenesis of living matter.

The mechanism works by the mediation of nucleic acids. It is well known that two nucleic acids, ribonucleic acid (RNA) and deoxyribonucleic acid (DNA), play the key role. In the second chapter we referred to the heuristic role played by symmetry considerations in the discovery of their precise structure, the double helix and the bonds connecting it. Watson describes the last phase of their lengthy research as follows:

"Not until the middle of the next week, however, did a nontrivial idea emerge. It came while I was drawing the fused rings of adenine on paper. Suddenly I realized the potentially profound implications of a DNA structure. [...] Most important, two symmetrical hydrogen bonds could also hold together pairs of guanine, cytosine or thymine. I thus started wondering whether each DNA molecule consisted of two chains with identical base sequences held together by hydrogen bonds between pairs of identical bases. [...] Despite the messy backbone, my pulse began to race. If this was the DNA, I should create a bombshell by announcing its discovery."

But this was not yet the final solution. Later:

"I [...] began shifting the bases in and out of various other pairing possibilities. Suddenly I became aware that an adenine-thymine pair held together by two hydrogen bonds was identical in shape to a guanine-cytosine pair held together by at least two hydrogen bonds. All the hydrogen bonds seemed to form naturally; no fudging was required to make the two types of base pairs identical in shape. [...] I suspected that we now had the answer to the riddle of why the number of purine residues exactly equaled the number of pyrimidine residues. Two irregular sequences of bases could be regularly packed in the center of a helix if a purine always hydrogen-bounded to a pyrimidine. [...] This type of double helix suggested a replication scheme much more satisfactory than my briefly considered like-with-like pairing. Always pairing adenine with thymine and guanine with cytosine meant that the base sequences of the two intertwined chains were complementary to each other. Given the base sequence of one chain, that of its partner was automatically determined.

[...] Thus the next several days were to be spent using a plumb line and a measuring stick to obtain the relative positions of all atoms in a single nucleotide. Because of the helical symmetry, the locations of the atoms in one nucleotide would automatically generate the other positions."

Thus both RNA and DNA are right-twisting macromolecules in the form of a double helix (Figure 12.17), that we have been able to admire for nearly five hundred years on the central staircase of the castle of Chambord by the Loire, whose projection is attributed to unknown architects under the profound influence of Leonardo da Vinci. The spines of the two threads run in opposing directions, however. The ribose and deoxyribose carbohydrate units forming the individual threads of the two helices are connected into chains by phosphodiester bonds. The two threads are connected by so-called base bonds: in DNA we find thymine-adenine and cytosine-guanine base pair bonds, while in RNA thymine is replaced by uracil. The RNA and DNA molecules characteristic of an individual are distinguished by the sequence of the base pairs.

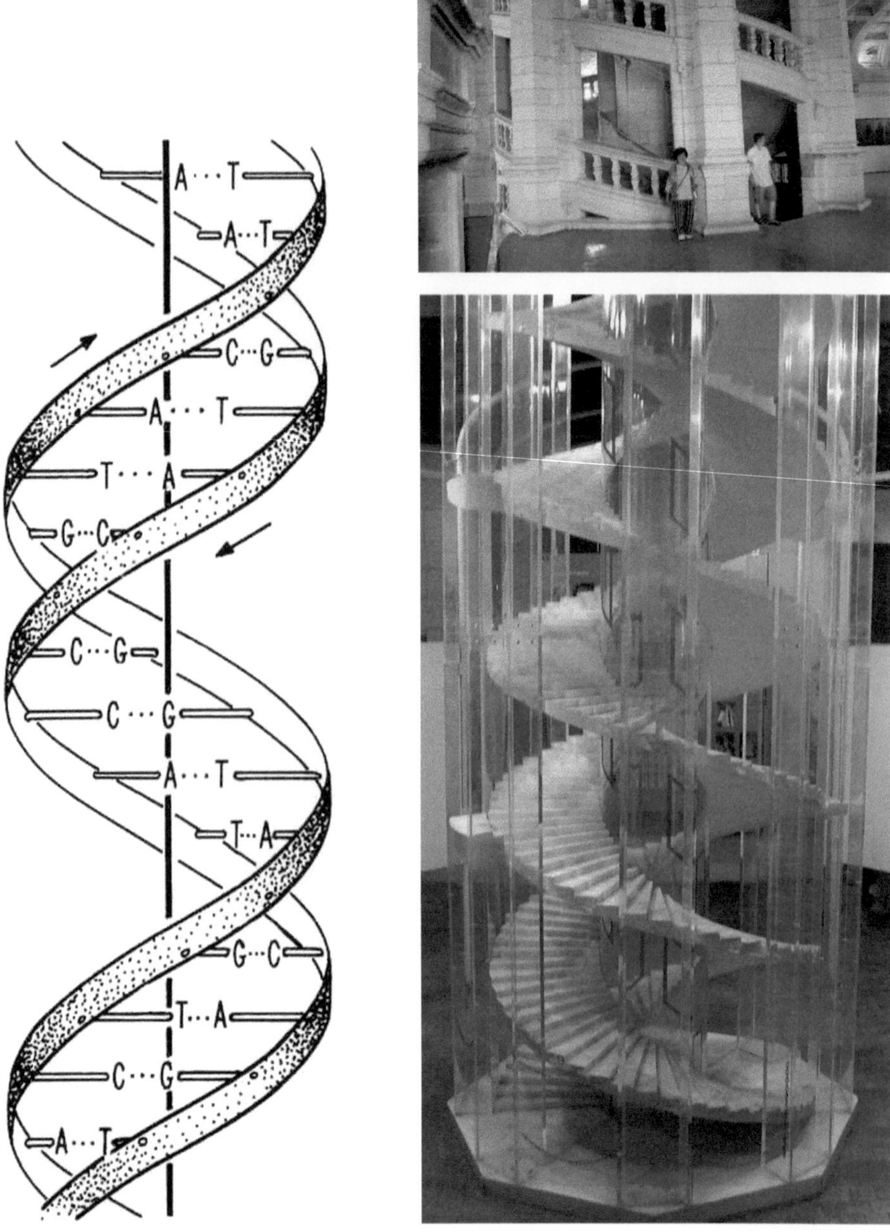

Figure 12.17. *The double helix of the DNA molecule (model), and the opposite-twisted double helical central staircase of the castle of Chambord and its structural model*

The characteristics of sugar molecules discussed above can help us to understand the symmetry breakings between organic molecules appearing in inanimate and living matter, for ribose, which makes up ribonucleic acid, is a similar molecule ($C_5H_{10}O_5$) to sugars, and, in particular, is structurally more closely related to *D*-glucose. Before the development of the first RNA, however, chiral asymmetry did not exist: the selection cannot have taken place with the help of living matter that did not yet exist. The causes, as we have mentioned, are to be found either in an accidental effect, or in the rise to dominance of the effect of a dissymmetry that has already been coded into inanimate matter. Ribose ($C_5H_{10}O_5$), which participates in the construction of both threads of the helix of RNA, is in all instances a right-handed (*D*-) carbohydrate (Figure 12.18).

Deoxyribose, which is the basis for the structure of the both helical threads of DNA, is a D-ribose which at the second carbon atom does not contain oxygen (Figure 12.19). It can be produced from *D*-glucose. No known DNA is produced from *L*-glucose, not even exceptionally.

```
    CHO
     |
H—C—OH
     |
H—C—OH
     |
H—C—OH
     |
    CH2OH
```

Figure 12.18. *The structure of nucleic acids. Ribose*

```
    CHO
     |
H—C—OH
     |
H—C—OH
     |
H—C—OH
     |
    CH2OH
```

Figure 12.19. *The structure of nucleic acids. Deoxyribose*

The base pairs connecting the two chains display a similar structure, and display symmetry in their bonds and their spatial arrangement. Their schematic structure can be seen in Figure 12.20.

The role of the nucleic acids that transmit inheritance is to determine the structure characteristic of the proteins that form the living material. On the one hand, when a new specimen is created from the DNA of the two parents, they determine the base sequence to be transmitted from the individual gene sections to the descendant. On the other hand, during the ontogenesis of the specimen, they ensure the preservation of the pro-

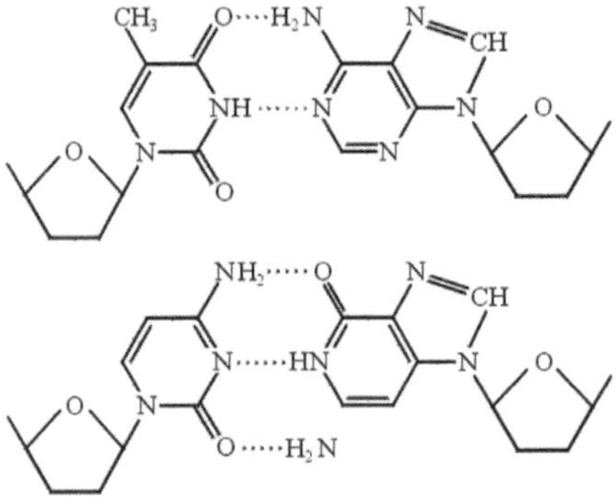

Figure 12.20. *The symmetries and asymmetries created by bond pairs* thymine-adenine *and* cytosine-guanine *on the bridges connecting the helices of DNA. In RNA* thymine *is replaced by* uracil.

tein structure, which is particular to the individual and which was created during the transmission of the DNA base sequence in the course of the propagation and replacement of cells.

Proteins are made up of amino acids (or parts of amino acids). The sequence of the amino acids in proteins is species-specific and inherited. There are a total of 20 amino acids used in protein synthesis. The amino acids that make up proteins are only present in all living creatures in left-handed form. In all probability the protein synthesis mechanism determines that with the mediation of RNA molecules with the same direction of rotation (*D*-), made up of riboses with the same direction of rotation (*D*-), only amino acids that are identically but complementarily twisted (*L*-) can participate in the construction of proteins. The scheme of the helix structure of the peptide chain that makes up proteins can be seen in Figure 12.21, depicted on the surface of an imaginary cylinder.

In the cell nucleus, the DNA, which carries the information on the sequence of amino acids of the protein, creates a messenger RNA (*m*-RNA) onto which the information is copied. The base sequence of the latter reflects the base sequence of the DNA (this is how the information is transmitted). The messenger RNA wanders from the cell nucleus to the location of the protein synthesis, and controls the sequence of the amino acids in the protein being generated. This it does by copying the information on the surface of the ribosomes, rather like a barcode scanner.

The activation of the amino acids in accordance with this information (the connection to the peptide chain) happens with the participation of a separate enzyme and different transfer RNA molecule (*t*-RNA) to acti-

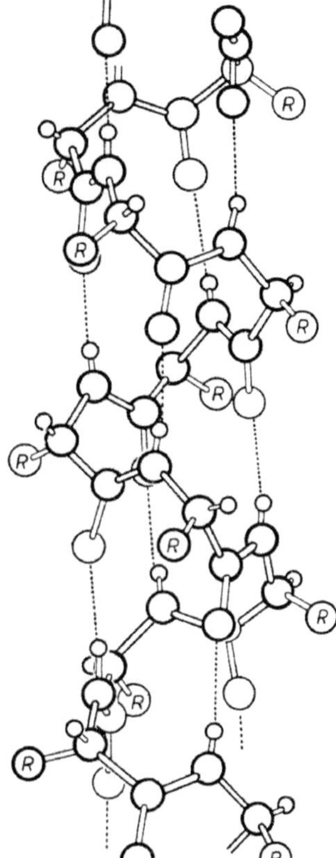

Figure 12.21. *The helix structure of a protein. R marks the side chains of the amino acid (after L. Pauling [1901–1994] and R.B. Corey [1897–1971])*

vate each of the 20 protein-forming amino acids. In the course of protein synthesis, the acid radical of the amino acid adjoins the sugar part at the end of the chain of a transfer RNA molecule. This is how the new protein molecule is created. The mechanism is presented in the schematic diagram in Figure 12.22.

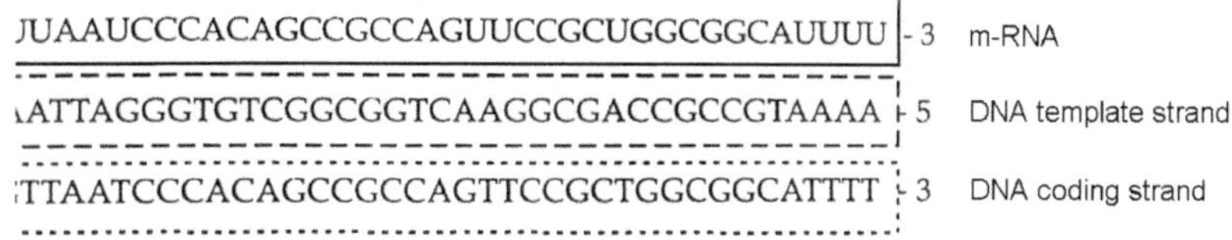

Figure 12.22. *Protein synthesis. In the RNA (*upper strand*), bonds are formed by four separate bases,* adenine, cytosine, guanine *and* uracil.

We can read off from the diagram that the corresponding base from the DNA coding strand is copied with *reflection* onto the DNA template strand, and then the corresponding complementary base for this in the RNA is copied onto the *m*-RNA strand. The base sequences of the three strands are, pair by pair, the "mirror images" of each other. Three of the four bases in the RNA form a base triplet (codon) which builds one of the 20 amino acids into the polypeptide chain. The chemical structure of three successive bases determines which amino acid will be built into the polypeptide chain. The mathematics of the combinations of the triplets that can be selected from the four types of base and the pairings between the 20 amino acids provides a number of further interesting symmetries, which also display a connection to the atomic number (proton number) of the atoms participating in the construction of the given molecules. (This question is discussed in greater detail in the 2001 volume by S. V. Petoukhov [1946–].)

The process as a whole clearly shows that asymmetric causes produce asymmetric effects. No mechanism is later capable of compensating an asymmetry that has emerged during the development of living matter. Were molecules of opposing chirality to appear in small quantities in living matter, it would perhaps be possible to assume a process intended to offset asymmetry. For lack of a biological molecule with chirality opposite to the majority, however, there is no chance of this. The asymmetry lies in every single molecule of living matter.

The following step in the self-organizing mechanism of living matter is cell division. A stand-alone cell usually displays spherical symmetry. Examples of things that are spherical in their symmetrical environment are single-cell creatures floating in water, pollen grains spreading in the air, and zygotes. H. Weyl discusses at length the question of whether there is an inherent median plane in the egg. If there is, this means there is an asymmetry coded into it from the outset, which determines the direction of division. If there is not, then it remains a mystery as to what determines the plane along which the cell splits in two. What is for sure is that the division occurs sooner or later, whether the (spontaneous) symmetry breaking occurs during the life of the cell, or whether it is carried in the material it is made of from the outset. We would probably cause great disappointment were we unambiguously to state that symmetry breaking is only and exclusively capable of taking place inside the cell, in spontaneous fashion, or as a result of random effects. For it follows from the above discussion that all the vital molecules making up the cell bear this asymmetry. We cannot, however, give a clear answer to the question of whether it is exclusively these causes that play a role in the symmetry-breaking process of cell division, or whether there is also a role played by new, qualitatively different symmetry-breaking elements not found at the lower levels. For new symmetries *can occur* at every level of the evolution of matter, and new symmetry breakings *do occur.* There are qualitative differences between them, and these cannot be explained mechanically with the properties of the lower levels. Yet what is certain is that the symmetry properties of the lower levels (including their violation) contribute to the development of the symmetry properties of the new level. This contribution can have the effect that symmetries characteristic of the lower levels are violated at each step in the course of the evolution of matter. We can observe this in the organization of inanimate matter just as in the living world. All this amounts to saying that while the violation of spherical symmetry in the course of cell division is not independent of the chiral asymmetry of the vital molecules that make up the cell, a qualitatively new asymmetry emerges in the course of its actualization.

Cell division (Figure 12.23) produces two apparently mirror symmetric cells. Biologically the two cells are equivalent. This equivalence means that equivalent, complete specimens can develop from each of them sep-

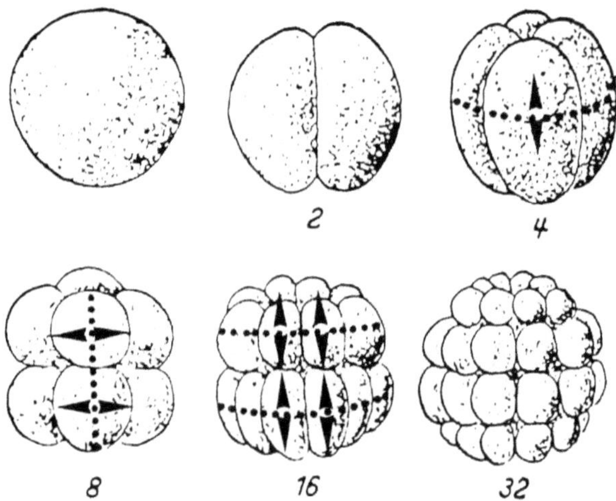

Figure 12.23. *Cell division*

arately. Another division leads to the morula state, in which one of the four cells is differentiated, but there are also instances where this differentiation only occurs after the following division. The existence of this differentiated cell means that a functional asymmetry appears between cells that externally appear to be symmetrical. The differentiated cell will thereafter no longer possess the properties needed for the development of all the organs of the specimen. This functional asymmetry then continues during further divisions. The blastula still strives to preserve spherical symmetry, and to this end the cells all locate themselves on a spherical surface (otherwise the cells located inside the sphere and those on its surface would not be equivalent), but the continued propagation of the cells demands both the sacrifice of this symmetry principle and also further differentiation. It is visible that functional asymmetry and morphological asymmetry develop in parallel. The blastula still displays spherical symmetry morphologically, but not genetically: its cells are not longer all equivalent. It is no surprise that sooner or later cell differentiation also manifests itself in morphological phenomena: during gastrulation, the appearance of the blastopore breaks the spherical symmetry, and at a certain point crinkling appears on the spherical surface. The building begins of the alimentary canal, which strives for cylindrical symmetry. Cylindrical symmetry then turns into multifold rotational symmetry and radial symmetry. Until, in the course of another differentiation, the rotational symmetry is replaced by left-right symmetry in echinoderms, followed by the appearance of even organs, then the breaking of dorsal-ventral symmetry. The development of uneven organs brings asymmetry once and for all to the morphological construction of the body (e.g. the twist of the intestinal canal, the development of the circulation system and the heart, the twist of a snail's shell). The series of breakings of morphological symmetries in living matter are presented schematically in the drawings in Figure 12.24.

The question is whether the chirality and the breaking of left-right symmetry that appears in the case of the heart, the intestines and a snail's shell, are of morphological or genetic origin. This can be answered with an examination of uniovular twins. If it is true that complete mirror symmetry prevails within the cell during that particular first division, which produces biologically equivalent cells, then every component of the cell

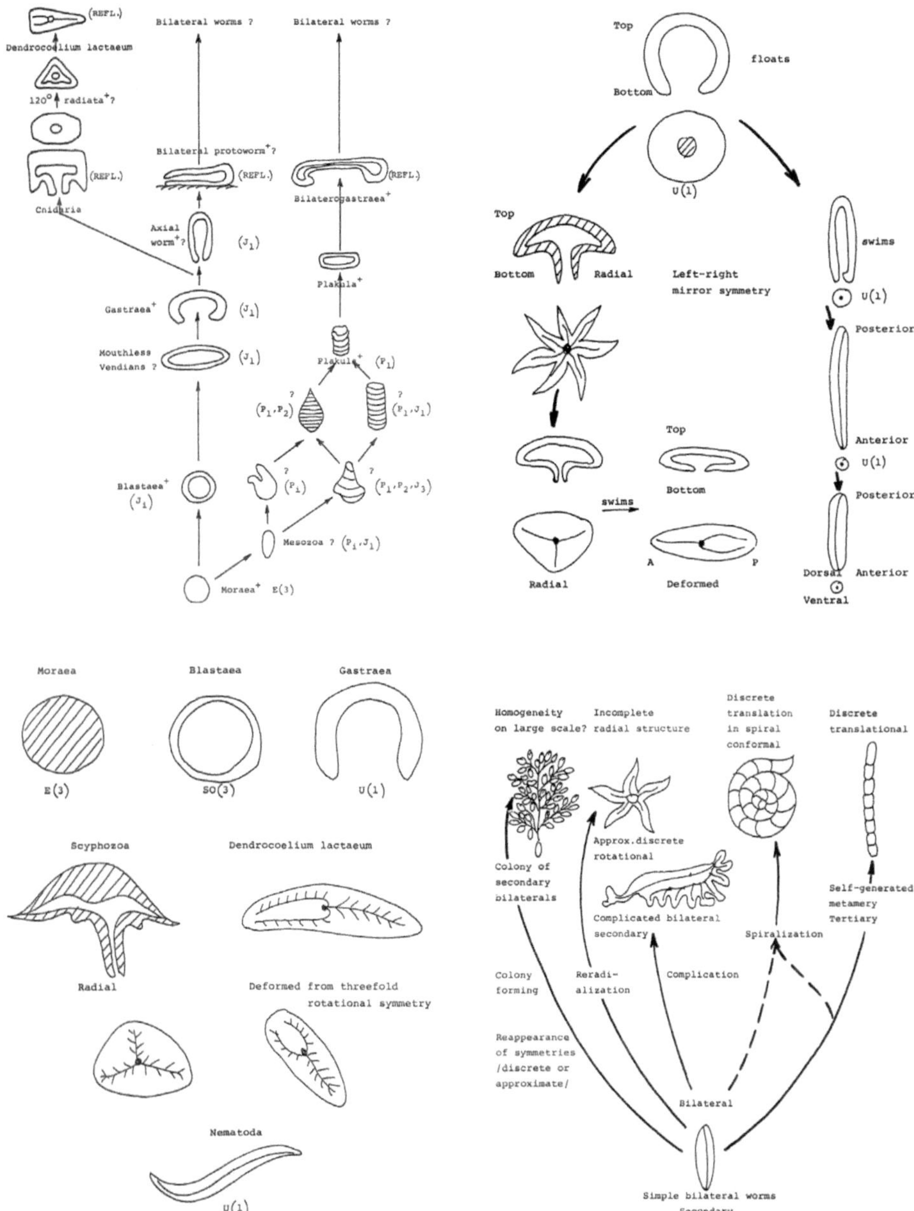

Figure 12.24. *Series of breakings of morphological symmetries* (after S. Bérczi, B. Lukács and I. Molnár)

would be mirrored. Mirror symmetric pairs of twins would develop from the two mirror symmetric cells, with uneven internal organs located and twisted mirror symmetrically. This is not what happens. For both of the twins, their heart is found shifted towards the left side of the chest, and its left ventricle is what pumps fresh blood into the body; twins are each just as likely to be right-handed, and their uneven organs are located in the same way as their twin's and indeed anyone else's. The chirality of nucleic acids and protein molecules determines the chirality of the construction of the body. We carry within us the asymmetry brought from lower levels; we cannot escape from it.

We can also observe another symmetry breaking in the course of the phylogenesis of living organisms: the declining ability to regenerate. The ability to regenerate means the capacity to reverse a locally created alteration, namely an injury. Regeneration is a particular, local form of time reflection.

As phylogenesis can be traced through the course of ontogenesis, the latter can be used to study certain phases of the process. At the outset the ability to regenerate is almost complete. Separating the first split cell from its pair, a normally smaller but fully functional specimen can develop from it. Experiments with transplanting buds of future limbs show that following the fixing of the anterior-posterior axis, the dorsal-ventral axis can still be inverted. The two claws of certain types of lobster are not of equal size. If the larger claw is cut off when the lobster is young, the other one can grow larger, while in the place of the removed claw a smaller one grows. This shows that, genetically, the potential for growing larger is there in the cells of the smaller claw, too. This potential is only realized in certain situations, but it is nevertheless realizable. The cell plasma contains the potential for the symmetrical variant. Unusual external conditions can result in the inversion of the genetically coded variant.

The property of regeneration, together with the capacity individually to adjust to the environment, tends to decline in the course of phylogenesis, and is weakest in the case of humans. We compensate for this biological property with our consciousness and the work that we consciously perform.

The development of living creatures is made up of a continuous series of minute phase alterations to their acclimatization to their environment. In his book *On Growth and Form*, first published in 1917, D'Arcy Thompson modelled the morphological development of vertebrates with similitude symmetry. Species that have undergone transformations are similar to their ancestors. Drawing a grid on a picture depicting the external appearance of particular species, the nodes of the grid transfer particular characteristic points of the body by means of similitude transformations on the corresponding points or cells of similar grids of specimens depicted in the various phases of the development of the species. Figure 12.25 shows some of the noteworthy similitude transformations described in his book.

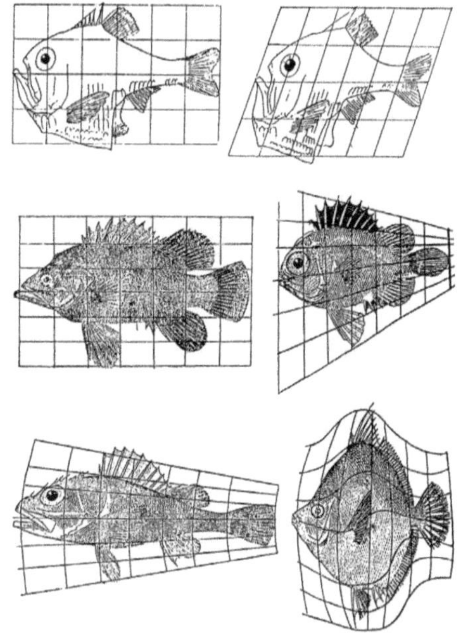

Figure 12.25. *Argyopelecus Olfersi → Sternoptyx diaphana; Polyprion → Pseudopriacanthus altus; Scorpaena sp. → Antigonia capros*

A metamorphosis performed with similitude transformations appears in the artistic drawings of M. C. Escher shown in Figures 12.26–12.28.

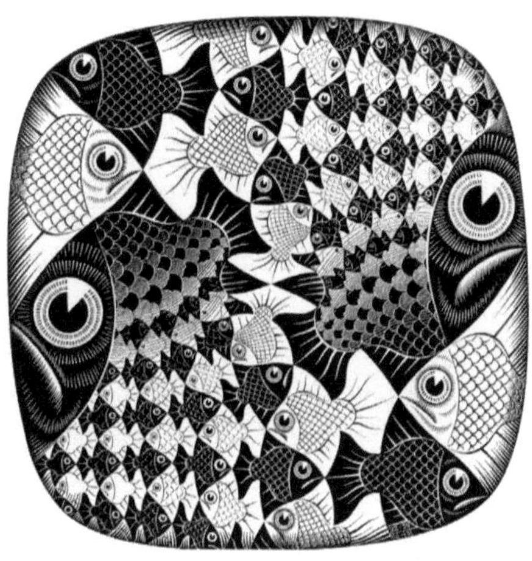

Figure 12.26. *M. C. Escher: Fish and Scales* (1959)

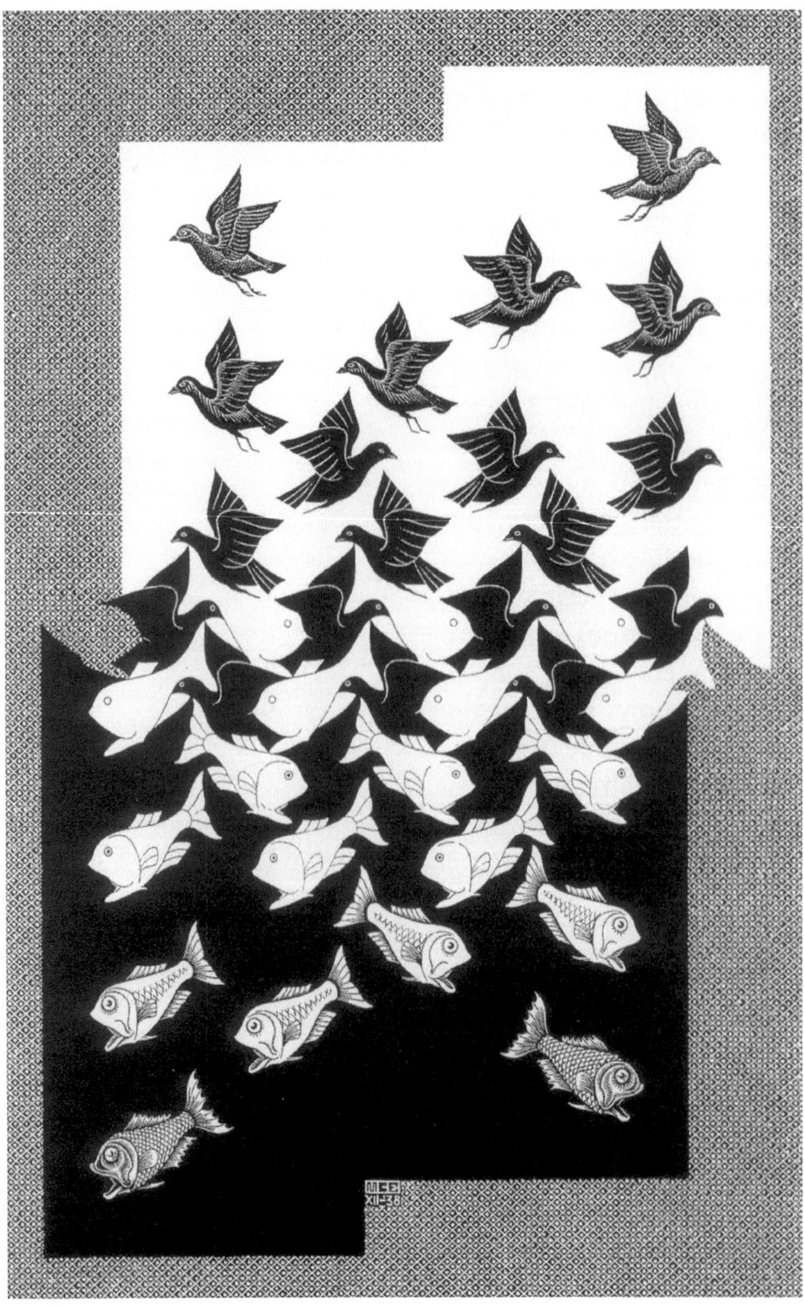

Figure 12.27. *Sky and Water II* (1938)

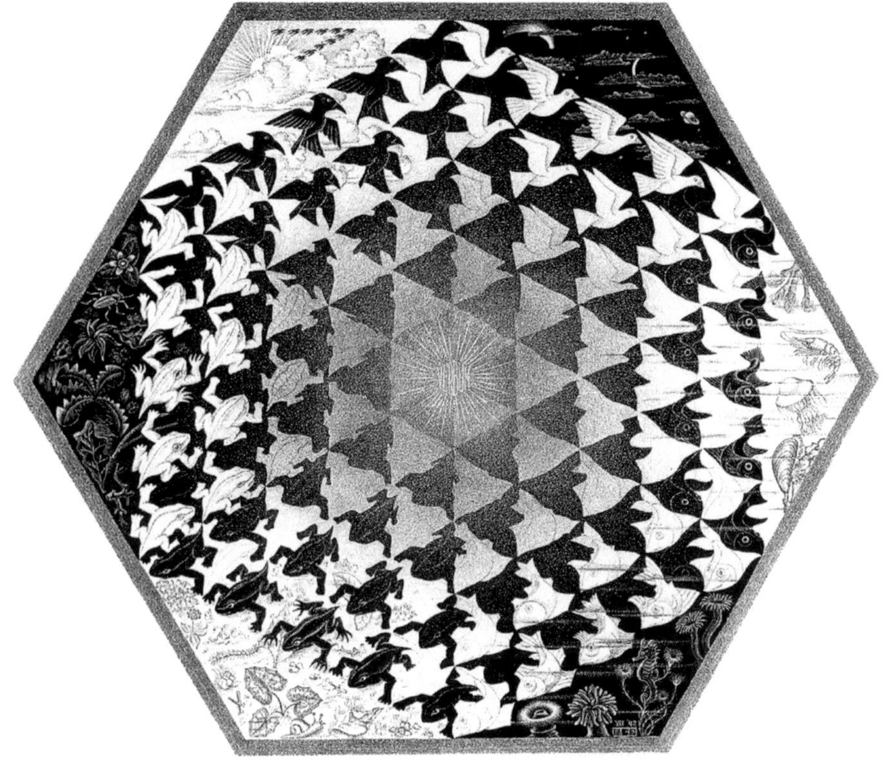

Figure 12.28. *Verbum (Earth, Sky and Water)* (1942)

It is in the case of human beings, at the pinnacle of phylogenesis, that an asymmetry has become dominant which at its early stages appeared in certain mammals, but in its developed form is specific to man. It can be shown to have been present in the early stages of human development. This asymmetry is the differentiation of the two hemispheres of the brain. The asymmetry of the brain is evident in both its morphology and its function. We deal with cerebral asymmetries and their consequences in the following chapter.

Chapter 13

Cerebral asymmetries

and their consequences

We have seen that the history of the evolution of matter is, in both the inanimate and living worlds, the history of a series of symmetry breakings. The chiral molecules built into the fundamental cornerstones of the living organism set the path of further breakings of symmetry in the course of phylogenesis. The asymmetry of the right-twisting DNA molecule was the precursor of a number of molecular consequences, and other symmetry breakings affected living creatures either indirectly or as the result of a combination of external influences.

If we examine our own bodies, we see that we are aware of a good number of morphological asymmetries. Most particularly that of the heart, whose asymmetric development in the course of phylogenesis has become the subject for school textbooks. In the course of its development, the heart did not, thanks to the function it was required to perform, have either rotational symmetric or mirror symmetric alternatives. Let us compare a modern, essentially schematic, anatomical textbook drawing with Leonardo's artistic drawing of the human body's internal organs (Figure 13.1).

According to Aristotle, human beings are much more beautifully formed than animals, because the symmetry of the various parts of the human body are more marked than those of animals. From the outside this certainly seems to be true.

The asymmetric location of the heart in our bodies also affects the location of other organs. The morphological asymmetry causes functional asymmetry in the blood supply. We can easily convince ourselves of this if we measure the blood pressure in each of our arms: we will reliably find that the value for our right arm is higher. The semicircular aorta does not bifurcate into the vessels symmetrically, and so the turbulent blood reaches the various arteries at different pressures. The situation is similar with the other artery pairs.

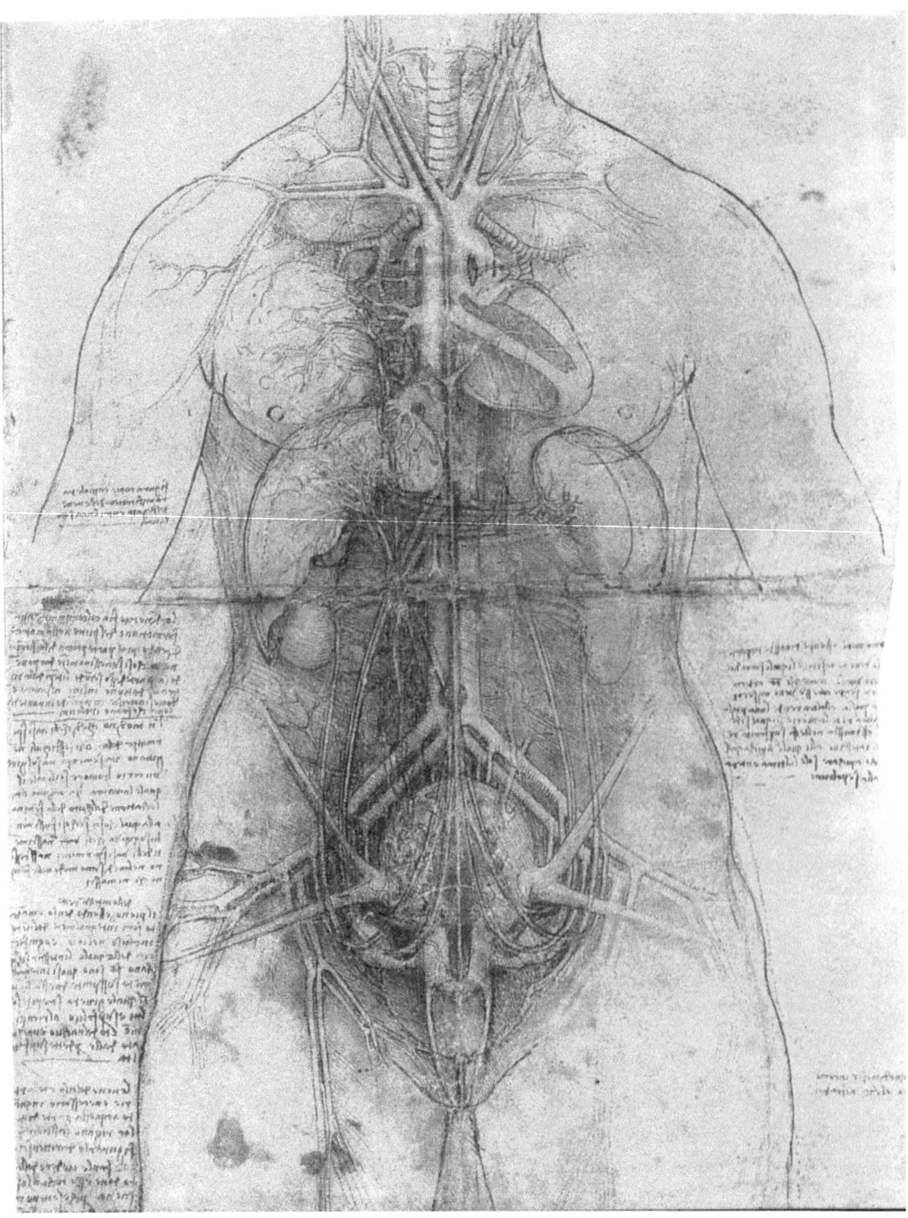

Figure 13.1. *Leonardo's drawing of the human body's internal organs*

It is possible that the blood supply played a role in the development of human cerebral asymmetries; it is possible that its role was only secondary. Today we are not able to decide for certain. Like many phenomena that we have long been aware of, the reason for the development of the cerebral asymmetries belongs to the more mysterious chapters of symmetry breakings.

The functional separation of the two hemispheres of the brain is one of the most well-known asymmetric phenomena. The most obvious sign of it is the dominance of right-handedness. Thanks to the analysis of injuries, we have also long known that the speech centre and the motor centre are located in the left frontal lobe. The first direct proof of this to achieve widespread renown was provided in 1861 by French doctor Paul Broca (1824–1880), when he reported to the French Anthropological Society on how he had discovered injuries in the frontal lobe of some of his patients who had lost their ability to speak after a stroke. Four years later, at the same place, Broca also announced that the speech motor centre is localized in the left hemisphere. We owe it to historical accuracy to mention that — although Broca was unaware of this — Marc Dax (1770–1837) had already reported on his observations in Montpellier in 1836, according to which the injury of his patients who lost their ability to speak was always to the left cerebral hemisphere, though it is true that he did not localize it more specifically. Thus since Dax we have known that the two hemispheres of the brain control different functions. It was Broca who went on to relate right-handedness to the left hemisphere. And we have known of the crossover of nerve fibres since Hippocrates (c. 460–377 BC).

In 1874, German neurologist Karl Wernicke (1848–1904) determined that it is not only the *motor* speech centre, but also the *auditory* speech centre that is in the left cerebral hemisphere — more precisely in the upper part of the left temporal lobe — and is thus found separate from the motor speech centre, which is located in the rear area of the frontal lobe. If the auditory centre is injured separately, the power of speech is retained, but loses its comprehensibility. Interestingly, it was another hundred years before the functions associated with the right cerebral hemisphere were identified.

The right and left hemispheres of the human cortex also show slight morphological deviation: we stress that is only true for human beings, and only for the cortex which developed in the later stages of evolution. All evidence suggests that the morphological deviations are the result rather than the cause of the functions specialized for one hemisphere or the other. The arrows on Figure 13.2 show certain such minor deviations.

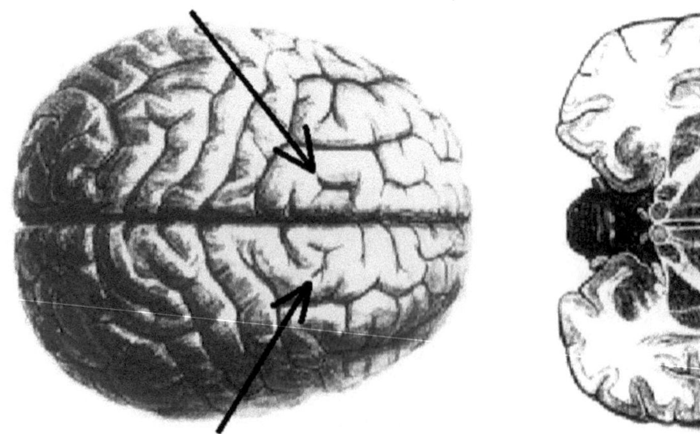
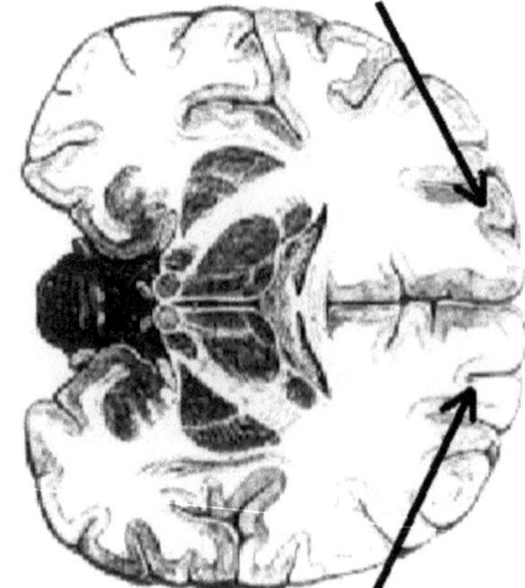

Figure 13.2.

About 90 per cent of people use their right hand to perform actions that require considerable skill. Various theories of differing scientific thoroughness have arisen to explain the development of right-handedness, but it would be early to make a final decision on the correctness of even for the most solid of these. The best-established theories agree, at least, that right-handedness developed in stages along the road to becoming human beings, and in connection to learning to walk on two legs. For apelike primates, the hand was freed, and eyes that were closer together made spatial awareness more refined. Manipulative activity combined with vision is a property exclusive to primates. The basis for the exploration of space with hands was provided in both cerebral hemispheres by the inferior parietal lobe. The preferred use of the right hand only appeared after this. Evidence from archaeological excavations shows that ancient man was already mostly using his right hand to make tools in the early stone age, around half a million years ago; indeed, other finds show that this was even the case for *Homo habilis*, 1.4–1.9 million years ago.

How did asymmetry develop in the use of hands, and why? One of the rival theories, that of Marian Annett, holds that the cause is genetic, that it became asymmetric in the course of an accidental mutation, and then became fixed that way. According to this view, the bilateral symmetry of the body, ever since it emerged in phylogenesis, together with handedness, is determined by a gene that appears in two forms (two alleles): an *RS+* shifting to the right and an *RS–* shifting to the left. Of these two, it was *RS+* which became dominant as the result of a mutation in the early stages of human evolution, and it was this which resulted in right-handedness for the majority of the predecessors of modern man. According to this same theory, the *RS–* formation does not automatically result in left-handedness, merely in a neutral state in which left- or right-handedness can equally arise. This theory can be used with reasonable accuracy to explain the approximately 90: 10 per cent proportion of left- and right-handed people in almost all human populations. It appears that a mechanism developing in such a way, even if the mutation happened previously, must have disappeared before man started moving on two legs, and this is why we do not encounter left- and right-handedness in earlier stages of evolution.

In and of itself, however, the dominance of right-handedness does not offer an explanation for the asymmetry of the cerebral hemispheres. All the evidence is that right-handedness can have developed earlier than comprehensible speech in the modern sense. In all probability, communication between human beings was helped with the hand. This may have had consequences for the localization of the communication centre and thus later the speech centre in the left inferior parietal lobe controlling the right hand. The development of the speech centre in the left cerebral hemisphere induced further functional asymmetries, and ultimately it was this that made possible the appearance of speaking, self-aware human beings.

A rival theory begins with the premise that the straightening of the human spine was not accompanied by a rearrangement of the valves of the vascular system. It had previously demanded lower pressure to pump blood to a brain that was at about the same height as the heart. Once humans straightened up and became erect, the vessels that became vertical (that had previously been horizontal) did not have the valves necessary to regulate flow. In the vessels leading upwards to the brain, the role of small pressure differentials increased. As a result of the asymmetry of the heart and the bifurcations of the aorta (Figure 13.3), the minimal difference in pressure at the outlets of the arteries leading to the two cerebral hemispheres meant that minimally less blood, and thus minimally less oxygen, was passed to the right hemisphere of an erect human than to the left one.

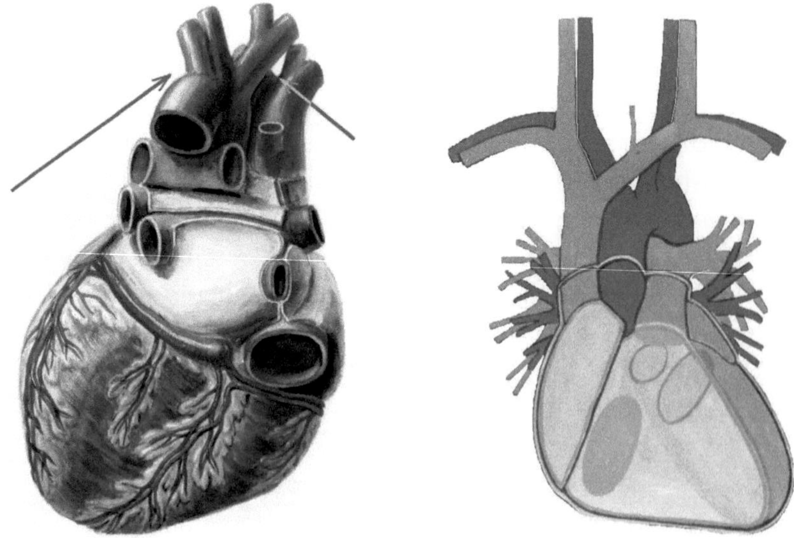

Figure 13.3. *The heart and the bifurcations of the aorta (above, in red), in detail and schematically*

As man's activities increased, so did his need for oxygen. For this reason as little as a one per cent difference in oxygen supply could multiply in its significance. This could, so the theory supposes, have caused a division of labour to emerge between the two cerebral hemispheres, according to which the tasks requiring more oxygen are performed by the left hemisphere, and those requiring relatively less by the right. The crossover of nerve fibres causes the left hemisphere to control the right of our body, and the right to control the left. This would have been the cause of the localization of both right-handedness and rational thought in the left cerebral hemisphere.

The latter theory claims that the asymmetry of the brain must have developed before right-handedness, while according to the former one this was the other way around. The latter theory gives an evolutionary explanation, while the former puts the dominance of the right hand down to an accidental mutation (albeit to the development of the preponderance of an asymmetric gene latent in us since the beginning of phylogenesis). Proponents of the former theory refer to measurements which do not bear out a sizeable enough difference between the oxygen supply to the two hemispheres for this to be an adequate explanation. The aim of this book is to provide an exposition rather than to take up a position between rival theories.

A division of functions has developed between the two hemispheres of the brain, most probably as the result of the mechanism of one of these theories. For example, the left cerebral hemisphere is commonly referred to as the *talking* hemisphere, and the right as the *silent* but *seeing* hemisphere. In a certain sense the two hemispheres function in different ways. In another sense, certain functions are dominantly directed by one hemisphere or the other. In the majority of humans, the dominance between the two hemispheres only causes small-scale difference (both hemispheres fulfil most functions, just in differing degrees), but it can be a qualitative difference whether the control of a given activity is initiated by one hemisphere or the other. Vital functions can be maintained by one hemisphere on its own, but in the case of healthy persons the two hemispheres communicate with one another. The corpus callosum connects the two hemispheres with nerve fibres, through which the communication between the two flows. From the 1930s onwards, the corpus callosum of serious epileptic patients was cut to prevent their attacks from occurring. The communication between the two hemispheres of the brain ceased. This did not trouble the everyday lives of the patients, but in certain situations generated right-left coordination problems, and difficulties with counting.

From the middle of the 1980s, the identification of functions with particular areas of the brain picked up speed, and to this day our knowledge of the workings of the brain are enriched with huge amounts of new information every year. In parallel with the mapping of the brain, so-called neural models have been successfully used to model a number of brain functions with computers. One of the most interesting results of this was the discovery that the functioning of our left cerebral hemisphere is more akin to that of a digital computer, while our right hemisphere is more like the workings of an analogue one. It was with the help of this observation that by the early 1990s Tamás Roska (1940–) and Leon O. Chua (1936–) developed the first working neural model, the artificial retina, which, mimicking the asymmetry of the brain, passes visual information to the brain with the help of digital and analogue processors integrated into a single chip. This same theoretical model helped find a rational explanation for the many observed distinctions in function-pairs between the two cerebral hemispheres.

We have reached the last known stage of the series of symmetry breakings that have occurred in the course of the evolution of living matter. The asymmetric functioning of the brain, like Kant's antinomies, ascribes antithetical pairs to the two hemispheres. These antithetical pairs have far-reaching effects, whose consequences have not yet all been discovered. They induce deviations, more significant than the motor functions mentioned above (hand movement, speech), in our thought, processes of cognition and learning, in our relations with each other and the world, and in our world-view.

With the knowledge we have today, we can display neither a correlation nor the lack of one between right-handedness and left hemispheric dominance in thinking. It is clearly the functioning of the left cerebral hemisphere that is responsible for rational thought. The right hemisphere is responsible for our emotional thought functions formed by feelings and impressions. In healthy people, both hemispheres are almost equally developed, and work in cooperation with each other; in general, dominance means that the functioning of one very slightly exceeds that of the other.

Scientific thought primarily requires rational brain functions. Schoolchildren with dominant left hemispheres will find it easier to learn natural sciences. Activities that require spatial manipulation more than a detailed understanding of our environment, like artistic disciplines, will be more successfully learned by pupils whose right hemisphere is dominant.

Experience shows that women are more susceptible to thought driven by emotions, while men are more driven by rational thinking. It is highly probable, but not proven beyond doubt, that for the majority of women it is the right hemisphere that is dominant, while for the majority of men it is the left. Our knowledge of the history of science suggests that the majority of the women most successful in mathematical discoveries were more masculine in nature than our customary picture of the average woman would suggest. There is no evidence in the history of art, however, that the majority of eminent artists might have been women. Another challenge to this categorization is the fact that girls learn to talk earlier than boys, and they retain this verbal advantage for a good while, while in childhood boys are better at spatial manipulation, which is associated with the right hemisphere. We have, therefore, to treat categorical classifications with caution.

As well as the pairs of properties associated with masculine and feminine thought, it is customary to mention the difference between Eastern and Western ways of thinking. The so-called Western way of thinking, and the Western-style science associated with it, is built on cold rationality. It is based on the formulation of propositions and their proof with the precision of mathematics. A good number of philosophical schools of thought and eminent philosophical figures have tried to transfer the methods of the exact sciences to the social sciences, ethics, and the world of human activity and relations (e.g. Spinoza). Some approaches, with only slight exaggeration, only regard as science that which can formulate and prove its propositions in exact form. According to this, impressions, heuristics and intuition can only be given rights in science if we can prove them on the basis of earlier, axiomatically constructed knowledge. In contrast, the so-called Eastern or Far Eastern way of thinking is much more inclined to be visual and provide examples. Impressions and analogies play a much more significant role in the process of acquiring knowledge. The science of nature and that of society are less clearly separated. The criteria for the exactness of natural science are more relaxed than in Western-style thinking, which forces the logical rules of Euclidean geometry onto all branches of science. For example, the method of proof of *wasan*, traditional Japanese geometry, does not follow the criteria demanded by Euclidean geometry, and has nevertheless succeeded in recognizing and proving many important theorems.

Where categorical antithetical pairs have to be outlined, even in sciences of the spirit (*Geisteswissenschaft*), in contrast to the left-right mirror world of the West, in the East it is the antisymmetry of the yin-yang and rotational symmetry blurring clear boundaries that come to the fore. In the East, asymmetry plays a larger role in visual depictions. This displays itself in decorative art, in the culture of the environment and in music just as much as in calligraphy and prose styles. Writing with pictures, from the outset, reflects a different associational world. A world of beliefs which is less categorical, and less apt to record the knowledge of a particular era in dogmas, can more easily build new knowledge into itself, and more easily remain closer to nature and the knowledge we have gained of it. The people of the East, who grow up in this visual and conceptual world, are less inclined to make categorical black-white, yes-no judgments, not only in their everyday communication, but also in their scientific thought. We have no indubitable proof that the majority of 'the' people from the East would *ab ovo* have a dominant right hemisphere. Yet those who grow up in this cultural world carry this way of seeing things with them, even if it is not a genetic characteristic, just like the preference for rational thought in the Western cultural world. As we saw in physics, the cooperation of these two ways of thinking with their roots in different cultures has helped to accept violations of symmetry into our scientific world-view. The result is that today we can see the world as a system that unites symmetries, antisymmetries, chiral symmetries and asymmetries.

It is no accident that the functioning of the left cerebral hemisphere responsible for rational thought is compared to the workings of a digital chip, and that the right hemisphere directing emotional functions is considered to be analogue in its operation. In the following we survey a few antisymmetrical pairs of functions with regards their correspondence to one hemisphere or the other.

The *rational* left hemisphere derives *logical* conclusions. This hemisphere is *detail-oriented,* comprehending the phenomena of the outside world, the stimuli that reach it, as separate objects. It builds on *elements.* In contrast, the right hemisphere, of itself, thinks *intuitively,* feeding on *impressions.* The right hemisphere grasps the phenomena of the outside world as a whole, *holistically.* This hemisphere detects *continuous* phenomena. It senses a *set* as a whole, not as made up of its elements. The left hemisphere directs so-called *intellectual* activities, while the right guides actions better described as *instinctive.* The left hemisphere is more characterized by *abstract* thinking (belonging to schemes), while it is *object-centred* thinking that is more typical of the right hemisphere. The left hemisphere primarily guides us to new knowledge by means of *conclusions;* the right

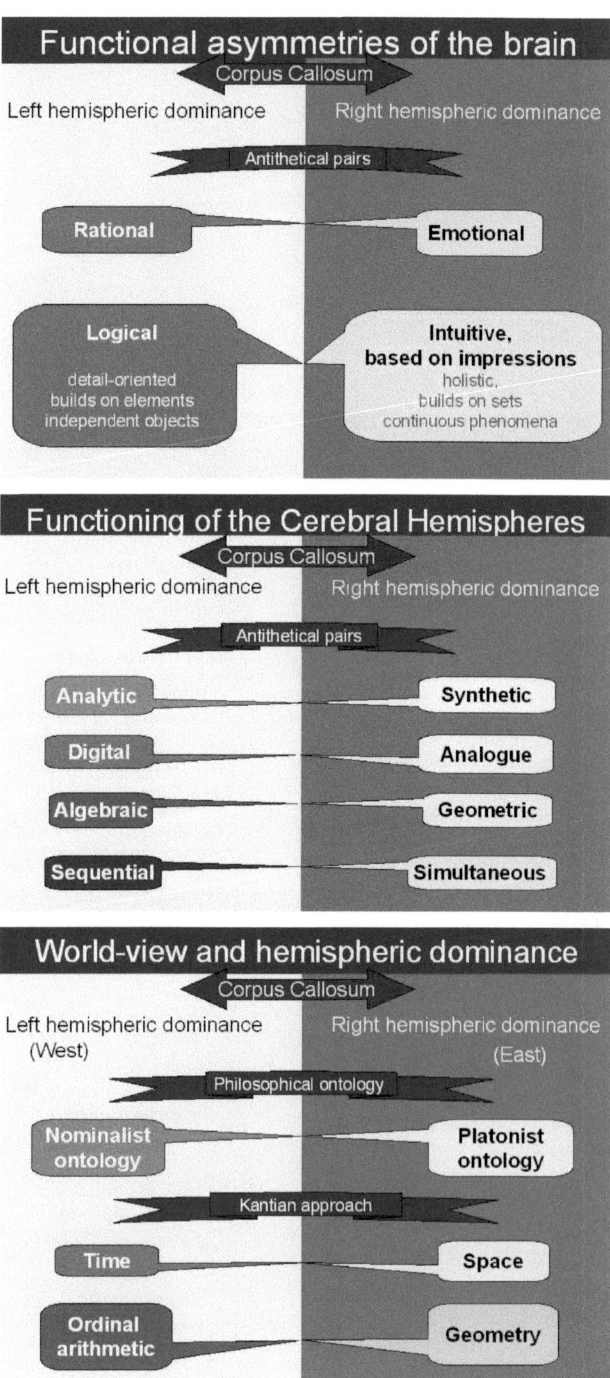

Figure 13.4. *Tables of right and left hemispheric functions*

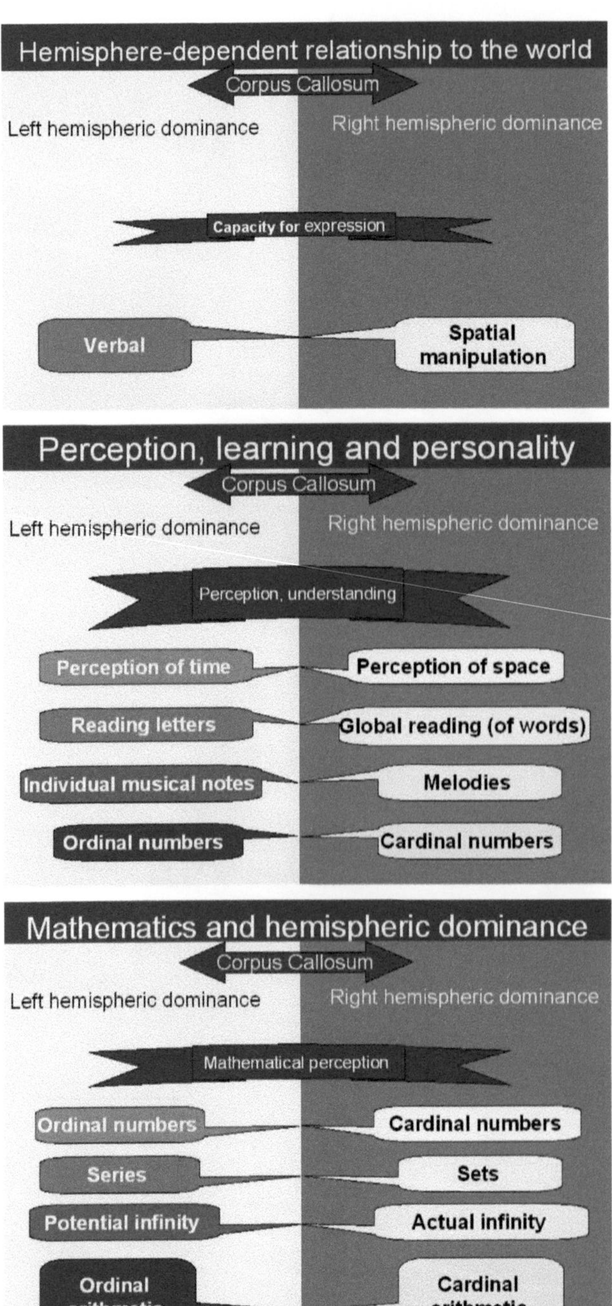

Figure 13.4. (cont.)

one does so more via our creative powers, our *imagination*. A result of the latter is that the left hemisphere attributes meaning according to a single, strictly set logical order of the information perceived; it comprehends the world in its *objective* reality, cannot really give something meaning brought by an association, and thereby *has no sense of humour*. The right hemisphere, in contrast, is more impulsive, interprets the world more *subjectively*, associates more freely, and thereby *has a sense of humour*.

The left hemisphere processes the information interpreted as being made up of elements *analytically*. It treats the stimuli arriving from the outside world element by element, in a *digital* fashion. The right hemisphere deals with the world grasped as a whole *synthetically*, and processes the pieces of information received simultaneously, in an *analogue* fashion.

In the course of mathematical thinking, the left hemisphere is receptive to *algebraic* solutions, which analyze details, while the right hemisphere prefers *geometric* solutions, which cover the phenomenon as a whole and present it in a visual fashion. The following example is often used: if a tourist asks on a London street corner how to get to Big Ben, she will be told to go straight on, then take the second left, then take a right after house number 37, and after three streets she will be where she wants to be; if a tourist inquires about the emperor's palace on a street corner in Kyoto, the local will draw him a geometrical drawing on a slip of paper, with roughly the right proportions, marking the directions to take. We experience the same difference between the visiting cards our English and Japanese friends give us: the former has a street name and house number, while the latter has a drawing.

It follows from the hemispheric differences in perception and processing of data that the right hemisphere, sensing better in *space*, grasps events in one go, i.e. *simultaneously*. In contrast, the left hemisphere, which captures details, the moment, interprets particular events in their *temporal* order, i.e. *sequentially*. The right hemisphere, therefore, is responsible for our *spatial vision*, and the left for our *sense of time*. The left hemisphere grasps a single moment in each instant, then another one in the next instant, and so on: this is how it builds up its own picture of the world. In a given instant, the right hemisphere captures the whole of the perceptible environment, recording this as the picture it has formed of the world, but is not receptive to the chronological order of images recorded at different times.

This extreme picture is not borne out by our everyday experiences. The reason is that we sense with both our cerebral hemispheres at the same time. The two hemispheres exchange their information through the corpus callosum. It is the synthesized image of the worlds separately sensed by the two hemispheres that appears in our consciousness. There are many consequences of the elemental asymmetrical properties associated with each hemisphere whose conscious application is only now developing or becoming widespread.

In the world of our philosophical thought, for example, we can in retrospect reinterpret a number of factors. Left hemispheric dominance and building upon elements is in this case more likely to be the verification of a *nominalist ontology*, while left hemispheric dominance, which captures things in their entirety, is more likely that of a *Platonist ontology*. Similar categorization can be introduced between the teachings of the various logical schools. The pairs of Kant's antinomies can likewise be linked to the two cerebral hemispheres. Of these four antithetical pairs, the first two deserve particular mention: the second, which on the one hand sees things as being made up of simple, indivisible parts, and on the other hand as complex and holistically unified, and the first, which relates to the spatial and temporal finiteness or infiniteness of the world. The latter is of interest to us in two aspects. It grasps the concepts of the finite world and of infinity in their potentiality: infinity can be approached by a series of finite things, but not reached. This *potential* concept of the *infinite* is the left hemispheric approach. From a right-hemispheric perspective there exists the concept of *actual infinity*. The other aspect is that *space* is associated with the right hemisphere and *time* with the left. Of the pairs of the third antinomy, *freedom* corresponds to the thinking of the right hemisphere, while the left hemisphere is characterized more by being directed or controlled. The *necessity* of the fourth antinomy is associated with the logical rationality of the left hemisphere, and *contingency* by the spontaneity of the right one.

Man's capacity for verbal expression is linked to the left hemisphere. Someone who is particularly good at verbal expression, however, is not necessarily creative in manipulative activities. Spatial manipulation and spatial coordination of hand movements are connected to the right hemisphere. To avoid any misunderstandings, this does not mean that manipulative creativity goes together with left-handedness. Spatial coordination is performed by the right hemisphere in the right-handed, too. This is why a talent for drawing only really manifests itself after the age of seven or eight, when the myelination of the fibres in the corpus callosum is complete, and communication and the transfer of information between the two hemispheres becomes fully developed. Right hemispheric dominance is more likely to mean a predestination for artistic activity. So it often happens that excellent artists have difficulty expressing themselves verbally; they prefer to put what they have to express into a certain form, or to draw it.

The differences between the two hemispheres have a role in sensation, in the process of perception, understanding and gaining knowledge, and consequently in the course of learning.

The sensation of time is linked to the left hemisphere, the sensation of space to the right. Let us at this point quote Lorentz' observation that space and time transformations can be summarized in a single invariance. The two types of functioning of the two hemispheres of the human brain are given meaning precisely by the fact that the two hemispheres communicate with one another, and, working in cooperation, are capable of grasping and perceiving the world in its fullness and its reality. Spatiality and temporality equally belong to the fullness of the world. This is reflected in the physical description of the world in the assertion that space and time transform together, according to a unified symmetry principle. This is how the objectivity of the physical world that surrounds us blends with the structure and functioning of our brain as it forms an awareness of it into a system that follows the same, unified, symmetry (Lorentz invariance, the symmetries of space-time) or asymmetry (the difference between space and time and that between the cerebral hemispheres). This is what established harmony between objective reality, the mode of acquiring knowledge of it, and the picture of the world that this knowledge brought to our consciousness.

An interesting example of the cooperation of the cerebral hemispheres is the mechanism of reading. In the course of letter reading, the left hemisphere digitally detects the individual letters one after another (sequentially). It passes the information, the sequence of letters, through the corpus callosum to the right hemisphere, which synthesizes it into an image of a word, which it then sends back to the left hemisphere. The left hemisphere analyzes the word image — this is the comprehension phase. Once it has detected a number of consecutive words, it again passes on this sequence to the right hemisphere, which synthesizes them into a sentence before passing it back to the left hemisphere. The left hemisphere analyzes and interprets the sentence. Once it has detected a number of sentences, it furthers them to the right hemisphere, which synthesizes them into a narrative. This again is returned to the left hemisphere, which analyzes and interprets it. The entire mechanism is an iterated process of to and fro between the two hemispheres. This sort of reading presupposes left hemispheric dominance: this dominance takes the form of the left hemisphere initiating the process. Not all children find it easy to learn to read with this method.

With the so-called global reading method, we first recognize word images. The word image as a whole is sensed by the right hemisphere and furthered to the left hemisphere. The left hemisphere analyzes it, breaking it down into its elements, its letters, while at the same interpreting the word image. Once it has interpreted and analyzed a number of words, it returns the sequence of them to the right hemisphere, which synthesizes them into a sentence. From this point onwards the iterative process continues in the same way, with the participation of both hemispheres. In this process the analysis of letters one by one has little significance. Text comprehension can equally succeed without registering the individual letters: breaking down words into letters can come at a later stage of learning to read. Writing, however, which is inevitably sequential, demands it. Alongside the capacity to read gained with the global method, the capacity to write can be learned later, as the analytical stage takes hold. Right-hemispheric dominance is here seen in the way in which the first initiative begins from the right hemisphere, which plays the leading role. There are children, however, for whom this method is the harder way of learning to read.

The conclusion from a comparison of the two reading methods is that there is no single best method in the teaching of reading. For the lion's share of children, the functioning of the two hemispheres is essentially balanced, and they can learn to read with either method. For those children, however, for whom one hemisphere or the other is dominant, learning to read is only easy with the method that is suitable for them.

We sense musical sounds in a similar way to letters. The individual musical notes reach our ears sequentially. In those with left hemispheric dominance, individual musical notes are passed from the left hemisphere to the right one, which synthesizes them and turns them into a melody, then passes this back to the left hemisphere, which, after analyzing a few musical units one after the other, sends this back to the right hemisphere for synthesis, which is then sent back again, until the entire work is put together. The whole process is like that of verbal comprehension. Those with right hemispheric dominance find it hard to distinguish the various notes in a song. They sense the melody first, which is what their right hemisphere passes on to the left for analysis, and so on. Those with left dominance easily solmizate a melody as soon as they hear it, but have more difficulty singing it right away; those with right dominance can immediately hum what they have heard, but find it hard instantly to break it down into its constituent parts and solmizate it or play it on a piano. Sat at a concert, the former group can almost see the score note by note, while the latter group appreciate the music for its overall impression.

Something similar happens with the learning of mathematics. The left cerebral hemisphere is the algebraic, the arithmetic one, that builds upon individual numbers. It grasps sets in terms of their individual members: numbers one by one, one after another, in turn. The right hemisphere is the global, the geometric one, for which the set appears as a single unit, and numbers are registered as sets. In practice, this begins in early childhood, with the beginning of the development of the notion of numbers. If we place four counters in front of a child with left hemispheric dominance, it will count them by showing each around in turn: this is the first counter, this the second, this is the third, this is the fourth. Then we can asked the child how many counters it saw, and the child will reply. A child with right dominance senses the set of counters as a whole, detecting that there are four without counting them separately. If in kindergarten they play a game in which the toy dolls go in a line to have breakfast, then if there are already five in the line, what the left-hemispheric child registers is that the fifth doll was the last to join it. The right-hemispheric child does not detect that the doll joined the line as the fifth, simply that there happen to be five dolls in total. The latter does not concentrate on the last doll to join the line, but the line as a whole, while the former is only capable of deciding how many dolls there are in total through information about one particular doll or another.

Put in precise terms, we can say that the left hemisphere thinks in ordinal numbers, the right one in cardinal numbers. The two types of number concept come into harmony with one another in the consciousness in the course of the change of developmental level around the age of six, one of the so-called changes in Piaget-levels. We originally have two types of number concept, however, on which two different types of arithmetic can be built: an ordinal arithmetic, and a cardinal arithmetic. In ordinal mathematics, the infinite is a potential concept. In cardinal mathematics, which thinks in terms of sets (of points and of numbers), the concept of the infinite can be actual. Both types of arithmetic are capable of interpreting all mathematical operations and relations. The difference is only in the way they are approached. Hilbert's formalism, for example, which is built upon discrete, individual objects, is based on ordinal mathematics. An example of cardinal mathematics is the logicism of Frege (1848–1925), which is built on sets. (Russell's type theory represents a dualist approach in-between these.) The two types of formulation of the theory of quantum mechanics, which developed in parallel, present an example of how the same physical theory can equally be described with either of the two mathematical approaches. The matrix mechanics of Heisenberg (1901–1976) works with matrices made up of *discrete* elements (a left-hemispheric, ordinal theory). The wave mechanics of Schrödinger and de Broglie (1892–1987) reaches the same results by searching for the discrete eigenvalues of *continuous* functions (a right-hemispheric, cardinal theory).

Mathematics textbooks in current use are based on the cardinal concept of numbers. Arithmetic textbooks that build on ordinal numbers only appear rarely and in experimental form. Nevertheless we can say that in education it is primarily left-hemispheric rationality and logic that is prevalent. Schools primarily teach children laws of nature and the learning of proven facts, and try to point to the logical method of their proof. It is no accident that one of the first right-hemispheric teaching experiments was initiated by the Israeli sculptor Yaacov Agam (1928–). As an artist of international renown, he understood that school had for him been a series of failures because the teaching method which had been forced upon him had been designed for children with left hemispheric dominance. In Hungary the teaching experiment of József Zsolnai (1935-) in Törökbálint set itself the goal of establishing a school where in teaching as a whole, and in the distribution of the daily timetable, the children's left and right cerebral hemispheres are burdened alternately and in roughly equal measure. To achieve this, the classes primarily teaching logical knowledge are followed in turn by activities serving aesthetic and emotional development and those demanding physical effort and manipulative creativity. This is how asymmetries became consciously implemented in everyday teaching practice.

We gained our information about asymmetries of the brain initially from the post-mortem data of those affected by strokes, then later from psychological tests and experiments. Our knowledge of brain asymmetries was greatly helped by the observation of schizophrenic patients, and by Piaget's observations from experiments on children. Both directions of experiment displayed a strong correlation with the attitude of experimental subjects to symmetry. In early childhood, the attitude to symmetry plays a role in determining the so-called Piaget levels of development. The observation of patients helped, for example, in the elaboration of the so-called Rorschach test. In the course of this the psychologist analyses the associations evoked by the figureless and slightly dissymmetrical shape, depending in part on whether the participant in the experiment puts greater emphasis on the symmetry of the figure or on the violation of this symmetry.

It is not only in this context that psychology makes use of symmetry. In contrast to mirror symmetry, which is based on opposites, it is well known in the psychology of advertising that the audience targeted by an advertisement is most susceptible to rotational symmetric emblems and logos. Their repetition, together with the permanence of the way in which they are rotated back to their original position, suggest stability, dependability and a sense of security, and strengthen the trust placed in the advertiser. It is as a combination of these qualities that the yin-yang, blending timelessness and the harmony of opposites (and which today also serves as Korea's national symbol), can have become the symbol of various philosophical explanations and interpretations over thousands of years.

With certain illnesses, the presence or lack of a sense of humour can help determine the preservation or decline of the healthy balance between the two cerebral hemispheres. The observation of epileptic patients with a severed corpus callosum helped recognize the coordination or counting disorders that can appear: how the left hemisphere becomes dominant in a situation where, in the absence of communication, a spontaneous decision occurs between the execution of instructions separately directed by the two hemispheres, and which functions are unaffected by the absence of the connecting role of the corpus callosum. Nowadays there are a number of non-invasive methods of examination at our disposal which allow us precisely to determine which areas, or even which groups of cells, are active in the brain during particular operations.

Some functions of the two hemispheres are clearly antithetically opposed (antisymmetrical), such as digital or analogue operation, or the location of the motor centre or the speech centre. The distribution of the majority of functions, however, is to be understood in terms of the dominance of one hemisphere or the other: both perform the function, but one plays a slightly more emphatic role than the other (dissymmetry). It is important to stress this, lest someone might think of belittling the emotional life of a scientist on the basis of left hemispheric dominance, or the capacity of an artist for rational reasoning, citing dominance of the right hemisphere.

Indeed, artists do not all display right hemispheric dominance. The members of the Brueghel family, for example, painted meticulous pictures that were worked to the tiniest detail, which suggests left hemispheric dominance. The majority of impressionist and cubist painters did not attribute significance to minute detail: they were more guided by the overall impression and by emphasizing the more important characteristics of the theme being depicted, which suggests a right-dominant approach. Figure 13.5 shows two paintings. The same theme is depicted by two artists with two different attitudes and two different approaches. The upper painting, *The Maids of Honour* (*Las Meninas*) by Velazquez (1599–1660), is a minutely-elaborated work striving to paint details accurately. Picasso (1881–1973) made a "copy" of this same painting, as he saw it. We see this picture in the

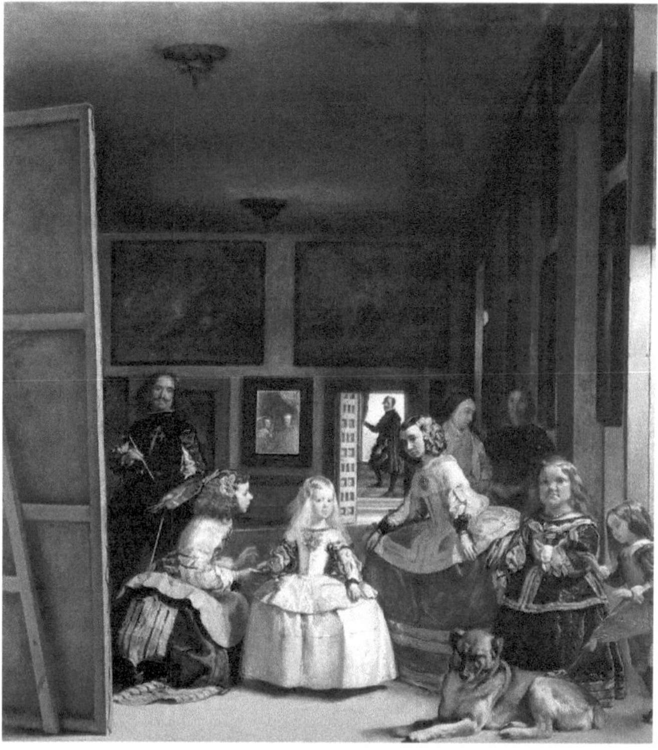

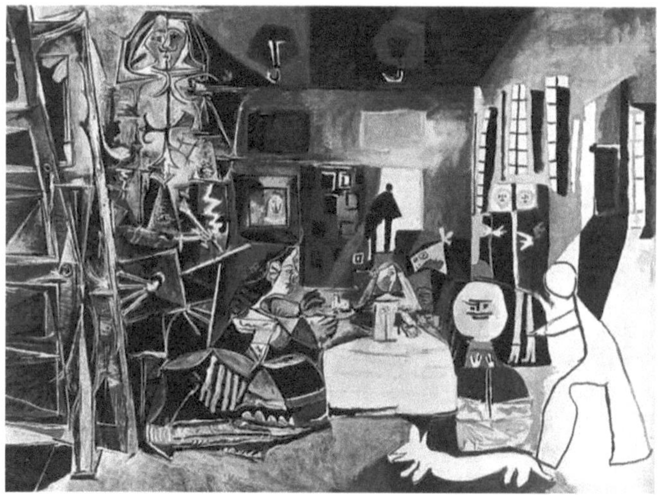

Figure 13.5. *The Maids of Honour (Las Meninas),* as seen by Velazquez and Picasso

lower part of the figure. Picasso was not interested in small details: what remained for him was an overall impression of the painting, in which only the picture's general proportions, centres of gravity and main protagonists were significant, could be symbolized even by triangles or rectangles, and the rest was not important. The two paintings hold up a mirror to the relative asymmetric hemispheric dominance of the two painters.

Human Creativity

Chapter 14

Beauty and truth

The emotional and rational functions of the human brain: art, science, technology

In the preceding chapter we encountered the asymmetries at the highest point of evolution, the human brain, and their consequences. We saw that asymmetry not only governs life functions –like the use of the hand, the motor regulation of speech –but also cognitive and emotional functions, like speech comprehension, the processing of written and visual information, counting, the appreciation of music, the learning process, the decision between the dictates of reason and of the emotions, creativity, aesthetic sensibility, a sense of humour, and so on. We do not yet understand the precise mechanism of most of these, but we know of their existence. This alone is enough to have an effect on the way in which we interpret and group our *knowledge*, on the one hand, and on the *route by which we gain it*, on the other.

In the historical introduction we mentioned that we are thus far aware of three so-called golden ages of European culture in which the arts and the sciences flourished at the same time. This can be said of the golden age in Greece in the 6th–5th centuries BC, of the Renaissance, and of the current era. In the intervening periods the progress of the arts and the sciences followed separate paths. The scientific and the artistic *acquisition of knowledge* took place in their own particular ways, isolated from one another, just as *science* painted a different picture of the world around us from that of *art*; the passages between the two were few and far between, and it was not the done thing to point them out.

Today we see this all in a different light. We do not deny the particularities of scientific and artistic acquisition of knowledge, and the two different reflections of the world, and yet at the same time we recognize that there are points of connection between them, that they influence one another. We know, or in the case of some details we guess, that this interaction takes place thanks to cooperation between the two cerebral hemispheres. The difference –as well as the interaction between the different –is a result of their asymmetry. Put more precisely: in some senses, the two cerebral hemispheres, like an organ performing certain operations on the basis of antithetical pairs, are, despite the term commonly used to describe them, not such much asymmetric as *anti*symmetric. This antisymmetry is the result of a long period of development; it is the consequence of symmetry breakings which occurred at lower levels. The symmetry breakings at lower levels, however, are only responsible for the fact that symmetry breaking necessarily occurs at the higher levels, too. It is the qualitative changes playing a role in the creation of each level that are responsible for the concrete ways in which symmetry breaking manifests itself. We discussed the reasons and possible mechanism of the development of the antisymmetries of the cerebral hemispheres in the preceding chapter. Now we explore what role they play in the scientific and artistic acquisition of knowledge about the world, and the way in which these are reflected.

Science is the art of searching for the *truth.* It attempts to collect and compose our knowledge concerning nature and man, and their relationship, in the form of true propositions. Its method is rationality. It tries to use the tools of logic to put its results into the most perfect form. Its mode of expression is mainly verbal. Scientific research is primarily left-hemispheric creative work, both as an activity intended to acquire knowledge, and as the systematization of knowledge already acquired.

Art is the craft of searching for *beauty*. It tries to portray the picture we have developed of nature and man, and of the relationship between the two, primarily in a way that affects our feelings. Its method is the deployment of the emotions. It make its results of public value by putting them in a form that is grasped by our sense and generates an emotional response. Its mode of expression is non-verbal depiction. Other art forms, like literature (poems, drama, etc.), which are expressed in words, also affect our emotions. Artistic depiction is primarily a right-hemispheric activity.

The asymmetry of the brain (or rather, as we just described more precisely above, its antisymmetry) manifests itself in the course of human activity in the difference between the particular characteristics of science and of art. Various philosophical systems have tried to express this difference in a variety of ways. Most of all, in a way that fitted their system, they discussed the way in which, in the object-subject relation, in science it is the objective element which is dominant, while in art it is the subjective one. With our knowledge of the a(nti)symmetrical functioning of the brain, today we can be more precise. In the case of science, the object-subject relation is predominantly realized with the mediation of the left hemisphere. More precisely, we could say that in the scientific acquisition of knowledge and reflection, it is the left hemisphere which takes the initiative and controls the process on behalf of the subject. In the case of the arts, this same object-subject relation is predominantly created with the mediation of the right hemisphere. Again, put more precisely, in artistic depiction and acquisition of knowledge, it is the right hemisphere that takes the initiative and controls the process on behalf of the subject.

Rationality is characterized by the adherence to logical rules. It is a chain of consecutive (distinct) events connected by causal relations that form the subject of science, while the goal of scientific activity itself is the search for *truth* (in the logical sense).

Emotion grasps the entirety of things as one. This is the characteristic of holistic thinking. In emotion, the general impression gained of particular things, people or the world as a whole play a more important role than individual details. "Gut feelings" or intuition are prevalent. To this group belong activities predominantly influenced by our feelings. Emotional activity (in contrast to the sequential nature of the rational) takes into account the complete horizon of events at the given moment. Together, these moments determine the subject of art, whose objective is the search for *beauty*.

These observations are polarized ones, of course. There is no purely rationally-thinking individual, nor one completely driven by emotion. For healthy people, left-hemispheric and right-hemispheric effects are mixed – to a degree that varies from one person to the next. Both cerebral hemispheres play a part in both types of activity. It is the initiative, the centre of control, and the effect that occurs predominantly that tips the balance to one side or the other.

The search for *good* is realized through the synthesis of rational and emotional effects. The search for the right way to act, the right way to behave and the right decision to make is the subject of *ethics*. The search for good is partly emotional, because what is right is also beautiful, and partly rational, that is the following of the *laws* created by human society on the basis of *moral* principles: the rational logic of the law (which allows justice to prevail).

In the history of philosophical thinking, the search for *beauty* and *truth* have often been complemented by the search for *good* to form the above trinity. While noting that this trinity is also present in Eastern cultures — indeed its role is more on a par with dichotomous systems than in the West — we have chosen our examples from Western cultures. It was often precisely symmetry that represented the connection between the beautiful and the true.

It was Socrates who was the first to draw our attention to this role: "If we cannot hunt down the good under a single form, let us secure it by the conjunction of three: Beauty, Symmetry, and Truth."

It was no accident that Plato, who made note of all this, paid such attention to these thoughts of his master. As we have explained in greater depth, he himself attributed a central role in his work to the search for perfection. And, for him, perfection was associated with symmetry, more precisely geometrical symmetry.

Aristotle, Plato's student and the great synthesizer of Greek philosophy, extended the idea of perfection to include *the form and content of our thoughts*. This characterizes the way his logic is constructed, as well as his views on art. The perfection of logical propositions lies in harmony with the search for truth (cf. logical truth tables), while the perfection of artistic works is in harmony with the search for beauty. The perfection of geometrical presentation, which is associated with proportionality, harmony, that is symmetry, played a common role in both cases. This is how symmetry became the golden middle way for Aristotle between the true and the beautiful. Truth and beauty — complemented with the golden middle way (also known as proportionality or measuredness) — had already in Aristotle's ethics extended to include the perfection of our actions, that is the search for good (Figure 2.17).

Truth, beauty and good fitted into scholastics in an organic fashion. Medieval Christian thought easily took trichotomy on board –we need only mention the Holy Trinity –though it is true that this existed in parallel with certain dichotomous elements (nominalism-realism, faith-knowledge). Mature scholastics had three sources. From Jewish culture via Maimonides (1135–1204), from Greek culture via Avicenna (1135–1204) and Averroës (1126–1198), all three reached back primarily to the teachings of Aristotle, and partly to those of Plato, through the mediation of Arab culture. They translated and studied Aristotle's *Physics, Metaphysics* and *Nichomachean Ethics.* In scholastics, three elements were embedded in Aristotle's scheme: faith, kind acts that pleased God, and knowledge, acquired through experience and then elaborated with logic. All three confronted one another in order to form as perfect system as possible.

The symbolic, trichotomous structure appeared in the trilogy of Dante (1265–1321). With the perfection of geometrical presentation as a starting point, Spinoza poured his ethics into an axiomatic form borrowed from geometry. Kant also followed the trichotomy of Socrates and Aristotle in the structure of his three main philosophical works:

The Critique of Pure Reason	(science — truth)
The Critique of Practical Reason	(ethics — good)
The Critique of Judgement	(aesthetics — beauty)

The whole of Kant's oeuvre, of which these three volumes were the high point, embodies the unity of the true, the beautiful and the good, yet devotes little space to their interaction. In line with the spirit of the times, Kant's transcendental logic and transcendental aesthetics hold that the sciences can only make use of pure rationality, keeping their distance from the intuitive arts and the influence of beauty.

Hegel constructed his complete philosophical system based on the principle of the trinity, from which the order of thesis, antithesis and synthesis, strictly built one upon the other, came to enter public consciousness.

The use of symmetry as an intermediary notion to connect beauty and truth is not only found in the works of philosophers as the personificators of rational science. It can also be found in literature. A conscious awareness of the connection was as important to artists as to scientists.

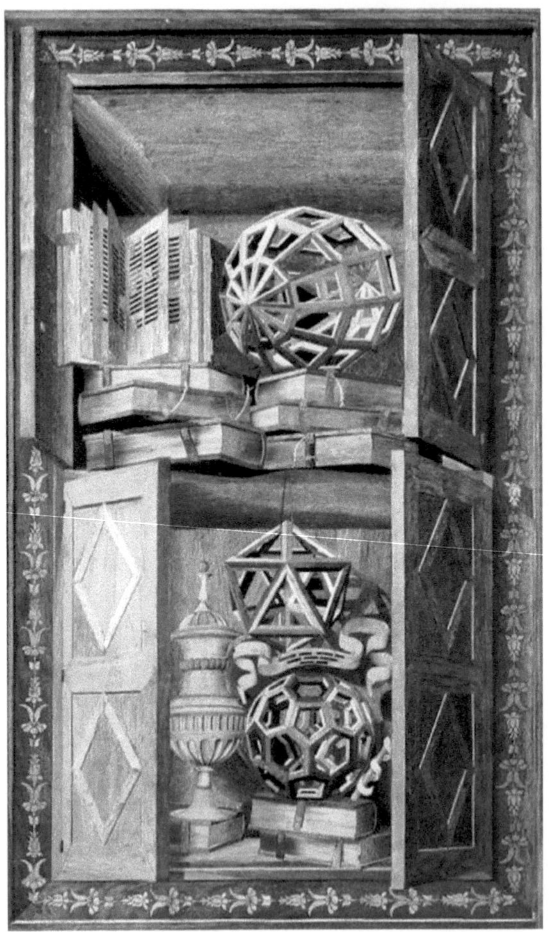

Figure 14.1. *An intarsia of Fra Giovanni*

In the Renaissance, beauty and scientific perfection still went hand in hand. This is what we saw from Alberti through P. della Francesca to Pacioli, from P. Uccello (1397–1475) through Leonardo and Dürer to Kepler and to C. Ripa's *Iconology*, which brought the Renaissance to a close. The meeting of the two can clearly be seen in the masterfully beautiful intarsia in Figure 14.1, a work of Fra Giovanni of Verona (?–1525/6) from around 1520.

In the upper part of the planar intarsia that suggests perspective spatiality, we see a sphere covered with 72 planar faces in the company of books symbolizing science. The "circles of latitude" approximately represent the

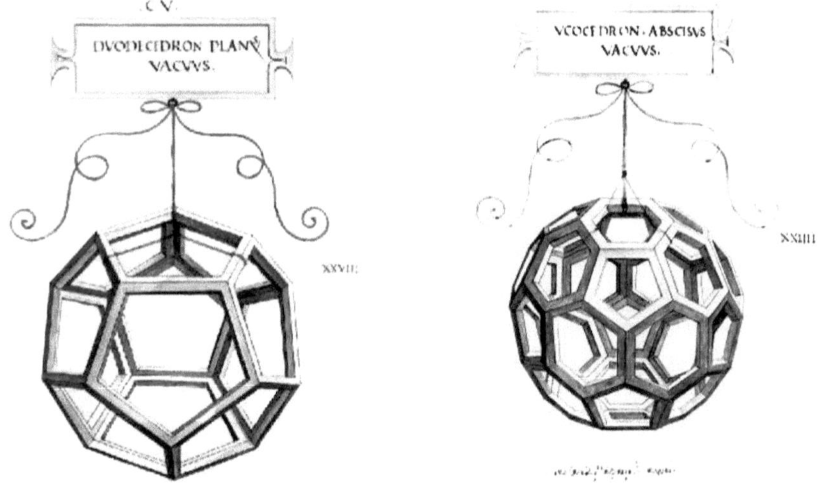

Figure 14.2. *The dodecahedron and the truncated dodecahedron or truncated icosahedron in Leonardo's drawings*

well-known parallels of the Earth: the equator, the polar circles, and the tropics. Beneath it we find a classical icosahedron, then its truncated version below, which we today refer to as fullerene, and which in a previous chapter we saw in Leonardo's drawing in Pacioli's book published more than a decade earlier (Figure 14.2).

Shakespeare (1564–1616), who represented the culmination of late Renaissance poetry, connects beauty and truth in the following oft-quoted form. He begins Sonnet LIV with these lines:

"O! How much more doth beauty beauteous seem
By that sweet ornament which truth doth give!"

Furthermore, as Shakespeare connects beauty and truth, the ornament is given the attribute 'sweet', which symbolizes good.

In twentieth-century Hungarian literature, the use of sweet as a symbol of good reappears in the poetry of Attila József, who links the good, the beautiful and the true in the following way:

"My mother's mouth gave me the sweetest nectar,
my father's mouth, the truth as beautiful."
(*By the Danube,* trans. Zsuzsanna Ozsváth and Frederick Turner)

Of the three, the closer connection between beauty and truth is merged through the persona of his father, while good (which for József, too, is symbolized by the "sweetness" of nectar) is related to the truth-beauty pair, along the antisymmetry of the mother-father relation, rather like the antithetical pair of the yin-yang. That he receives the nourishment from his mother's "mouth" like the words from the mouth of his father underlines the symmetry between these two relations.

Returning to the early modern period after this diversion of many centuries: in this age, too, scientists open to the other half of culture were able to learn from art. We saw an example of this in Kepler, who not only constructed an artistically perfect model of the solar system from Plato's perfect bodies, and not only made reference to Aristotle, but also referred to the Pythagoreans and the music of the spheres. A proportion of artists, meanwhile, were not left unaffected by the achievements of science. Let us quote the Frenchman R. de Piles in this regard, who wrote the following on the beauty of painting in 1708: "There are three kinds of truth in painting: Simple Truth, Ideal Truth, and Complex or Perfect Truth." (*Course de peinture*, p. 30.) According to de Piles, "the Simple Truth [...] is the simple and faithful imitation" (cf. Plato's mimesis), while "Ideal Truth is a selection of perfections which are never to be found simultaneously [...]". De Piles gives us less information about the Complex or Perfect Truth, but we do learn that it represents the combination of the aforementioned two. Perfect Truth is the middle way between the Simple and the Ideal, which strives on the one hand for perfection, and on the other for the artistic reflection of the truth. According to its objective, then, art also depicts the truth, the real world, except that it finds artistic perfection with the differently proportioned combination of the simple and the ideal, along a different route from that which science follows in order to find logical perfection.

Art and science progressed hand in hand, just on the two opposite banks of the culture's increasingly wide river, observing one another from a discreet distance. Intellectuals with an open mind looked for bridges between the two sides. Again and again, they would find a pillar or two of the bridge by looking back to ancient cultures, in which art and science, beauty and truth had not yet split apart. As an illustration of this, let us have a glance at the last two lines of the best-known poem by John Keats:

" 'Beauty is truth, truth beauty,' –that is all
Ye know on earth, and all ye need to know."
(*Ode on a Grecian Urn*)

The theme of reaching back to the classical period of antiquity can only fully be understood in the light of Plato's Sun Myth, in which Plato's morphology (cf. the perfect bodies) is explained. In this system the Sun embodies the idea of good being at the top of the pyramid, and it is from this idea that beauty and truth are derived. The light of the Sun is represented by the fire burning in front of the entrance to the Cave, discussed in Book VII of *The Republic*, and by the shadows it projects onto the cave walls, which we mortals are able to watch. To understand the relationship with the forms, we can read the following in the Ode, four lines above:

"Thou, silent form! dost tease us out of thought
As doth eternity"

In the Ode, the urn represents the perfect form, and the urn expresses the warning intended as a conclusion for the reader: "Beauty is truth, truth beauty". The rest of the words –"that is all / Ye know on earth, and all ye need to know" –are added to the urn's message by the poet.

These two lines of poetry reveal almost everything about the unity of the culture. The summarizing words that end the poem were, for Keats, not merely poetic tools with which he could make an effective end to the Ode. Nor was it just a question of posterity imposing its interpretation on the poem. Keats himself attributed particular significance to this sentence, which can clearly be traced in his correspondence in November and December 1817, the months after the Ode was written (cf. his letters to his friend B. Bailey, and to his brothers George and Tom). In these letters Keats himself mentions the last lines of the Ode, analyzing and interpreting his own text, which shows that what he was trying to express meant more to him than in the case of his other poems. What a coincidence that it was in the same year, and also in England, that the crystallographer D. Brewster patented his technical-scientific novelty so spectacularly displaying the beauty of symmetry: the rotatable kaleidoscope!

Since then, literary scholars have dedicated hundreds of pages to the Ode, with considerable emphasis on the analysis of the last two lines, not to mention the attention that Keats himself drew to his own words. These lines are quoted in a wide variety of fields, thus not only by literary critics, but, for example, by the mathematician László Fejes Tóth. We should add that it is no accident that this book devotes more space to them, as, from Brewster onwards, it was symmetry that heeded the warning of the urn with the most devotion. It was research into symmetry that implemented Keats' dream, the reunification of our knowledge of beauty and of truth.

Arthur Koestler (1905–1983) wrote: "the artist and scientist do not inhabit separate universes, merely different regions of a continuous spectrum –a rainbow stretching from the infra-red of poetry to the ultra-violet of physics, with many intermediate ranges –such hybrid vocations as architecture, photography, chess playing, cooking, psychiatry, science fiction or the potter's craft. But to avoid oversimplification, after emphasising the affinities, I must briefly discuss the differences –some apparent, some real – between the opposite ends of the continuum. The most obvious difference seems to lie in the nature of the criteria by which we judge scientific and artistic achievement" (Koestler, 1978, p. 152.) What else could these criteria be than those that evaluate the attainment of truth and beauty?

The paths followed by scientists and artists sometimes cross. When researchers formulate a scientific truth, they try to do so in an attractive form. This makes their results easier to understand for their peers. A beautiful mathematical proof, for example, offers an aesthetic experience. Almost all mathematicians know this feeling. Paul Erdős (1913–1996) was of the opinion that God had collected together the most beautiful, most perfect proofs into a separate volume. In his words: "you need not believe in God, as a mathematician, you should believe in THE BOOK". It is to H. Poincaré to whom the claim is attributed that it was mathematical beauty, the harmony of numbers and forms, and the sensing of geometrical elegance that subconsciously drove him to the successful combinations that resulted in new discoveries. It is said of P. Dirac (1902–1984), one of the greatest figures in twentieth-century physics, that he went even further than this, stating that it is more important that an equation be beautiful than that it fit the experimental results. How reminiscent this is of the construction of the Herodotean world-view!

When artists give a form to the beauty of their work, they try to depict reality in a faithful, that is realistic way. Many artists have been inspired by theories motivated by science: the golden section, the geometry of perspective and foreshortening, and the final laws of the Renaissance masters on the perfect proportion. Here we can include the theorem of Cézanne (1839–1906), according to which all natural forms can be reduced to spheres, cylinders and cones. We can also include reflection, repetition and counterpoint in musical composition, the harmony of colours and the combination of the utility, simplicity, beauty and symmetry in the principles of the Bauhaus, and the works of graphic artists and sculptors inspired by higher geometrical dimensions, non-Euclidean spaces, quasi-crystals or fractals.

There are many examples to prove the mutual influence of scientists and artists. This interaction is permitted by the fact that it is not possible completely to separate truth and beauty. At least not to the extent that both parties operate within the same social environment, and it is this same system which appraises the outcome of their work.

Koestler, for example, analyzes Keats' lines quoted above as follows: "This is [...] a touching profession of faith in the essential unity of the two cultures, artificially separated by the quirks in our educational and social system. In the unprejudiced mind, any original scientific discovery gives rise to aesthetic satisfaction, because the solution of a vexing problem creates harmony out of dissonance; and vice versa, the experience of beauty can only arise if the intellect endorses the validity of the operation –whatever its nature –designed to elicit the experience. Intellectual illumination and emotional catharsis are the twin rewards of the act of creation, and its re-creative echo in the beholder. The first constitutes the moment of truth, the *Aha* reaction, the second provides the *Ah...* reaction of the aesthetic experience. The two are complementary aspects of an indivisible process." (Koestler, 1978, p. 155.)

Beauty and truth together laid the foundation for the cultic function of the arts. Cave paintings already reflected this dual function, followed by the decorations of the chambers of the pyramids. They expressed some truth of importance to their creators, in a beautiful form. Later it was these functions which motivated the decoration of practical appliances: they had to be both usable (suitable for their function, that is real, genuine) and beautiful at the same time. We see examples of this in the decoration

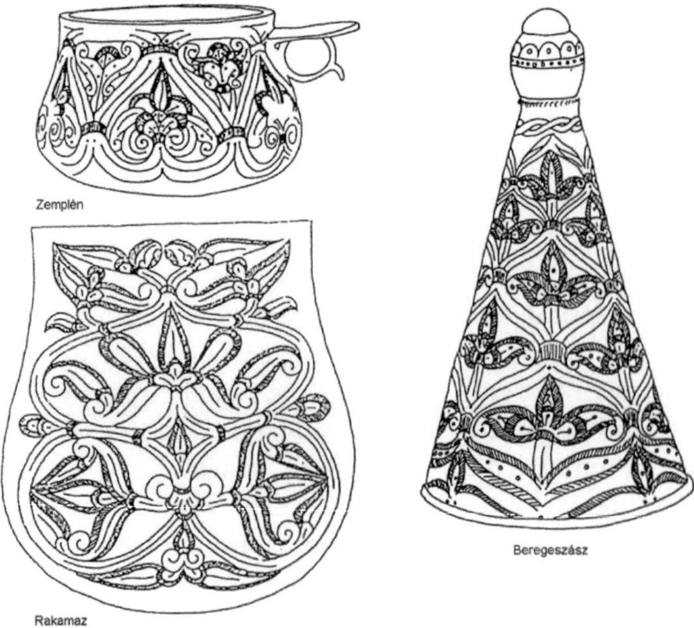

Figure 14.3. *Decorations of practical objects from the time of the Magyar conquest of Hungary (9th century AD) (from the motif collection of Szaniszló Bérczi* [1950–])

of weapons, and that of objects and appliances for use in the home, like the form and decorative motifs of tools and bowls (Figure 14.3).

It was these applied objects that would later go their separate ways, according to their function: separate works of art on the one hand, and the decoration of everyday objects on the other. It was from the former that fine art separated to become a distinct field, leaving architecture in the middle ground; the latter in turn gave birth to applied arts and design. Some middle ground, some connection, always remained between the creation of distinct works of art and the decoration of practical objects. In the twentieth century these connections gained new content. It is this content to which the next chapter is devoted.

Chapter 15

Rationality and impression

Function and art in the works of the twentieth century

It transpired that nature is not perfectly symmetrical. Neither are human beings. Can art achieve perfect symmetry?

Assume we are able to construct a geometrically perfect shrine to symmetry. We enter it, and at once our presence ruins it. Symmetry turns into dissymmetry. Why are we, humans, asymmetrical? As we saw, this is a result of our biological development, our phylogenesis. The asymmetries of our physiology are derived from the asymmetries of biologically active molecules, which became asymmetric as a result of the fine structure of matter at the most detailed level. The shrine was constructed of these same elementary building blocks. There flitting around us are the neutrinos, which only exist in one chiral form, while individual atoms in the crystals of the marble have their angular and own (intrinsic) momentums, and others may also have magnetic momentum, and so on, which all determine an arrow in space. On the other hand, the glass in a window is transparent precisely because the molecules in it are distributed in complete spatially disordered directions. A building can be constructed of bricks that are each geometrically entirely symmetrical, but which contains asymmetry at a much smaller scale. On the other hand, a building can also be constructed of rough stones that are all individually asymmetrical, and yet be geometrically symmetrical at a larger scale. Even if it is not visible to the naked eye, the shrine contains the dissymmetry even in our absence.

From other points of view, we can also ask whether, if an artistic depiction is symmetrical — in its geometry, at least — it is at the same time a faithful reflection of the real world (or the part of the world it is intended to depict), or just an approximation of it?

Furthermore, if a depiction is geometrically symmetrical, and if possible a faithful reproduction of the world, does it automatically bear the hallmarks of harmony?

We cannot address all of these considerations at once. A good example of this is the history of perspective in the art of painting. Perspective set as its goal the most faithful possible reflection of the reality we see. Its methods were similarity and (affine) projection. We saw in the introduction that these belong to the elementary geometrical symmetry operations. Yet things were not so simple. The richness of the artistic way of seeing cannot be realized with any single technique, however geometrically perfectly created it might be. The image and the what we see become separate. The application of perspective has followed a lengthy path from Brunelleschi (1377–1446) to Miklós Barabás (1810–1898), and its interpretation has progressed from Pliny's ancient notions about painting, through Alberti in the Renaissance, to Panofsky (1892–1968) in the twentieth century, and the contemporary R. Berger (1915–), and from Leonardo's description of perspective, through his *sfumato* and *camera obscura,* to the "isms" of the twentieth century. Alongside central perspective with a single vanishing point appeared depiction using two, then three vanishing points (Figures 15.1, 15.2, 15.3 and 15.4).

For now, we will ignore extensions of a non-geometric nature, like aerial and colour perspective, instead drawing the reader's attention to the fact that if the vanishing points of perspective could be extended to three, then there is no reason why they could not be extended even further (Figure 15.5 and 15.6).

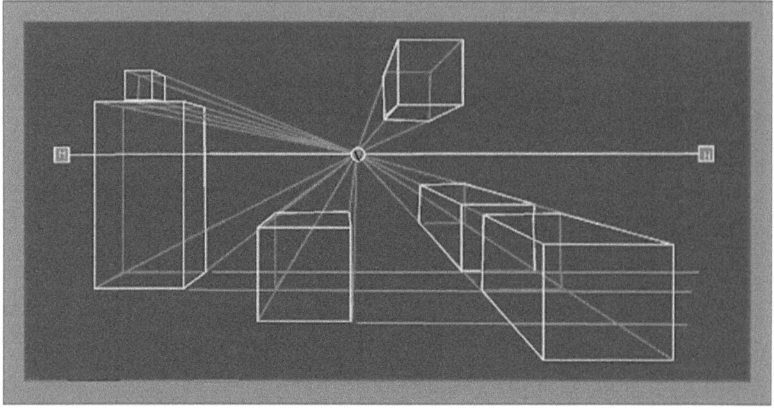

Figure 15.1. *Perspective with a single vanishing point*

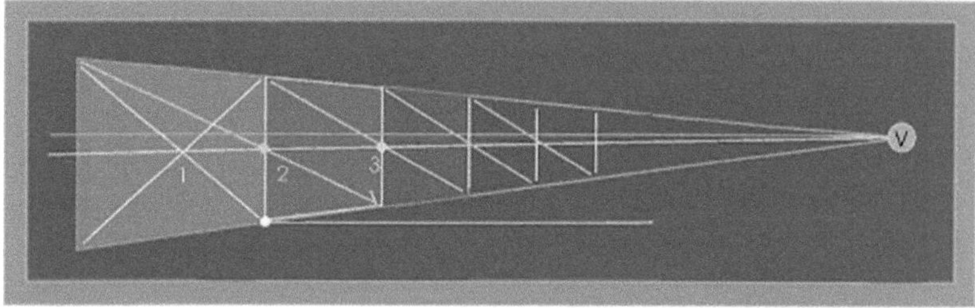

Figure 15.2. *The proportions of perspective, in the course of similarity and affine projection*

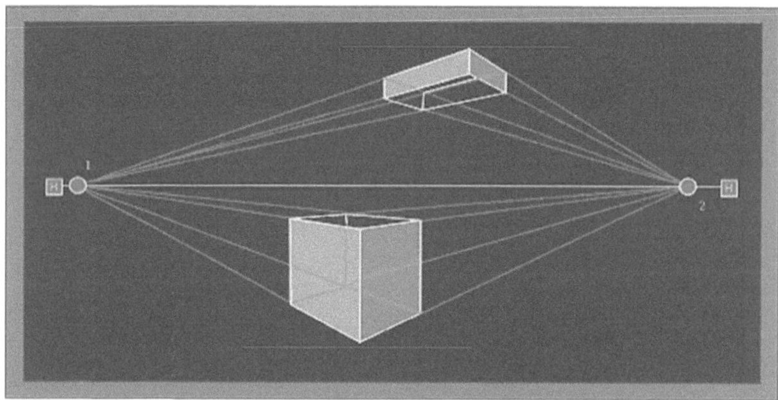

Figure 15.3. *Perspective with two vanishing points*

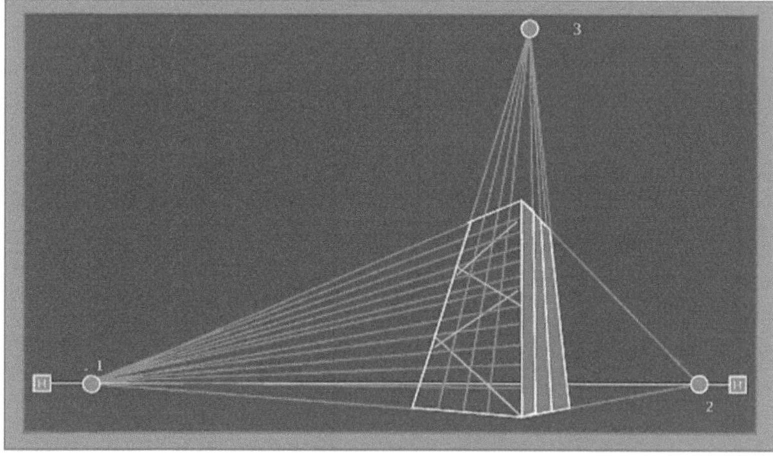

Figure 15.4. *Perspective with three vanishing points*

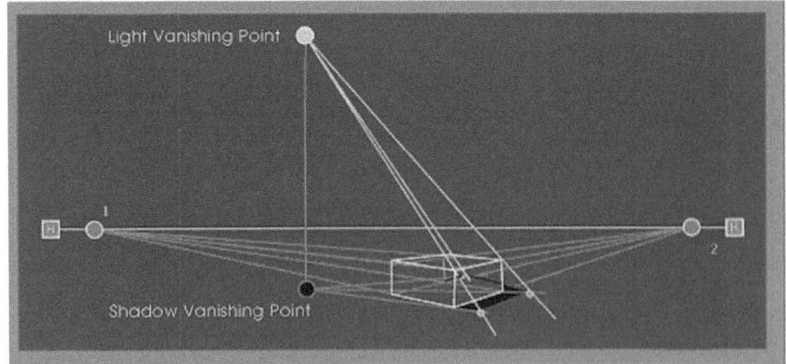

Figure 15.5. *Perspective with multiple vanishing points*

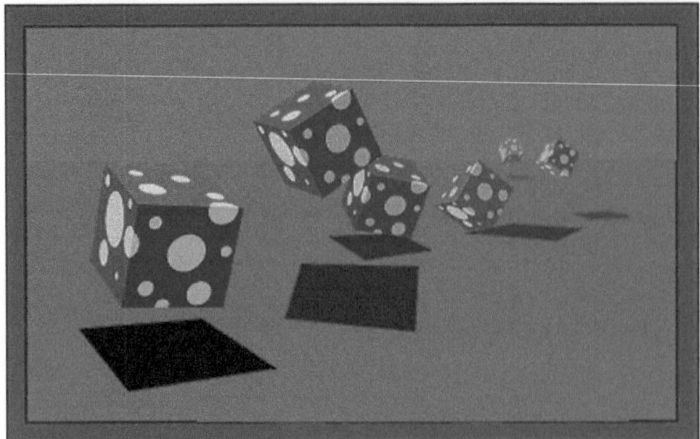

Figure 15.6. *Aerial and colour perspective*

Perspective with multiple vanishing points could also be paired with the sacrifice of the limits of traditional spatial vision. The widening and enriching of science's concept of space freed up artistic thinking. If new invariance operations and new transformational bases could be introduced to science (Lorentz invariance, Minkowski [1864–1909] space, Riemann [1826–1866] space), why could the same not be done in the arts? A good example of the escape from traditions is the art of Escher (e.g. his works presented in Figure 15.7).

The old interpretation of perspective came to a close with Cézanne. In modern painting, the multiplication of vanishing points began at the beginning of the twentieth century. And if it was possible for vanishing

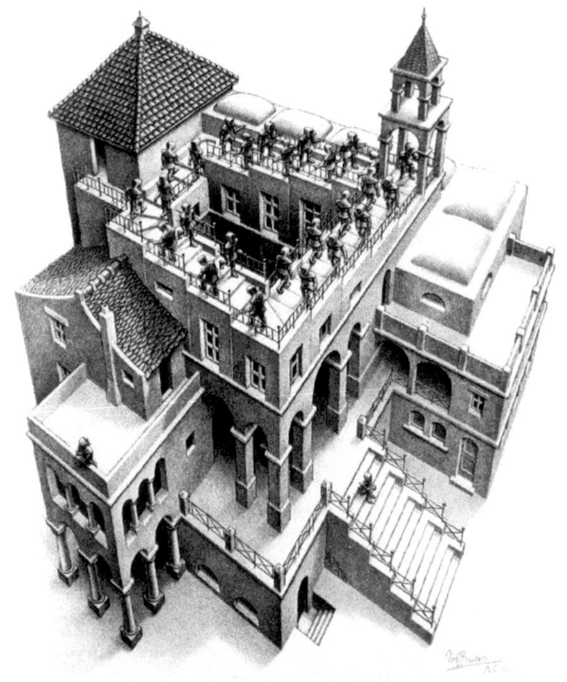

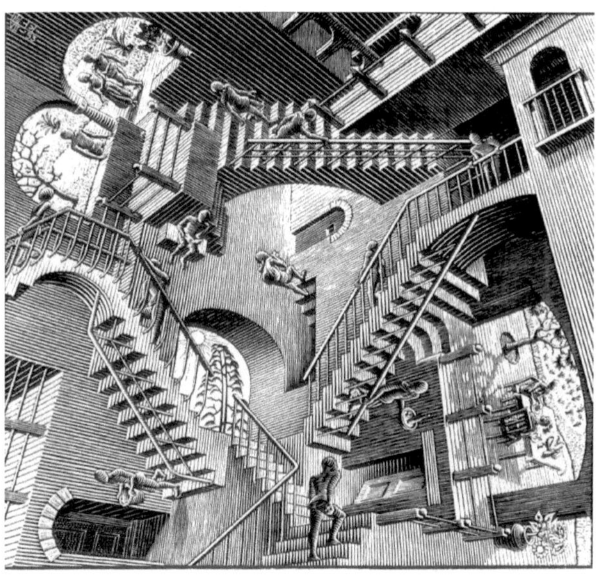

Figure 15.7. *M. C. Escher:* Ascending and Descending *(1960) and* Relativity *(1953)*

points to multiply, what obstacle was there to the point of view of the painter no longer being a single point? Why could our own viewpoint not be multiplied? If the point of projection of similarity as a symmetry operation can be expanded along a horizon, first in two directions, then in three, these points of projection can be lifted from the line of horizon Leonardo still considered to be so sacred. The canvas is stretched between the depicted perspective and the viewer like a mirror plane. This allows us to extend the abstraction further; if the projection can be directed at our eyes from a number of points behind the plane of the canvas, similarity projective lines that are directed from our eyes towards the depicted object can also originate from various different viewpoints in front of the canvas. Third, if the vanishing points of the similarity symmetry can be moved or become multiplied, why can we not also transform the plane of affine projection? We can tip the plane of projection, indeed we can multiply it in line with the multiple points of projection. It was this that Picasso, Braque (1882–1963), the pioneers of cubism, realized in their art. Perspective with multiple vanishing points, with multiple viewpoints, with multiple planes of projection, became a symmetry operation of equal merit, an equally acceptable mode of depicting reality. There is no doubt that, within the realm of the tools it uses, it paints just as faithful picture of the world it wishes to depict as its predecessors, in a way that is unusual in comparison with the past, which gives the artist much more freedom, and which demands compliance from the beholder to the way of seeing chosen by the artist, requiring an acclimatization to and acceptance of new ways of seeing.

The process can be understood in a number of stages. First the vanishing points remained fixed, and the artist's viewpoint moved. In Picasso's *Nude on a Beach* (Figure 15.8), the artist sees the subject from a number of different angles.

Spending a little time looking at the painting, with our eyes we can mark the viewpoints from which the nude's face has been tripled, while the angles under which her nipples appear are fixed. In the next step, it was the both the vanishing points of perspective and the painter's viewpoints that moved. In Picasso's *Portrait of Marie-Thérèse* (Figure 15.9), for example, the former can be observed in the (distorted [sic!] perspective of the) edges of the room, while the latter can be seen in the woman's face.

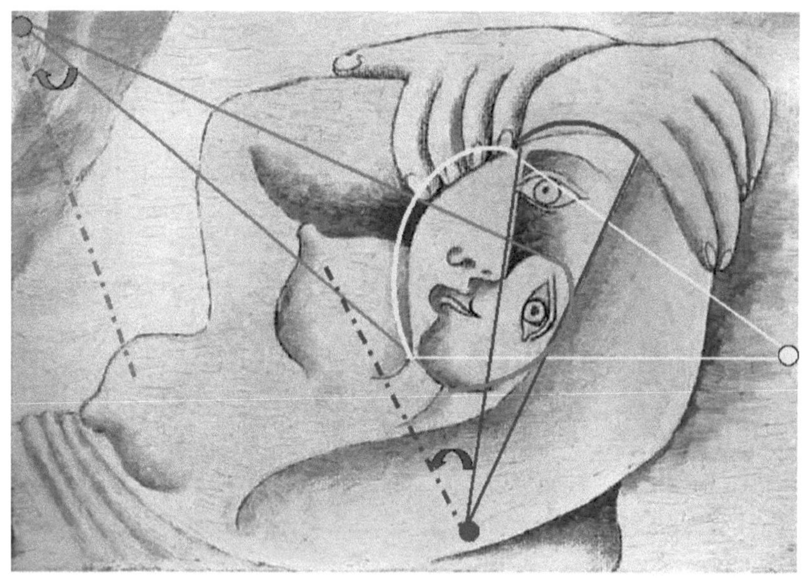

Figure 15.8. *Picasso:* Nude on a Beach (1929)

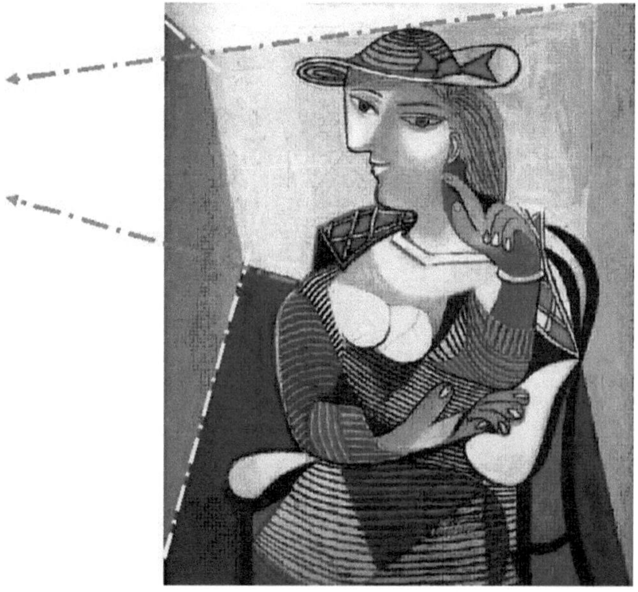

Figure 15.9. *Picasso:* Portrait of Marie-Thérèse (1937)

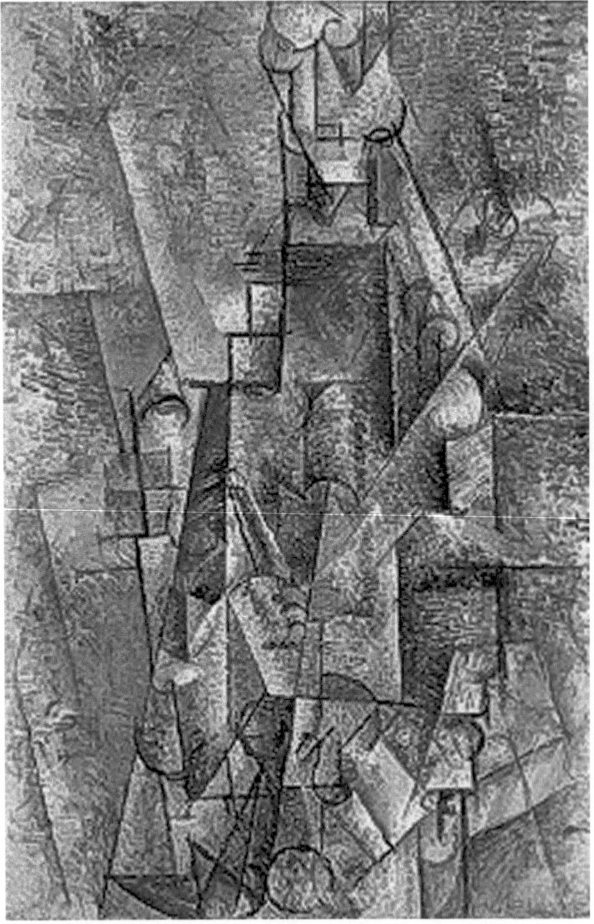

Figure 15.10. *Picasso:* Man with a Clarinet (1911–12)

The point movements and multiplications were multiply repeatable. This is illustrated by two paintings (although of earlier origin), Picasso's *Man with a Clarinet* (Figure 15.10) and Braque's *Girl with a Mandolin* (Figure 15.11).

Both paintings, however, retained the horizontal perspective of traditional depiction. The horizontal view of the subject, regarded as absolute in Renaissance perspective, was sacrificed by futurism when it located its viewpoint above the subject. It viewed the objects to be depicted at an

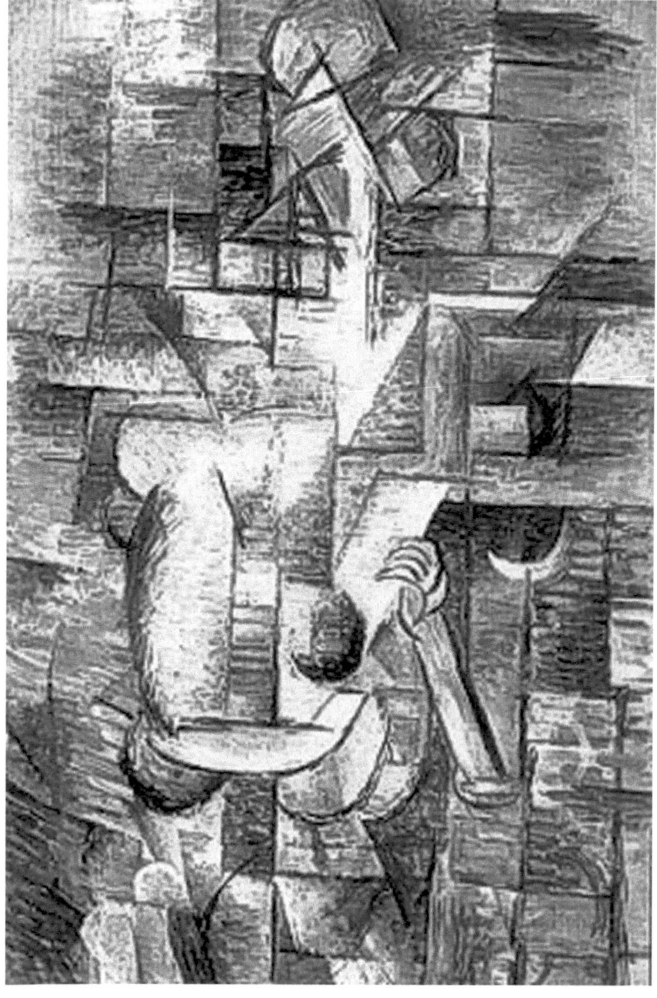

Figure 15.11. *Braque:* Girl with a Mandolin (1910)

angle, or even vertically. The perspective multiplication seen in the paintings in Figures 15.8–15.11 can be observed from a vertical perspective in the futurist picture in Figure 15.12.

The question is more complicated than it appears at first glance, i.e. than simply rotating the direction of our vision through 90°. The horizontal and vertical directions do not play equal roles in our vision (sense of space), and so perspectives in the two directions are not equal, either.

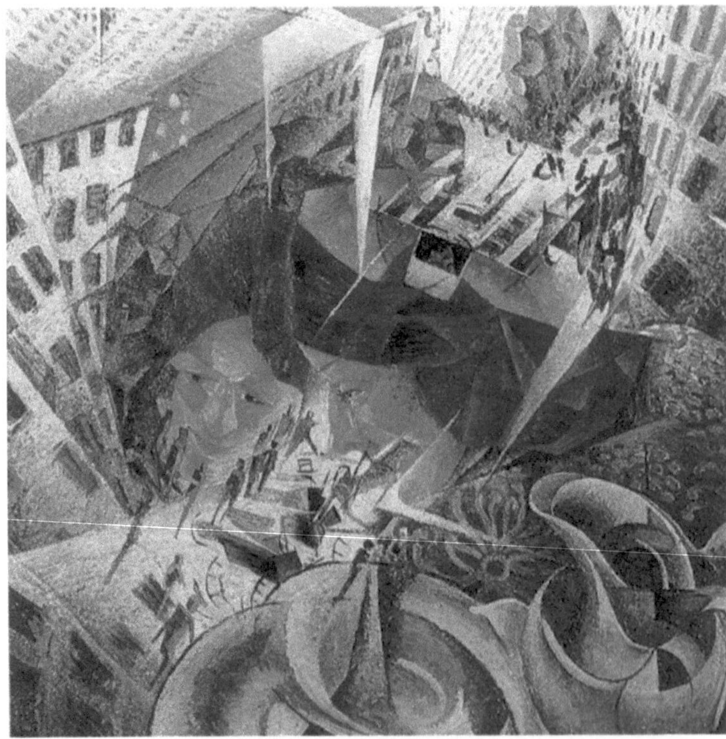

Figure 15.12. *Boccioni:* Simultaneous Visions (1911)

We do not sense a straight line segment as being of the same length if it is vertical rather than horizontal. (For example, let us imagine that there is a 10 metre-high tree about 300 metres away from us, and a fence of the same length [10m] at the same distance, but running horizontally. We sense the tree as being taller than the fence is wide.) As the roles of the horizontal and vertical directions in our sense of space cannot be commuted, we can say that their roles are not symmetrical (not invariant under the commutation of the two directions). As a result, the perspective will also be different if we view the object of our examination from above and not horizontally. It is this viewpoint looking down from above that characterizes the perspective of the futurists. If we can disregard this dependence on direction, however, perspective with multiplied vanishing points appears in a similar way in both schools, as is clearly demonstrated by the

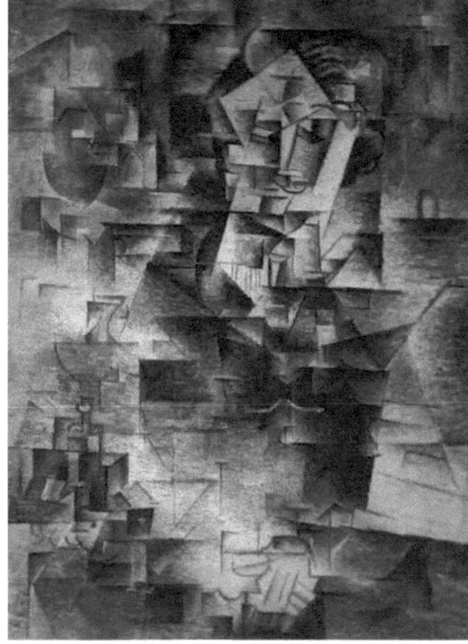

Figure 15.13. *Picasso:* Portrait of D-H Kahnweiler (1910)

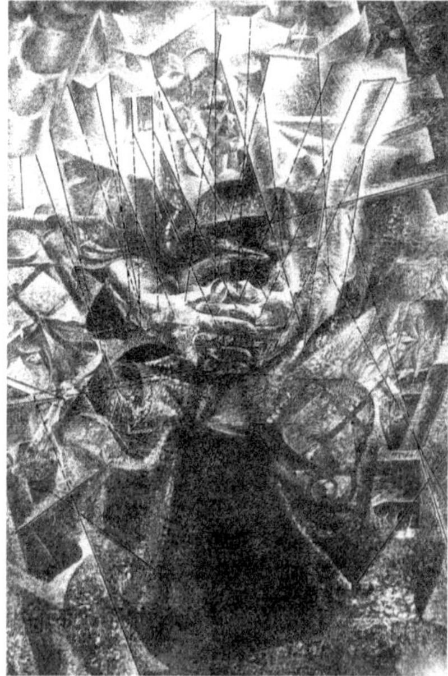

Figure 15.14. *Boccioni:* Materia (1911–12)

cubist Picasso's *Portrait of D-H. Kahnweiler* (Figure 15.13) and the futurist Boccioni's (1882–1916) *Materia* (Figure 15.14).

In was not only in terms of perspective that twentieth-century science liberated the artist's way of seeing, but also in a few other ways. Examples were the more flexible treatment of views on *dimensionality*, the acceptance of the real constituents of non-Euclidean *spaces*, the reinterpretation of the relationship between *space and time*, the changing relationship between *part and whole*, and scale dependence. Artistic depiction did not necessarily apply all of these in the same sense that scientists understood them, but it was able to change traditional interpretations in a more flexible way (e.g. Magritte 1898–1967). Artistic depiction liberated the dogmatic obligations, a step previously considered taboo. We learned about the relationship between the aspects listed above and invariances in the chapter on physical symmetries. These individual aspects influenced art in different ways; moreover, the same aspect could have different effects in different artistic disciplines.

Art, thus freed from the absolute nature of three-*dimensional* space, sometimes realized 'real' spaces with four or more dimensions, and sometimes filled with artistic content spaces that were inconceivable in three dimensions, as in the works of Tamás F. Farkas (Figures 15.15 and 15.16).

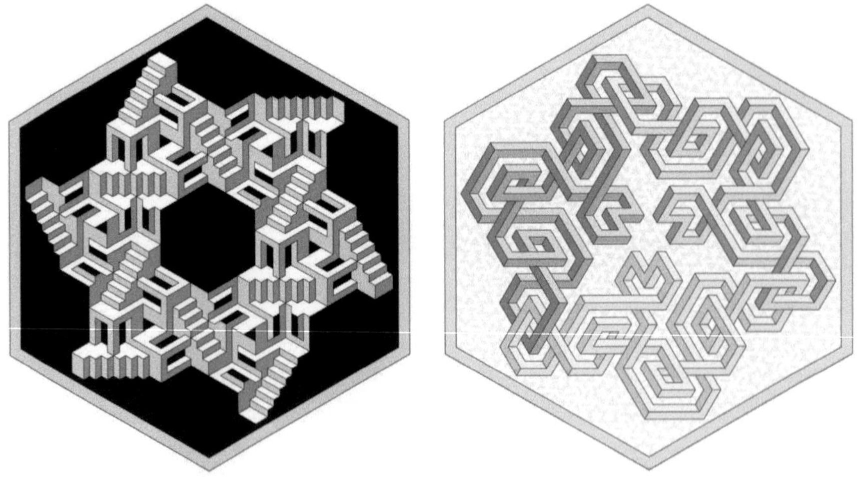

Figure 15.15. *Tamás F. Farkas:* Geocity I (2003)

Figure 15.16. *Tamás F. Farkas:* Genetix IV (2002)

Escher's spaces represent the liberation of *spatial vision*, beyond the reinterpretation of perspective (Figure 15.17 and Figure 15.18).

The *relationship between space and time* moved the artistic imagination in new directions, encouraging it to depict time in spatial dimensions. It was partly this dynamism that futurism set as its goal, and partly the depiction of this which motivated MADI artists (cf. *"M"*). On the one hand, they tried to express motion, changes in time, through fixed, immobile artistic objects; on the other hand, they created mobile works of art. (Mobile space plastics as works of art had already appeared as 'forerunners' to the works of recent or contemporary artists close to or belonging to the MADI movement, like Nicholas Schöffer [1912–1992] and István Harasztÿ [1934–]. The earliest mobiles appeared in works by Naum Gabo [1890–1977], László Moholy-Nagy [1895–1946] and Marcel Duchamp [1887–1968], followed by Alexander Calder [1898–1976], who perhaps became the best-known in this regard. Today, Y. Agam, mentioned above, uses his large-scale art works to combine play with musical sound and light with a restrained

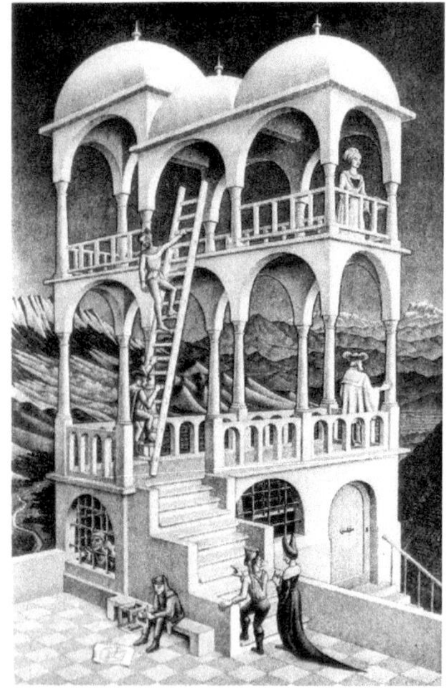

Figure 15.17. *M. C. Escher:* Belvedere (1958)

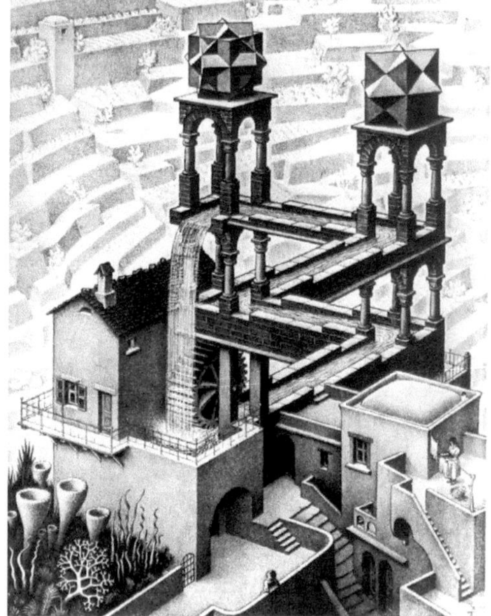

Figure 15.18. *M. C. Escher:* Waterfall (1961)

use of motion. Previously it had been said of music that it was the art of time (we saw in the chapter on cerebral asymmetries that this was not entirely without justification), until the arrival of the motion picture at the turn of the twentieth century, which was capable of depicting spatial scenes in temporal succession. Even before the twentieth century, however, music was not exclusively the art of one-dimensional time. Firstly, emphasis had been placed on harmonics before; in orchestral works, operas and oratoria, spatial orchestration had been given a role, in a similar way to the spatial distribution of organ pipes in large spaces. Secondly, themes arranged in space were depicted with the methods of music (e.g. *The Moldau* by Smetana [1824–1884] and *Pictures at an Exhibition* by Mussorgsky [1839–1881]).

Following the new topology of atomic sizes, the *relation between part and whole* was again reinterpreted once self-similar fractals became an increasingly important part of mathematics from the 1970s onwards. In the latter instance, however, art preceded scientific discussion. Some of Escher's graphic works had already pioneered the use of fractals as an artistic tool (Figure 15.19), while from another aspect these same works can, for example, be considered as the artistic projection of the representations of non-Euclidean spaces like the Klein model and the model proposed by Poincaré.

Although in this instance there was no mention of any direct influencing effect, we saw, in the cases of fullerenes and quasi-crystals, examples of how art had an inspirational effect on science.

We concluded the previous chapter by separating stand-alone decorative art and applied arts. With its special space-creating function, and its monumental character, architecture was as good as a connecting element between them. Architectural styles were always the functions of the opportunities provided by the technology available in the given era. New technology widened the possibilities for the formation of space, decoration, and the formation of style, but did not restrict the use of previous techniques, either. This is what led to opportunities becoming richer and richer.

Technology was always related to the possibility for static solutions, and statics generally favoured symmetrical solutions. Vitruvius had already become aware of this when he elucidated the classical styles. The need for monumentality played a style-forming role in architecture from the outset, and architecture appeared to be a predestined art form for its implementation. Our main public buildings were constructed in line with this requirement, including the Parliament in Budapest, the National Gallery in London, other important museums, the palaces at Versailles, German and Austrian royal castles, bridges over the Thames, the Tiber and the Danube, the headquarters of the Hungarian Academy of Sciences and the Académie Française, our universities, cathedrals, theatres and public spaces. Monumentality itself necessitated the use of symmetrical elements.

As technology increasingly freed architecture from its static limitations, newer viewpoints could emerge, and these new viewpoints mixed new symmetry elements into the picture. Le Corbusier's *Modulor* introduced the human proportions of Renaissance artists. Similar principles based on human scale — ones that interpreted practical considerations in a broader, more multidisciplinary fashion — were introduced by the *Bauhaus*, which was primarily founded by architects. These artists included Walter Gropius

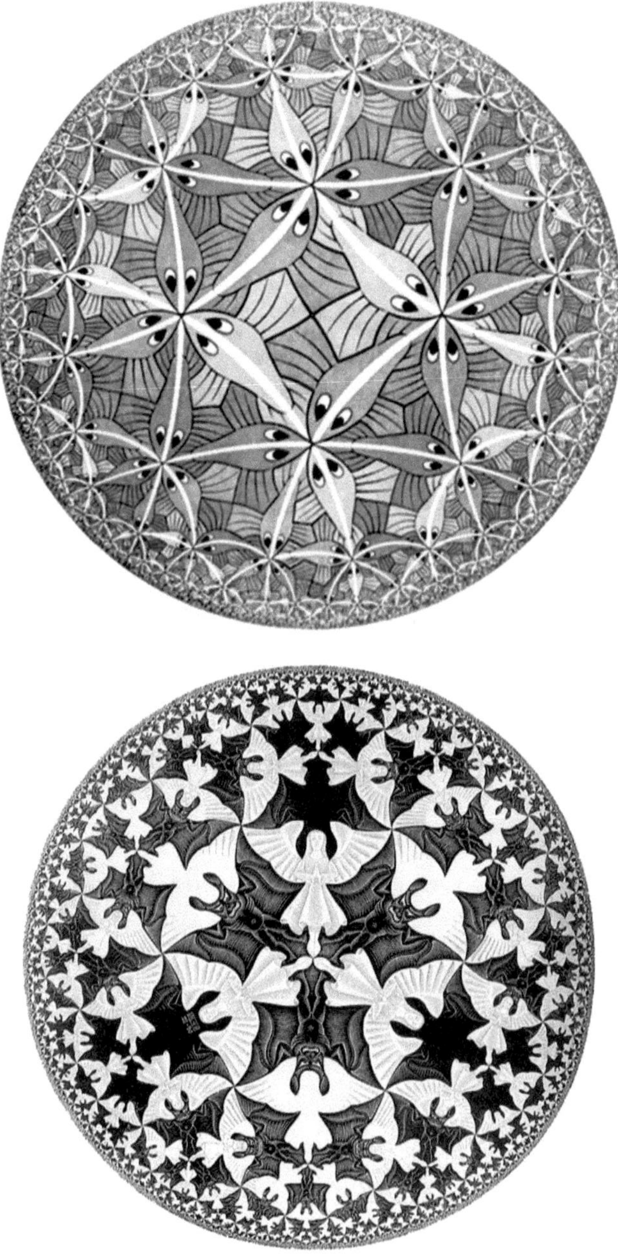

Figure 15.19. *M. C. Escher: Circle Limit III and IV*

(1883–1969), L. Mies van der Rohe (1886–1969), Marcell Breuer (1902–1981), in some measure László Moholy-Nagy, and Max Bill, who joined them later, as well as architects outside Germany who implemented their ideas. Architects by training, they projected their key principles (to which we will return below) onto other associated art forms.

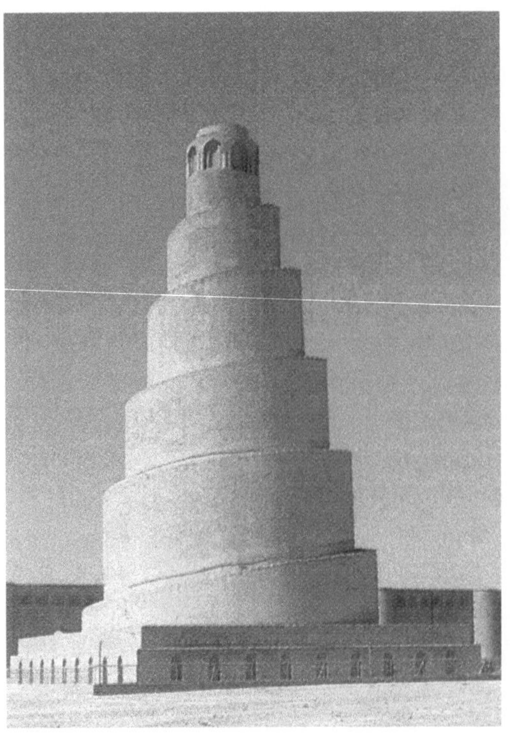

Figure 15.20. *The Great Mosque of Samarra (9th century)*

The relationship between architecture and symmetry stretches beyond individual buildings, making its imprint on collections of buildings, on town design, even urban planning, in which new functional considerations had to be taken into account (transport, communal services, public administration, organization, etc.). Figuratively speaking, these considerations introduced new spatial dimensions to thinking, reviving the notion of ideal cities from earlier ages. In Middle Eastern architecture, from the fifteenth and sixteenth centuries onwards — and further developing the tiered tower construction of the Babylonian *Ziggurat*, with its rectangular ground-plan — tower buildings started to appear and spread with round ground-plans, and a gradually rising surrounding wall running around the building in a spiral helix. (This is how P. Brueghel [c. 1525–1569] depicts the tower of Babel in his famous painting.)

In the spiral plans of Le Corbusier, it is first the rectangular spiral of the Ziggurat that appears, which he regards as the organic form of arrangement (cf. his 1939 plans for the museum in Philippeville shown at the end of Chapter Five). The spiral order of the organic arrangement, which he combines with the golden section, outstripped the boundaries set by the walls of the building, and also influenced his ideas for urban planning.

His principle of constructing a city in an organically developing spiral, complemented by some Indian cult inspirations (the integrated yoga represented by Sri Aurobindo [1872–1950]), was fulfilled in the plan for the city of Auroville by French architect Roger Anger (1923–). Auroville is a new town near Pondicherry in the southern Indian state of Madras. Its original plans were laid down in the 1960s, and they conceived of a central space with spiral arms growing out of it, like those of a galaxy, with the increasingly wide spaces encompassed by the spiral arms that stretch out into the green belt being reserved for special functions (Figure 15.21).

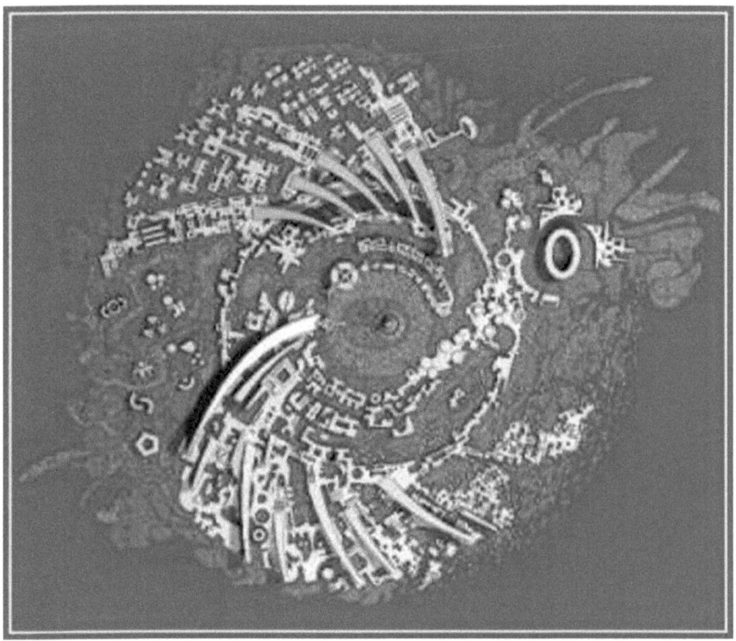

Figure 15.21. *The plan for the city of Auroville*

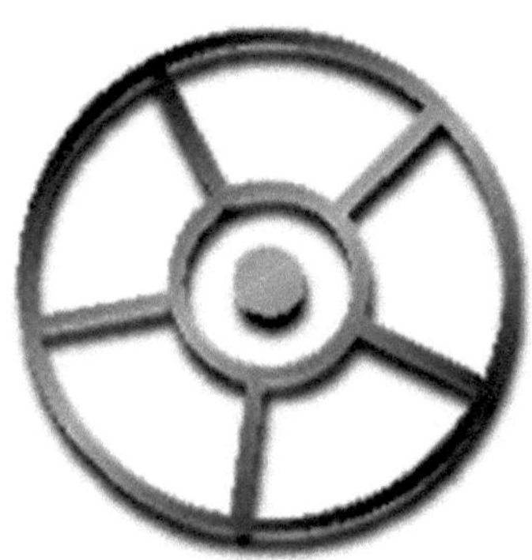

Figure 15.22. *The symbol of the city of Auroville*

The symbol of the city expresses the principle of connecting together the functions of the different districts (residential, cultural, industrial, etc.) with an emblem that displays not only fivefold rotational symmetry, but also mirror symmetry (Figure 15.22).

Another reinventor of the concept of spiral-helical construction is Kisho Kurokawa (1934–), whose helical city design, accommodating as it does to the restricted space for construction in Japan, can be seen in Figure 15.23. Although neither of the plans were realized in the form detailed here, the ideas behind them are worthy of note.

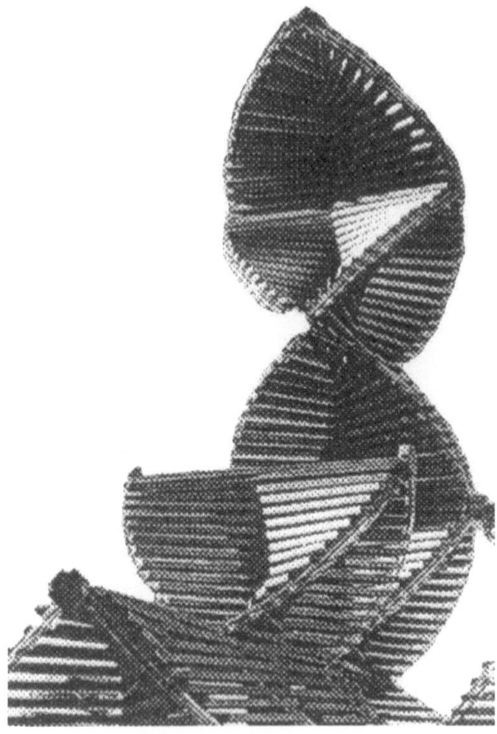

Figure 15.23. *The model for Kurokawa's helical city plan* (1961)

The architectural appearance of the helix in Europe was in the form of stairs, like the well-known ducal palace in Urbino. In the castle of Chambord by the Loire — where in 1539 French king François I (ruled 1515–47) welcomed German-Roman Emperor Charles V (ruled 1519–56) — we find a staircase running up in a double helix (Figure 12.17), which could almost be a predecessor of four centuries of the DNA model constructed by Francis Crick and James Watson based on the X-ray diffraction fibre diagrams of Maurice Wilkins (1916–2004).

The functional architectural schools of thought in twentieth-century Europe also found followers in the United States. Louis Kahn followed principles that were essentially similar to those of the Bauhaus, and, like Le Corbusier and Anger, he also worked outside his homeland for Indian cities that at the time were open to accepting and implementing new plans. Frank Lloyd Wright (1869–1959) formulated his own principles, partly emphasizing the building's harmony with nature by dint of its horizontal lines (e.g. *The Waterfall House*, Figure 15.24), partly functional considerations, like the corridor of the *Guggenheim Museum* in New York, helical in

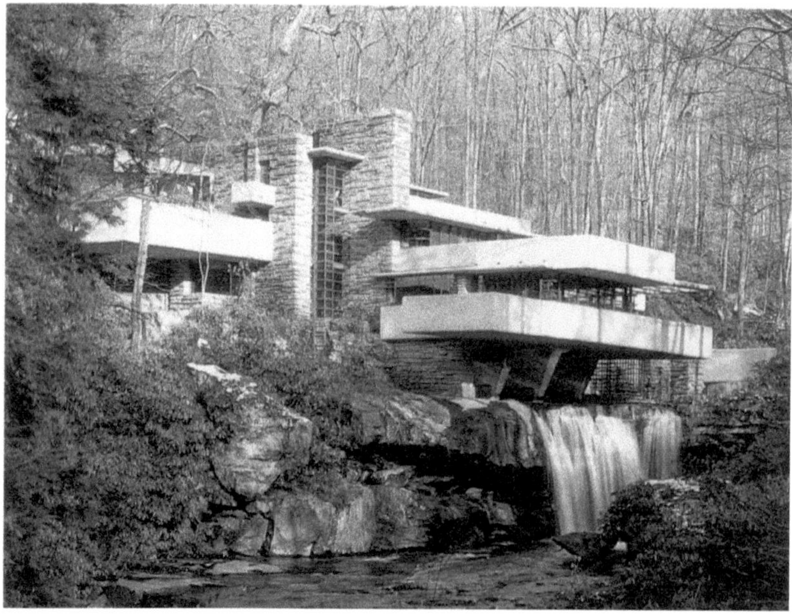

Figure 15.24. *The Waterfall House* (1935–39)

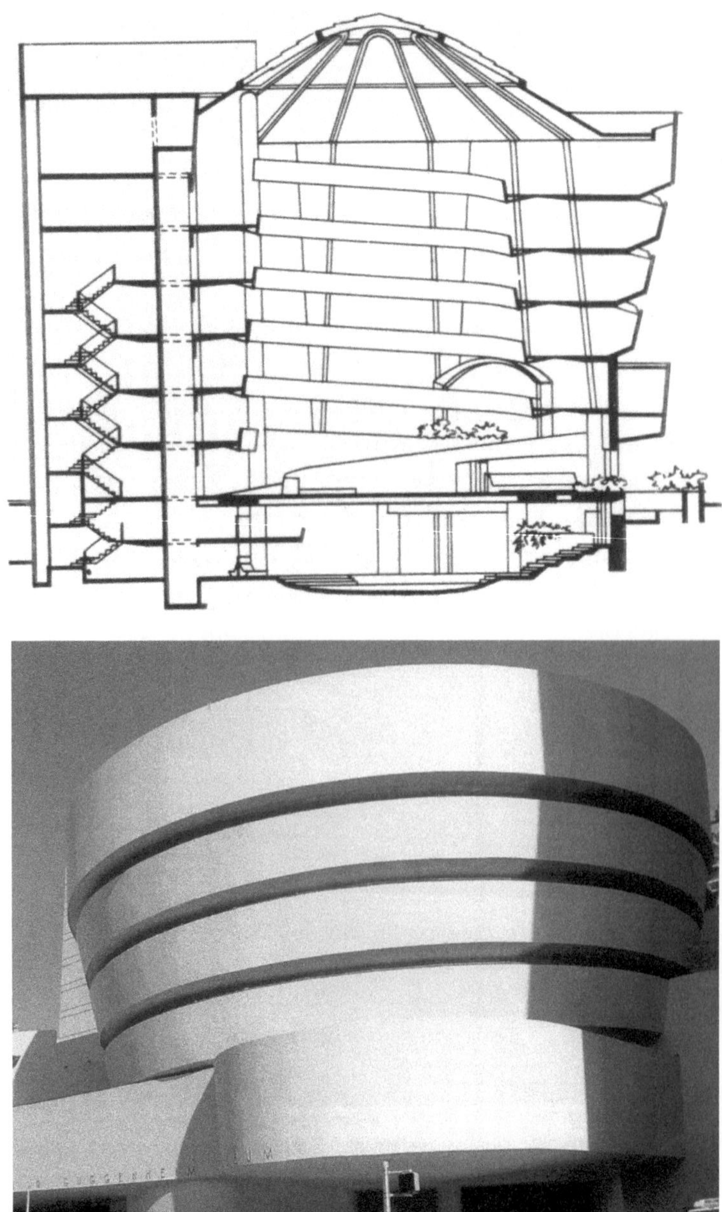

Figure 15.25. *Cross-section drawing of the New York Guggenheim museum, and a view of it from outside*

shape and slightly leaning and touching the inner surface of the vault of the truncated upside-down cone-shaped dome, which equally serves the symmetrically even ceiling lighting of the gallery and the comfort of the visitor.

This building is almost the inverse of the Ziggurat-style buildings that went on to become circularly concentric, seen above in the form of the Great Mosque of Samarra in Figure 15.20. In Budapest, on a smaller scale, it was on this basis that the central space of the Millenary Park was constructed, which connected what originally were the only two halls into a single exhibition building.

We have had the opportunity to become acquainted with the synergetic principle of R. Buckminster Fuller, which originated from architecture but whose scope went well beyond it. In Japan, the new contours of the supporting structures of Kenzo Tange (1913–2005) embraced new, more softly arced lines of symmetry (e.g. the Yoyogi National Gymnasium constructed for the Tokyo Olympics in 1964).

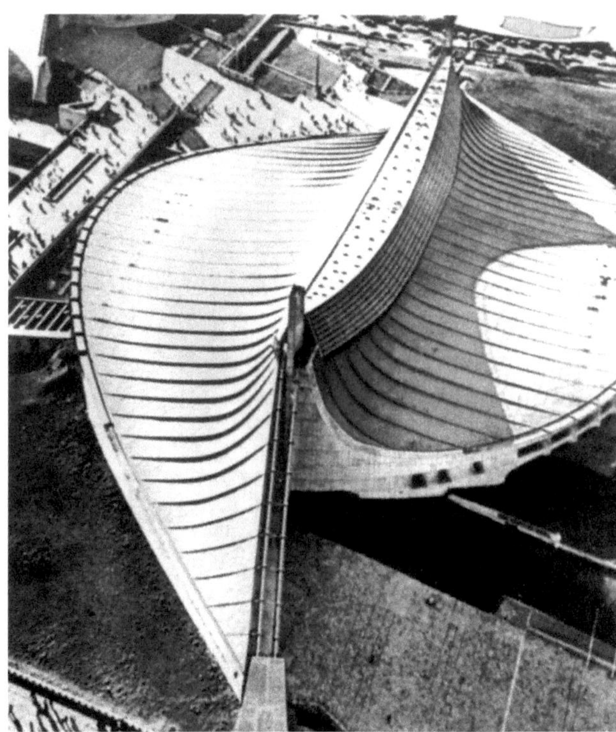

Figure 15.26. *The Yoyogi National Gymnasium in Tokyo*

New creations also come about on a scale smaller than that of buildings, depending on the technology that the science of the age makes possible. These have to be put into an aesthetically acceptable form. This was achieved by the handicraft of the age, then by design and applied art. It generates harmony between the requirements for material used, function, form and usability on the one hand, and the user on the other. It generates consonance between the possibilities given by function and aesthetic considerations, and between function and form. Its most important creative areas are industrial machinery, precision instruments, manual devices, vehicles, clothes and domestic textiles, household furniture, fixtures in public places, surfaces, insulation, cultic objects, bowls and plates, other everyday items, e.g. books and writing instruments.

A number of considerations have played a role in creating harmony: one such consideration was synergetics. It was a synergetic principle, for example, that played a role in how, through the research of Koyro Mirua, using the principles and technique of traditional Japanese origami (paper folding), it was possible to go from an umbrella that can be opened in a single movement to much larger and heavier parabolic antennas in packed form, which could be unfolded with the smallest number of operations, thus reducing to a minimum the technical risk facing satellites once shot up into space. Previously, origami and space travel can have seemed like applications as remote as the design of geodesic domes, the surface structure of viruses and the chemistry of carbon molecules, and yet it is a similar principle that established a relationship between them.

In fine art, symmetry transformations not only played a role in enriching perspective. The transformations themselves became a means of expression, making it possible to simplify forms and for them to become more geometrical. A series of geometrical fine art movements emerged. Alongside form, some attributed greater significance to colours, while all of them put greater emphasis on proportions. The various manifestations of symmetry became of greater significance. This applies equally to geometrical and generalized symmetries (related to various transformations, objects, properties).

From the beginning of the twentieth century, artistic schools with an interest in geometry spread like wildfire. We cannot address all of them, neither is this the intention, but some, which declared and used certain symmetry principles, we are obliged to mention. We previously touched upon cubism and futurism. The Russia of the 1910s saw the onset of *suprematism* — whose leading figure was Kazimir Malevich (1878–1935) — some of whose proponents later joined the constructivists in the West. At the same time, *De Stijl* was being formed in the Netherlands — Piet Mondrian (1872–1944), Theo van Doesburg (1883–1931) — which constructivism and then concrete art both heralded as their predecessor. The most influential movement, which exists to this day, was that of *constructivism*. Malevich and Mondrian were part of the movement, as was the Hungarian Lajos Kassák (1887–1967), but it also included V. Kandinsky (1866–1944), N. Gabo, V. Tatlin (1885–1953), El Lissitsky (1890–1941) and P. Klee (1879–1940), as well as many of our contemporaries. In addition to fine art, it also had an influence on functionalist architecture. Constructivism was claimed as its own by the *Bauhaus*, where many of its representatives taught.

Constructivism consciously proclaimed and continues to proclaim the adherence to the formal order of geometry, and the primacy of the structural components of depiction. In 1930, some of them formed the group of *concrete* artists, which professed to even more abstract artistic principles than the previous movements. A few alterations notwithstanding, it also exists to this day.

After World War II, partly in Paris, and partly in Buenos Aires, *MADI* (founded by C. Arden Quinn [1913–]) became a movement, its name being an acronym from the keywords *M*ovement, *A*bstraction, *D*imension and *I*nvention. Its formal world was delimited by even stricter geometrical restrictions, but it kept a door open to rounded forms and the tools of a world of more distinct colours. It freed itself of the limits set by rectangles, and opened up towards polygonality. Invariance transformations determined by right-angles were extended by manifold rotation transformations and continuous ones provided by circles, and the invariance of forms was replaced by the invariant commutation of colours. Today it is enjoying a revival.

Alongside concrete (fine) art grew concrete poetry. The formal presentation of poetry conveys meaning, and this is reflected by the typography used. It has an effect together with the verbal meaning.

In performing arts, pantomime and dance apply the techniques of movement to make use of the symmetric appearance of the human body. Since ancient times, folk dances have preserved the symmetries of movement. Modern ballet has taken over this tradition. Choreographical notation, similar to a musical score, expresses the connection between the two art forms, while the way it is written employs the symmetries of parts of the human body. In Laban (1879–1958) notation, which is the most widespread, markings correspond to even and odd parts of the body (hands and feet; head). The Eshkol-Wachman notation goes even further: it marks movement down to every single movable part (arm–lower arm–hand–fingers), breaking down the symmetry of even parts of the body as far as the human bone system. The harmony of music, poetry and movement is created by eurhythmy.

Le Corbusier's Modulor (Figure 15.27) is rather similar in taking the construction of the human body as its starting point. The modulor is a systematically elaborated system of measures and proportions. It is based on the proportions of the body. The division of full height as reached by stretching one's arm up into the air determines a set of measures that is in line with the golden section. It is to this that furniture and other fixtures used by human beings, and even the sizes of buildings and their harmonically designed proportions, are, depending on their functions, made to correspond.

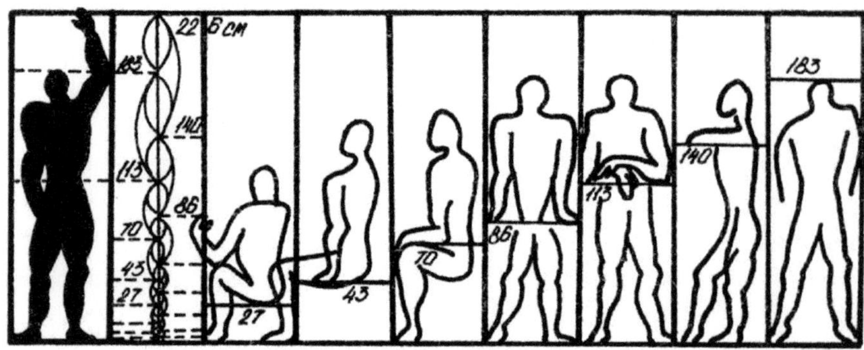

Figure 15.27. *Modulor*

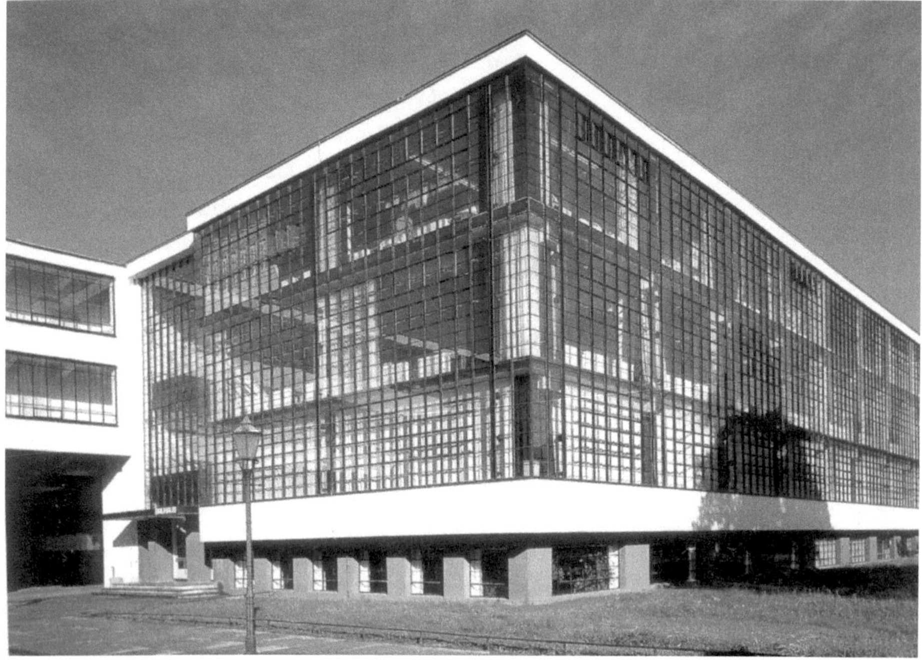

Figure 15.28. *W. Gropius: The Bauhaus in Dessau (1925–26)*

The *Bauhaus*, founded (1919–) and operated (–1933) by Le Corbusier's contemporaries — W. Gropius in particular — established its fundamental principles in its manifestos. In its day, the Bauhaus was a pioneering institution for creativity, research and education, which was the first to implement integrated principles in its activity. These fundamental principles, which gradually developed over the years, also gave a role to symmetry. The starting points were simplicity, usability, and the application of modern technology. They declared and taught the harmony between the various modern techniques, science and the arts, and the need to establish connections between them. In addition to architecture, they used similar fundamental principles for household furniture, the applied art of everyday appliances, fine arts, the theory of colours, typography, stage arts, and photography.

Although technology has since progressed radically, indeed a quantitatively new element has appeared in the form of the computer, the funda-

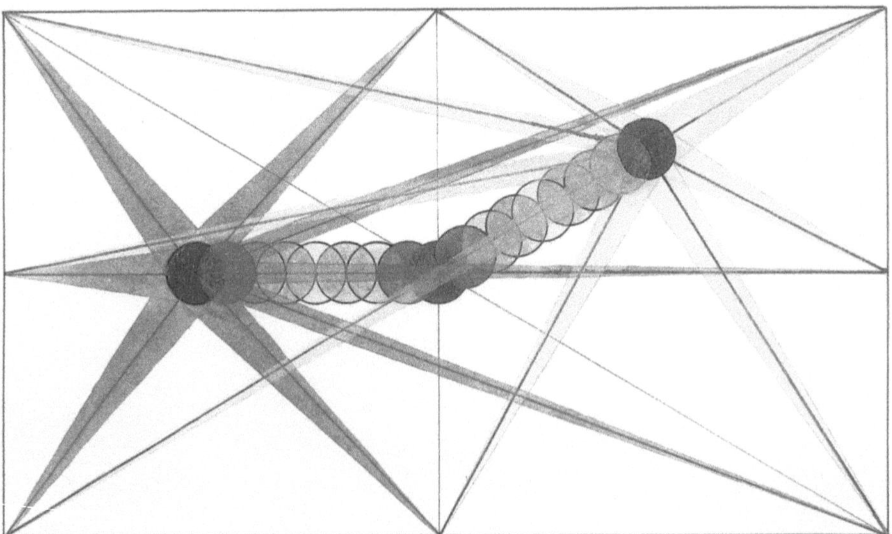

Figure 15.29. *A study examining perspective, colours and proportions, prepared by student L. Meyer-Bergner for a class held by his teacher Paul Klee (1927)*

mental principles declared in the manifestos of the Bauhaus have stood the test of time. The residences in Budapest considered to be the most inhabitable are to this day those in buildings designed in the 1930s by architects who studied at the Bauhaus, primarily Farkas Molnár (1897–1945). Marcell Breuer's steel pipe furniture is now attributed classic status, and has enjoyed renown on a par with the career of the (Thonet) bentwood chair. László Moholy-Nagy is considered to be the father of photography as a separate art form. Many are unaware of the fact that the designs for their practical serving bowls, which are still in use to this day, emerged from the workshops of the Bauhaus. In many quarters, their theory of colours remains textbook knowledge (J. Itten 1888–1967), which is no small acknowledgement of the school established in Weimar, coincidentally the city of Goethe, who, many years earlier, was the one to create the first colour theory.

The fundamental principles of the Bauhaus may hold true today, but need to be broadened to include the virtual reality created by computers. Although virtual reality is only one of the many applications of the computer, let us pick out just one single element. Just as art doubles the real world, so a computer maps within itself the objective reality treated by science, but also much from the (man-made) world of technology. It is more than merely a tool for technical design: it is a tool for the creation of precious products of the human spirit. It makes human beings capable of creating works they would not otherwise be able to create. Just as the computer as a tool and the virtual reality it creates are no substitute either for the participation of the human brain or for objective reality, likewise it cannot replace man in the creation of artistic works, and it is no substitute for the world depicted by man, i.e. his works of art. But it does and can help it to come about. The computer, with the unique opportunities and functions it offers, has appeared as a contributor alongside thinking and creative man, the objective world, and its scientific and artistic reflection. We must neither overestimate nor underestimate its role and its value. It is no longer possible to ignore its significance.

In 1933, the teachers at the Bauhaus, and most of their students, had no choice but to scatter from Germany, and some of them went on to found post-Bauhaus schools. They attempted to implement the spirit of the Bauhaus in its entirety. László Moholy-Nagy founded a school in Chicago, while W. Gropius and the others worked on the East Coast of the United States. In Cambridge, near Boston, Marcell Breuer established an institute at Harvard University. (György Kepes [1906–2002] later did likewise, also in Cambridge, at the Massachusetts Institute of Technology.) In the decades after the war, Max Bill set up an institute in Ulm, Germany. Sadly, not of these organizations survived for long, at least not in their original incarnations. It was the *Hochschule für Gestaltung* (1953–1968) in Ulm that was most able to inherit the principles of Weimar, Dessau and Berlin in their fullness. It was here, under the direction of Tomas Maldonado (1922–) that symmetry first became a separate subject of study. Moholy-Nagy's students from Chicago, scattered around the world, are still working and still stick together, just like the handful of alumni of the HfG. The Kepes Institute in Cambridge, Massachusetts, safeguarding the principles of its founder of combining arts and sciences, operates with a narrower profile, while at Harvard it is primarily the building of Carpenter Center which bears the signature of a great personality, and today only a few scraps remain of the spirit of the original institution's teaching.

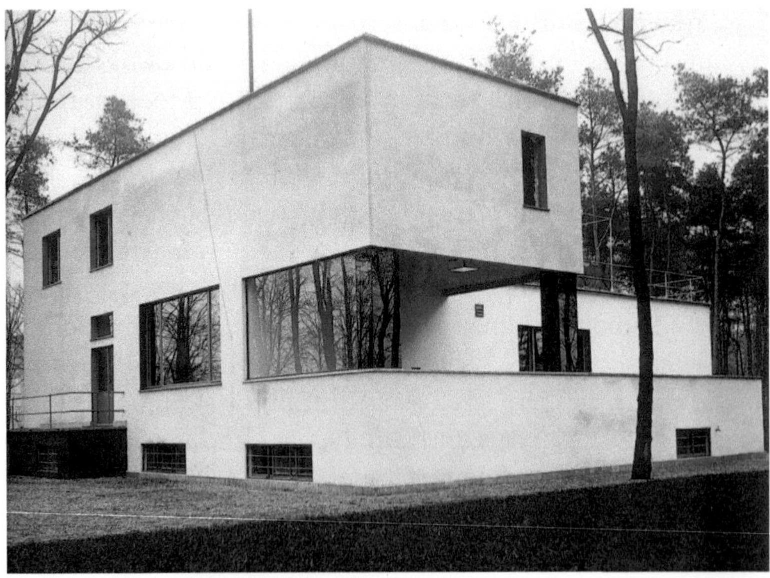

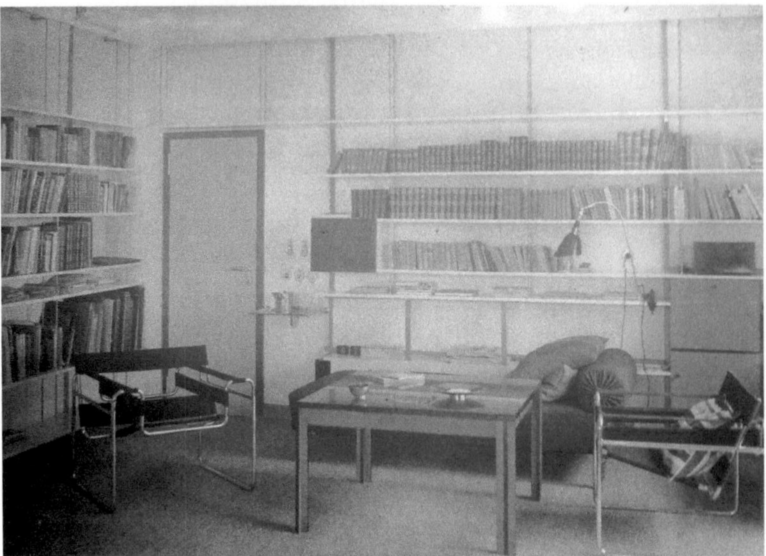

Figure 15.30. *The house of W. Gropius in Dessau, which he designed himself (1925–26), and his own room with furniture and fittings designed by Marcell Breuer. (The photographs were taken by* Lucia Moholy-Nagy.*) After emigrating, Gropius and Breuer taught design together at Harvard, then for a while jointly operated an architectural office of their own.*

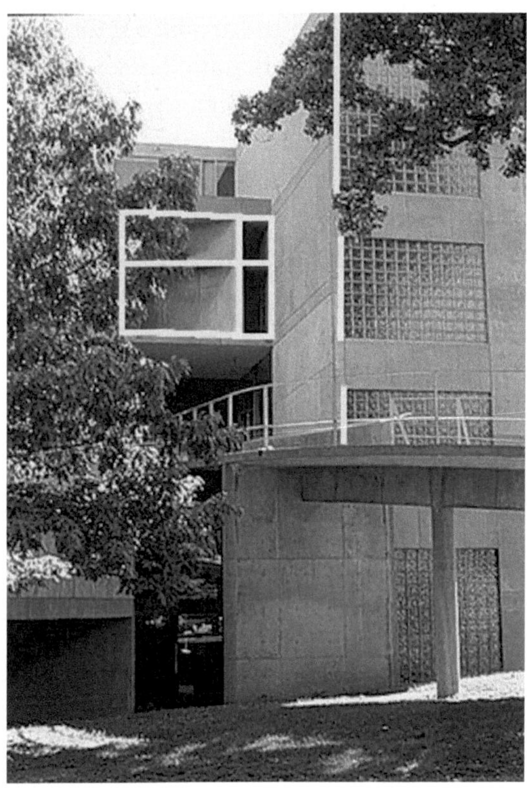

Figure 15.31. *In the details of the façade of Harvard University's Carpenter Center for the Visual Arts, designed by Le Corbusier, the proportions previously discussed in this book are clearly visible*

The influence of their design concepts can equally strongly be felt in the collection of buildings constructed for the Barcelona World Expo as in the UNESCO headquarters in Paris. Their students and their students pass on what they have learned, from Europe to Japan, from India to California. Perhaps it is no surprise that the reason for mentioning them here is that, of the principles that bind them together, symmetry is by no means the least significant.

— * —

These principles were elaborated in sciences and in the arts in parallel. We have tried to present their common features in a systematic fashion, from their conceptual foundations and appearance in nature, to a survey of some results of modern science and some directions in contemporary art, always with an eye on the path of this ontological evolution and historical considerations. We have attempted to trace the process of human thought and creativity, together with some of its results, from this unique perspective, grouped around a central guiding principle — that of symmetry or the lack of it.

Bibliography

Abas, S. J. and Salman, A. S. (1995) *Symmetries of Islamic Geometrical Patterns*, Singapore: World Scientific, xxii + 396 pp, 16 colour plates.

Adair, R. K. (1988) A flaw in a universal mirror, Scientific American, February, p. 30; In Hungarian: A "világtükör" sérülése, *Tudomány* (április) 20–26.

Adler, I. (1990) Symmetry in phyllotaxis, *Symmetry: Culture and Science*, 1, 2, 171–183.

Agazzi, E., Darvas, G., eds., (1997) *Philosophy of Mathematics Today*, Episteme 22, Dordrecht: Kluwer, xxix + 361 p.

Akopyan, I. D. (1980) *Simmetriya i asimmetriya v poznanii*, [Symmetry in Asymmetry in learning, in Russian], Erevan, Armenia: Izdatel'stvo Akademii Nauk Armyanskoi SSR, 132 pp.

Alberti, L. B. (1435) *De pictura.*

Alberti, L. B. (1436) *Della pittura.*

Alberti, L. B. (1956) *On Painting*, New Haven: Yale University Press.

Andrews, E. (1990) *Markedness Theory: The Union of Asymmetry and Semiosis in Language, Sound and Meaning*: The Roman Jakobson Series in Linguistics and Poetics, Durham, N.C.: Duke University Press, viii + 220 pp.

Annett, M. (1985) *Left, Right, Hand and Brain: The Right Shift Theory*, Hillsdale, NJ: Lawrence Erlbaum.

Annett, M. (1992) Spatial ability in subgroupt of left- and right-handers, *British Journal of Psychology*, 83 493–962.

Avnir, D., et al. (1998) The measurment of symmetry: Bridging order and disorder, *Symmetry: Culture and Science*, 9, 131–134.

Baggott, J. E. (1994) *Perfect Symmetry: The Accidental Discovery of Buckminsterfullerene*, Oxford: Oxford University Press., ix + 315 pp.

Baglivo, J. A. and Graver, J. E. (1983) *Incidence and Symmetry in Design and Architecture*, Cambridge: Cambridge University Press, xi + 306 pp.

Balogh T. (1999) A szimmetria személyes és szociális struktúrái, [The personal and social structures of symmetry, in Hungarian], *Magyar Tudomány*, 3, 311–316.

Balogh T., szerk. (1992) *Szimmetria – Aszimmetria,* [Symmetry – Asymmetry, in Hungarian], Szeged: MTA SZAB, 59 p.

Balogh T., szerk. (1999) *Szimmetria – Aszimmetria II.,* [Symmetry – Asymmetry 2, in Hungarian], Szeged: MTA SZAB, 83 p.

Barabás, M. (1859) A festészeti távlatról [On perspective in painting, in Hungarian] *Magyar Akadémiai Értesítő, III.*

Barlow, W. Z. (1894) *Z. Krist.*

Beke, L. (2000) Horison line and vanishing point, 479–485 p., in: Peternák, M., Erős, N. szerk. *Perspektíva.*

Beke, Zs. (2000) Perspective, overhead view, abstraction: The space of the image and the image of space in the Italian futurist aeropittura, 205–217. p., in: Peternák, M., Erős, N. eds. *Perspective.*

Bérczi Sz. (1991) *Szimmetria és Struktúraépítés,* [Symmetry and Structure Building, in Hungarian], egyetemi jegyzet/textbook, Budapest: Tankönyvkiadó, 260 p.

Bérczi, Sz. (1980) Cyclicity in the history of matter and its application to the elementary structure-building processes of the solar system, *Acta Geologica Academia Scientiarum Hungaricae,* 23 (1–4) 163–171.

Bérczi, Sz. (1986–87) *Szimmetria és techné a magyar, avar és hanti díszítőművészetben,* [Symmetry and techné in Hungarian, Avar and Hanti decorative art, in Hungarian], Budapest: Eötvös Loránd Tudományegyetem, Általános Technika Tanszék; Leuven, Belgium: Leuveni Katolikus Egyetem, Collegium Hungaricum, 59 pp.

Bérczi, S. (1989) Symmetry and technology in ornamental art of old Hungarians and Avar-Ongurians from the archaeological finds of the Carpathian basin, seventh to tenth century A.D., in: *Symmetry 2: Unifying Human Understanding* (ed. I. Hargittai) 715–730 p.

Bérczi, S., Papp, L. (1992) A unique fivefold symmetrical building: A Calvinist Church in Szeged, Hungary, in: *Fivefold Symmetry* (ed. I. Hargittai) 235–244.

Bérczi, Sz. Lukács, B. guest eds. (1993) Symmetry and Topology in Evolution 1–2, Special issues, *Symmetry Culture and Science,* 4, 1–2, 1–224.

Bernal, I., Hamilton, W. C., and Ricci, J. S. (1972) *Symmetry: A Stereoscopic Guide for Chemists,* San Francisco: Freeman, viii + 180 pp.

Beyer, J. (2002) Symmetry in quilt and fabric design, in: *Symmetry 2000* (eds. I Hargittai and T. C. Laurent) 561–575.

Bhagavantam, S. (1966) *Crystal Symmetry and Physical Properties*, London: Academic Press, x + 230 pp.

Bir, G. L. and Pikus, G. E. (1972) *Simmetriya i deformatsionnye effekty v poluprovodnikakh*, [Symmetry and Deformation Effects in Semiconductors, in Russian], Moskva: Nauka, 584 pp.

Birss, R. R. (1964) *Symmetry and Magnetism*, Amsterdam: Norht Holland Publ. Co.

Bochvar, D. A., Galpern, E. G. (1973) *Dokdlady Akademii Nauk SSSR*, 209, 610–612.

Bodnar, O. Ja. (1994) *Zolotoe sechenie i neevklidova geometrija v prirode i iskusstve*, [Golden Section and Non-Euclidean Geometry in Nature and Arts, in Ukrainian], Lvov: Izd. Svit, 205 str.

Bonpunt, L. (1999) The emergence of symmetry concepts by the way of the study of crystals (1600–1900), *Symmetry: Culture and Science*, 10, 127–242.

Brading, K. A. (2002) Which Symmetry? Noether, Weyl and conservation of electric charge, *Studies in Hist. and Phil. of Mod. Phys.* 33, 3–22.

Brandmüller, J. (1992) Fivefold symmetry in mathematics, physics, chemistry, biolog and beyond, in: *Fivefold Symmetry* (ed. I. Hargittai) 11–32.

Braun, T. (1996) *A káprázatos C_{60} molkeula*, [The Wonderful C_{60} Molecule, in Hungarian], Budapest: Akadémiai Kiadó, 170 p.

Brewster, D. (1819) *The Kaleidoscope: Its History anbd Construction with its Application to the Fine and Useful Arts*, Edinburgh: A. Constable & Co.

Broca, P. (1865) Sur la siège de la facultè du langage articulè, [in French], *Bulletins de la Societe d'Anthropologie de Paris*, 6 377–393.

Bruhn, S. (1992) Symmetry and irreversibility in the musical language(s) of the twentieth century, *Symmetry: Culture and Science*, 3, 2, 187–200.

Bruhn, S., guest ed. (1996) Symmetry in Music, Cance and Poetry, Special issue, *Symmetry: Culture and Science*, 7, 2, 113–224.

Bruhn, S., guest ed. (1996) Symmetry in Music, In memoriam Ernő Lendvai 2, Special issue, *Symmetry: Culture and Science*, 7, 4, 337–448.

Bruhn, S., Nagy, D. guest eds. (1994) Symmetry in Music, In memoriam Ernő Lendvai, *Symmetry: Culture and Science*, 5, 3, 225–336.

Bruno, E. (1976) *The Magic Mirror of M. C. Escher*, New York: Random House.

Burckhardt, J. J. (1988) *Die Symmetrie der Kristalle: Von René-Just Haüy zur kristallographischen Schule in Zürich*, [in German], Basel: Birkhäuser, 195 pp.

Burt, M. (1966) *Spatial Arrangement and Polyhedra withe Curved Surfaces and their Architectural Applications*, Haifa: Technion.

Burt, M. (1996) *The Periodic Table of the Polyhedral Universe*, Haifa: Technion.

Burton, R., Cavendish, R. (1992) *A világ száz csodája: Barangolás az építészet lenyűgöző alkotásai között*, [100 Great Wonders of the World, in Hungarian], Budapest: Magyar Könyvklub, 240 p.

Caglioti, G. (1983) *Simmetrie infrante nella scienza e nell'arte*, Milano: CLUP, 182 pp. *The Dynamics of Ambiguity*, Berlin: Springer, 1992, xx + 170 pp.

Carrol, L. (1962) *Alice's Adventures in Wonderland* and *Through the Looking Glass*, London: Puffin Books, 346 p.

Cesariano, C. (1521) *M. Vitruvio Pollione De Architectura*, Como.

Chomsky, N. (1980) On cognitive structures and their development: A reply to Piaget, in: M. Piattelli-Palmarini, ed., *Language and learning: The debate between Jean Paigaet and Noam Chomsky*, Cambridge, MA: Harvard University Press.

Chorbachi, W. K. (1989) In the tower of Babel: Beyond symmetry in Islamic design, in: Symmetry 2: *Unifying Human Understanding* (ed. I. Hargittai) 751–789.

Chorbachi, W. K., Loeb, A. L. (1992) An Islamic pnetagonal seal from scientific manuscripts of the geometry of design, in: *Fivefold Symmetry* (ed. I. Hargittai) 283–305.

Coleman, S. (1985) *Aspects of Symmetry: Selected Erice Lectures of Sidney Coleman*, Cambridge: Cambridge University Press, xiv + 402 pp.

Corballis, M. C. (1990) Symmetry and asymmetry in psychology, *Symmetry: Culture and Science*, 1, 2, 183–194.

Coxeter, H. S. M. (1948) *Regular Polytopes*, London: Methuen.

Coxeter, H. S. M. (1961) *Introduction to Geometry*, New York: Wiley.

Coxeter, H. S. M. (1975) *Complex Polytopes*, Cambridge University Press.

Coxeter, H. S. M., Emmer, M., Penrose, R. and Teuber, M. L., eds. (1986) *M. C. Escher: Art and Science*, Amsterdam: Elservier.

Crowe, D. W. (1992) Albrecht Dürer and the regular pentagon, in: *Fivefold Symmetry* (ed. I. Hargittai) 465–487.

Crowe, D. W. (1992) Symmetry in the aboriginal art of Australia, *Symmetry: Culture and Science*, 3, 3, 227–240.

Curie, P. (1894) Sur la symétrie dans les phénomenes physiques, symétrie d'un champ eléctrique et d'un champ magnétique, [in French], *J. Phys.* (3rd ser) 3, 393–415.

Curie, P. (1908) *Oeuvres de Pierre Curie*, [in French], Paris: Gauthier–Villars.

Czernohaus, K. (1988) *Delphindarstellungen von der minoischen bis zur geometrischen Zeit*, [in German], Göteborg: P. Aströms Förlag, cxxi + 111 p.

Dalai Emiliani, M. (2000) "We place the viewer into the centre of teh picture" A reinterpretation of the futurist space of Umberto Boccioni, 219–232. p., in: Peternák, M., Erõs, N. eds. *Perspective*.

Darvas, G. (1995) ISIS-Symmetry: Looking for truth and beauty, *Symmetry: Culture and Science*, Special issue: Symmetry: Natural and Artificial, 1, 9–17.

Darvas, G. (1995) Symmetry — A concept and phenomenon linking arts and sciences, p. 5, In: *Einstein meets Magritte, An Interdisciplinary Reflection on Science, Nature, Human Action and Society*, Abstracts (of the International conference to mark the 25th anniversary of the Vrije Universiteit Brüssel, June, 1995), 52 p.

Darvas, G. (1996) Dichotomies and trichotimies in the Eastern and Western cultures, in: *Katachi U Symmetry* (eds. Ogawa, T., Miura, K., Masunari T, and Nagy, D.) 91–100.

Darvas, G. (1996) Symmetry: A concept and phenomenon linking mathematics and art, p. 63, In: Koptsik, V. A., Riznichenko, G. Yu., eds., Mezhdunarodnaia konferencia: *Matematika i Isskusstvo*, Suzdal, 23–27 sentiabria 1996, Tezisü, Moskva, 1996, 100 p.

Darvas, G. (1997) Symmetry in mathematics and world view, 319–334 pp., in: Agazzi, E., Darvas, G., eds., (1997) *Philosophy of Mathematics Today*, Episteme 22, Dordrecht: Kluwer, xxix + 361 p.

Darvas, G. (1998) Laws of symmetry breaking, *Symmetry: Culture and Science*, 9, 179–188.

Darvas, G. (1998) Ontological levels and symmetry breaking, *Paideia*, Philosophy of Science, http://www.bu.edu/wcp/Papers/Scie/ScieDarv.htm

Darvas Gy. (1999) Disszimmetria, [Dissymmetry, in Hungarian], *Élet és Tudomány*, 30, 935–937.

Darvas Gy. (1999) Szimmetria a tudományban és a művészetben, [Symmetry in science and art, in Hungarian], *Magyar Tudomány*, 3, 257–265.

Darvas, G., (1999) Tudomány a művészetben — Művészet a tudományban, Disszimmetrikus előszó, pp. 6–8, In: Darvas, G. ed., (1999) *Ars (Dis) Symmetrica '99; A Tudomány a Művészet Tükrében — Művészet a Tudományban, Science Reflected in the Arts — Art in the Sciences*, Institute for Research Organisation, Hungarian Academy of Sciences and Symmetrion, Budapest, 1999, 144 p.

Darvas, G. (2001) Symmetry and asymmetry in our surroundings; Aspects of symmetry in the phenomena of nature, physical laws and human perception, , 136–149 pp., in: *Surroundings Surrounded* (ed. P. Weibel) ZKM Karlsruhe, 703 p.

Darvas, G. (2003) Perspective as a symmetry transformation, *Nexus Network Journal*, vol. 5 no. 1 (Spring), http://www.nexusjournal.com/Darvas.html

Darvas, G., Nagy, D., eds., (1989) *Symmetry of Structure*, Extended Abstracts, Budapest: International Society for the Interdisciplinary Study of Symmetry, Vols. I and II, 656 p.

Darvas, G., Nagy, D. (1992) Symmetry of patterns and patterns of symmetry, *Symmetry: Culture and Science*, Special issue: Symmetry of Patterns, Extended abstracts of the 2nd Interdisciplinary Symmetry Symposium and Exhibition, August 17–23, 1992, Hiroshima, 1, 6–8.

Darvas, G., Nagy, D. (1993) Symmetry as an organising principle in science(s), P3–7, P9, 25, In: Hormigon, M., Ausejo, E., Dhombres, J., eds., *Book of Abstracts — Scientific Sections of the XIXth International Congress of History of Science*, 22–29 August 1993, Zaragoza: ICHS.

Darvas, G., Nagy, D. (1993) Symmetry: Bridging east and west, art and science, Abstract, presented at the Conference held by the Institut für Integrative Musikpädagogik und Polyästhetische Erziehung and the Mozarteum Salzburg, in Mittersill, Austria, July 1993 on *Mathematics and Music*, p. 2.

Darvas, G., Nagy, D. (1994) Wasan versus western mathematics: Asymmetry of the brain and education, pp. 321–324, In: Ogawa, T., ed., *Katachi U Symmetry, Extended Abstracts*, International Symposium, University of Tsukuba, 21–25 November, 1994, 360 p.

Darvas, G., Nagy, D., Pardavi-Horvath, M., eds., (1995) *Symmetry: Natural and Artificial*, Extended Abstracts, Budapest: Symmetrion, 734 p.

Debolini, F. (1998) *A reneszánsz géniusz: Leonardo, a tudós és a művész*, [The Renaissance Genius: Leonardo the Scientist and Artist, in Hungarian], Artbook, Pécs: Alexandra, 143 p.

Depenchuk, N. P. (1963) *Simmetriya i asimmetriya v zhivoi prirode*, [Symmetry and Asymmetry in Nature, in Russian], Kiev: Izdatel'stvo Akademii Nauk Ukrainskoi SSR, 174 pp.

Deregowski, J. B. Perception of symmetry: Some psychological observations, *Symmetry: Culture and Science*, 3, 3, 263–280.

Dixon, R. (1992) Green spirals, in: *Spiral Symmetry* (eds. I. Hargittai and C. A. Pickover) 353–368.

Dóczi, Gy. (1981) *The Power of Limits*, Shambhala Publ. Inc., Boulder, Co, 167 p.

Donáth, T. (1991) *Anatómiai atlasz*, [Atlas of Anatomy, in Hungarian], Budapest: Medicina Könyvkiadó, 212 p.

Droste, M. (1991) *Bauhaus, 1919–1933*, Berlin: Benedikt Taschen Verlag, 256 p.

Dunham, D. (1998) Morphological transformation of hyperbolic patterns, *Symmetry: Culture and Science*, 9, 199–205.

Dürer, A (1525) *Underweysung der Messung Mit dem Zirckel un Richtscheyt, in Linien Ebnen, und gantzen Corporen* [Treatise on measurement with compass and ruler in lines, planes and whole bodies, in German].

Dürer, A (1528) *Vier Bücher von menschlicher Proportion*, [Four Books on Human Proportion, in German].

Elliott, J. P. and Dawber P. G. (1979) *Symmetry in Physics*, Vol. 1, Principles and Simple Applications, Vol. 2, Further Applications, London: Macmillan, xix + xviii + 557 pp.

Esaki, L. (1997) The birth of the tunnel diode and the semiconductor superlattice, *Symmetry: Culture and Science*, 7, 3–4, 227–238.

Euklid, *Elements*.

Family, F., Vicsek, T. (1991) *Dynamics of Fractal Surfaces*, Singapore: World Scientific.

Fedorov, E. S. (1885) *Trans. Imp. St. Petersburg Min. Soc.* 21, 1.

Fedorov, E. S. (1949) *Simmetriya i struktura kristallov: Osnovnye raboty*, [Symmetry and Structure of Crystals: Basic Works, in Russian], Ed. by A. V. Shubnikov and I. I. Shafranovskii, Klassiki nauki, Moskva: Izdatel'stvo Akademii Nauk SSSR, 630 pp.

Fedorov, E. S. (1971) *Symmetry of Crystals*, ACA Monograph, No. 7, New York: American Crystallographic Association, x + 315 pp.

Fejes Tóth L. (1964) *Regular Figures*, New York: Macmillan.

Fejes Tóth, L. (1964) What the bees know and what they do not know, *Bulletin of the American Mathematical Society*, 20, 468–481.

Feynman, R. P. (1965) *The character of Physical Law*, Cambridge, MA: MIT Press.

Feynman, R. P. (1997) *Six Not-So-Easy Pieces: Einstein's Relativity, Symmetry, and Space-Time*, Reading, Massachusetts: Addison-Wesley, xxvii + 152 pp.

Field, M. (1996) *Lectures on bifurcation, dynamics and symmetry*, Pitman Research Notes in Mathematics, Vol. 356, Harlow: Longman, 159 pp.

Field, M. and Golubitsky, M. (1992) *Symmetry in Chaos: A Search for Pattern in Mathematics, Art and Nature*, Oxford: Oxford University Press, xii + 218 pp.

Freitas, L. de, (1992) Notes on some pentagonal "mysteries" in Egyptian and Christian iconography, in: *Fivefold Symmetry* (ed. I. Hargittai) 307–332.

Fujita, S. (1991) *Symmetry and Combinatorial Enumeration in Chemistry*, Berlin: Springer, ix + 368 pp.

Fuller, R. Buckminster, with Applewhite E. J. and Loeb, A. L. (1975) *Synergetics*, New York: Macmillan.

Fuller, R. Buckminster, with Applewhite E. J., ed. (1979) *Synergetics 2*, New York: Macmillan.

Gafurio, F. (1480, 1492, 1508) *Theorica musicae*.

García, S. (1681/1991) *Compendio de architectura y simetría de los templos: Conforme a la medida del cuerpo humano, con algunas demostraziones de geometría*, [in Spanish], año de 1681 Valladolid: Colegio Oficial de Arquitectos, Vols. 1–2.

Gardner, M (1988) *Penrose Tiles to Trapdoor Codes*, New York: W. H. Freeman.

Gardner, M. (1964) *The Ambidextrous Universe: Left, Right, and the Fall of Parity*, New York: Basic Books, x + 294 pp.

Gardner, M. (1977) Mathematical Games: Extraordinary nonperiodic tiling that enriches the theory of tiles, *Scientific American*, January, 236, 110.

Gardner, M. (1990) *The Ambidextrous Universe: Symmetry and Asymmetry from Mirror Reflections to Superstrings*, New York: W. H. Freeman.

Gardner, M. (1997) Penrose csempézés, [Penrose tiling, in Hungareian], *Természet Világa*, 8–9.

Gazzaniga, M. S. (1995) Principles of human brain organization derived from split brain studies, *Neuron* 14, 217–228.

Genz, H. (1987) *Symmetrie — Bauplan der Natur*, [Symmetry — Blueprint of Nature, in German], München: Piper, 464 pp.

Genz, H. and Decker, R. (1991) *Symmetrie und Symmetriebrechung in der Physik*, [Symmetry and Symmetry Breaking in Physic, in German], Braunschweig: Vieweg, 376 pp.

Gerdes, P. (1990) On ethnomathematical research and symmetry, *Symmetry: Culture and Science*, 1, 2, 154–170.

Gerdes, P. (1992) Fovefold symmetry and (basket) weaving in various cultures, in*: Fivefold Symmetry* (ed. I. Hargittai) 245–262.

Gerdes, P. (1996) *Lunda Geometry: Designs, Polyominoes, Patterns, Symmetries*, Maputo, Mozambique: Universidade Pedagógica, 152 pp.

Gerő L. (1955) *Magyarországi várépítészet*, [Castle Architecture in Hungary, in Hungarian], Művelt Nép, 511 p.

Gévay G. (1999) A kaleidoszkóp tanulságai, [Lessons of the kaleidoscope, in Hungarian], *Magyar Tudomány*, 3, 289–294.

Gévay, G. (1992) Icosahedral morphology, in *Fivefold Symmetry* (ed. I. Hargittai) 177–204.

Ghiberti, L. (1455) *Commentarii.*

Gombrich, E. H. (1984) *The Sense of Order*, Ithaca: Cornell University Press.

Gortari, E. de (1970) Symmetry as a scientific method, pp. 30–34, in*: La Symetrie. Comme principe heuristique dans les differentes sciences*, Bruxelles: Office international de libraire, 135 p.

Gruber, B, et al., eds (1980, 1986, 1989, 1990, 1991, 1993) *Symmetries in Science I–VI*, New York: Plenum.

Grünbaum, B. (1967) *Convex Polytopes*, New York: Interscience.

Grünbaum, B., Shepard, G. C. (1987) *Tilings and Patterns*, New York: W. H. Freeman.

Gympel, J. (1997) *Az építészet története az ókortól napjainkig*, [The Story of Architecture from Antiquity to the Present (Konemann, 1996), in Hungarian], Kulturtrade Kiadó, 120 p.

Haake, A. (2002) XX-TEX — Textiles and crystallography, in: *Symmetry 2000* (eds. I Hargittai and T. C. Laurent) 223–230.

Haeckel, E. H. P. A. (1899–1904) *Kunstformen der Natur* [in German].

Haeckel, E. H. P. A. (1880–95) *Report on the Radiolaria collected by H. M. S. Challenger during the years 1873–76*, Edinburgh.

Haeckel, E. H. P. A. (1887) *Challenger Monograph.*

Hahn, Th, ed. (1985 and 1988) *International Tables for Crystallography*, Boston: Reidel.

Hahn, W. (1989) *Symmetrie als Entwicklungsprinzip in Natur und Kunst*, Königstein: Langewiesche, 320 S.

Hahn, W. (1998) *Symmery as a Developmental Principle in Nature and Art*, Singapore: World Scientific, 510 p.

Hajnóczi G. (1999) A 'symmetria' és a művészetelméleti terminológia kezdetei a reneszánszban, ['Symmetria' and the beginnings of the terminology of art theory in the Renaissance, in Hungarian], *Magyar Tudomány*, 3, 266–273.

Hajnóczi, G. (1999) The concept of symmetry in early Renaissance art theory. Leon Battista Alberti, De Pictura, *Symmetry: Culture and Science*, 10, 7–16.

Haken, H. (1978) *Synergetics — an Introduction*, Berlin-Heidelberg-New York: Springer Verlag.

Haldane, J. B. S. (1960) Pasteur and cosmic asymmetry, *Nature*, 185, 87.

Hambidge, J. (1920) *Dynamic Symmetry: The Greek Vase*, New Haven, Connecticut: Yale University Press, 161 p.

Hambidge, J. (1923) *Dynamic Symmetry in Composition as Used by the Artists*, New York: Brentano, 83 p.

Hambidge, J. (1924) *The Parthenon and Other Greek Temples: Their Dynamic Symmetry*, New Haven, Connecticut: Yale University Press, xxii + 103 pp.

Hambidge, J. (1926) *The Elements of Dynamic Symmetry*, New Haven, Connecticut: Yale University Press and New York: Brentano, xx + 140 pp.

Hambidge, J. (1932) *Practical Applications of Dynamic Symmetry*, Ed. by M. C. Hambidge, New Haven, Connecticut: Yale University Press and London: Milford and Oxford University Press, xviii + 109 pp.

Hámori J. (1999) Aszimmetriák a biológiában, [Asymmetries in biology, in Hungarian], *Magyar Tudomány*, 3, 302–310.

Hámori J. (1999) *Az emberi agy aszimmetriái*, [The Asymmetries of the Human Brain, in Hungarian], Budapest-Pécs: Dialóg Campus Kiadó, 140 p.

Harenberg, B., ed. (1990) *Az emberiség krónikája*, [The Chronicle of Mankind, in Hungarian], Budapest: Officina Nova, 1223 p.

Hargittai I., ed. (1986) *Symmety: Unifying Human Understanding*, New York: Pergamon Press.

Hargittai I., ed. (1989) *Symmety: Unifying Human Understanding 2*, Oxford: Pergamon Press.

Hargittai I., Pickover C. A., eds. (1992) *Spiral Symmetry*, Singapore: World Scientific, xvii + 449 pp.

Hargittai, I. (1983) *Szimmetria egy kémikus szemével*, [Symmetry with the Eyes of a Chemist, in Hungarian], Budapest: Akadémiai Kiadó, 292 p.

Hargittai, I. (2002) The unreasonable effectiveness of the symmetry concept (homage to Eugene P. Wigner), in: *Symmetry 2000* (eds. I. Hargittai and T. C. Laurent) 609–620.

Hargittai, I. and Hargittai, M. (1994) *Symmetry: A Unifying Concept*, Bolinas, California: Shelter Publications, xviii + 222 pp.

Hargittai, I., ed. (1992) *Fivefold Symmetry*, Singapore: World Scientific, xvi + 561 pp.

Hargittai, I., Laurent T. C., eds. (2002) *Symmetry 2000* (Vols. 1 and 2), Venner Grem Intnl. series, London: Portland Press, xix + 627 pp.

Hargittai, M. and Hargittai, I. (1989) *Fedezzük föl a szimmetriát!*, [Discover Symmetry!, in Hungarian], Budapest: Tankönyvkiadó, 148 p.

Hargittai I., Hargittai M. (2003) *Szimmetriák a Felfedezésben*, [Symmetries in Discovery, in Hungarian], Tudomány-Egyetem sorozat, Budapest: Vince Kiadó, 264 p.

Hartmann, E. (1998) *An Introduction to Crystal Physics (Description of the Physical Properties of Crystals)*, International Union of Crystallography, http://www.iucr.org/iucr-top/comm/cteach/pamphlets/18/18.html

Hay, D. R. (1846) *First Principles of Symmetrical Beauty*, Edinburgh: Blackwood, viii + 88 pp., 100 plates.

Hecker, Z. (1976) *Polyhedral Architecture*, Jerusalem: The Israel Museum.

Hegstrom, R. A, Kondepudi, D. K. (1990) Jobb- vagy balkezes-e a Világegyetem? [in Hungarian], *Tudomány* (március) 46–53; The handedness of the Universe, *Sci. Amer.* 262 (1) 108–115.

Hellige, J. B. (1993) *Hemispheric Asymmetry: What's Right and What's Left*, Cambridge, Massachusetts: Harvard University Press, xiii + 396 pp.

Heracleitos (1983) *Múzsái*, vagy *A természetről*, [Musea, or On Nature, in Hungarian], Budapest: Helikon, 78 p.

Herodotos (1893) *Történeti könyvei* [Historical works, in Hungarian].

Hilton, P. and Pedersen, J. (1988) *Build your own Polyhedra*, Reading, MA: Addison-Wesley.

Hilton, P., Pedersen, J. (1997) Symmetry in theory: Mathematics and aesthetics, *Symmetry: Culture and Science*, 8, 3–4, 239–264.

Hochstrasser, R. M. (1966) *Molecular Aspects of Symmetry*, New York: Benjamin, xiii + 355 pp.

Hoffman, W. C. (1992) Symmetry and geometric psychology, *Symmetry: Culture and Science*, 3, 3, 305–310.

Hofman-Jablan, J. (1995) *Simetrija muzickog dela*, [Symmetry of Musical Works, in Serbian], Beograd: Zaduzbina Andrejevic, 159 pp.

Hofstadter, D. R. (1979) *Gödel, Escher, Bach*, New York: Basic Books.

Horwich, P. (1987) *Asymmetries in Time: Problems in the Philosophy of Science*, Cambridge, Massachusetts: MIT Press, xiii + 218 pp.

Husimi, K. (1996) Symmetry in Mon and Mon-yo: Artisans preceded Mathematicians, in: *Katachi U Symmetry* (eds. Ogawa, T., Miura, K., Masunari T, and Nagy, D.) 75–80.

Huybers, P., Ende, P. v. d. (1990) Polyhedral geometry as a basis of building structures and other materials systems, *Symmetry: Culture and Science*, 1, 4, 387–422.

Ivanov, V. V. (1986) *Páros és páratlan: Aszimmetria az agyban és a jelrendszerben*, [Even and Odd: Asymmetry in the Brain and in the Symbol System, in Hungarian], Budapest: Kozmosz, 263 p.

Jablan, S. V. (1985) *Theory of Symmetry and Ornament*, Beograd: Matematicki institut, 1995, 331 p.

Jacobi, B. (1968) *Beszélő kövek*, [Story-telling Stones, in Hungarian; original edition: Verweht und ausgegraben Archäologische Forschungen der letzten fünfzig Jahre, Leipzig: Prisma-Verlag, 1964], Budapest: Gondolat, 394. p.

Jean, R. V. (1992) On the origins of spiral symmetry in plants, in: *Spiral Symmetry* (eds. I. Hargittai and C. A. Pickover) 323–351.

Jean, R. V. (1994) *Phyllotaxis: A Systematic Study in Plant Morphogenesis*, Cambridge University Press.

Jean, R. V. (1992) Nomothetical modelling of spiral symmetry in biology, in: *Fivefold Symmetry* (ed. I. Hargittai) 505–528.

Johnson, D. L. (2001) *Symmetries*, London: Springer, 198 p.

József A. (late 1930s) *Összes versei és és műfordításai*, [Collected poems and translations, in Hungarian], Budapest: Cserépfalvi, 544 p.

Jung, C. G. (1977) *A szellem szimbolikája*, [Symbolics of the Spirit, two essays, in Hungarian], Budapest: Európa Könyvkiadó.

Kalayan, H. (1988) *The Architectural Information through Symmetry*, Amman: Department of Antiquities, The Hashimite Kingdom of Jordan, 89 pp.

Kálmán A. (1999) Kitüntetett, gyakori, ritka és rendhagyó szimmetriák, [Distinguished, frequent, rare, and irregular symmetries, in Hungarian], *Magyar Tudomány*, 3, 280–288.

Kambartel, W. (1972) *Symmetrie und Schönheit: Über mögliche Voraussetzungen des neueren Kunstbewusstseins in der Architekturtheorie Claude Perraults*, [Symmetry and Beauty, in German], München: Fink, 179 p.

Kappraff, J. (1990) *Connections, the Geometric Bridge between Art and Science*, New York: McGraw Hill.

Kappraff, J. (1992) The relationship between mathematics and mysticism of the golden mean through history in: *Fivefold Symmetry* (ed. I. Hargittai) 33–66.

Kepes, G., ed. (1965) *The Module, Proportion, Symmetry and Rhythm*, New York: Braziller.

Kepler, J. (1596) *Mysterium Cosmographicum*.

Kepler, J. (1609) *Astronomia nova seu Physica coelestis*.

Kepler, J. (1619) *De harmonice mundi*

Kholodnyi, V. A. and Price, J. V. (1998) *Foreign Exchange Option Symmetry*, Singapore: World Scientific, xvii + 134 pp.

Kinsey, L. Ch, Moore T. E. (2002) *Symmetry, Shape, and Space: An Introduction to Mathematics through Geometry*, Key College Publ. and Springer, 494 p.

Kiss D., Horváth Á., Kiss Á. (1998) *Kísérleti Atomfizika*, [Experimental Particle Physics, in Hungarian], Budapest: ELTE Eötvös Kiadó, 472 p.

Kittel, Ch. (1966) *Bevezetés a szilárdtest-fizikába*, [Introduction to Solid State Physics, in Hungarian], Budapest: Műszaki Könyvkiadó, 699 p.

Klein, P. (1990) On symmetry in science education, *Symmetry: Culture and Science*, 1, 1, 77–91.

Knight, T. W. (1994) *Transformations and Design: A Formal Approach to Stylistic Change and Innovation in the Visual Arts*, Cambridge University Press, xvii + 258 p.

Koch, A. S., Tarnai, T (1988) The aesthetics of viruses, *Leonardo*, 21, 161–166.

Koestler, A. (1978) *Janus. A Summing Up*, New York: Random House, 354 p.

Kondepudi, D. K., Nelson, G. W. (1983) *Phys. Rev. Lett.* 50, 1023–1026.

Koptsik, V. A. (1966) *Shubnikovskie gruppy: Spravochnik po simmetrii i fizicheskim svoistvam cristallicheskikh struktur*, [Shubnikov Groups: Guide to Symmetry and Physical Properties of Crystal Structures, in Russian], Moskva: Izdatel'stvo Moskovskogo universiteta, 723 p.

Korobko, V. I., Korobko, G. N. (1995) *The Golden Section and a Man*, Stavropol: Caucasian Library Publ. House, 174 p.

Korobko, V. I., Primak, G. N. (1992) *Zolotaia proporcija i cselovek*, [Golden Section and Man, in Russian], Stavropol: Kavkazszkaia Biblioteka, 173 p.

Kovács J. (1999) *Sejttan*, [Cells, in Hungarian], egyetemi tankönyv, Budapest: ELTE Eötvös Kiadó, 81 p.

Krätschmer, W., Lamb, L. D., Fostiropoulos, K., Huffman, D. R. (1990) Solid C_{60} a new form of carbon, *Nature*, 347, 354–358.

Kroto, H. W., Heath, J. R., O'Brien, S. C., Curl, R. F., Smalley, R. E. (1985) C_{60} Buckminsterfullerene, *Nature*, 318, 162–163.

Kroto, H. W. (1989) C_{60} Buckminsterfullerene, other fullerenes and the icospiral shell, in: *Symmetry 2: Unifying Human Understanding* (ed. I. Hargittai) 417–423.

Kultermann, U. (1977) *Die Architektur im 20. Jahrhundert*, [Architecture in the 20th c., in German], Köln: DuMont.

Kuroda, R. (2002) Chirality and achirality in life, in: *Symmetry 2000* (eds. I Hargittai and T. C. Laurent) 279–294.

Lalvani, H. (1977) *Structures and Hyperstructures*, Brooklyn: Pratt Institute.

Lalvani, H. (1977) *Transpolyhedra*, Brooklyn: Pratt Institute.

Lanczos, K. (1970) *Space through the Ages: The Evolution of Geometrical ideas from Pythagoras to Hilbert and Einstein*, London: Academic Press.

Le Corbusier (1953) *Der Modulor*, [in German], Stuttgart: DVA, 241 S.

Le Corbusier (1962) *Musee National D'Art Moderne*, [in French], Paris: Ministere d'etat affaires culturelles, 76 p.

Lee, T. D. (1966) Space inversion, time reversal and particle-antiparticle conjugation, *Phys. Today* 19 (3) 23–31.

Lee, T. D. (1988) *Symmetries, Asymmetries, and the World of Particles*, Seattle, Washington: University of Washington Press, xii + 66 pp.

Lee, T. D., Yang, C. N. (1956) Question of parity conservation in weak interactions, *Phys. Rev.* 104, 254–259.

Lendvai, E. (1993) *Symmetries of Music, An introduction to Semantics of Music*, Kecskemét: Kodály Institute, 1993, 155 p.

Leonardo, da Vinci (1492–1510) *Codex Urbinax Latinus 1270.*

Leonardo, da Vinci (1498?) *Trattato della pittura,* [On painting, in Italian].

Leyton, M. (1992) *Symmetry, Causality, Mind*, Cambridge, Massachusetts: MIT Press, vii + 630 pp.

Loeb, A. L. (1971) *Color and Symmetry*, Wiley Monographs in Crystallography, New York: Wiley, xiii + 179 pp.

Loeb, A. L. (1989) The magic of the pentangle: Dynamic symmetry from Merlin to Penrose, in: *Symmetry 2: Unifying Human Understanding* (ed. I. Hargittai) 33–48.

Loeb, A. L. (1990) Reflections on rotations connections and notations, *Symmetry: Culture and Science*, 1, 39–56.

Loeb, A. L. (1991) *Space Structures: Their Harmony and Counterpoint*, Boston: Birkhäuser, 188 p.

Loeb, A. L. (1993) *Concepts & Images: Visual Mathematics*, Boston: Birkhäuser, 228 p.

Loeb, A. L., Varney, W. (1992) Does the golden spiral exist, and if not, where is its center? in: *Spiral Symmetry* (eds. I. Hargittai and C. A. Pickover) 47–61.

Lucich, P. (1987) *Genealogical Symmetry: Rational Foundations of Australian Kinship*, Armidale, New South Wales: Light Stone Publications, xxxi + 684 pp.

Ludwig, W. and Falter, C. (1988) *Symmetries in Physics: Group Theory Applied to Physical Problems*, Springer Series in Solid-State Sciences, Vol. 64, Berlin: Springer, xi + 461 pp.

MacGillavry, C. H. (1965) *Symmetry Aspects of M. C. Escher's Periodic Drawings*, Utrecht: Oosthoek, xi + 84 pp.

Mackay, A (1982) Crystallography and the Penrose Pattern, *Physics* 114A, 609–613.

Mackay, A. (1996) Growth and From, in: *Katachi U Symmetry* (eds. Ogawa, T., Miura, K., Masunari T, and Nagy, D.) 47–62.

Mainzer, K. (1988) *Symmetrien der Natur: Ein Handbuch zur Natur- und Wissenschaftsphilosophie*, [Symmetries of Nature: a Manual to Philosophy of Nature and Science, in German], Berlin: de Gruyter, xii + 739 pp.

Mainzer, K. (1990) Symmetry in philosophy and history of science, *Symmetry: Culture and Science*, 1, 3, 319–328.

Makovicky, E. (1992) 800-year-old pentagonal tiling from Maragha, Iran, and the new varieties of aperiodic tiling it inspired, in *Fivefold Symmetry* (ed. I. Hargittai) 67–86.

Mandelbrot, B. B. (1983) *The Fractal Geometry of Nature*, New York: Freeman.

Mann, Th. (1924) *The Magic Mountain*.

Marsden, J. E., Montgomery R., and Ratiu, T. (1990) *Reduction, Symmetry, and Phases in Mechanics*, Memoirs of the American Mathematical Society, No. 436, Providence, Rhode Island: American Mathematical Society, iv + 110 pp.

Marsden, J. E., Ratiu, T. S. (1999) *Introduction to Mechanics and Symmetry*, New York: Springer, 582 p.

Matsuno, K., Marijuán, P. C., guest eds. (1996) Symmetry and Information, Special issue, *Symmetry: Culture and Science*, 7, 3, 225–336.

Matsuno, K., Marijuán, P. C., guest eds. (1997) Symmetry and Information 2, Special issue, *Symmetry: Culture and Science*, 8, 2, 113–224.

Mezei, O. (1981) *A Bauhaus magyar vonatkozásai: Előzmények, együttműködés, kisugárzás*, [The Hungarian Connections with the Bauhaus, in Hungarian], Budapest: Népművelési Intézet és Művelődéskutató Intézet, 169 p.

Mikasa, T. (1996) Development and modification of the "Holy symmetrical design" — West and East of the "silk-road", in: *Katachi U Symmetry* (eds. Ogawa, T., Miura, K., Masunari T, and Nagy, D.) 81–90.

Miller, W., Jr. (1972) *Symmetry Groups and Their Applications*, Pure and Applied Mathematics, Vol. 50, New York: Academic Press, x + 432 pp.

Mills, R. (1989) Gauge fields, *Am. J. Phys.* 57 (6), June, 493–507.

Miyazaki, K. (1986) *Form and Space, Polygons, Polyhedra and Polytopes*, New York: Wiley.

Miyazaki, K. A mystic history of fivefold symmetry in Japan, in: *Fivefold Symmetry* (ed. I. Hargittai) 361–395.

Moller, A. P. and Swaddle, J. P. (1997) *Asymmetry, Developmental Stability, and Evolution*, Oxford: Oxford University Press, xi + 291 pp.

Mozrzymas, J. (1992) *Ewolucja symetrii*, [Evolution of Symmetry, in Polish], Wroclaw: Uniwersytet Wroclawski, 96 p.

Neumann, F. E. (1833) Die thermischen, optischen und krystallographischen Axen des Krystallsystems des Gypses, *Poggendorff Ann. Phys.* 27, 240–274.

Ne'eman, Y. (1961) *Nucl. Phys.* 26, 222–229.

Ne'eman, Y. (1979) *Symétries gaujes et variétés de groupe*, [Gauge Symmetries and Variations of Groups, in French], Montréal: Presses de l'Université de Montréal, 141 p.

Ne'eman, Y. (1990) The interplay of symmetry, order, and information in physics and the impact of gauge symmetry on algebraic topology, *Symmetry: Culture and Science* 1, 3, 229–255.

Ne'eman, Y. (1999) Symmetry as the leitmotif at the fundamental level in the twentieth century physics, *Symmetry: Culture and Science*, 10, 143–162.

Ne'eman, Y. (2002) Pythagoreanism in atomic, nuclear and particle physics, in: *Symmetry 2000* (eds. I Hargittai and T. C. Laurent) 265–278.

Ne'eman, Y., Kirsh, Y. (1986) *The Particle Hunters*, Cambridge University Press, 272 p.

Niggli, P. (1919) *Geometrische Kristallographie des Disconinuums*, Leipzig: Bontaraeger.

Noether, E. A. (1918) Invariante Variationsprobleme, *Nachrichten von der Königlichen Gesellschaft der Wissenschaften zu Göttingen: Mathematisch-physikalische Klasse*, 235–257.

Nonnenmacher, T. F., Losa, G. A. and Weibel, E. R., eds. (1994) *Fractals in Biology and Medicine*, Basel: Birkhäuser, 397 p.

Ogawa, T. (1996) Katachi and symmetry: Towards interdisciplinary and intercultural cooperation, in: *Katachi U Symmetry* (eds. Ogawa, T., Miura, K., Masunari T, and Nagy, D.) 7–26.

Ogawa, T., Miura K., Masunari, T. and Nagy, D., eds.(1996) *Katachi U Symmetry*, Tokyo: Springer, xv + 417 pp.

Ogawa, T., Teshima, Y., Watanabe, Y. (1996) Geometry and crystallography of self-supporting rod structures, in: *Katachi U Symmetry* (eds. Ogawa, T., Miura, K., Masunari T, and Nagy, D.) 239–246.

Olovsson, I. (2002) Periodic and non-periodic patterns (tessellation), in: *Symmetry 2000* (eds. I Hargittai and T. C. Laurent) 185–200.

Osawa, E. (1970) *Kagaku*, 25, 854–863.

Otto, B. (1976) *Geometrische Ornamente auf anatolischer Keramik: Symmetrien frühester Schmuckformen im nahen Osten und in der Ägäis*, [Geometric Ornaments on Anatolic Ceramics, in German], Mainz am Rhein: von Zabern, xiii + 203 pp.

Pacioli, L. (1494) *Summa de arithmetica, geometria, proportioni et proportionalita,* Venetiis.

Pacioli, L. (1509) *De Divina Proportione.*

Palladio, A. (1570) *Quattro libri dell'architettura.*

Panofsky, E. (1921) Die Entwicklung der Proportionslehre als Abbild der Stilentwicklung, *Monatshefte für Kunstwissenschaft,* 14, pp. 188–219

Panofsky, E. (1960) *Renaissance and Renascences in Western Art,* Stockholm: Almquist and Wiksell

Panofsky, E. (1984) A perspektíva, mint szimbolikus forma, [Perspective as a symbolic form, in Hungarian], in: Panofsky, E. *A jelentés a vizuális művészetekben,* [Meaning in Visual Arts, in Hungarian], Budapest: Gondolat.

Pasteur, L. (1848) *Compt. Rend.*, Paris, 26, 535–538.

Paul, I., Zsille, Zs. (1976) *Keresztszemes hímzésminták*, [Crosswoven Needlework Ornaments, in Hungarian], 179 p.

Pauli, W. (1941) Relativistic field theories, *Rev. of Mod. Phys.* 13, 203–232.

Penrose, R. (1974) The role of aesthetics in pure and applied mathematics research, *J. Inst. Mathematics and its Applications*, 10, 266–271.

Perjés Z. (1999) Szimmetriák a fizikában, [Symmetries in physics, in Hungarian], *Magyar Tudomány*, 3, 274–279.

Perry, Ch. O. (2002) Symmetrical sculpture, in: *Symmetry 2000* (eds. I Hargittai and T. C. Laurent) 543–549.

Peternák, M., Erõs, N., eds. (2000) *Perspective*, [in English and Hungarian], Budapest: C^3 Alapítvány and Műcsarnok, 514 p.

Petoukhov, S. V. (1990) Symmetric-algorithmic properties of regular biostructures, *Symmetry: Culture and Science*, 1, 3, 295–312.

Petoukhov, S. V. (1999) Genetic code and the ancient Chinese Book of Changes, *Symmetry: Culture and Science*, 10, 211–226.

Petoukhov, S. V. (2001) *The Bi-periodic Table of Genetic Code and the Number of Protons* (in Russian) Moscow, 258 p.

Petoukhov, S. V. (1981) *Biomekhanika, bionika i simmetriya*, [Biomechanics, Bionics and Symmetry, in Russian], Moscow: Nauka, 238 p.

Piero, della Francesca (1998) *De la perspective en peinture*, [in French], Paris: In Medias Res, 351 p.

Piles, R. de (1708) *Cours de Peinture* [in French].

Plato (1983) *Válogatott művei*, [Collected works, in Hungarian], Bratislava: Madách, 667 p.

Plato (1994) *Az állam*, [The Republic, in Hungarian], Budapest: Gondolat, 468 p.

Plato (2001) *Philebosz*, [Philebos, in Hungarian], Budapest: Atlantisz, 138 p.

Pogány F. (1967) *Róma*, [Rome, in Hungarian], Budapest: Corvina, 390 p.

Polyclitus, (. . .) *Canon*.

Popper, K. R. (1979) *Objektive Knowledge: An Evolutionary Approach*, Revised Edition, Oxford: Clarendon Press, x + 395 pp.

Prigogine, I, Stengers, I, (1989) *Enre le temps et l'éternité*, [in French], Paris: Fayard.

Radman, M., Wagner, R. (1988) A DNS megkettőződésének rendkívüli pontossága, [The unusual accuracy of doubling DNA, in Hungarian], *Tudomány* (október) 20–26.

Renaud, P. (1935) Sur une généralisation du principe de symeétrie de Curie, [On the generalisation of the Curie symmetry principle, in French], *C. R. Acad. Sci. Paris*, 200, 531–534.

Ripa, C. (1997) *Iconologia*, [in Hungarian], Budapest: Balassi; Magyar Képzőművészeti Főiskola, 693 p.

Roero, C. S. (1999) Mean, proportion and symmetry in Greek and Renaissance art, *Symmetry: Culture and Science*, 10, 17–48.

Rollins, H. E., ed. (1958) *The Letters of John Keats 1814–1821*, Vol. 1, Cambridge University Press.

Rosen, J. (1975) *Symmetry Discovered: Concepts and Applications in Nature and Science*, Cambridge University Press.

Rosen, J. (1980) Symmetry and conservation: Inverse Noether's theorem and general formalism. *J. Phys.* A 13, 803–813.

Rosen, J. (1981) Extended Mach principle, *Amer. J. Phys.*, 49, 258–264.

Rosen, J. (1983) *Symmetry Primer for Scientists*, New York: Wiley, xiv + 192 pp.

Rosen, J. (1990) Fundamental manifestations of symmetry in physics, *Found. Phys.* 20, 283–307.

Rosen, J. (1991) Symmetry at the foundations of quantum theory, *Found. Phys.*, 21, 1297–1304.

Rosen, J. (1992) Reduction as symmetry, *Symmetry: Culture and Science*, 3, 2, 201–210.

Rosen, J. (1995) *Symmetry in Science: An Introduction to the General Theory*, New York: Springer, xv + 213 p.

Rosen, J., ed. (1982) *Symmetries in Physics: Selected Reprints*, Stony Brook, N.Y.: Amer. Assoc. of Phys. Teachers.

Rosen, J., Freundlich (1978) Symmetry and conservation, *Amer. J. Phys*, 46, 1030–1041.

Rothen, F. (1994) Phyllotaxis or self-similarity in plant morphogenesis, in: *Fractals in Biology and Medicine* (eds. Nonnenmacher, T. F., Losa, G. A. and Weibel, E. R.) 141–157.

Salam, A. (1966) *Symmetry Concepts in Modern Physics: Iqbal Memorial Lectures*, Lahore, Pakistan: Atomic Energy Centre, 55 p.

Sato, R. and Ramachandran, R. V. (1998) *Symmetry and Economic Invariance: An Introduction*, Research Monographs in Japan-U.S. Business and Economics, Boston: Kluwer, viii + 134 pp.

Schattschneider, D. (1980) Tiling the plane with Congruent pentagons, *Math. Magazine*, 51, 29–44.

Schattschneider, D. (1990) *Visions of Symmetry: Notebooks, Periodic Drawings, and Related Works of M. C. Escher*, New York: Freeman, xiv + 354 pp.

Schattschneider, D. (2002) The many faces of symmetry in the work of M. C. Escher, in: *Symmetry 2000* (eds. I Hargittai and T. C. Laurent) 173–184.

Schattschneider, D., Wallace, W. (1977) *M. C. Escher: Kaleidocycles*, New York: Ballantine.

Scherer J. (1999) *100 év formatan: Forma a természetben, a tervezésben és a művészetekben. A formatan oktatásának története a Magyar Iparművészeti Főiskolán*, [Form in Nature, Design, and Arts. Hundred Years of Form Studies at the Hungarian Institute of Design, in Hungarian], Budapest: Magyar Iparművészeti Főiskola, 108 p.

Schmutzer, E. (1972) *Symmetrien und Erhaltungssätze der Physik*, [in German], Braunschweig: Vieweg and Berlin: Akademie-Verlag, 165 pp.

Schönflies, A. (1891) *Kristallsysteme und Kristallstruktur*, [Crystal Systems and Crystal Structures, in German], Leipzig: B. G. Taubner

Schwabe, C. (1993) Natürliche Magie, *Turicum: Schweizer Kultur und Wissenschaft*, April/May, 26–33.

Senechal, M. (1981) Which tetrahedra fill space? *Mathematics Magazine*, 54, 227–243.

Senechal, M. (1990) *Crystalline Symmetries: An Informal Mathematical Introduction*, Bristol: Hilger, xii + 137 pp.

Senechal, M. (2002) Reflections on the kaleidoscope, in: *Symmetry 2000* (eds. I Hargittai and T. C. Laurent) 101–108.

Senechal, M. Fleck, G., eds. (1988) *Shaping Space: A Polyhedral Approach*, Boston: Birkhäuser Design Collection (series ed. A. L. Loeb).

Senechal, M., Taylor, J. (1990) Quasicrystals: The view from Les Houches, *The mathematical Intelligencer*, 12 (2) 54–64.

Shafranovskii, I. I. (1968) *Simmetriya v prirode*, [Symmetry in Nature, in Russian], Leningrad: Nedra, 184 p.

Shafranovskii, I. I. and Plotnikov, L. M. (1975) *Simmetriya v geologii*, [Symmetry in Geology, in Russian], Leningrad: Nedra, 144 p.

Shakespeare, W. (1975) *The Complete Works*, New York: Random House, 1227 p.

Shapiro, M. (1976) *Asymmetry: An Inquiry into the Linguistic Structure of Poetry*, North-Holland Linguistic Series, Vol. 26, Amsterdam: North-Holland, xiv + 231 pp.

Shechtman, D., Blech, I., Gratias, D. and Cahn, J. (1984) Metallic phase with long-range orientational order and no translational symmetry, *Phys. Rev. Letters*, 53, 1951–1954.

Sheikov, N. (1987) *Élet és szimmetria,* [Life and Symmetry, in Hungarian], Budapest: Gondolat, 127 p.

Shevelev, J. (2000) *Metalanguage of the Animate Nature,* Moszkva: Voskresenye, 348. p.

Shubnikov, A. V. (1940) *Simmetriya (Zakony simmetrii i ih primenenie v nauke, tekhnike i prikladnom iskusstve),* [Symmetry: Laws of Symmetry and Their Applications in Science, Technology and Applied Arts, in Russian], Moskva: Izdatel'stvo Akademii Nauk SSSR, 176 p.

Shubnikov, A. V. (1951) *Simmetriya i antisimmetriya konechnykh figur,* [Symmetry and Antisymmetry of Finite Forms, in Russian], Moskva: Izdatel'stvo Akademii Nauk SSSR, 172 p.

Shubnikov, A. V. (1961) *Problema dissimmetrii material'nykh ob'ektov,* [Problems of Dissymmetry of Material Objects, in Russian], Moskva: Izdatel'stvo Akademii Nauk SSSR, 55 p.

Shubnikov, A. V. and Koptsik, V. A. (1974) *Symmetry in Science and Art,* New York: Plenum Press, xxv + 420 pp.

Shubnikov, A. V., Belov, N. V., et al. (1964) *Colored Symmetry,* Ed. by W. T. Holser, New York: Macmillan and Oxford: Pergamon Press, xxv + 263 pp.

Simon T., Csapody V. (1987) *Kis növényhatározó,* [Little Plant Identification Handbook, in Hungarian], Budapest: Tankönyvkiadó, 205 p.

Simonyi K. (1981) *A fizika kultúrtörténete,* [*The Cultural History of Physics,* in Hungarian] 2. ed., Gondolat, 491 p.

Smith, C. S. (1980) *A Search for Structure,* Cambridge: MIT Press.

Smith, C. S. (1980) *From Art to Science,* Cambridge: MIT Press.

Smolina, N. I. (1990) *Traditsii simmetrii v arkhitekture,* [Traditions of Symmetry in Architecture, in Russian], Moskva: Stroiizdat, 343 p.

Speiser, A. (1923) *Theorie der Gruppen von Endlicher Ordnung, mit Anwendungen auf Algebraische Zahlen un Gleichungen sowie auf die Kristallographie,* [Theroy of Finite Order Groups with Applications to Algebraic Figures and Comparisions as well as to Crystalography, in German], Berlin: Springer, ciii + 194 p.

Sperry, P. W. (1974) Lateral specialization in the surgically separated hemispheres, in: F. O. Schmitt, F. G. Worden, eds., *The Neurosciences: Third Study Program,* Cambridge, MA: MIT Press.

Spinadel, V. W. (1999) The family of metellic means, *Symmetry: Culture and Science*, 10, 317–338.

Staar, Gy. (1997) A nagy csempéző: Beszélgetés Roger Penrose-zal, az Oxfordi Egyetem professzorával, ["The great tiler": an interview with R. Penrose, in Hungarian], *Természet Világa*, 12.

Stewart, I. and Golubitsky, M. (1992) *Fearful Symmetry: Is God a Geometer?*, Oxford: Blackwell, xix + 287 pp.

Stocklas, K., ed., (1986) *Symmetrie und Asymmetrie in der Architektur*, [Symmetry in Architecture, in German], IRB-Literaturauslese, Fachbibliographie, No. 1192, Stuttgart: IRB, 64 p.

Sugiura, K., *Lectures in the N.H.K. TV*, Tokyo: N.H.K.

Szabó, Á. (1997) *A görög matematika*, [Greek Mathematics, in Hungarian], Magyar Tudománytörténeti Szemle könyvtára; 4, 195 p.

Szabó, Á. (1998) *Antik csillagászati világkép: árnyék, naptár, földrajz, geometria*, [Astronomical World View in the Antiquity: Shadow, Calendar, Geography, Geometry, in Hungarian], Budapest: Typotex, 233 p.

Széles, K. (1999) Symmetry and infinity in the work of art, *Symmetry: Culture and Science*, 10, 255–268.

Szoboszlai M. (1999) Ábrázolás, konstrukció és CAD programhasználat, [Representation, construction and the use of CAD software programs, in Hungarian], *Magyar Tudomány*, 3, 295–301.

Szoboszlai, M. (1996) Heliocids and their projection, in: *Katachi U Symmetry* (eds. Ogawa, T., Miura, K., Masunari T, and Nagy, D.) 185–191.

Tamari, B. (1997) *Conservation and Symmetry Laws and Stabilization Programs in Economics*, Jerusalem: Ecometry, 58 p.

Tarnai, T. (1974) Spherical grids of triangular network, *Acta Technica*, Hungarian Academy of Sciences, Budapest.

Tarnai, T. (1989) Buckling patterns of shells and spherical honeycomb structures, in: *Symmetry 2: Unifying Human Understanding* (ed. I. Hargittai) 639–652.

Tarnai, T. (1996) Symmetry of golf balls, in: *Katachi U Symmetry* (eds. Ogawa, T., Miura, K., Masunari T, and Nagy, D.) 207–214.

Tarnai, T. (2002) Optimum packing of circles in a circle, in: *Symmetry 2000* (eds. I Hargittai and T. C. Laurent) 121–132.

Tayler, K. (1989) *Symmetry and Antisymmetry*, Cambridge: Haslingfield Press, 111 p.

Taylor, E. F., Wheeler, J. A. (1963) *Spacetime Physics*, San Franciscot: W. H. Freeman And Co.

Thompson, d'Arcy W. (1942) *On Growth and Form*, Cambridge University Press.

Tímár, L. (1979) *Albrecht Dürer: Della simmetria dei corpi humani*, Budapest: Chemical Works of Gedeon Richter, 64 pp.

Toth, N. (1990) The prehistoric roots of a human concept of symmetry, *Symmetry: Culture and Science*, 1, 3, 257–281.

Tsvetkov, V. D. (1997) *Serdtse, Zolotoe Sechenie i Simmetria*, [Heart, Golden Section and Symmetry, in Russian], Pushchino: PNTs RAN.

Trivedi, K. (1990) Symmetry in Hindu philosophy, *Symmetry: Culture and Science*, 1, 4, 369–386.

Tusa E. (1999) Szimmetria a zenében, [Symmetry in Music, in Hungarian], *Magyar Tudomány*, 3, 317–321.

Urmantsev, Yu. A. (1974) *Simmetriya prirody i priroda simmetrii (Filosofskie i estestvennonauchnye aspekty)*, [Symmetry of Nature and the Nature of Symmetry (Philosophical and scientific aspects), in Russian], Moskva: Mysl', 229 p.

Utiyama, R. (1956) Invariant theoretical interpretation of interaction, *Phys. Rev*. 101 (5), 1597–1607.

Utiyama, R. (1959) Theory of invariant variation and the generalised canonical dynamics, *Progr. of Theor. Phys. Suppl*. 9, 19–44.

Vainshtein, B. K. (1979) *Sovremennaya kristallografiya*, Tom 1: Simmetriya kristallov, Metody strukturnoi kristallografii, Moskva: Nauka, 383 p. English trans., *Modern Crystallography* 1: Symmetry of Crystals, Methods of Structural Crystallography, Springer Series in Solid-State Sciences, Vol. 15, Berlin: Springer, 1981, xvii + 399 pp.

Van Fraassen, B. C. (1989) *Laws and Symmetry*, Oxford: Oxford University Press, xv + 395 pp.

Vendler, H. V. (1983) *TheOodes of John Keats*, Cambridge, MA: The Belknap Press of Harvard University Press, viii + 330 p.

Verheyen, H. (1981) *An Alternative Generation and Transformational Theory of Uniform Polyhedra*, Antwerp: Universitaire Instelling.

Verheyen, H. (1996) *Symmetry Orbits, Design Science Collection*, Boston: Birkhäuser, vi + 236 p.

Verheyen, H. F. (1992) The icosahedral design of the Great Piramid, in: *Fivefold Symmetry* (ed. I. Hargittai) 333–359.

Vicsek M., Vicsek, T. guest eds. (1993) Fractals, Special issue, *Symmetry: Culture and Science,* 3, 3 225–336.

Vitruvius, P. (1960) *The Ten Books on Architecture* [translated by Morris H. Morgan, 1914] New York: Dover Publ.

Wachman, A., Burt, M. and Kleinmann (1974) *Infinite Polyhedra,* Haifa: Technion.

Walberg, G. (1987) *Kamares: A Study of the Character of Palatial Middle Minoan Pottery,* Göteborg: P. Aströms Förlag, 203 p.

Washburn, D. (2002) The cultural salience of symmetry, in: *Symmetry 2000* (eds. I Hargittai and T. C. Laurent) 429–444 pp.

Washburn, D. K. (1977) A Symmetry Analysis of Upper Gila Area Ceramic Design, *Papers of the Peabody Museum of Archaeology and Ethnology,* Vol. 68, Cambridge, Massachusetts: Harvard University Press, xiii + 193 pp.

Washburn, D. K. and Crowe, D. W. (1988) *Symmetries of Culture: Theory and Practice of Plane Pattern Analysis,* Seattle, Wash.: University of Washington Press, x + 299 pp.

Watson, J. D., Crick, F. H. C. (1953) Molecular structure of nucleic acids: A structure of Deoxyribose Nucleic Acid, *Nature,* 171, 737–738.

Watson, J. D. (1968) *The Double Helix; A Personal Account of the Discovery of the Structure of DNA,* New York: Atheneum.

Weibel, P. (2000) *Perspective, Toward the principle of construction: A geometry of the imagination,* 233–245 pp. in: Peternák, M., Erõs, N. eds. Perspective.

Wenninger, M. (1971) *Polyhedron Models,* Cambridge University Press.

Wenninger, M. (1978) *Spherical Models,* Cambridge University Press.

Wenninger, M. (1983) *Dual Models,* Cambridge University Press.

Wenninger, M. J. (2002) Symmetrical patterns on a sphere, in: *Symmetry 2000* (eds. I Hargittai and T. C. Laurent) 41–50 pp.

Wernicke, C. (1874) *Der Aphasische Symptomenkomplex,* [in German], Breslau: Cohn and Weigart.

Wester, T. (2002) Dual symmetry for plane- and point-based structures, in: *Symmetry 2000* (eds. I Hargittai and T. C. Laurent) 31–40 pp.

Weyl, H. (1918) Gravitation und Elektrizität, [Gravity and Electricity, in German], *Sitzungsberichte Preussische Akademie der Wissenschaften,* 465–480.

Weyl, H. (1929) Electron and gravitation (Engl. transl. in: O'Raifeartaigh (1997): *The Dawning of Gauge Theory*, Princeton Univ. Press).

Weyl, H. (1929) *Z. f. Physik*, 56, 330–352.

Weyl, H. (1931) *The Theory of Groups and Quantum Mechanics*, New York: Dover.

Weyl, H. (1946) *The Classical Groups, Their Invariants and Representations*, Princeton University Press.

Weyl, H. (1952) *Symmetry*, Princeton, NJ: Princeton University Press, 168 p.

Weyl, H. (1982) *Szimmetria*, [Symmetry, in Hungarian, translated and appended by Sz. Bérczi] Budapest: Gondolat, 226 p.

Wigner, E. P. (1959) *Group Theory and its Application to the Quantum Mechanics of Atomic Spectra*, [English translation], New York: Academic Press.

Wigner, E. P. (1965) Violations of symmetry in physics, *Sci. Amer.* 213 (6) 28–36.

Wigner, E. P. (1967) *Symmetries and Reflections, Scientific Essays of Eugene P. Wigner*, Bloomington: Indiana University Press.

Williams, K. (1999) Symmetry in Architecture, *Symmetry: Culture and Science*, 10, 269–282.

Witherspoon, G. and Peterson, G. (1995) *Dynamic Symmetry and Holistic Asymmetry in Navajo and Western Art and Cosmology*, American Indian Studies, Vol. 5, New York: Lang, x + 191 pp.

Wu, C. S., Ambler, E., Hayward, R. W., Hoppes, D. D. and Hudson, R. P. (1957) Experimental test of parity conservation in beta decay, *Phys. Rev.* 105, 1413–1415.

Yaglom, I. M. (1962, 1968, 1973) *Geometric Transformations*, Vols. I, II, III, New York: Random House.

Yaglom, I. M. (1988) *Felix Klein and Sophus Lie: Evolution of the Idea of Symmetry in the Nineteenth Century*, Boston: Birkhäuser, vii + 237 pp.

Yang, C.N., Mills, R.L. (1954) Conservation of isotopic spin and isotopic gauge invariance, *Phys. Rev.* 96 (1), 191–195.

Yushkin, N. P., Shafranovskii, I. I., and Yanulov, K. P. (1987) *Zakony simmetrii v mineralogii*, [Laws of Symmetry in Minerology, in Russian], Leningrad: Nauka, 333 p.

Zamorzaev, A. M. (1976) *Teoriya prostoi i kratnoi antisimmetrii*, [The Theory of Simple and Multiple Antisymmetry, in Russian], Kishinev: Shtiintsa, 283 p.

Zamorzaev, A. M., Galyarskii, E. I. and Palistrant, A. F. (1978) *Tsvetnaya simmetriya, ee obobshcheniya i prilozheniya,* [Colour Symmetry, Its Generalisations and Applications, in Russian], Kishinev: Shtiintsa, 275 p.

Zamorzaev, A. M., Karpova, Yu. S., Lungu, A. P., and Palistrant, A. F. (1986) *P-simmetriya i ee dal'neishee razvitie,* [Parity Symmetry and Its Evolution, in Russian], Kishinev: Shtiintsa, 154 p.

Zaslow, B. (1990) On the use of symmetry as an archaeological and anthropological tool, *Symmetry: Culture and Science,* 1, 1 49–56.

Zee, A. (1986) *Fearful Symmetry: The Search for Beauty in Modern Physics,* New York: Macmillan, xiv + 322 pp.

Zheludev, I. S. (1976) *Simmetriya i ee prilozheniya,* [Symmetry and Its Applications, in Russian], Moskva: Atomizdat, 286 p.

Zheludev, I. S. (1987) *Fizika kristallov i simmetriya,* [Crystal Physics and Symmetry, in Russian], Moskva: Nauka, 187 p.

Sources of illustrations

1.1 The *Allegory of Symmetry*. D. Calvaert (1540–1619), Bologna. (Graphical archive of the Museum of Fine Arts, Budapest, K.66.25). Reproduced courtesy of the Museum of Fine Arts, Budapest.

1.2a *Chenonceau Castle (France)*. Photo by G. Darvas.

1.3 M.C. Escher: *Reptiles*. Reproduced courtesy of the M.C. Escher Foundation. M.C. Escher works © 2003 Cordon Art – Holland. All rights reserved.

1.4 M.C. Escher: *Symmetrical Drawing 21* (1938). Reproduced courtesy of the M.C. Escher Foundation. M.C. Escher works © 2003 Cordon Art – Holland. All rights reserved.

1.5 *The translational symmetry of the columns of the Mezquita in Cordoba*. Photo by G. Darvas.

1.8 *Leonardo's drawing of the proportions of the human body*. Taylor, R.A. (1928) Leonardo the Florentine, N.Y.: Harper (5. edition) XXXI + 580 p., 72 t.

1.11 Drawing produced by the author based on H. Weyl: *Symmetry* (Princeton University Press, 1952).

1.12 *Reflection of electric current and a compass*. Produced courtesy of the authors, after Y. Ne'eman, Y. Kirsh: *The Particle Hunters* (Cambridge University Press).

1.14 *The* new Danube bridge *in Bratislava*. Photo by G. Darvas.

1.15 *Kyo Takenouchi's asymmetric bridge. Tokyo, Japan.* Reproduced courtesy of the journal *Symmetry: Culture and Science*.

1.16 *Dissymmetrical wall details. Alhambra monastery, Granada, Spain.* Details of photographs by the author.

1.17 *Dissymmetrical columns*. M.C. Escher: *Doric columns* (1945). Reproduced courtesy of the M.C. Escher Foundation. M.C. Escher works © 2003 Cordon Art – Holland. All rights reserved.

1.18 *Stonehenge's column order* (aerial photograph and ground-plan). Adapted by the author from Jacobi, B. *Beszélő kövek*, [Story-telling Stones, in Hungarian] (Gondolat, 1968).

1.22 *Parallels in meander and related designs between Europe and China*. Compilation, based on Shubnikov, A.V.: *Simmetriya* (Moscow, 1940) and drawings received by the author in private correspondence.

1.23 M.C. Escher: *Whirlpools* (1957). Reproduced courtesy of the M.C. Escher Foundation. M.C. Escher works © 2003 Cordon Art – Holland. All rights reserved.

1.24–1.25 Symmetric and antisymmetric images produced using the method patented by the author (P0302597 Hungarian Patent Office, 2003).

1.26 Reconstruction of a copy of the original image from the 1.24–1.25 symmetric and antisymmetric image components. C. Monet: *Alice Hoschedé in the Garden* (1881). Produced by the author.

2.1 *Neolithic bowl.* Yang-shao culture, c. 2000 BC. Simonyi K.: *A fizika kultúrtörténete* [The cultural history of physics, in Hungarian]. Reproduced courtesy of the copyright owners of the K. Simonyi estate.

2.2 *Fresco*, Megarón, Tiryns, c. 1200 BC. Czernohaus, K. (1988) *Delphindarstellungen von der minoischen bis zur geometrischen Zeit*, Tafel LXXXVI/3 (P. Aströms Förlag Göteborg). Reproduced courtesy of the publisher.

2.3 *The appearance of the dolphin motif in the Tripolis culture in North Africa c. 4000–3500 BC.* Jablan, S. V. (1985) *Theory of Symmetry and Ornament*, Beograd: Matematicki institut, 1995, 152. p. Fig. 2.89 (c). Reproduced courtesy of the author.

2.4 *Ancient Chinese dish, with two fish biting each other's tails.* T. Ogawa, private correspondence, photocopy.

2.5 *Design of dolphins biting each other's tails, as it appears on the decoration of Cretan ceramic dishes from the 17th-14th centuries BC.* Czernohaus, K. (1988) *Delphindarstellungen von der minoischen bis zur geometrischen Zeit*, Tafel XXXII/2, és Walberg, G. (1987) *Kamares: A Study of the Character of Palatial Middle Minoan Pottery*, 187. old. Fig. 42. 11/10. (P. Aströms Förlag Göteborg). Reproduced courtesy of the publisher.

2.6 *Ceramic pformer with two fish biting each other's tails. Basin of Mexico, 10th-6th century BC.* Anthropological Museum of Mexico City. Photo by G. Darvas.

2.7 *Some stages in the development, according to Chinese sources, of the fishes biting each other's tails into the yin-yang.* (Sugiura, K: *Lectures in the N.H.K. TV*) T. Ogawa, private correspondence. Compilation by the author.

2.8 *The Chinese predecessor of the threefold tomoe.* T. Ogawa, private correspondence, photocopy.

2.9 Kamares. Walberg, G. (1987) *Kamares: A Study of the Character of Palatial Middle Minoan Pottery*, 183. old. Fig 38. 6/1-7 182. old. Fig.37 5/1-6 (P. Aströms Förlag Göteborg). Reproduced courtesy of the publisher.

2.10 *Greek vase with meander designs from the early geometric period.* Knight, T.W. (1994) *Transformations and Design: A Formal Approach to Stylistic Change and Innovation in the Visual Arts* p. 114. Fig. 6.1. Adapted by the author from an original courtesy of Cambridge University Press.

2.11 *Sumerian decoration on silver bowl of King Entemena.* Lagash, around 2700 BC. Based on the Hungarian edition of H. Weyl: Symmetry (*Szimmetria*, Budapest: Gondolat, 1982).

2.12 *Sumerian picture.* Based on the Hungarian edition of H. Weyl: Symmetry (*Szimmetria*, Budapest: Gondolat, 1982).

2.13 *Settling a contract.* Ugaritic stele, 14th–13th Century BC. Adapted by the author from Jacobi, B. *Beszélő kövek*, [Story-telling Stones, in Hungarian] (Gondolat, 1968).

2.14 *Glazed sphinxes from the palace of Dareios at Susa.* Persia, c. 490 BC. Louvre, Paris. Photo by G. Darvas.

2.15 *The lions on the gate of the citadel of Mycenae.* Photo by G. Darvas.

2.18 *Pythagoras (c. 560-c. 480 BC) and music.* Miniature, c. 1200. Simonyi K.: *A fizika kultúrtörténete*, [The cultural history of physics, in Hungarian.], Reproduced courtesy of the copyright owners of the K. Simonyi estate.

2.19 *Gaffurio* (1480) *Opera teorica della disciplina musicale.* S. Roero: Mean, Proportion and Symmetry in Greek and Renaissance Art (1999). Reproduced courtesy of the journal *Symmetry: Culture and Science.*

2.20 *The world as Herodotus saw it.* Herodotus: *Opera II.*

2.21 *The Pantheon in Rome.* After Smolina, N. I. (1990) *Traditsii simmetrii v arkhitekture,* Moskva: Stroiizdat p. 153.

2.22 *The duomo of Pisa.* Burton, R., Cavendish, R. (1992) A világ száz csodája: Barangolás az építészet lenyűgöző alkotásai között [A Hundred Great Wonders of the World, in Hungarian] p. 25. Reproduced courtesy of the publisher, Magyar Könyvklub.

2.23 *The wood-engraved illustration of the Gothic duomo of Milan used in the translation of Vitruvius (1521) by C. Cesariano (1483–1543).* Gympel, J. (1997) Az építészet története az ókortól napjainkig [The History of Architecture from Ancient Times to the Present Day, in Hungarian]. Reproduced courtesy of the copyright owners of the former Kulturtrade Kiadó publishing house.

2.24 (a) *Luca Pacioli and the truncated icosahedron.* Simonyi, K.: *A fizika kultúrtörténete* [The cultural history of physics, in Hungarian, 2nd edition, 1981]. Reproduced courtesy of the copyright owners of the K. Simonyi estate. (b) Drawings by Leonardo.

2.25 *Dürer's proportions of the human body.* S. Roero: Mean, Proportion and Symmetry in Greek and Renaissance Art (1999). Reproduced courtesy of the journal *Symmetry: Culture and Science.*

2.26 *The spheres drawn around Plato's regular bodies, in Kepler's* Mysterium Cosmographicum (1596).

2.27 *The harmony of the universe and the Platonic regular bodies.* Simonyi K.: *A fizika kultúrtörténete,* [The cultural history of physics, in Hungarian, 2nd edition, 1981]. Reproduced courtesy of the copyright owners of the K. Simonyi estate.

2.28 *Kepler's music for the spheres.* Simonyi K.: *A fizika kultúrtörténete,* [The cultural history of physics, in Hungarian, 2nd edition, 1981]. Reproduced courtesy of the copyright owners of the K. Simonyi estate.

2.29 *Borromini: the ground-plan for the Church of S. Carlo alle Quattro Fontane in Rome* (1638–41). Adapted by the author.

2.30 *Drawing of a few radiolaria from a page of Haeckel's Challenger Monograph.*

2.31 *One of D'Arcy Thompson's drawings of the metamorphosis of living creatures* (519-520. *Scarus* sp. → *Pomacanthus*), Thompson, d'Arcy W. (1917) *On Growth and Form.* Reproduced courtesy of Cambridge University Press.

2.32 *The diffraction picture of a quasicrystal, and its electron microscope image.* Reproduced courtesy of the journal *Symmetry: Culture and Science* (Vol. 3, 1992, No. 1).

2.33 *The structure of the 60 carbon atomic fullerene molecule.* Braun T.: *Az a káprázatos fullerén molekula,* [The Wonderful C_{60} Molecule, in Hungarian], (Budapest: Akadémiai Kiadó, 1996). Reproduced courtesy of T. Braun.

3.3–3.9 *Examples of the frieze groups.* Based on Shubnikov, A.V.: *Simmetriya* (Moscow, 1940).

3.10 *The seven basic friezes illustrated with footprints.* Kinsey, L.C., Moore, T.E. (2002) *Symmetry, Shape and Space.* pp. 148–150, Springer & Key College Publishing. Reproduced courtesy of the authors.

3.11 *The seven basic friezes: on the left, in natural phenomena, and on the right, in elements of Hungarian decorative art from the time of the Hungarian conquest.* Reproduced courtesy of S. Bérczi©.

3.12–3.15 *Plane symmetry transformations.* Produced by G. Darvas based on the website by D. Joyce.

3.16 *The ten possible two-dimensional point groups, the symmetry elements and the equivalent points.* Drawn by the author.

3.17 *A presentation of the 17 wallpaper groups.* Produced by Gergely Darvas and extended by György Darvas based on the website by D. Joyce, and on Shubnikov, A.V. and Koptsik, V.A. (1974) *Symmetry in Science and Art.*

3.18 *Details from the tile mosaics of the Alhambra.* Photos by G. Darvas.

3.19 *Cross stitch embroideries from Hungarian folk art.* Paul, I., Zsille, Zs. *Keresztszemes hímzésminták*, [Crosswoven Needlework Ornaments, in Hungarian, without publisher, 1976].

4.1 *Common sunflower.* Photo by G. Darvas.

4.2 *Evergreen myrtle.* Photo by G. Darvas.

4.3 *Echinoderm (*suborder Ophidia*) displaying fivefold symmetry.* Haeckel: *Challenger Monograph.*

4.4 *Discomedusa.* Haeckel: *Challenger Monograph.*

5.1 *The order of the branching of a plant's stalks follows the Fibonacci sequence.* R.V. Jean: On the origins of spiral symmetry in plants. In: Hargittai I. – Pickover C.A.: *Spiral Symmetry*, World Scientific, Fig. 9.1, p. 338, Reproduced courtesy of the publisher and the author.

5.2 (a) *The order of the division of cells on the stalk.* R. V. Jean: On the origins of spiral symmetry in plants. In: Hargittai I. – Pickover C.A.: *Spiral Symmetry*, World Scientific, Fig. 3.1, p. 330 (Original source: "Strasbourger's Textbook of Botany", London: Longman, 1971.) Reproduced courtesy of World Scientific and R. V. Jean. (b) *The resulting growth of leaves*: R. Dixon: Green spirals. In: Hargittai I. – Pickover C.A.: *Spiral Symmetry*, World Scientific p. 356. Reproduced courtesy of World Scientific.

5.3 *Galaxy.* Newspaper illustration from the nineteen sixties, from the author's collection.

5.4 *Logarithmic spiral drawn around a golden rectangle.* A. L. Loeb and W. Varney: Does the golden spiral exist and if not, where is its center? In: Hargittai I. – Pickover C.A.: *Spiral Symmetry*, World Scientific. Reproduced courtesy of A. Loeb.

5.5 *Logarithmic spirals in the natural world. The sunflower.* (a) Photo by G. Darvas. (b) R. Dixon: Green spirals. In: Hargittai I. – Pickover C.A.: *Spiral Symmetry*, World Scientific, p. 364. Reproduced courtesy of World Scientific.

5.6 *The spiral positioning of the seeds on a sunflower.* L.A. Bursill, J.L. Rouse, A. Needham: Sunflower quasicrystallography. In: Hargittai I. – Pickover C.A.: *Spiral Symmetry*, World Scientific, p. 298. Reproduced courtesy of World Scientific.

5.7. *Pinecones, pineapple scales.* Photo by G. Darvas.

5.10 *Snails' shells (after R. Hooke and M. Cortie)* K.L. Cope: Spinning Descartes into Blake. In: Hargittai I. – Pickover C.A.: *Spiral Symmetry*, World Scientific p. 414. Reproduced courtesy of World Scientific., and M. Cortie: The form, function and synthesis of the Molluscan shell, ibid. p. 378, Reproduced courtesy of the publisher and the author.

5.11 *The parameters of a snail's shell.* Bérczi S., *Szimmetria és Struktúraépítés*, [Symmetry and Structure Building, in Hungarian, 1991]. Reproduced courtesy of S. Bérczi.

5.12 *Broccoli flowers.* D. Friedman: Determination of spiral symmetry in plants and polymers. In: Hargittai I. – Pickover C.A.: *Spiral Symmetry*, World Scientific, p. 260. Reproduced courtesy of World Scientific.

5.13 *Self-repeating* broken-dimensional *(fractal) structure* K. Wicks: Spiral-based self-similar sets. In: Hargittai I. – Pickover C.A.: *Spiral Symmetry*, World Scientific, p. 113. Reproduced courtesy of the publisher and the author.

5.14 *The spiral arrangement of cherry leaves.* D. Friedman: Determination of spiral symmetry in plants and polymers. In: Hargittai I. – Pickover C.A.: *Spiral Symmetry*, World Scientific, p. 258. Reproduced courtesy of World Scientific.

5.15 *Cherry flower.* Photo by G. Darvas.

5.16 *The common origins of floral lattices that can be characterized with Fibonacci numbers.* Image by S. Bérczi, Addendum 1 to the Hungarian edition of H. Weyl: Symmetry (*Szimmetria*, Budapest: Gondolat, 1982). Reproduced courtesy of S. Bérczi.

5.17 *(1) The rotation symmetry of the plant organ, (2) Lattice order made up of similar elements on the surface of the rotation body; (3) The arrangement of the lattice order in Fibonacci-numbered ribbons.* Bérczi Sz.: *Szimmetria és Struktúraépítés*, [Symmetry and Structure Building, in Hungarian, 1991]. Reproduced courtesy of S. Bérczi.

5.18 *Rose petals.* Bérczi Sz.: *Szimmetria és Struktúraépítés*, [Symmetry and Structure Building, in Hungarian, 1991]. Reproduced courtesy of S. Bérczi.

5.19 *Rose petals.* Photos by G. Darvas.

5.20 *Michelangelo's mosaic at the Capitolium in Rome* (Du Perac's engraving, 1569) In: Pogány F.: *Róma* [Rome, in Hungarian]. Budapest: Corvina. Reproduced courtesy of the Corvina Kiadó publ. house.

5.21 *Leonardo's engraving* (Accademia Vinciana, c. 1495), Taylor, R.A. (1928) Leonardo the Florentine, N.Y.: Harper (5th edition) XXXI + 580 p., 72 t., *and a design by Dürer* (1507).

5.22 *Renaissance dome seen from below.* After Smolina, N.I. (1990) *Traditsii simmetrii v arkhitekture*, Moskva: Stroiizdat.

5.23 *The floor mosaic in the basilica in Seville.* Photo by G. Darvas.

5.24 *The Parthenon on the Acropolis.* Adapted by the author from Gympel, J. (1997) Az építészet története az ókortól napjainkig [The History of Architecture from Ancient Times to the Present Day, in Hungarian], p. 11. Reproduced courtesy of the copyright owners of the former Kulturtrade Kiadó publishing house.

5.25 *The proportions of the Parthenon and of Saint Peter's in Rome follow the golden section.* Images produced by the author.

5.26 *The proportions of the Gothic minster in Ulm.* After Korobko, V.I., Korobko, G.N. (1995) *The Golden Section and a Man*, Stavropol: Caucasian Library Publ. House,

5.27 Polykleitos: *The Lance Bearer.* Produced by the author after Dóczi, G. (1981) *The Power of Limits*, Shambhala Publ. Inc., Boulder, Co.

5.28 *The Baroque gardens of the Castle of Versailles* (La Nôtre, 1661). After Smolina, N.I. (1990) *Traditsii simmetrii v arkhitekture*, Moskva: Stroiizdat.

5.29 *The gardens of Schönbrunn Palace* (1759, contemporary picture). Az emberiség krónikája, [The Chronicle of Mankind, in Hungarian], Budapest: Officina Nova. Reproduced courtesy of the Officina Nova publishing house.

5.30 *Plan of the city of Karlsruhe* (1715) Az építészet története az ókortól napjainkig [The History of Architecture from Ancient Times to the Present Day, in Hungarian], p. 59. Reproduced courtesy of the copyright owners of the former Kulturtrade Kiadó publishing house.

5.31 Le Corbusier's *modulor.* Drawings.

5.32 L.B. Alberti *Della Statua* (1435). S. Roero: Mean, Proportion and Symmetry in Greek and Renaissance Art (1999). Reproduced courtesy of the journal *Symmetry: Culture and Science.*

5.33 *The proportions of the human body in the 1511 Vitruvius edition by F. Giocondo.* From the article by S. Roero: Mean, Proportion and Symmetry in Greek and Renaissance Art (1999). Reproduced courtesy of the journal *Symmetry: Culture and Science.*

5.34 *Le Corbusier's museum design for Philippeville.* After Le Corbusier (1962) *Musee National D'Art Moderne,* Paris: Ministere d'etat affaires culturelles, and Le Corbusier Museum Design for Philippeville (1939). Le Corbusier (1962), *Musee National D'Art Moderne,* Paris: Ministere d'etat affaires culturelles.

5.35 *Bird's-eye view of Auroville.* From the website of the city.

6.1–6.5 and 6.7 Based on Shubnikov, A.V.: *Simmetriya* (Moscow, 1940).

6.6 *The golden section kaleidoscope: (a) construction from a square; (b) side view, the so-called Kepler star can be seen in the centre in blue (C. Schwabe, 1986); (c) images of the polyhedral shapes produced by passing light through the curved slits in the mirrors of the kaleidoscope;* (a) Reproduced courtesy of the journal *Symmetry: Culture and Science.* (b) C. Schwabe (1986), *Turicum,* Schweizer Kultur und Wissenschaft, April/May 1993, pp. 6–7, Reproduced courtesy of the editor of the journal *Turicum.* (c) Photo of the golden section kaleidoscope taken by G. Darvas.

6.9a *Etruscan dodecahedron.* Louvre, Paris. Photo by G. Darvas.

6.10–6.12 Bérczi S. *Szimmetria és Struktúraépítés,* [Symmetry and Structure Building, in Hungarian, 1991]. Reproduced courtesy of S. Bérczi.

6.14 *The stereograms of the three-dimensional point groups.* Redrawn after the *International Tables for Crystallography.*

6.16 *The fourteen Bravais lattices.* After Shubnikov, A.V.: *Simmetriya* (Moscow, 1940).

6.19–6.20 *Mandalas.* After K. Trivedi, *Symmetry: Culture and Science,* Vol. 1, 1990, No. 3. Reproduced courtesy of the journal *Symmetry: Culture and Science.*

6.22 *Pheidias: Knights from the western frieze of the Parthenon.*

6.23 *Primary substances.* Simonyi K.: *A fizika kultúrtörténete* [The cultural history of physics, in Hungarian]. Reproduced courtesy of the copyright owners of the K. Simonyi estate.

6.24 *Leonardo's illustrations.* Taylor, R.A. (1928) Leonardo the Florentine, N.Y.: Harper (5th edition) XXXI + 580 p., 72 t.

6.26 *The necropolis of Giza.* Gympel, J. (1997) *Az építészet története az ókortól napjainkig* [The History of Architecture from Ancient Times to the Present Day, in Hungarian], p. 59. Reproduced courtesy of the copyright owners of the former Kulturtrade Kiadó publishing house.

6.27 *Art Tower Mito (Japan).* (a) Photo courtesy of Anikó Darvas.

7.1 *Wild strawberry flower; tobacco flower.* Photo by G. Darvas, drawing.

7.2 *Complete set of regular bodies dating back to the Neolithic age.* J. Kappraff: The relationship between mathematics and mysticism of the golden mean through history Fig. 1. p. 35. In: Hargittai I.: *Fivefold Symmetry,* World Scientific. Reproduced courtesy of the publisher and the author.

7.3 *Details of gravestones from the Jewish cemetery in Tokaj, Hungary.* Photos by G. Darvas.

7.4 *Dürer's construction of a pentagon.* D. Crowe: Albrecht Dürer and the regular pentagon, Fig. 1. p. 473. In: Hargittai I.: *Fivefold Symmetry*, World Scientific. Reproduced courtesy of the publisher and the author.

7.5 *Folding of a regular pentagon.* P. Gerdes: Fivefold symmetry and weaving in various cultures, Fig. 11, p. 251. In: Hargittai I.: *Fivefold Symmetry*, World Scientific. Reproduced courtesy of the publisher and the author.

7.6 *Weaving of a bag in Mozambique and tying together the threads of a broom.* After P. Gerdes: Fivefold symmetry and weaving in various cultures, Fig. 18, p. 254. In: Hargittai I.: *Fivefold Symmetry*, World Scientific. Reproduced courtesy of the publisher and the author.

7.8 *C.A. Von Nettesheim's drawing of the proportions of the human body.* L. de Freitas: Some pentagonal "mysteries" on iconography. Fig. 3, p. 314. In: Hargittai I.: *Fivefold Symmetry*, World Scientific. Reproduced courtesy of the publisher and the copyright owner of the author's estate.

7.9 *'Golden triangle' in the (Great) pyramid of Kheops.* H. Verheyen: The icosahedral design of the Great Pyramid, Fig. 11. p. 350. In: Hargittai I.: *Fivefold Symmetry*, World Scientific. Reproduced courtesy of the publisher and the author.

7.10 *The relationship between the Egyptian triangle and the golden triangle.* H. Verheyen: The icosahedral design of the Great Pyramid , Fig. 7. p. 346. In: Hargittai I.: *Fivefold Symmetry*, World Scientific. Reproduced courtesy of the publisher and the author.

7.11 *The Egyptian triangle and the golden triangle in the* hall of the Kings *of the great pyramid.* H. Verheyen: The icosahedral design of the Great Pyramid , Fig. 10. pp. 349–350. In: Hargittai I.: *Fivefold Symmetry*, World Scientific. Reproduced courtesy of the publisher and the author.

7.12 *The proportions of the pyramid.* H. Verheyen: The icosahedral design of the Great Pyramid, Fig. 13. p. 352. In: Hargittai I.: *Fivefold Symmetry*, World Scientific. Reproduced courtesy of the publisher and the author.

7.13 *Finding Osiris' sarcophagus (flight of the swallow), and dividing up his corpse.* L. de Freitas: Some pentagonal "mysteries" on iconography. Fig. 7, p. 321. In: Hargittai I.: *Fivefold Symmetry*, World Scientific. Reproduced courtesy of the publisher and the copyright owner of the author's estate.

7.14 *The temple of Seti I, Abydos, c. 1300*BC. L. de Freitas: Some pentagonal "mysteries" on iconography. Fig. 9, p. 325 In: Hargittai I.: *Fivefold Symmetry*, World Scientific. Reproduced courtesy of the publisher and the copyright owner of the author's estate.

7.15 *From the tomb of Amenophis II, Thebes, c. 1450* BC. L. de Freitas: Some pentagonal "mysteries" on iconography. Fig. 10, p. 326 In: Hargittai I.: *Fivefold Symmetry*, World Scientific. Reproduced courtesy of the publisher and the copyright owner of the author's estate.

7.16 *Minoan and Mycenaean stamps.* J. Brandmüller: Fivefold symmetry in mathematics and science, Fig. 7. p. 23. In: Hargittai I.: *Fivefold Symmetry*, World Scientific. Reproduced courtesy of the publisher and the copyright owner of the author's estate.

7.17 *A page from the Persian translation of Abul-Wafa al Buzjani's geometry* (Bibliotheque Nationale Paris, ancient fond Persian Ms. 169, folio 180a). W.K. Corbachi, A.L. Loeb: An Islamic pentagonal seal, Fig. 1, p. 285. In: Hargittai I.: *Fivefold Symmetry*, World Scientific. Reproduced courtesy of the publisher and A. Loeb.

7.18 *The Blue Shrine and an enlarged detail of its ornamentation, and a reconstructed drawing of it (with the edge-columns shaded in).* (Gunbad i-Qubud, 1196–1197, Maragha, Western

Iran). E. Makovicky: 800-year-old pentagonal tiling, Fig. 1, p. 68 In: Hargittai I.: *Fivefold Symmetry*, World Scientific. Reproduced courtesy of the publisher and the author.

7.19 *Zvi Hecker's dodecahedral design for a housing estate (Ramot). Symmetry: Culture and Science*, Vol. 1, 1990, No. 3. Reproduced courtesy of the journal *Symmetry: Culture and Science.*

7.20 *The ground-plan of a Japanese building from the 2nd-3rd c.* AD (Harima, Hyogo prefecture). After K. Miyzaki: A mystic history of fivefold symmetry in Japan, Fig. 21, p. 378. In: Hargittai I.: *Fivefold Symmetry*, World Scientific. Reproduced courtesy of the publisher and the author.

7.21 *Puntone, 15th C., Italy.* After Gerő L.: *Magyarországi várépítészet* [Castle Architecture in Hungary, in Hungarian], Budapest: Művelt Nép, 1955, p. 64.

7.22 *The ground-plan of a system with bastions.* After Gerő L.: *Magyarországi várépítészet* [Castle Architecture in Hungary, in Hungarian], Budapest: Művelt Nép, 1955, p. 67.

7.23 *The Castel Sant'Angelo in Rome, early 16th C.* After Gerő L.: *Magyarországi várépítészet* [Castle Architecture in Hungary, in Hungarian], Budapest: Művelt Nép, 1955, p. 72.

7.24 *Palma Nuova (1593 plan by* Vincenzo Scamozzi). After Gerő L.: *Magyarországi várépítészet* [Castle Architecture in Hungary, in Hungarian], Budapest: Művelt Nép, 1955, p. 90.

7.25 *Spiral castle design with a pentagonal gate tower.* After Gerő L.: *Magyarországi várépítészet* [Castle Architecture in Hungary, in Hungarian], Budapest: Művelt Nép, 1955, p. 146.

7.26 *The ground-plan of Hollókő, and its pentagon-based tower.* Photos by G. Darvas.

7.27 *Nagykanizsa castle (1665).* After Gerő L.: *Magyarországi várépítészet* [Castle Architecture in Hungary, in Hungarian], Budapest: Művelt Nép, 1955, p. 373.

7.28 *Lenti castle.* After Gerő L.: *Magyarországi várépítészet* [Castle Architecture in Hungary, in Hungarian], Budapest: Művelt Nép, 1955, p. 396.

7.29 *Karánsebes castle (1690).* After Gerő L.: *Magyarországi várépítészet* [Castle Architecture in Hungary, in Hungarian], Budapest: Művelt Nép, 1955, p. 400.

7.30 *Nagyvárad castle (16th C.).* After Gerő L.: *Magyarországi várépítészet* [Castle Architecture in Hungary, in Hungarian], Budapest: Művelt Nép, 1955, p. 402.

7.31 *Szatmár castle (1666).* After Gerő L.: *Magyarországi várépítészet* [Castle Architecture in Hungary, in Hungarian], Budapest: Művelt Nép, 1955, p. 414.

7.32 *Goryokaku castle* (Hokkaido, Japan, 19th C.). Tourism prospectus and after K. Miyazaki. Reproduced courtesy of K. Miyazaki.

7.33 *Villa Farnese* (Caprarola). Smolina, N.I. (1990) *Traditsii simmetrii v arkhitekture*, Moskva: Stroiizdat.

7.34 *The Ministry of Defense* (Arlington, Virginia, USA). Photo from the in-flight magazine of an American air company. Private collection of the author.

7.35 *The ground-plan of the Calvinist church on Calvin square, Szeged.* Bérczi S. *Szimmetria és Struktúraépítés*, [Symmetry and Structure Building, in Hungarian, 1991]. Reproduced courtesy of S. Bérczi.

7.36 *Tiling of the plane without gaps, with congruent but irregular (non-equal-angled) pentagons. Symmetry: Culture and Science*, Vol. 2, 1991, Nos. 1–2. Reproduced courtesy of the journal *Symmetry: Culture and Science.*

7.37 Two pages from the republication of Dürer's *Underweysung der Messung* in 1600.

7.38 *Tiling of the plane with regular pentagons and two types of rhombus, and a design for a house* (B. Kirschenbaum, USA, 1956). *Symmetry: Culture and Science,* Vol. 7, 1996, No. 1. Reproduced courtesy of the journal *Symmetry: Culture and Science..*

7.39 *The construction of B. Kirschenbaum's design with golden rhombuses. Symmetry: Culture and Science,* Vol. 7, 1996, No. 1. Reproduced courtesy of the journal *Symmetry: Culture and Science.*

7.40 *Tamás F. Farkas' design made up of two types of rhombus and a square* (1972). Reproduced courtesy of the artist.

7.41 *Y. Watanabe: quasi-periodic design using a rhombus, a triangle and a square.* Kerchief presented by the RIKEN Institute. Design from the collection of G. Darvas.

7.42 M.C. Escher: *Birds.* Reproduced courtesy of the M.C. Escher Foundation. M.C. Escher works © 2003 Cordon Art – Holland. All rights reserved.

7.45, 7.47–48 Kinsey, L.C., Moore, T.E. (2002) *Symmetry, Shape and Space.* Reproduced courtesy of the authors.

7.50 *Comparison of the Maragha pattern and the Penrose arrangement.* E. Makovicky: 800-year-old pentagonal tiling, Figs. 9–10, p.81–82 In: Hargittai I.: *Fivefold Symmetry,* World Scientific. Reproduced courtesy of the publisher and the author.

7.52 M.C. Escher: *Fishes and Birds.* Reproduced courtesy of the M.C. Escher Foundation. M.C. Escher works © 2003 Cordon Art – Holland. All rights reserved.

7.53 *Penrose's non-periodic chicken.* Adapted by G. Darvas.

7.55 *Akio Hizume's model, made of bamboo.* Photo by A. Hizume. By kind permission of the artist.

7.56 *Penrose-type cells, constructed of rhombohedra arranged in space* (R. Dewar, 1989). Reproduced courtesy of the journal *Symmetry: Culture and Science.*

7.57 *The poster design for the 1995 congress of the International Symmetry Society, and the icosahedral model of the HTLV-2 (AIDS) virus.* (a) *Symmetry: Culture and Science,* Vol. 6, 1995 No. 1. Reproduced courtesy of the journal *Symmetry: Culture and Science.* (b) adapted by the author from J. Kappraff: The relationship between mathematics and mysticism of the golden mean through history, Fig. 10. p. 44. In: Hargittai I.: *Fivefold Symmetry,* World Scientific. Reproduced courtesy of the publisher and J. Kappraff.

7.58 *Cross-section of the DNA molecule, displaying tenfold symmetry.* J. Kappraff: The relationship between mathematics and mysticism of the golden mean through history Fig. 7. p. 42. In: Hargittai I.: *Fivefold Symmetry,* World Scientific. Reproduced courtesy of the publisher and J. Kappraff.

7.59 *T. Ogawa's bamboo and point lattice models. Symmetry: Culture and Science,* Vol. 3, 1992, No. 2. Reproduced courtesy of the journal *Symmetry: Culture and Science.*

7.60 *X-ray diffraction image displaying tenfold symmetry* ($Al_{65}Cu_{20}Fe_{15}$ quasicrystal, An Pang Tsai). *Symmetry: Culture and Science,* Vol. 3, 1992, No. 2. Reproduced courtesy of the journal *Symmetry: Culture and Science.*

7.61 *Electron microscope image of the alloy* $Al_{65}Cu_{20}Fe_{15}$ (An Pang Tsai). *Symmetry: Culture and Science,* Vol. 3, 1992, No. 2. Reproduced courtesy of the journal *Symmetry: Culture and Science.*

8.1 *Honeycombs.* Photo by the author.

8.3 *A cob of corn: the densest tiling of a cylindrical surface using circles.* Photo produced by the author.

8.4 *Close-packing of congruent spheres, in space and viewed from above.* Bérczi S. *Szimmetria és Struktúraépítés,* [Symmetry and Structure Building, in Hungarian, 1991]. Reproduced courtesy of S. Bérczi.

8.5 *Covering a spherical surface with circles.* Tarnai Tibor: Buckling patterns of shells and spherical honeycomb structures. In: Hargittai I.: *Symmetry 2: Unifying Human Understanding,* p. 642. Fig. 6a–b. Reproduced courtesy of T. Tarnai.

8.6 *The series of simple truncations of the octahedron or the cube.* Bérczi S. *Szimmetria és Struktúraépítés,* 1991. Bérczi S. *Szimmetria és Struktúraépítés,* [Symmetry and Structure Building, in Hungarian, 1991]. Reproduced courtesy of S. Bérczi.

8.7 Computer graphics by Sándor Kabai designed for this publication.

8.8 *Leonardo's drawings of a dodecahedron and a truncated dodecahedron,* from L. Pacioli's *Divina Proportione* (1509).

8.9 *Fullerene molecule of 60 carbon atoms.* H.W. Kroto: C60B buckminsterfullerene, other fullerenes and the icospiral shell, Fig. 1, p. 418. In: Hargittai I.: *Symmetry 2: Unifying Human Understanding,* 1989, Pergamon Press. Reproduced courtesy of Elsevier Publishing.

8.10 C_{60}, C_{240}, C_{540}, H.W. Kroto: C60B buckminsterfullerene, other fullerenes and the icospiral shell, Fig. 7, p. 422. In: Hargittai I.: *Symmetry 2: Unifying Human Understanding,* 1989, Pergamon Press. Reproduced courtesy of Elsevier Publishing.

8.11 *The weaving pattern of a basket from Mozambique; a Japanese basket seen from below* . Produced by G. Darvas after P. Gerdes: Fivefold symmetry and weaving in various cultures, Fig. 5, p. 248. In: Hargittai I.: *Fivefold Symmetry,* World Scientific. Reproduced courtesy of the publisher and the author.

8.12 *Tube fullerene.* NEC prospectus.

8.13 *Virus (Polyoma).* http://bilbo.bio.purdue.edu/~baker/projects/papova/papova.html. Reproduced courtesy of the Tim Baker research group at Purdue University, U.S.A.

8.14 *Filling an icosahedron with spheres.* Alan Mackay: Packing of spheres, Acta Crystallographica, Vol. 15. p. 916. Reproduced courtesy of A. Mackay.

8.15 *The arrangement of the tissue cells of the eye of* Calliphora erythrocephala. Reproduced courtesy of G. Savostyanov.

8.16 *The parenchyma of maize.* After H. Weyl.

8.17 *Filling space without gaps using tetrakaidecahedra.* After H. Weyl.

8.18 *Icosahedral surface of the pollen of* Sildalcea malviflora. Electron microscope image manipulated by the author.

8.19 *Radiolaria.* Haeckel: *Challenger Monograph.*

8.20 *The siliceous skeleton of radiolarian called Aulonia hexagona* (after Weyl) *and the pentagons and heptagons appearing on its surface.* Tarnai Tibor: Buckling patterns of shells and spherical honeycomb structures. In: Hargittai I.: *Symmetry 2: Unifying Human Understanding,* p. 646. Fig. 12a–b. Reproduced courtesy of T. Tarnai.

8.21 *Left, a scanning electron microscope image of a radiolarian; right, the pentagons and heptagons on its surface are emphasized.* Tarnai Tibor: Buckling patterns of shells and spherical honeycomb structures. In: Hargittai I.: *Symmetry 2: Unifying Human Understanding,* p. 647. Fig. 13a–b. Reproduced courtesy of T. Tarnai.

8.23b *Ancient Egyptian design tiling.* After Shubnikov, A.V.: *Simmetriya* (Moscow, 1940).

8.24 *Golf ball designs.* From the collection of T. Tarnai. In: Katachi U Symmetry, Springer, 1995. Reproduced courtesy of T. Tarnai.

8.25 *R.B. Fuller's patent design for the geodesic supporting structure* (1951).

8.26 *The structural framework of a geodesic dome. An arrow shows one of the pentagons.* Adapted by the author from an image from http://www.GreatBuildings.com/buildings/US_Pavilion_at_Expo_67.html, © 1994–2003 Kevin Matthews and Artifice, Inc.

8.27 *A geodesic structure being built.* Adapted by the author from an image from http://www.GreatBuildings.com/buildings/US_Pavilion_at_Expo_67.html, © 1994–2003 Kevin Matthews and Artifice, Inc.

8.28 *The first geodesic dome* (Walter Bauersfeld, Zeiss planetarium, Jena, 1922). http://www.planetarium-jena.de/hi.htm

8.29 *B. Fuller's pavilion for the world expo in Montreal* (1967). Photo courtesy of M. Pardavi-Horvath, adapted by the author.

8.30 *A detail of the Montreal dome.* Photo courtesy of M. Pardavi-Horvath.

8.31 *Geodesic dome in Disneyland, Florida, and the Dome Villette in Paris.* (a) Burton, R., Cavendish, R. (1992) A világ száz csodája: Barangolás az építészet lenyűgöző alkotásai között [A Hundred Great Wonders of the World, in Hungarian] p. 197. Reproduced courtesy of the publisher, Magyar Könyvklub. (b) Photo by G. Darvas.

9.2 *The phase changes of the Moon as depicted around 30,000 years ago.* Simonyi K.: *A fizika kultúrtörténete* [The cultural history of physics, in Hungarian]. Reproduced courtesy of the copyright owners of the K. Simonyi estate.

9.3 *Eratosthenes' measurement.* Simonyi K.: *A fizika kultúrtörténete* [The cultural history of physics, in Hungarian]. Reproduced courtesy of the copyright owners of the K. Simonyi estate.

9.6 *The sun's apparent yearly movement across the sky, and the approximate symmetries of the earth.* Simonyi K.: *A fizika kultúrtörténete* [The cultural history of physics, in Hungarian]. Reproduced courtesy of the copyright owners of the K. Simonyi estate.

9.10 *Kepler's drawing of the nesting of planetary orbits in regular bodies and the spheres that can be drawn around them.* Simonyi K.: *A fizika kultúrtörténete* [The cultural history of physics, in Hungarian]. Reproduced courtesy of the copyright owners of the K. Simonyi estate.

9.11 *The spiral arms of atmospheric cyclones.* Images taken by NASA astronauts; publication courtesy of NASA.
(1) STS109-E-6003 (10 March 2002) – The astronauts on board the Space Shuttle Columbia took this digital picture featuring a well-defined subtropical cyclone. The view looks southwestward over the Tasman Sea (between Australia and New Zealand). The image was recorded with a digital still camera.
(2) ISS009E20440 (27 August 2004) – This photo of Hurricane Frances was taken by Astronaut Mike Fincke aboard the International Space Station as he flew 230 statute miles above the storm at about 9 a.m. CDT Friday, Aug. 27, 2004. At the time, Frances was located 820 miles east of the Lesser Antilles in the Atlantic Ocean, moving west-northwest at 10 miles per hour, with maximum sustained winds of 105 miles per hour.
(3) STS069-731-031 (9 September 1995) – Hurricane Luis is captured on film in its latter days in the Caribbean in this 70 mm frame. The multifaceted mission carried astronauts David M. Walker, mission commander; Kenneth D. Cockrell, pilot; and James S. Voss (payload commander), James H. Newman, Michael L. Gernhardt, all mission specialists.

10.6 *A harp from the shrinetomb of Rekhmire the Vizier* (18th dynasty). Simonyi K.: *A fizika kultúrtörténete* [The cultural history of physics, in Hungarian]. Reproduced courtesy of the copyright owners of the K. Simonyi estate.

10.7 *The first quantitative law of nature.* Simonyi K.: *A fizika kultúrtörténete* [The cultural history of physics, in Hungarian]. Reproduced courtesy of the copyright owners of the K. Simonyi estate.

10.3 Table *The Pythagoreans' search for perfect proportions.* Simonyi K.: *A fizika kultúrtörténete* [The cultural history of physics, in Hungarian]. Reproduced courtesy of the copyright owners of the K. Simonyi estate.

10.8 *Orpheus taming wild animals.* Simonyi K.: *A fizika kultúrtörténete* [The cultural history of physics, in Hungarian]. Reproduced courtesy of the copyright owners of the K. Simonyi estate.

10.10 *Reflection, reversion and gliding in a line of* Bartók's 24th Concerto. After E. Lendvai and K. Fittler. Reproduced courtesy of the journal *Symmetry: Culture and Science.*

10.11 *The reversion and inversion of the basic melody, then the reversion of the inverted melody in* Weber's Second Cantata. After K. Fittler. Reproduced courtesy of the journal *Symmetry: Culture and Science.*

10.12 *The construction of* Bartók's Sonata for two pianos and percussion. After E. Lendvai. E. Tusa, *Magyar Tudomány* [in Hungarian], 1999, 3, p. 319. Figure 4.

11.1 Produced by G. Darvas after and courtesy of the authors Y. Ne'eman, Y. Kirsh: *The Particle Hunters* (Cambridge University Press).

11.5 *Alice's parity mirror.* Adapted by G. Darvas from the journal *Tudomány* [in Hungarian], April, 1988, p. 21.

11.6 *Spin mirror.* After Ne'eman, Y., Kirsh, Y. (1986) *The Particle Hunters*, Cambridge University Press. Reproduced courtesy of Y. Ne'eman.

11.7 *The two options for the location of the mirror.* Simonyi K.: *A fizika kultúrtörténete* [The cultural history of physics, in Hungarian]. Reproduced courtesy of the copyright owners of the K. Simonyi estate.

11.8 *The results of Wu's experiments.* Simonyi K.: *A fizika kultúrtörténete* [The cultural history of physics, in Hungarian]. Reproduced courtesy of the copyright owners of the K. Simonyi estate.

11.10 M.C. Escher*'s magic mirror* (1946). Reproduced courtesy of the M.C. Escher Foundation. M.C. Escher works © 2003 Cordon Art – Holland. All rights reserved.

11.11 M.C. Escher: *Symmetrical drawing 70* (1948). Reproduced courtesy of the M.C. Escher Foundation. M.C. Escher works © 2003 Cordon Art – Holland. All rights reserved.

11.13 *Triple tomoes from Japan and the fish motif with threefold symmetry from China.* After: Sugiura, K: *Lectures in the N.H.K. TV.* T. Ogawa, private correspondence, photocopies.

11.15 *The façade of the* Gtsug-lag-khang, *the main church of Lhasa.* Photo taken and adapted by the author.

11.16–17 Ne'eman, Y., Kirsh, Y. (1986) *The Particle Hunters*, Cambridge University Press. Reproduced courtesy of Y. Ne'eman.

11.4 Table *The generations of the most basic particles.* After Kiss Á, Horváth D, Kiss D, *Kísérleti atomfizika* [Experimental Atomic Physics, in Hungarian] p. 425. Table 24.I. Reproduced courtesy of D. Horváth.

12.4a–12.5a *Dining room,* Cheverny Castle, France. Photos taken and produced by G. Darvas.

12.7 *The creation of man. Detail of the fresco on the ceiling of the Sistine Chapel in the Vatican* (Michelangelo, 1508–12). Az emberiség krónikája, [The Chronicle of Mankind, in Hungarian], Budapest: Officina Nova. Reproduced courtesy of the Officina Nova publishing house.

12.10 *The creation of a quark-antiquark pair (meson).* After Ne'eman, Y., Kirsh, Y. (1986) *The Particle Hunters,* Cambridge University Press. Reproduced courtesy of Y. Ne'eman.

12.11 *Chirality schematically and in contrarily twisting pairs of columns. The columns of the Karlskirche in Vienna, and Leonardo da Vinci's bed in Clos Luce castle by the Loire, in which he died.* (a) Journal *Tudomány* [in Hungarian], March, 1990. (b) and (c) Photos by G. Darvas.

12.12 *Mirror symmetric group of atoms. Entaniomorph pair.* Adapted by the author from R.N. Bracewell: *The Galactic Club? Intelligent life in outer space,* 1975, Freeman and Co., New York.

12.17a *The double helix of DNA.* The image is a courtesy of J.D. Watson.

12.17b-c *The double helix staircase in the Chambord castle, France.* Photos by G. Darvas.

12.21 *The helix structure of a protein.* Adapted from L. Pauling and R.B. Corey).

12.22 *Protein synthesis.* After Kovács J.: *Sejttan* [Cells, in Hungarian, ELTE, 1999.

12.23 *Cell division.* W. Hahn: *Symmetrie als Entwicklungsprinzip ...* p. 116, Fig. 271. Reproduced courtesy of W. Hahn.

12.24 *Series of breakings of morphological symmetries.* After S. Bérczi, B. Lukács and I. Molnár. *Symmetry: Culture and Science,* Vol. 4, 1993, No. 2. Reproduced courtesy of the journal *Symmetry: Culture and Science.*

12.25 *Argyopelecus Olfersi → Sternoptyx diaphana; Polyprion → Pseudopriacanthus altus; Scorpaena sp. → Antigonia capros.* Thompson, d'Arcy W. (1917, 2nd ed. 1942) *On Growth and Form.* Reproduced courtesy of Cambridge University Press.

12.26 M.C. Escher: *Fish and Scales* (1959). Reproduced courtesy of the M.C. Escher Foundation. M.C. Escher works © 2003 Cordon Art – Holland. All rights reserved.

12.27 M.C. Escher: *Sky and Water II* (1938). Reproduced courtesy of the M.C. Escher Foundation. M.C. Escher works © 2003 Cordon Art – Holland. All rights reserved.

12.28 M.C. Escher: *Verbum (Earth, Sky and Water),* (1942). Reproduced courtesy of the M.C. Escher Foundation. M.C. Escher works © 2003 Cordon Art – Holland. All rights reserved.

13.1 *Leonardo's drawing of the human body's internal organs.* Taylor, R.A. (1928) Leonardo the Florentine, N.Y.: Harper (5th ed.) XXXI + 580 p., 72 t.

13.2 *Human brain.* Produced by the author.

13.3 *The heart and the bifurcations of the aorta, in detail and schematically.* Adapted by G. Darvas from Donáth: *Anatómiai atlasz* [Atlas of Anatomy, in Hungarian, Medicina Kiadó] p. 126.

13.5 *The Maids of Honour (*Las Meninas*), as seen by* Velazquez *and* Picasso. Reproduced courtesy of Dr Olga Mataev, from her website.

14.1 *An intarsia of Fra Giovanni.*

14.2 *The dodecahedron and the truncated dodecahedron or truncated icosahedron in Leonardo's drawings.* Taylor, R.A. (1928) Leonardo the Florentine, N.Y.: Harper (5th ed.) XXXI + 580 p., 72 t.

14.3 *Decorations of practical objects from the time of the Magyar conquest of Hungary (9th century* AD), From the motif collection of Szaniszló Bérczi. In: Hargittai I.: *Symmetry 2: Unifying Human Understanding.* Reproduced courtesy of S. Bérczi.

15.1–5 *Vanishing points.* After G. Darvas: Perspective as a symmetry transformation. *Nexus Network Journal,* Vol. 5, No. 1 (2003) pp. 9–21. Drawings originally produced by G. Darvas using illustrations on the website by Ralph Larmann "Art Studio Chalkboard" http://www2.evansville.edu/studiochalkboard/draw.html.

15.6 *Aerial and colour perspective.* After G. Darvas: Perspective as a symmetry transformation. *Nexus Network Journal,* Vol. 5, No. 1 (2003) pp. 9–21. Drawings originally produced by G. Darvas using illustrations on the website by Michael Delahunt http://www.artlex.com/ArtLex/a/aerialperspective.html.

15.7 M.C. Escher: *Ascending and Descending* (1960) *and Relativity* (1953). Reproduced courtesy of the M.C. Escher Foundation. M.C. Escher works © 2003 Cordon Art – Holland. All rights reserved.

15.8 Picasso: *Nude on a Beach* (1929). Picasso Museum, Paris. Photo taken, processed and produced by G. Darvas.

15.9 Picasso: *Portrait of Marie-Thérèse* (1937). Adapted by G. Darvas. Reproduced courtesy of Dr Olga Mataev, from her website.

15.10 Picasso: *Man with a Clarinet* (1911–12). Reproduced courtesy of Dr Olga Mataev, from her website.

15.11 Braque: *Girl with a Mandolin* (1910). Reproduced courtesy of Dr Olga Mataev, from her website.

15.12 Boccioni: *Simultaneous Visions* (1911). Peternák M. et al. (eds.), *Perspective,* 2000. Reproduced courtesy of the editor.

15.13 Picasso: *Portrait of D.-H. Kahnweiler* (1910). Reproduced courtesy of Dr Olga Mataev, from her website.

15.14 Boccioni: *Materia* (1911–12). Peternák M. et al. (eds.), *Perspective,* 2000. Reproduced courtesy of the editor.

15.15 F. Farkas Tamás: *Geocity I* (2003). Reproduced courtesy of the artist.

15.16 F. Farkas Tamás: *Genetix IV.* (2002). Reproduced courtesy of the artist.

15.17 M.C. Escher: *Belvedere* (1958). Reproduced courtesy of the M.C. Escher Foundation. M.C. Escher works © 2003 Cordon Art – Holland. All rights reserved.

15.18 M.C. Escher: *Waterfall* (1961). Reproduced courtesy of the M.C. Escher Foundation. M.C. Escher works © 2003 Cordon Art – Holland. All rights reserved.

15.19 M.C. Escher: *Circle limit III. and IV.* Reproduced courtesy of the M.C. Escher Foundation. M.C. Escher works ©2003 Cordon Art – Holland. All rights reserved.

15.20 *The Great Mosque of Samarra (9th century).* Adapted by G. Darvas from Gympel, J. (1997) *Az építészet története az ókortól napjainkig* [The History of Architecture from Ancient Times to the Present Day, in Hungarian] p. 17. Reproduced courtesy of the copyright owners of the former Kulturtrade Kiadó publishing house.

15.21 *The plan for the city of Auroville.* From the website of the city.

15.22 *The symbol of the city of Auroville.* From the website of the city.

15.23 *The model for Kurokawa's helical city plan* (1961). Produced by G. Darvas after http://parole.aporee.org/files/ga12/helix_city.jpg

15.24 *The Waterfall House* (1935–39). Gympel, J. (1997) *Az építészet története az ókortól napjainkig* [The History of Architecture from Ancient Times to the Present Day, in Hungarian] p. 95. Reproduced courtesy of the copyright owners of the former Kulturtrade Kiadó publishing house.

15.25 *Cross-section drawing of the New York Guggenheim museum, and a view of it from outside.* (a) Smolina, N.I. (1990) *Traditsii simmetrii v arkhitekture,* Moskva: Stroiizdat. (b) Gympel, J. (1997) *Az építészet története az ókortól napjainkig* [The History of Architecture from Ancient times to the Present Day, in Hungarian] p. 102. Reproduced courtesy of the copyright owners of the former Kulturtrade Kiadó publishing house.

15.26 *The Yoyogi National Gymnasium in Tokyo.* Adapted by G. Darvas from Smolina, N.I. (1990) *Traditsii simmetrii v arkhitekture,* Moskva: Stroiizdat.

15.27 *Modulor.* Le Corbusier: *Modulor.*

15.28 *W. Gropius: The Bauhaus in Dessau* (1925–26). Adapted by G. Darvas from Gympel, J. (1997) *Az építészet története az ókortól napjainkig* [The History of Architecture from Ancient Times to the Present Day, in Hungarian] p. 89. Reproduced courtesy of the copyright owners of the former Kulturtrade Kiadó publishing house.

15.29 *A study examining perspective, colours and proportions, produced by student* L. Meyer-Bergner *for a class held by his teacher* Paul Klee (1927). M. Droste: *Bauhaus.* Reproduced courtesy of the Bauhaus Archive Berlin).

15.30 *The house of W. Gropius in Dessau* (1925–26), *and his own room with furniture and fittings designed by* Marcell Breuer. Photos by Lucia Moholy-Nagy. M. Droste: *Bauhaus.* Reproduced courtesy of the Bauhaus Archive (Berlin).

15.31 *The façade of Harvard University's Carpenter Center for the Visual Arts, designed by* Le Corbusier. Photo taken and produced by G. Darvas.

Subject index

Index of names

Colour plates

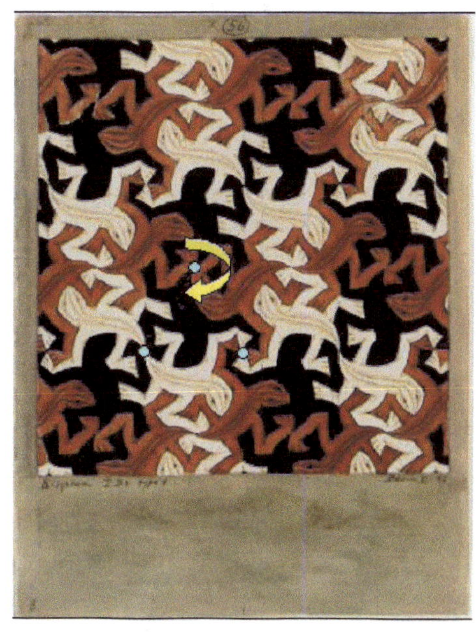

Figure 1.3. (page 6) *M. C. Escher: Reptiles*

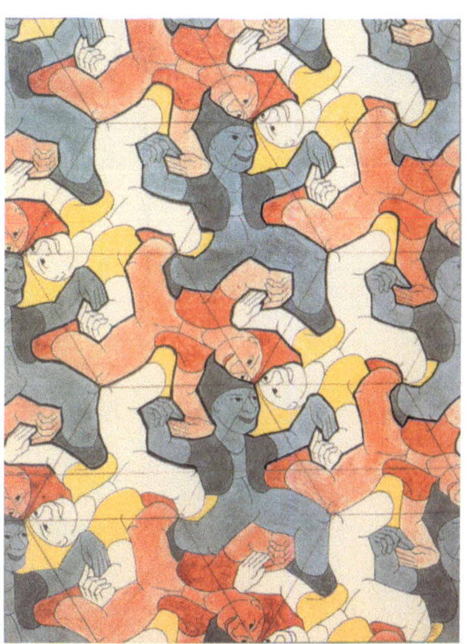

Figure 1.4. (page 7) *M. C. Escher: Symmetrical Drawing 21 (1938)*

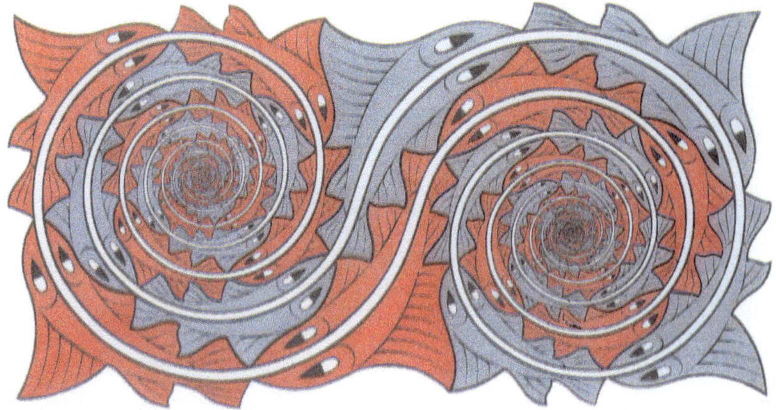

Figure 1.23. (page 25) *M. C. Escher: Whirlpools (1957)*

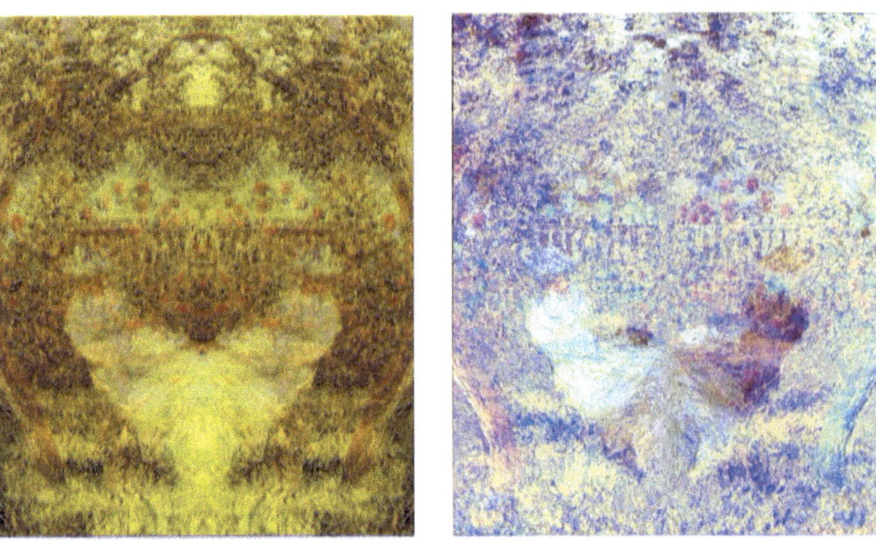

Figure 1.24. Symmetrical image **Figure 1.25.** Antisymmetrical image on the two sides of the perpendicular mirror axis, the colour of the respective mirror points is

identical *complementary*

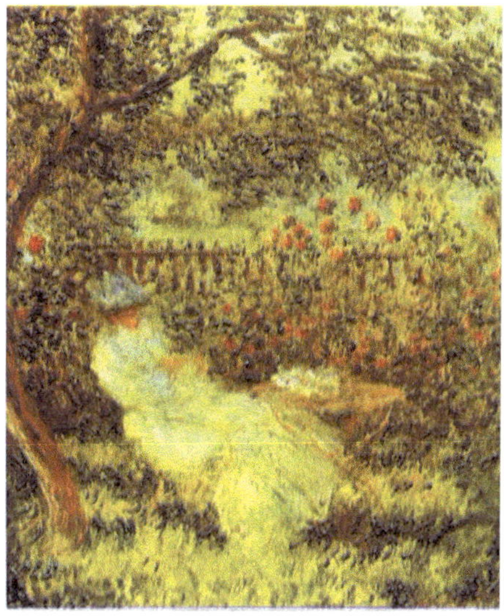

Figure 1.26. (page 28) *C. Monet: Alice Hoschedé in the Garden* (1881)

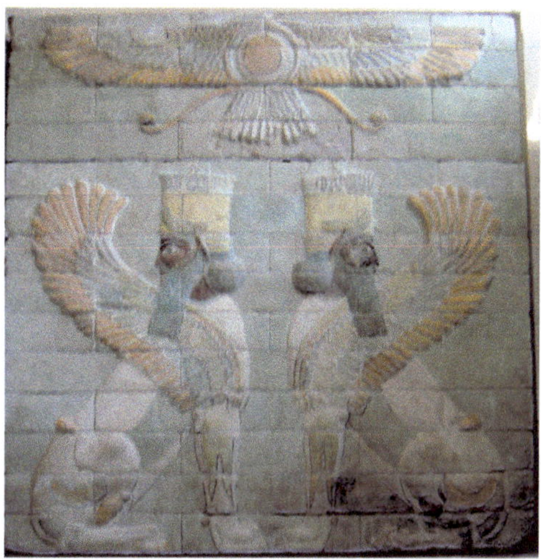

Figure 2.14. (page 42) *Glazed sphinxes from the palace of Dareios at Susa.* Persia, c. 490 BC

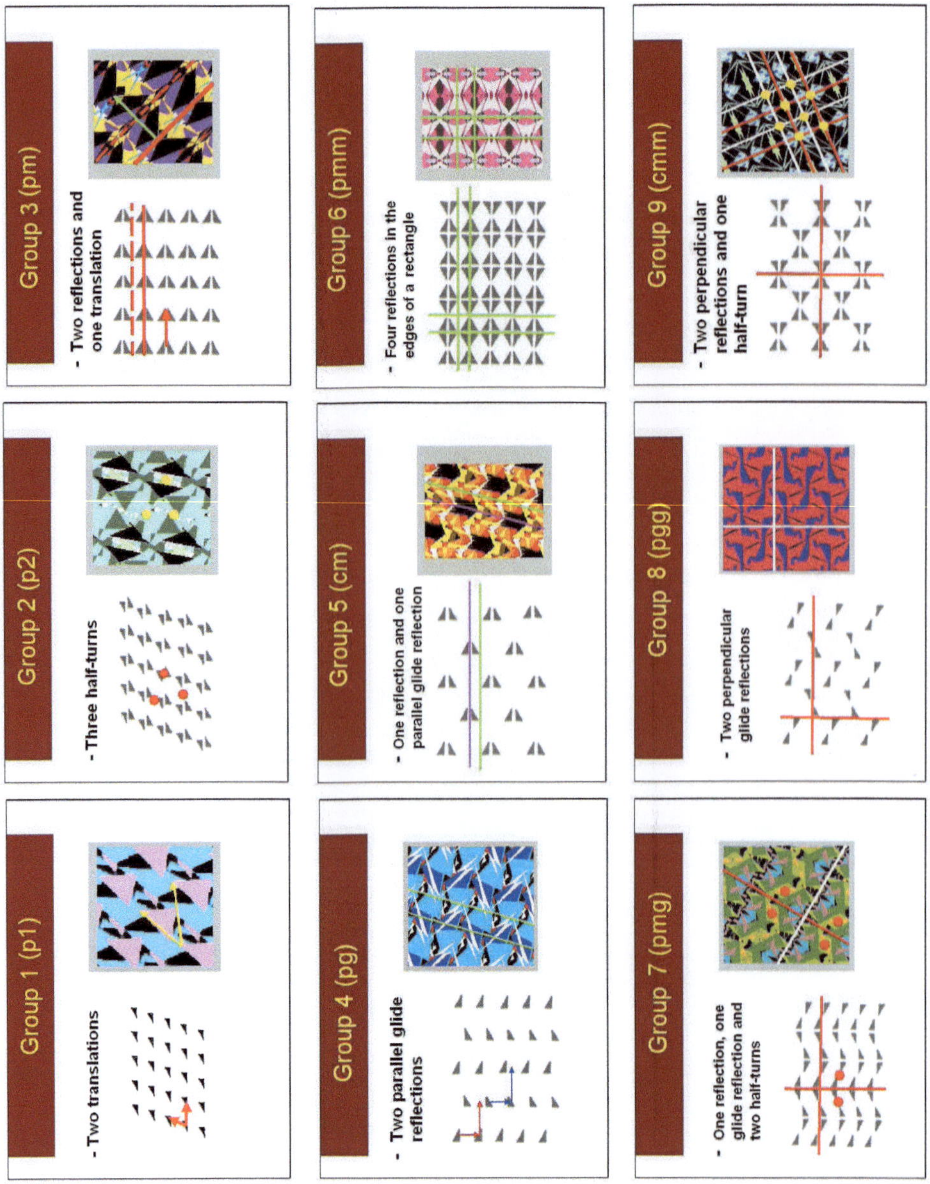

Group 1 (p1)

- Two translations

Group 2 (p2)

- Three half-turns

Group 3 (pm)

- Two reflections and one translation

Group 4 (pg)

- Two parallel glide reflections

Group 5 (cm)

- One reflection and one parallel glide reflection

Group 6 (pmm)

- Four reflections in the edges of a rectangle

Group 7 (pmg)

- One reflection, one glide reflection and two half-turns

Group 8 (pgg)

- Two perpendicular glide reflections

Group 9 (cmm)

- Two perpendicular reflections and one half-turn

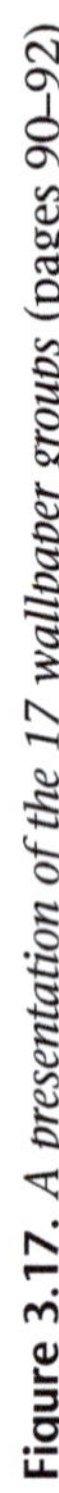

Figure 3.17. *A presentation of the 17 wallpaper groups* (pages 90–92)

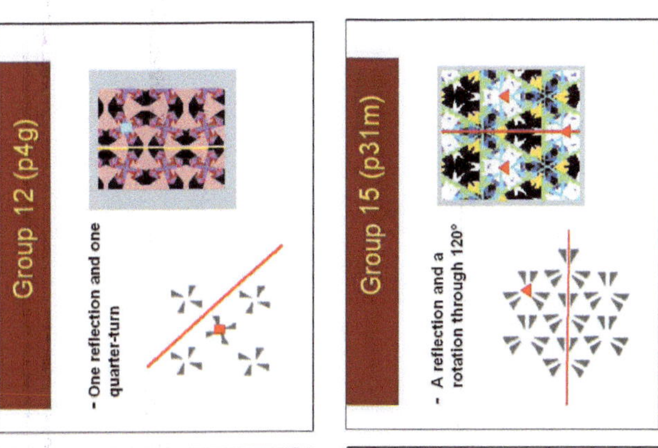

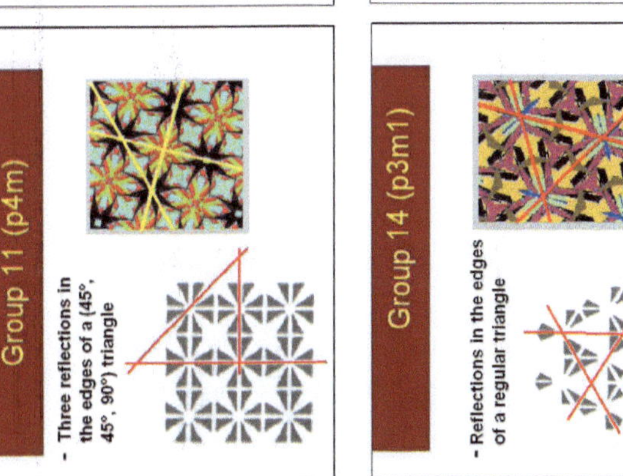

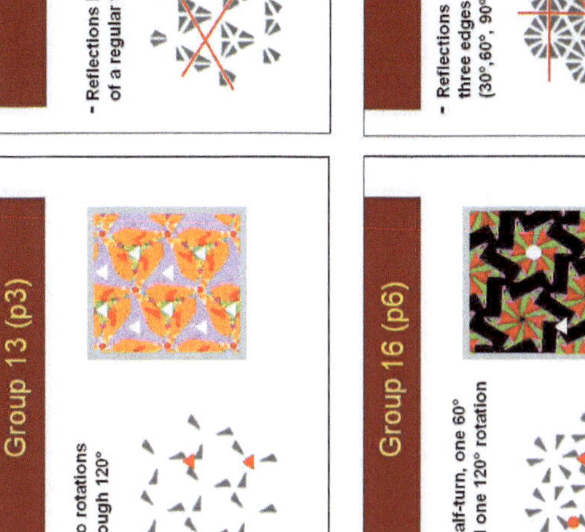

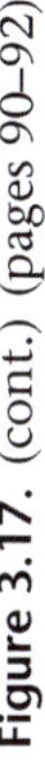
Figure 3.17. (cont.) (pages 90–92)

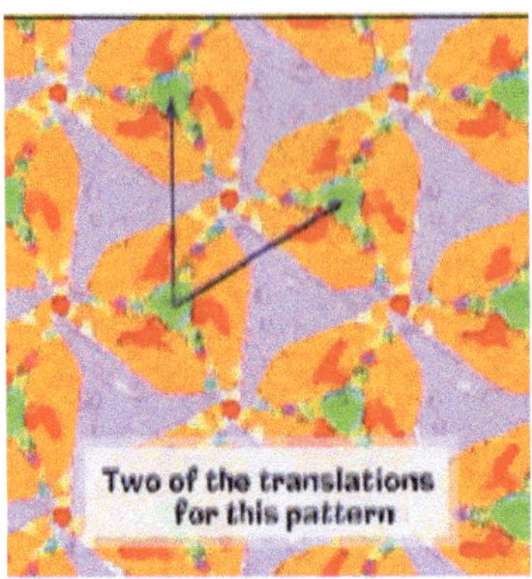

Figure 3.12. (page 87) *Translation. We have marked two possible translations on the figure.*

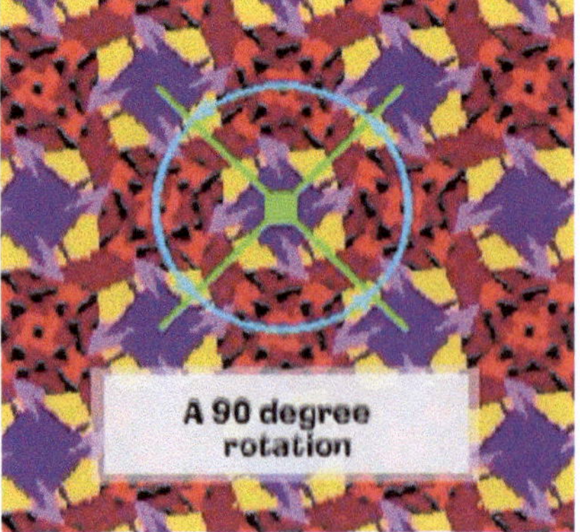

Figure 3.13. (page 87) *Rotation. We have marked a rotation through 90° on the figure.*

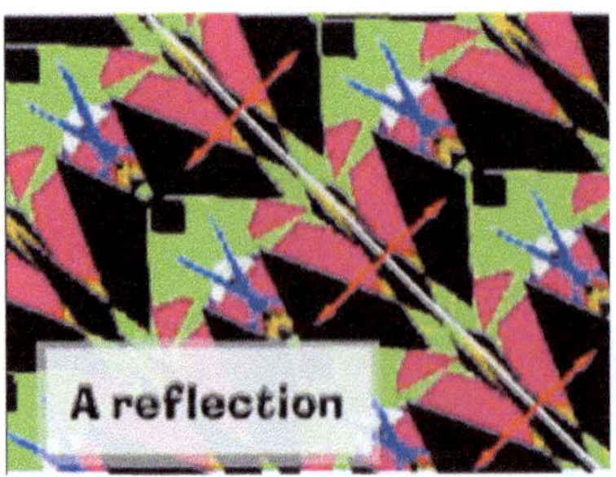

Figure 3.14. (page 87) *Reflection. We have marked the axis of a reflection on the figure.*

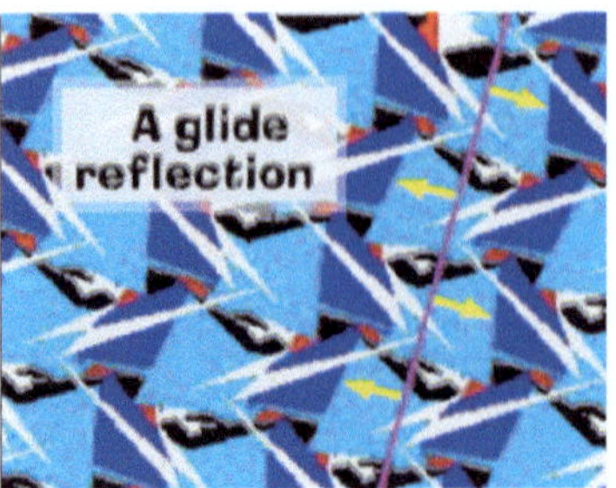

Figure 3.15. (page 87) *Glide reflection. We have marked the glide reflection on the figure with an axis, and marked the place of the glide-reflected elements with arrows*

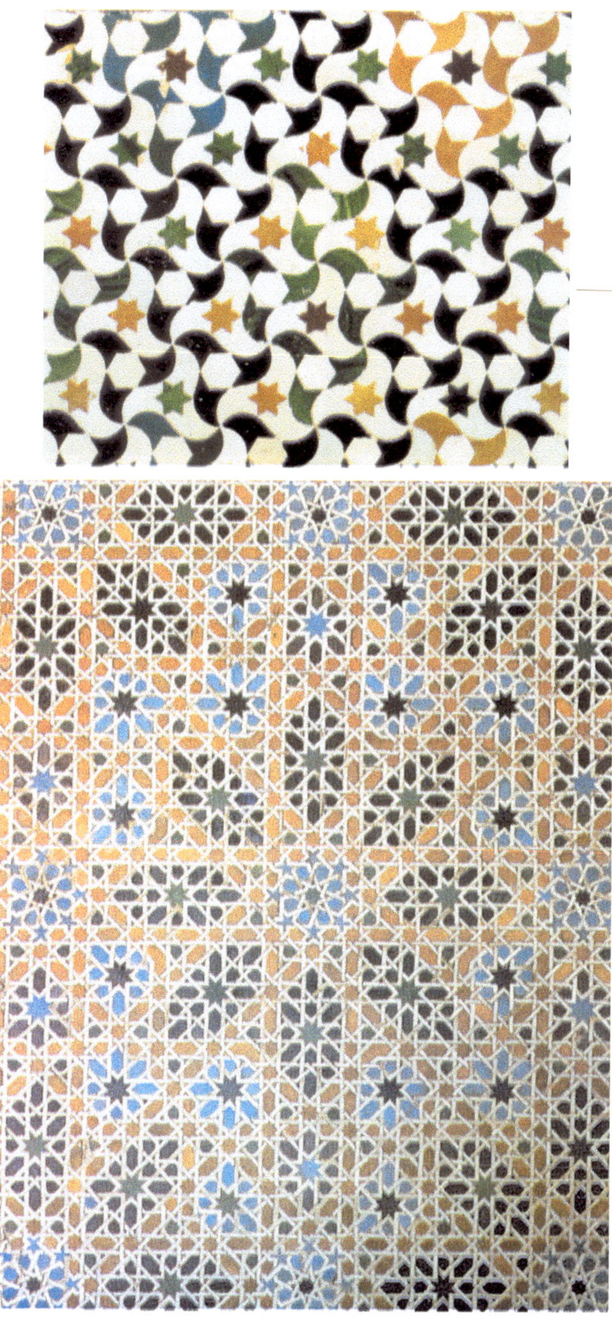

Figure 3.18. (page 93) *Details from the tile mosaics of the Alhambra*

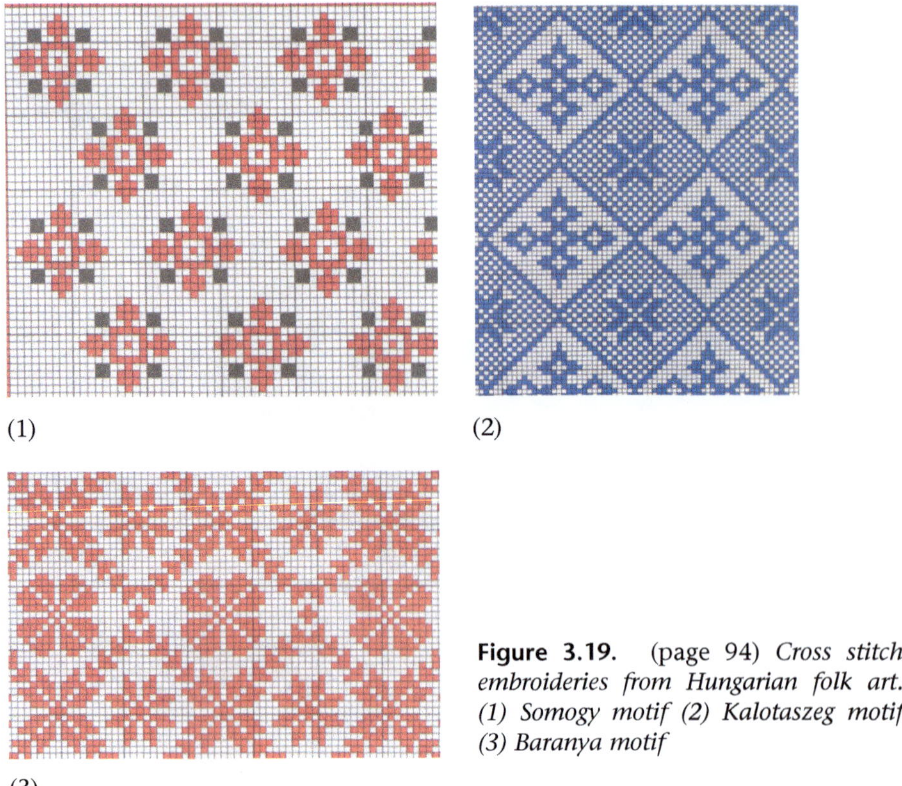

Figure 3.19. (page 94) *Cross stitch embroideries from Hungarian folk art. (1) Somogy motif (2) Kalotaszeg motif (3) Baranya motif*

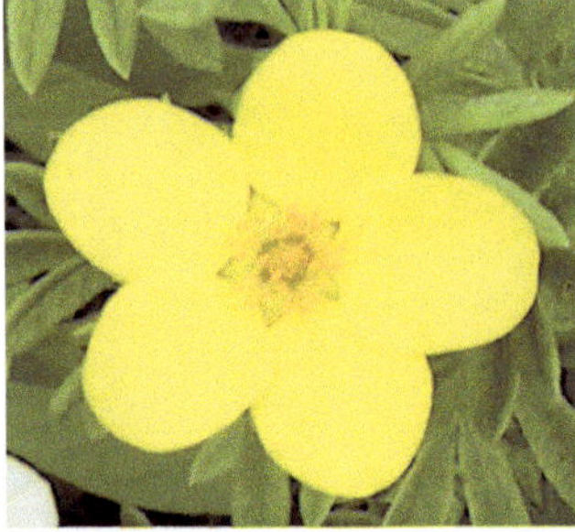

Figure 4.1. (page 98) Common sunflower

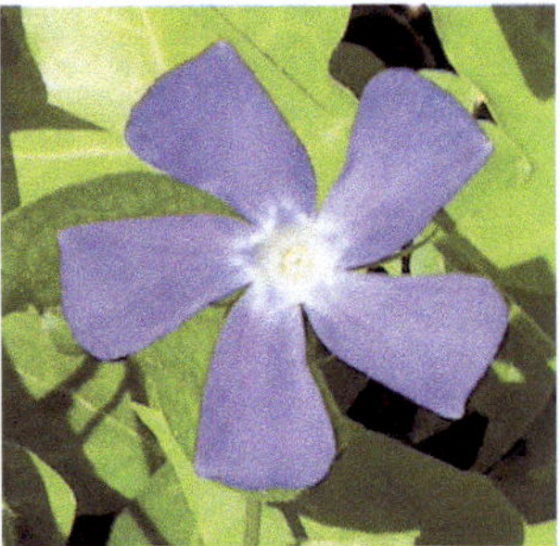

Figure 4.2. (page 98) Evergreen myrtle

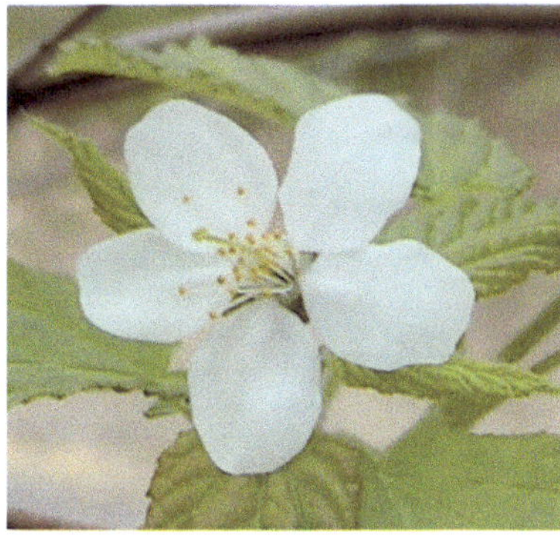

Figure 5.15. (page 119) *Cherry flower. Petals arranged in a single plane, with fivefold rotation and mirror symmetry*

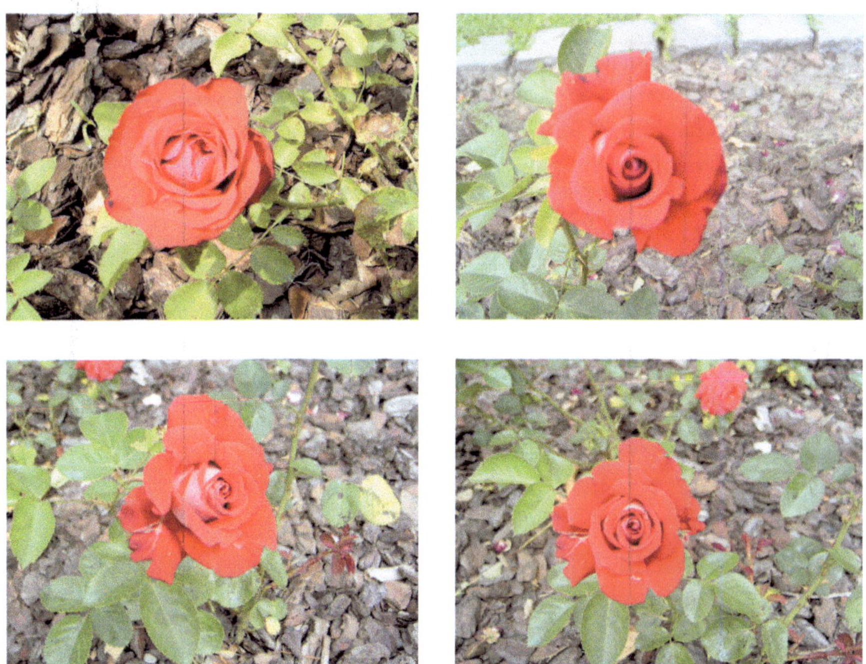

Figure 5.19. (page 122) *Rose petals that have blossomed*

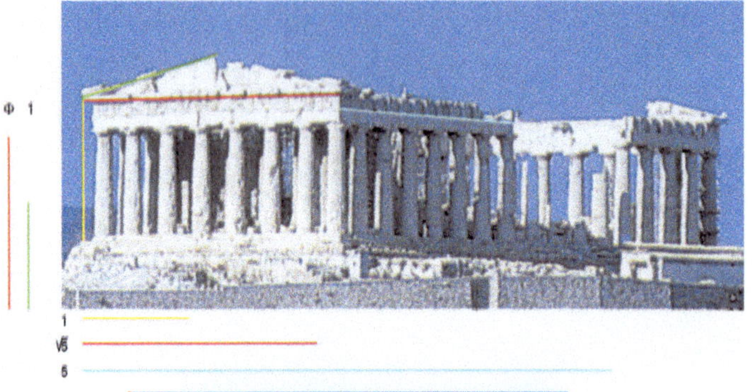

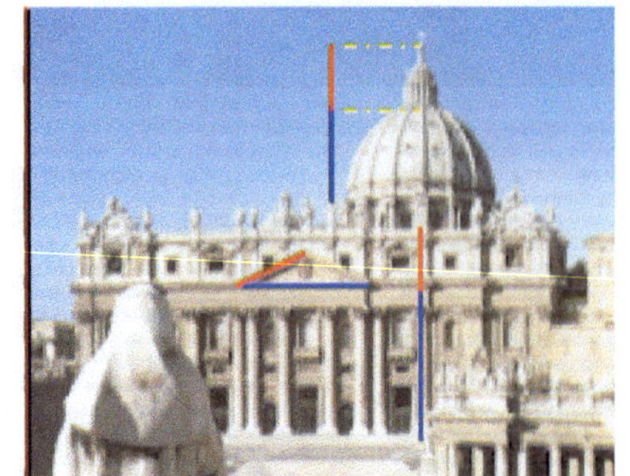

Figures 5.24, 5.25. (page 125) *The proportions of the Parthenon and of Saint Peter's in Rome follow the golden section.*
The Parthenon on the Acropolis (top)
Saint Peter's in Rome (left)

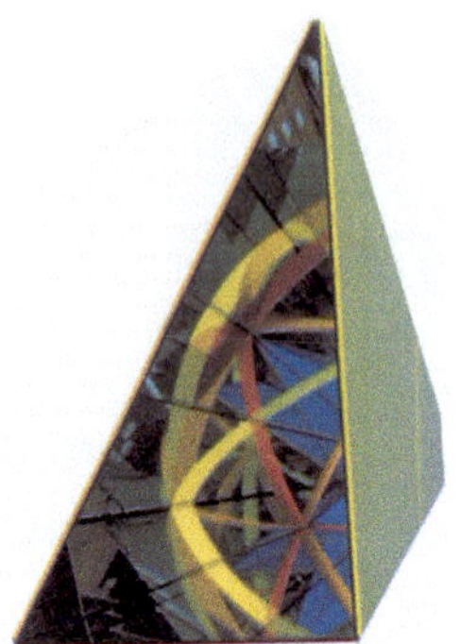

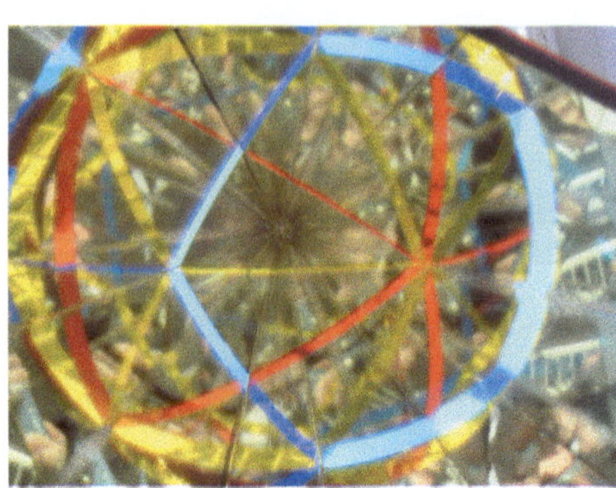

Figure 6.6b,c. (page 137) *The golden section kaleidoscope: (b) side view; the so-called Kepler star can be seen in the centre in blue (C. Schwabe, 1986); (c) images of the polyhedral shapes produced by passing light through the curved slits in the mirrors of the kaleidoscope (G. Darvas, 2003)*

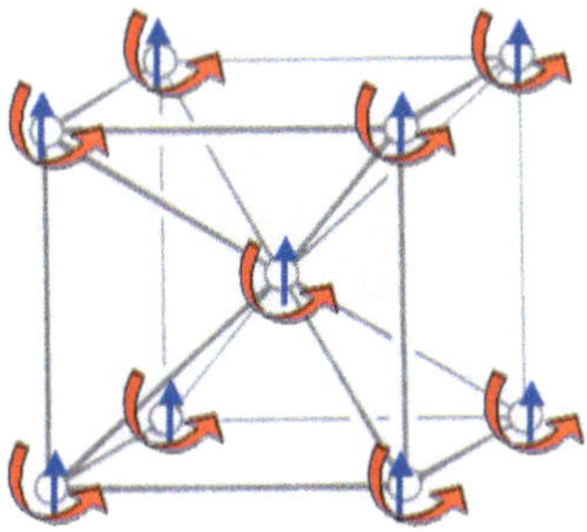

Figure 6.17. (page 157) *A detail of a crystal lattice, with atoms with internal structures*

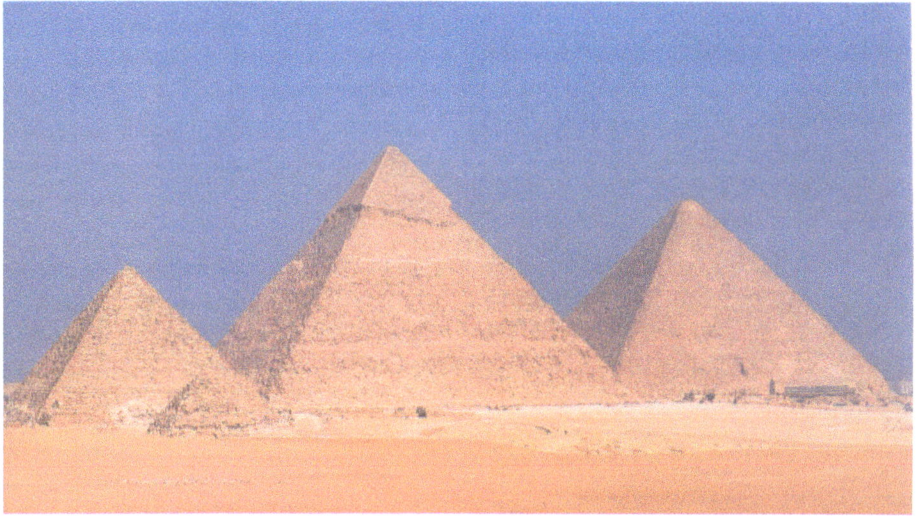

Figure 6.26. (page 168) *The necropolis of Giza. The pyramids of Mycerinos, Khephren and Kheops, with a smaller pyramid of a queen in the foreground*

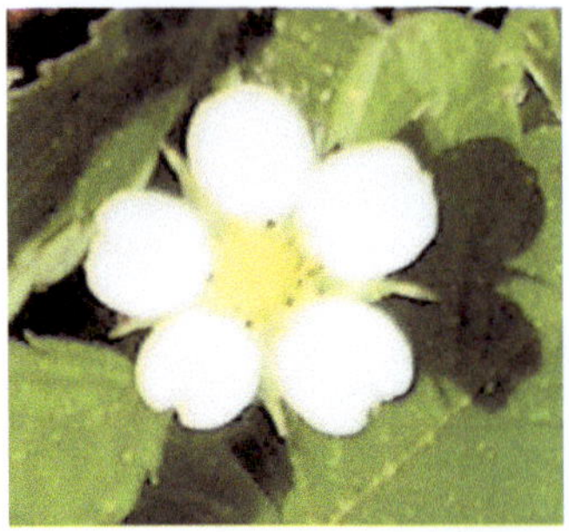

Figure 7.1. (page 171) *Wild strawberry flower; tobacco flower*

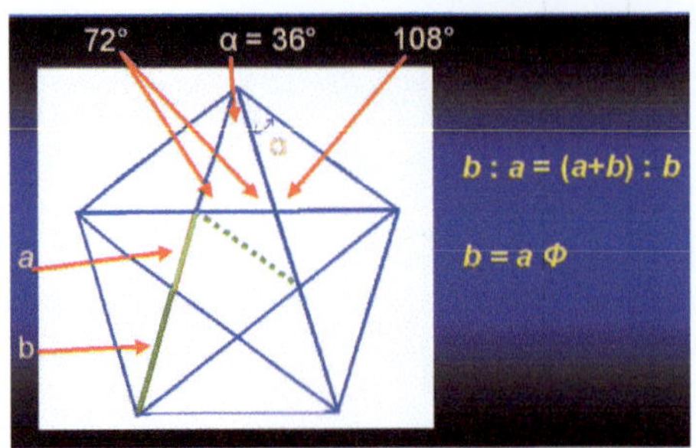

Figure 7.7. (page 174) *Properties of the regular pentagon*

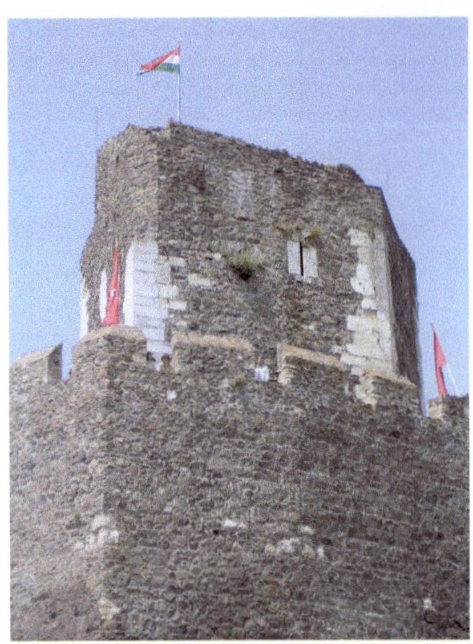

Figure 7.26b. (page 188) *The pentagon-based tower of Hollókő*

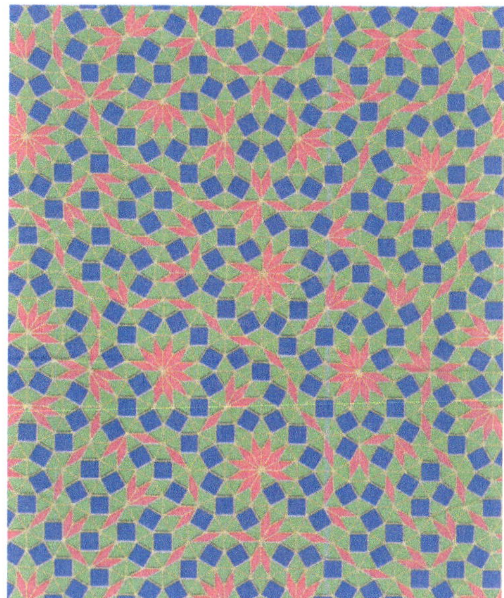

Figure 7.41. (page 198) *Y. Watanabe: quasi-periodic design using a rhombus, a triangle and a square*

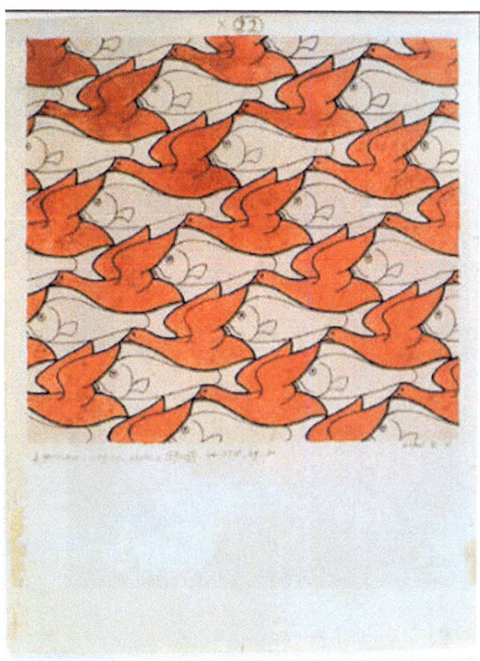

Figure 7.52. (page 206) *M. C. Escher: Fishes and Birds*

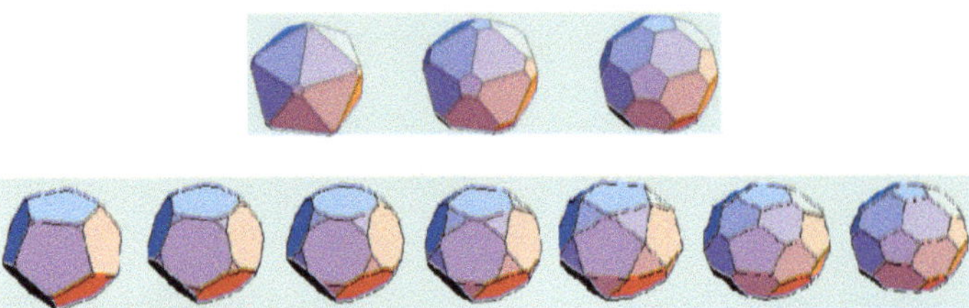

Figure 8.7. (page 221) *The truncation of the icosahedron or the dodecahedron* (computer graphics by Sándor Kabai)

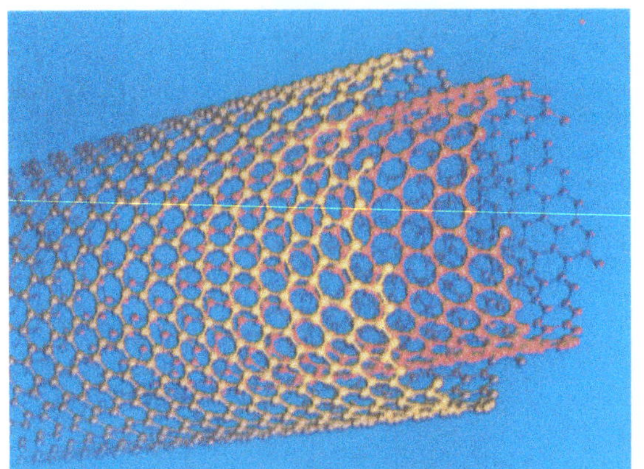

Figure 8.12. (page 226) *Tube fullerene*

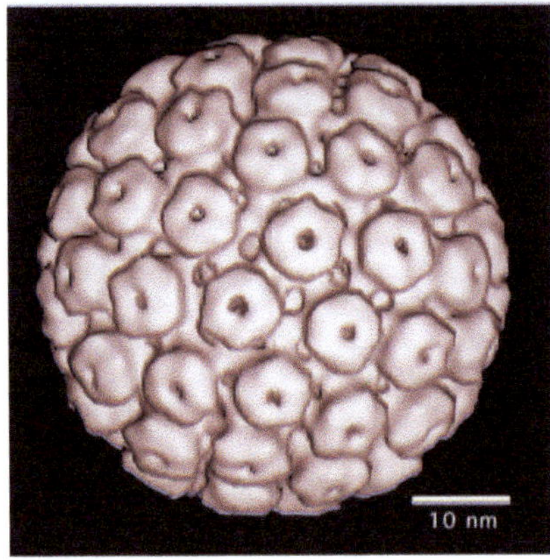

Figure 8.13. (page 226) *Virus (Polyoma) with an icosahedral surface structure*

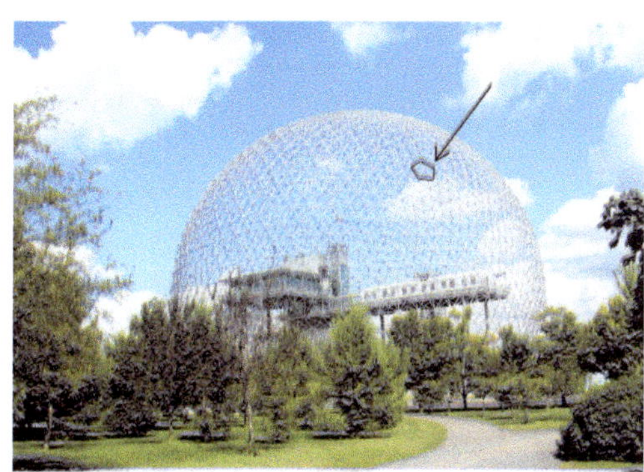

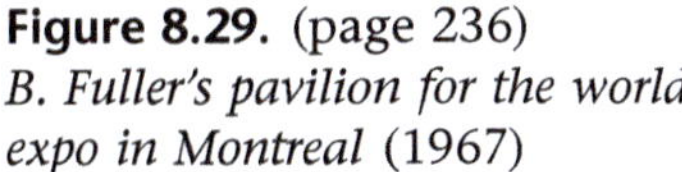

Figure 8.29. (page 236)
B. Fuller's pavilion for the world expo in Montreal (1967)

(a)

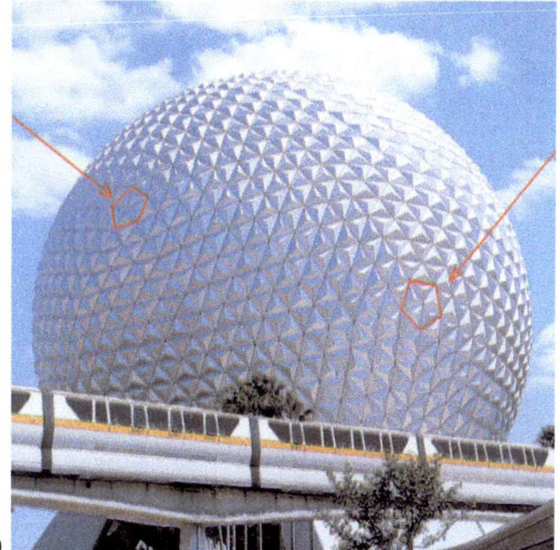

(b)

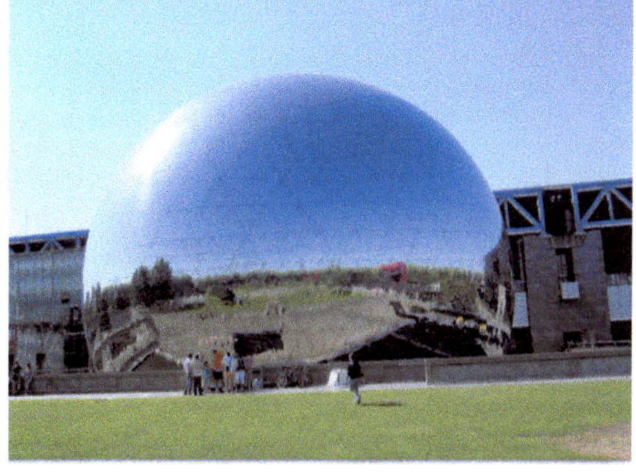

Figure 8.31. (page 237)
(a) Geodesic dome in Disneyland, Florida, and (b) the Dome Villette in Paris

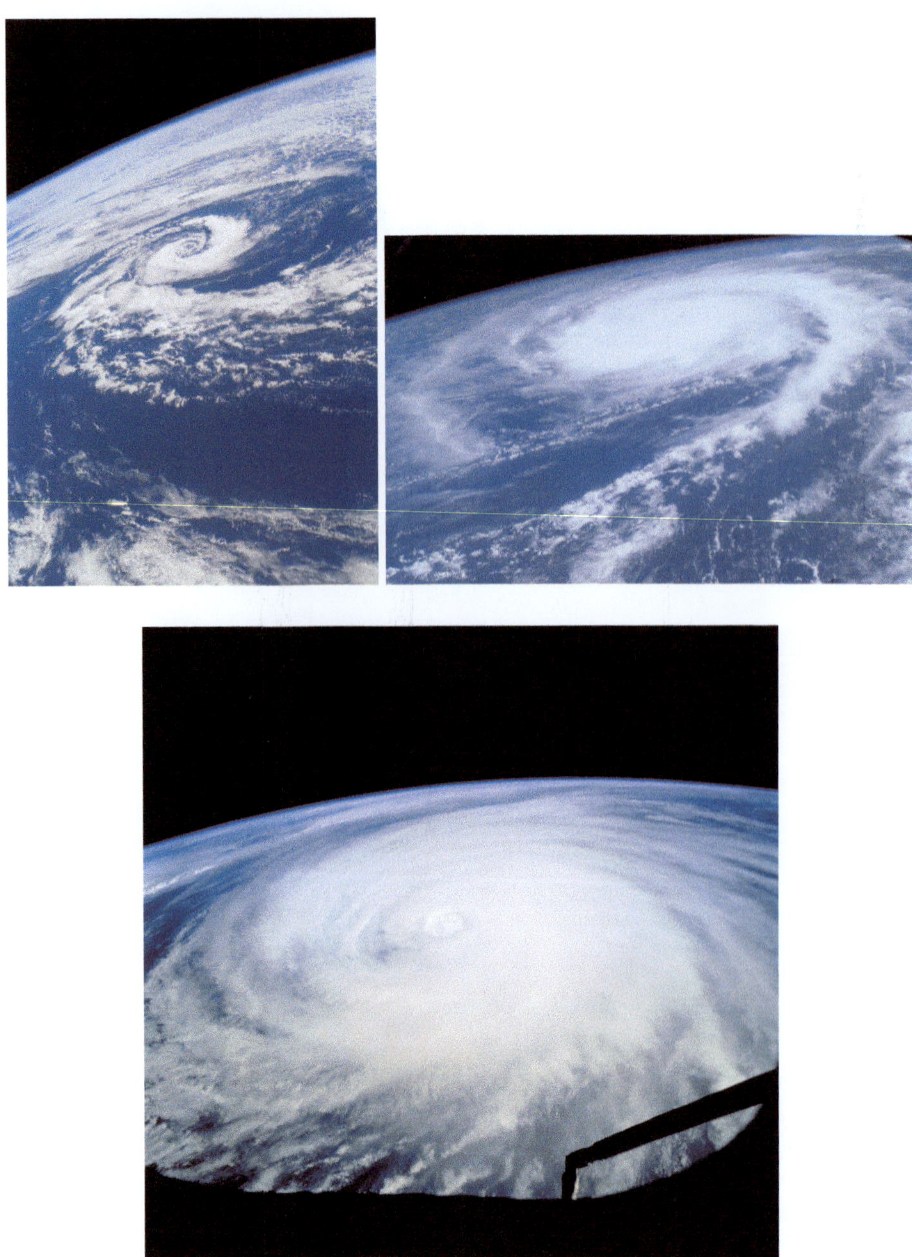

Figure 9.11. (pages 252–253) *The spiral arms of atmospheric cyclones*

The spectrum of visible light

from violet (380mμ) to red (760mμ)

multiplying wavelength by a factor of two,
and dividing frequency by a factor of two:

1 octave

with 7 basic colours in-between:

red, orange, yellow, green, cyan, blue, violet

the transition is continuous

Figure 10.2. (page 258) *The spectrum of visible light*

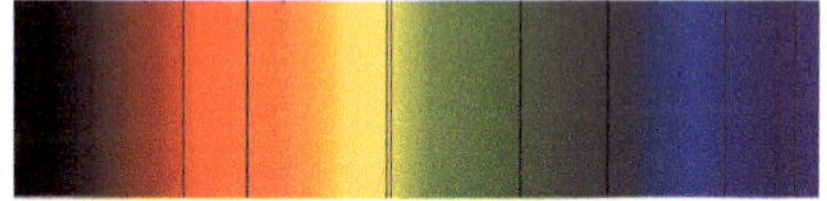

Figure 10.3. (page 259) *The spectrum of the rainbow*

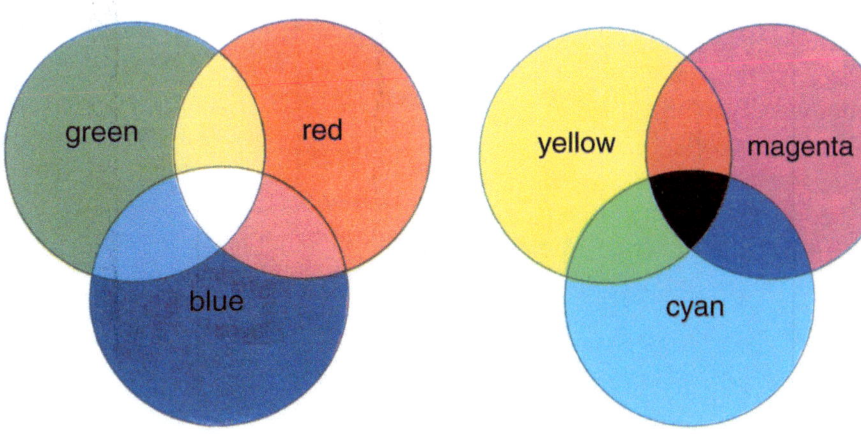

Figure 10.4. (page 262)
Colour synthesis with addition

Figure 10.5. (page 262)
Colour subtraction

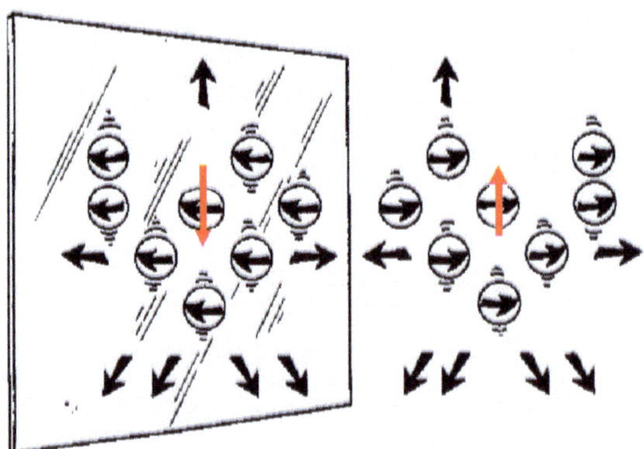

Figure 11.6. (page 297) *The mirror image of atoms with spin in the same direction (the upward direction is marked with a red arrow). The upward spin corresponds to a rightward direction of rotation. The mirror image of this will be rotation to the left, which — assuming the mirror image preserves right-handedness, i.e. parity — will correspond to a downward spin. If in an experiment a higher number of particles are emitted in a direction opposite to the spin, the mirror image of this will demonstrate that there are more particles emitted in the direction coinciding with that of the spin (after Ne'eman and Kirsh)*

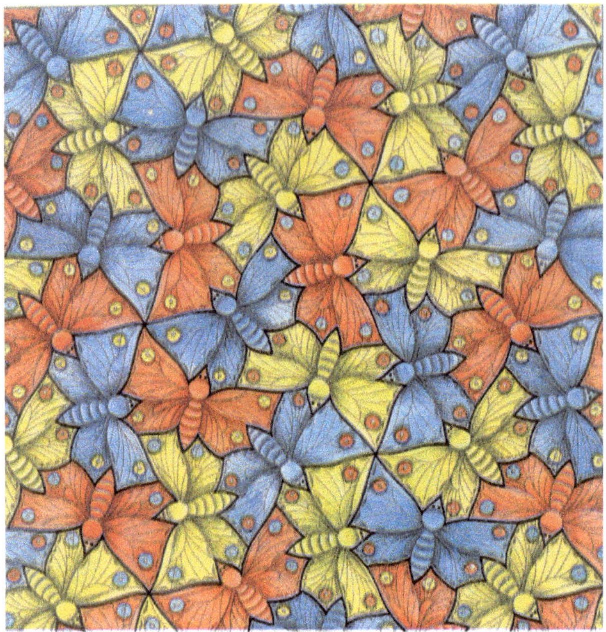

Figure 11.11. (page 304) *M. C. Escher: Symmetrical drawing 70* (1948)

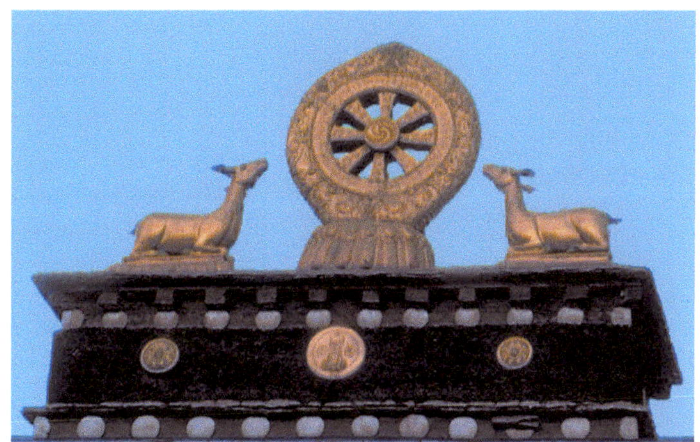

Figure 11.15. (page 310) *The combination of the eightfold way and threefold rotational symmetry: holy symbols made of gold on the façade of the* Gtsug-lag-khang, *the main church of Lhasa*

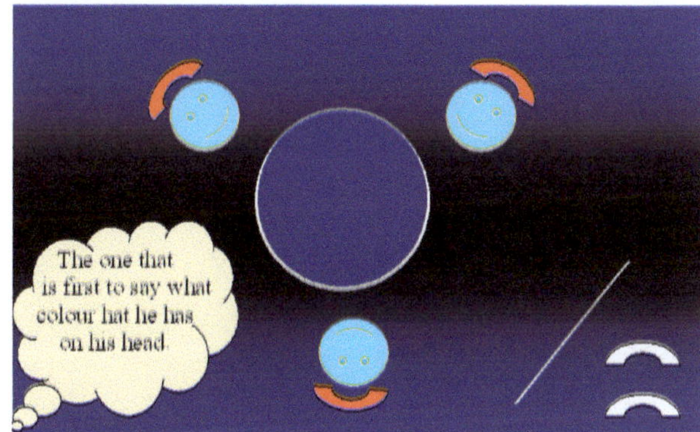

Figure 12.2. (page 319) *Which is the cleverest dwarf?*

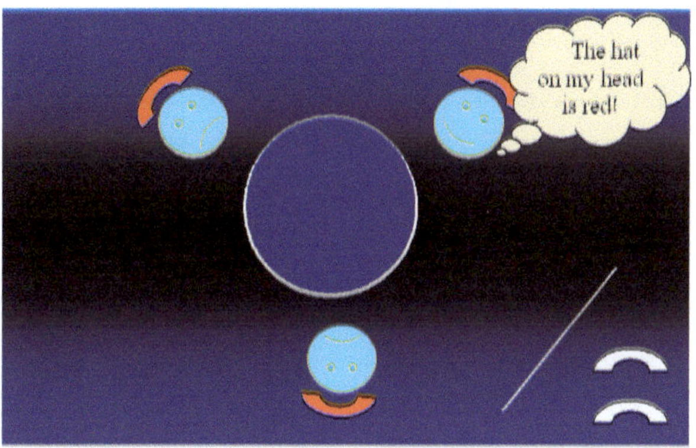

Figure 12.3. (page 319) *The cleverest dwarf*

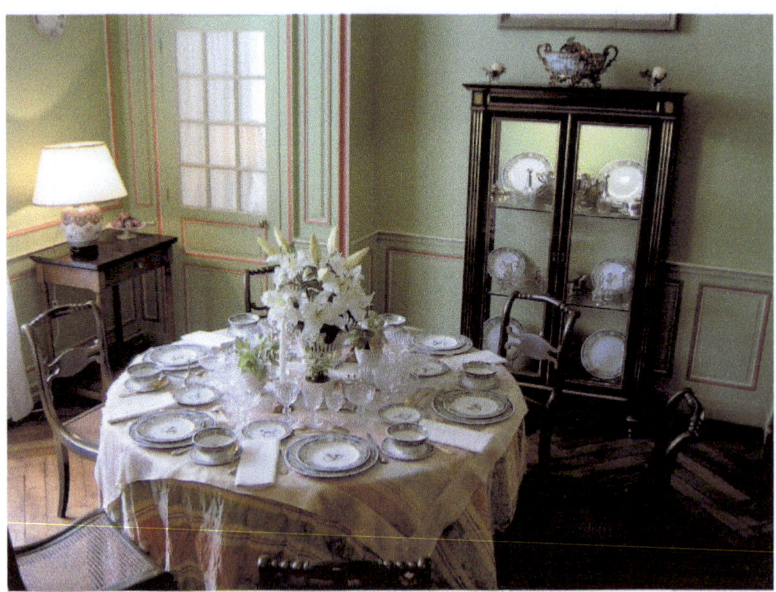

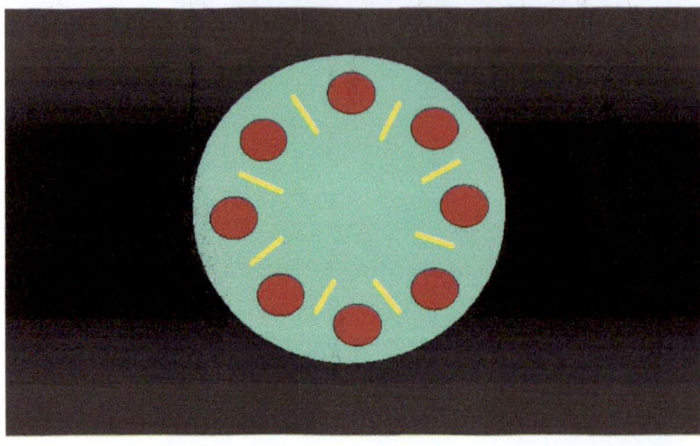

Figure 12.4. (page 320)
S. Weinberg's example: symmetrical setting of the table

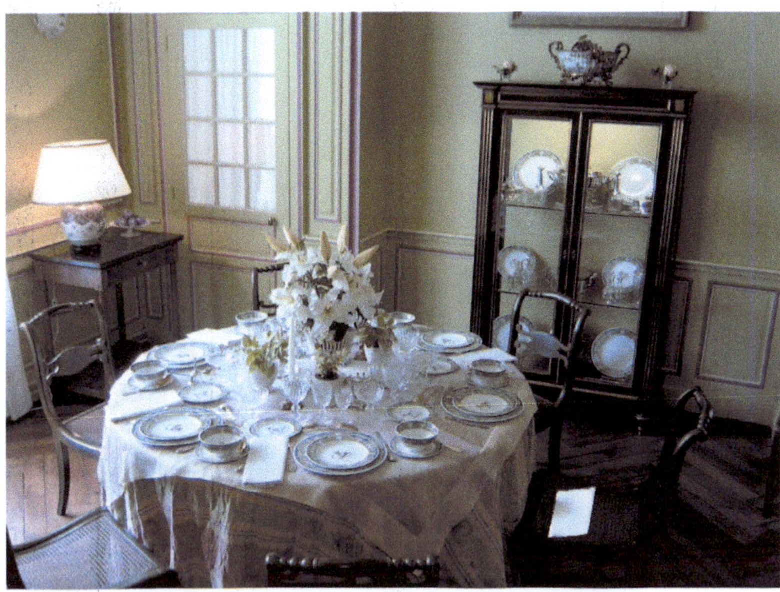

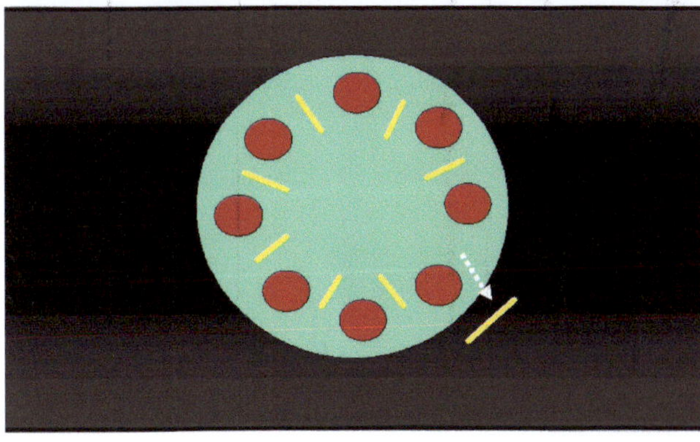

Figure 12.5. (page 321) *Spontaneous symmetry breaking*

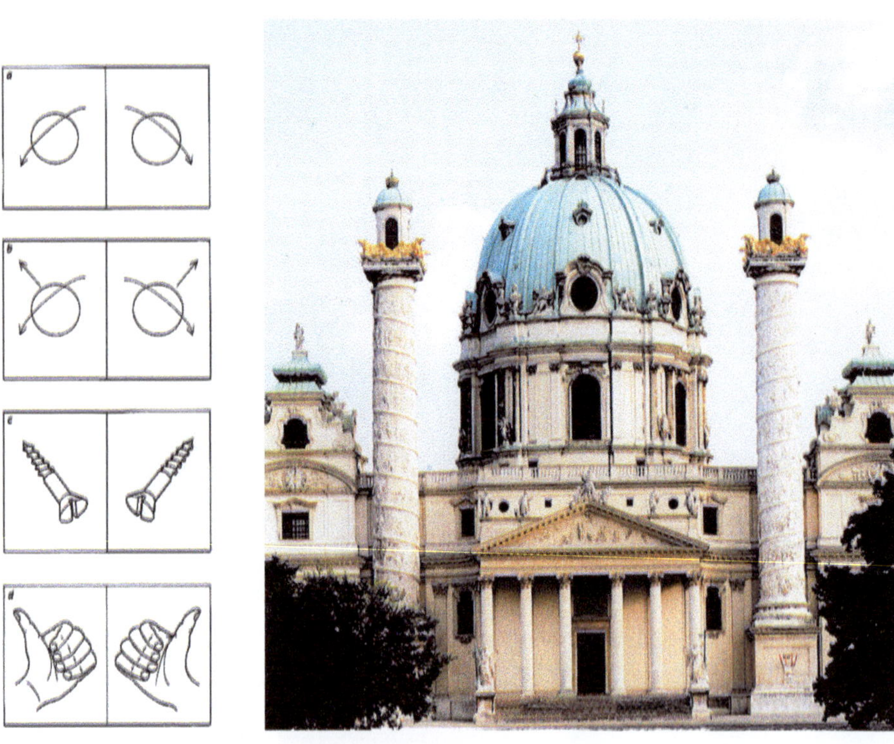

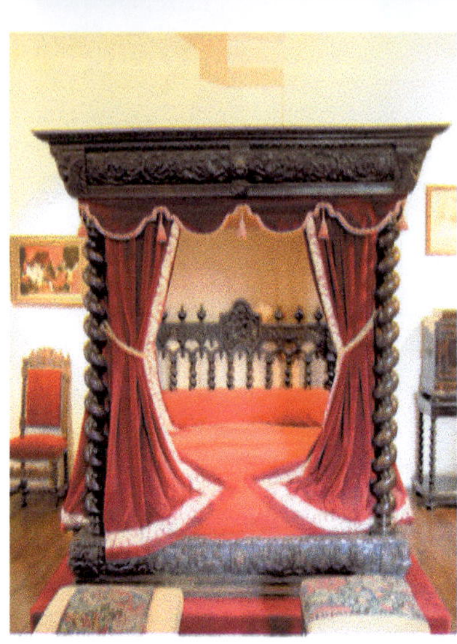

Figure 12.11. (page 330) *Chirality schematically and in contrarily twisting pairs of columns. The columns of the Karlskirche in Vienna, and Leonardo da Vinci's bed in Clos Luce castle by the Loire, in which he died*

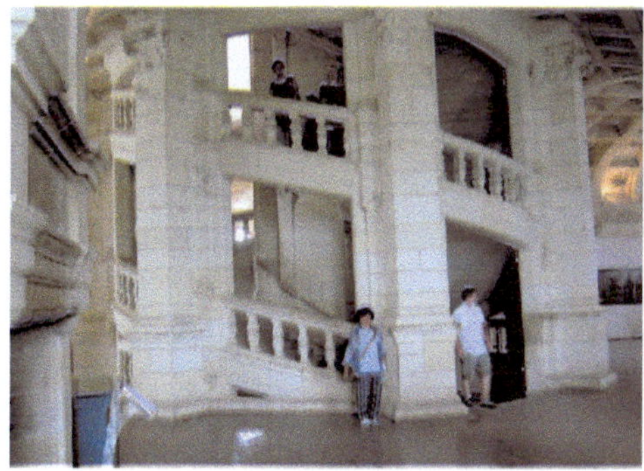

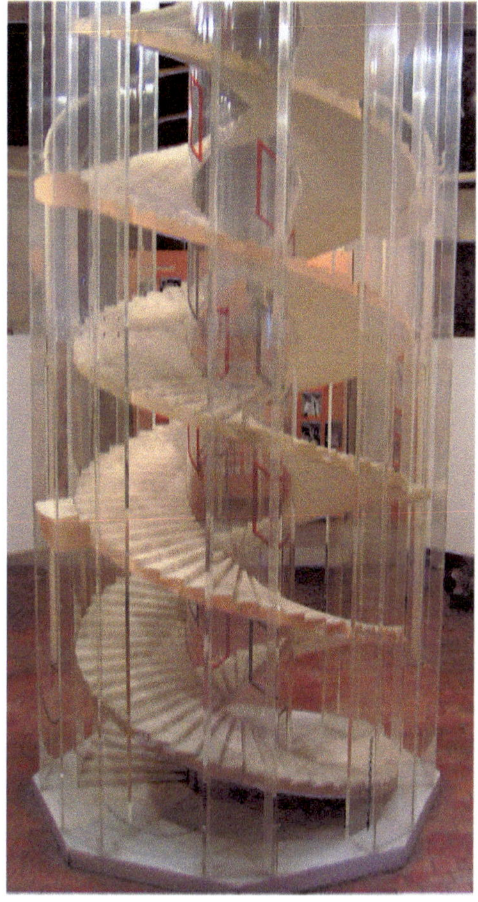

Figure 12.17b. (page 338) *The double helical central staircase of the castle of Chambord and its structural model*

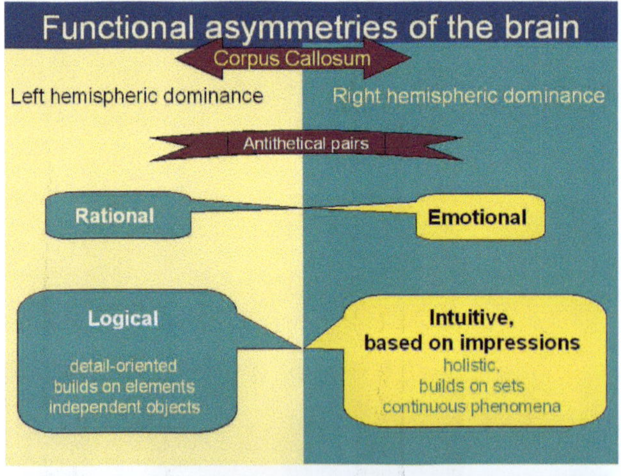

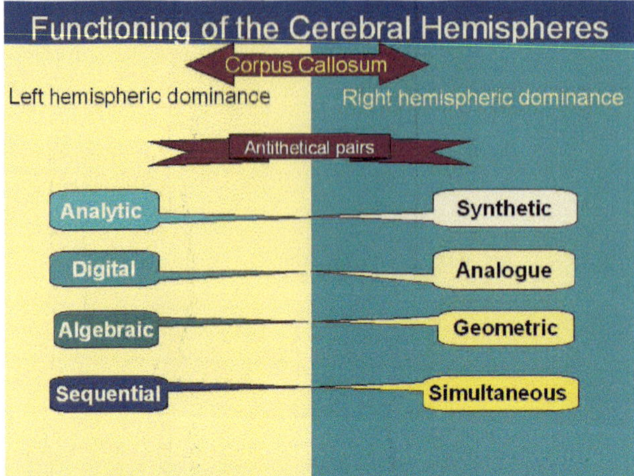

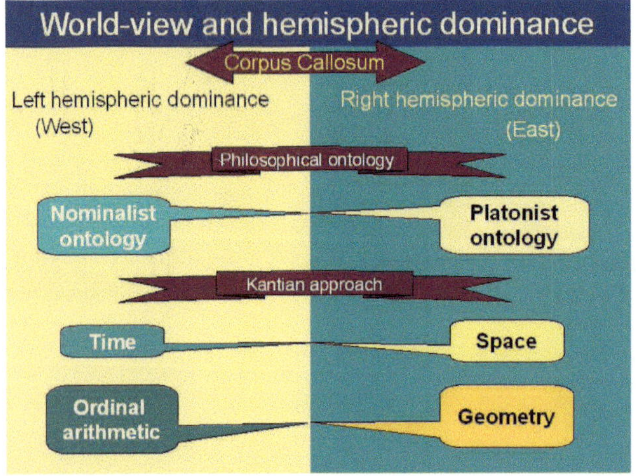

Figure 13.4. (page 361) *Tables of right and left hemispheric functions*

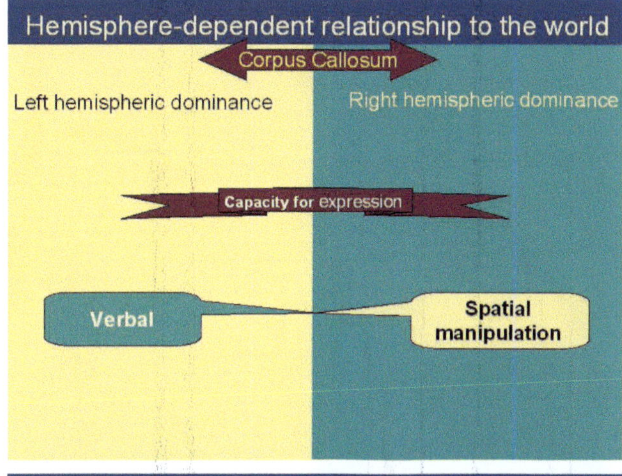

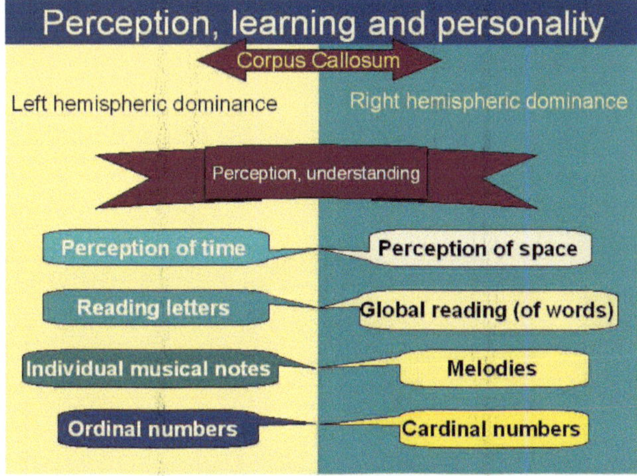

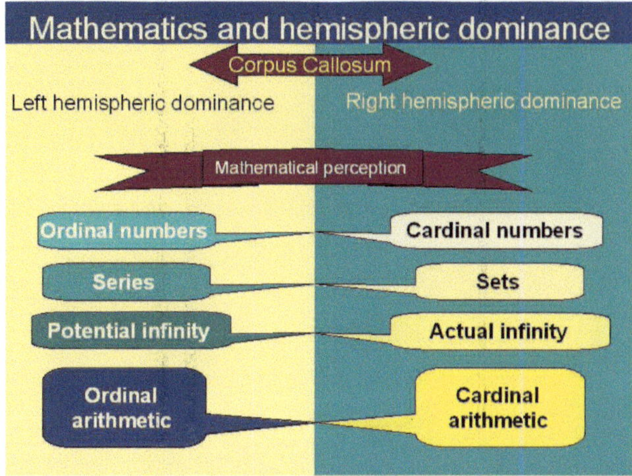

Figure 13.4.
(page 362)
(cont.)

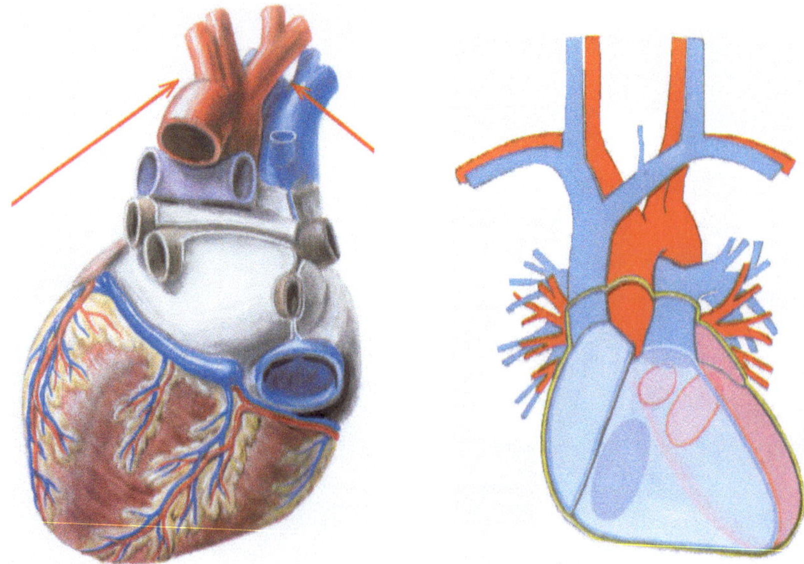

Figure 13.3. (page 356) *The heart and the bifurcations of the aorta (in red), in detail and schematically*

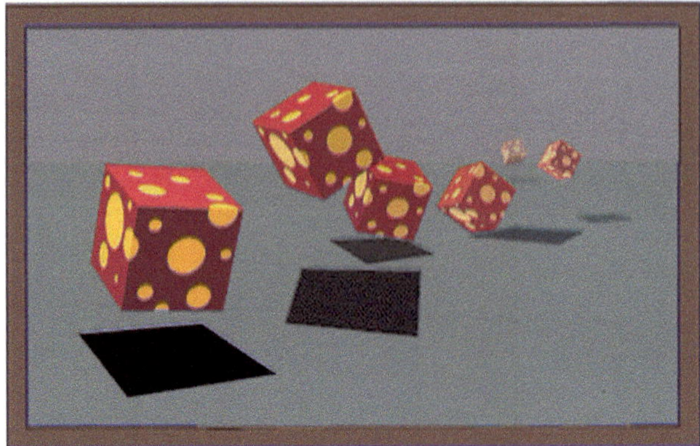

Figure 15.6. (page 388) *Aerial and colour perspective*

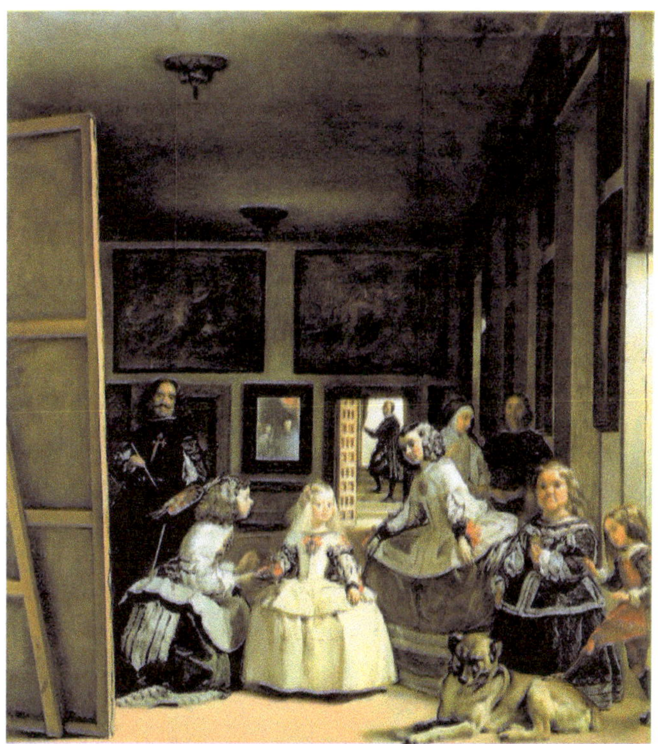

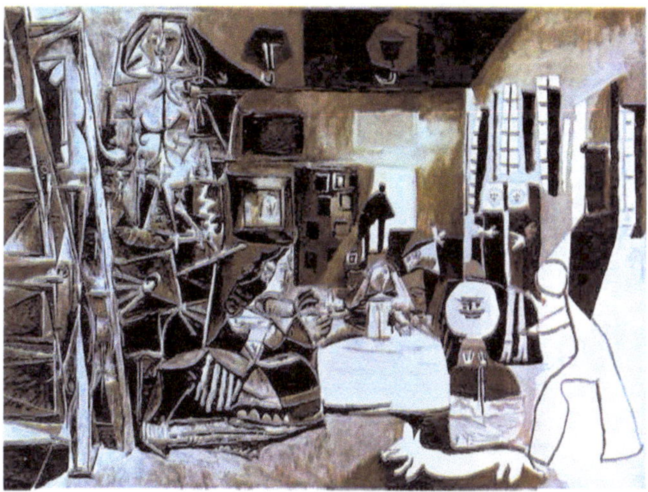

Figure 15.7. (page 371) *The Maids of Honour (Las Meninas),* as seen by Velazquez and Picasso

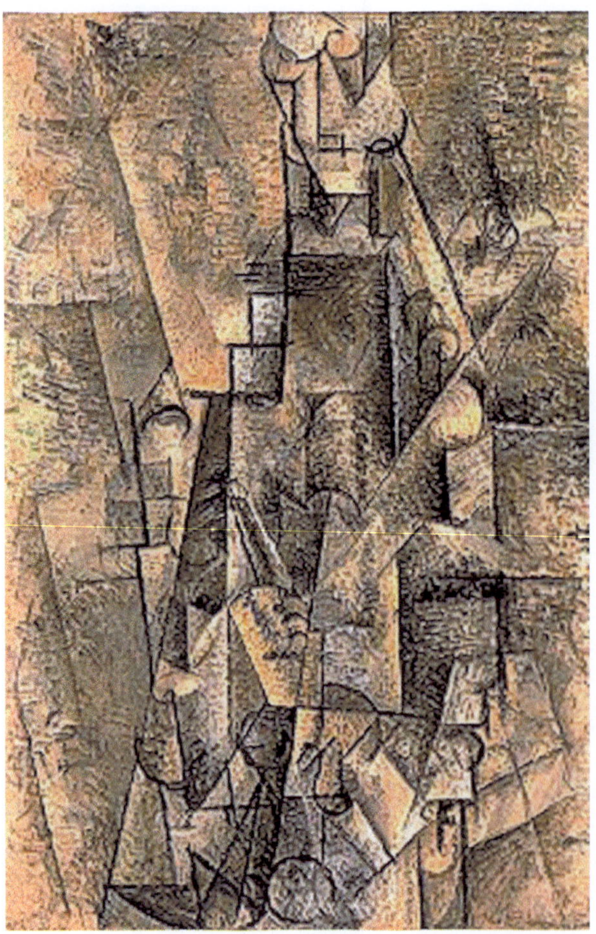

Figure 15.10. (page 392) *Picasso:* Man with a Clarinet (1911–12)

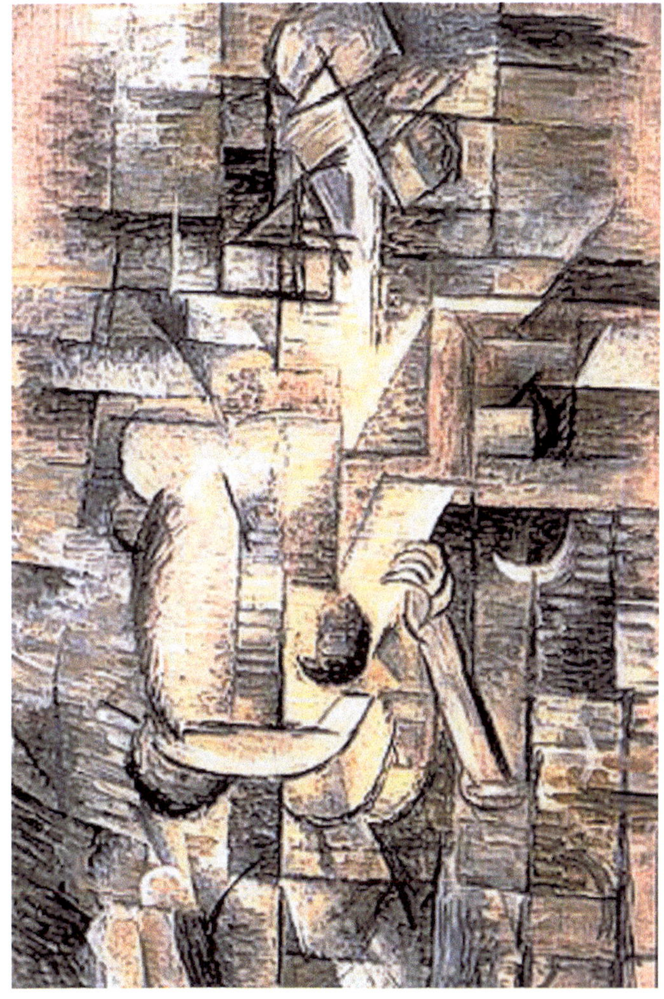

Figure 15.11. (page 393) *Braque:* Girl with a Mandolin (1910)

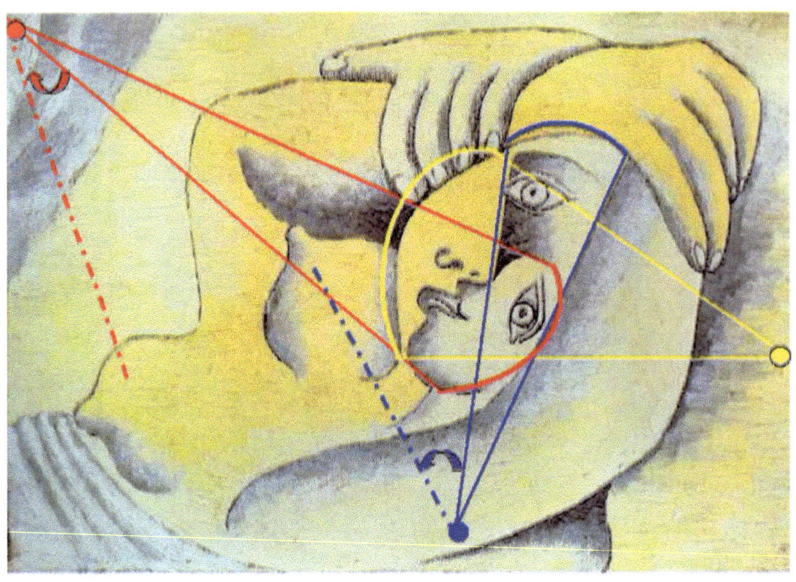

Figure 15.8. (page 391) *Picasso:* Nude on a Beach (1929)

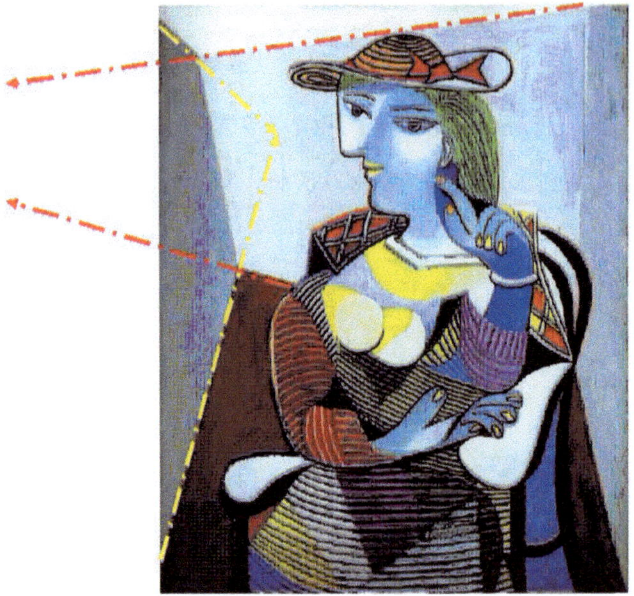

Figure 15.9. (page 391) *Picasso:* Portrait of Marie-Thérèse (1937)

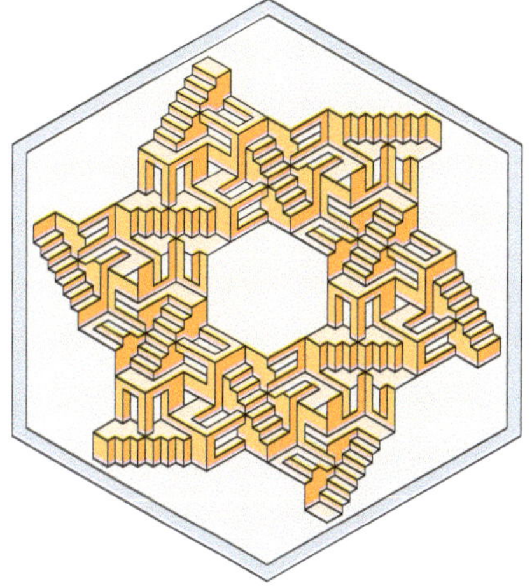

Figure 15.15. (page 396) *Tamás F. Farkas:* Geocity I (2003)

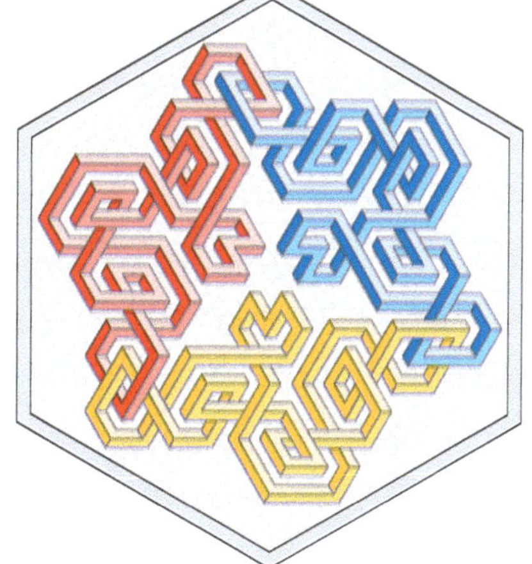

Figure 15.16. (page 396) *Tamás F. Farkas:* Genetix IV (2002)

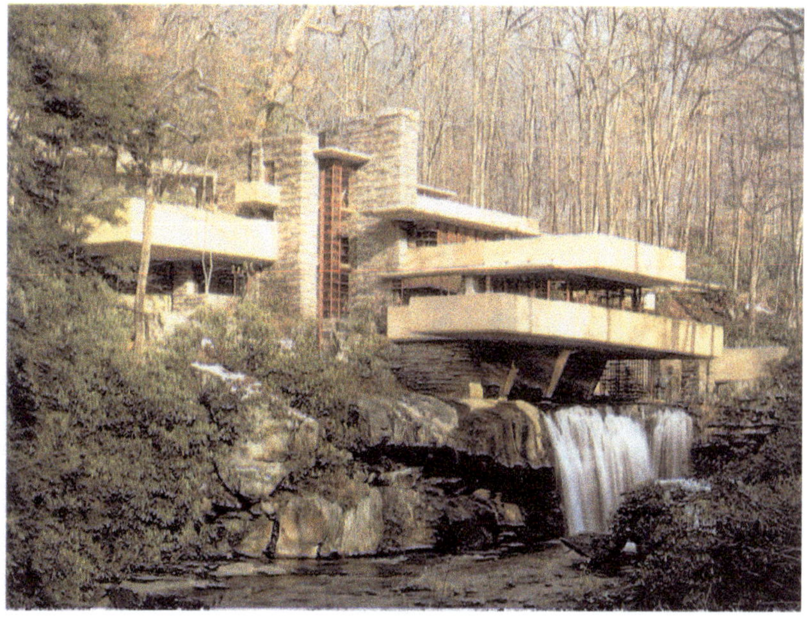

Figure 15.24. (page 403) *The Waterfall House* (1935–39)

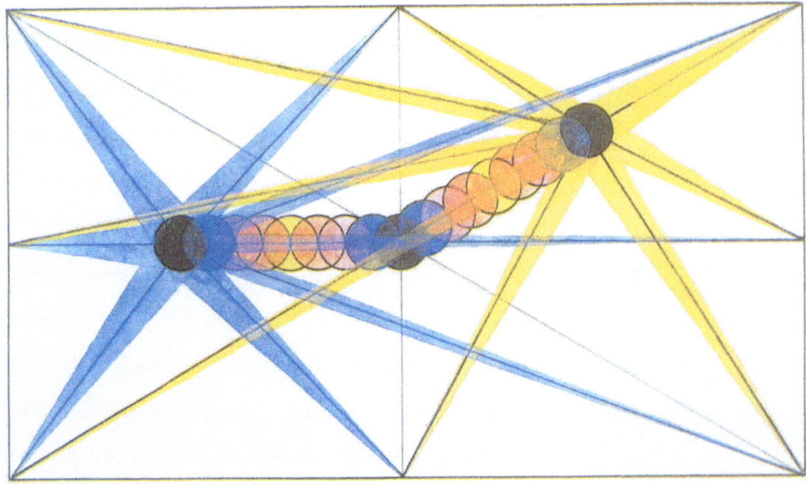

Figure 15.29. (page 236) *A study examining perspective, colours and proportions, prepared by student L. Meyer-Bergner for a class held by his teacher Paul Klee (1927)*

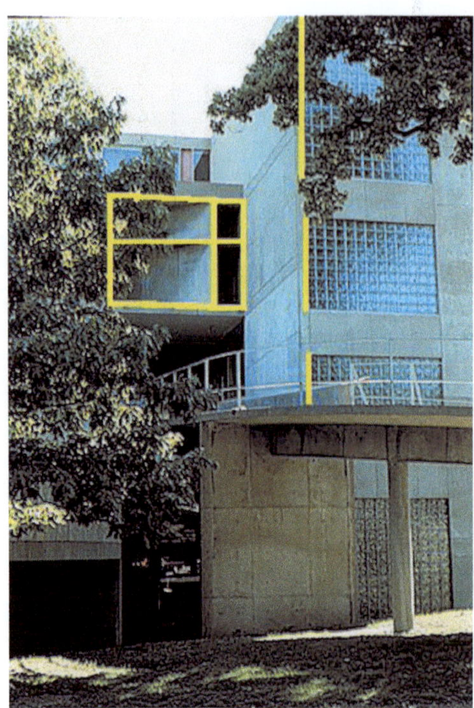

Figure 15.31. (page 413) *In the details of the façade of Harvard University's Carpenter Center for the Visual Arts, designed by Le Corbusier, the proportions previously discussed in this book are clearly visible*

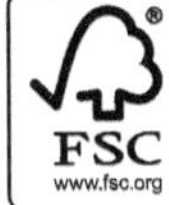

MIX
Papier aus verantwortungsvollen Quellen
Paper from responsible sources
FSC® C105338

If you have any concerns about our products,
you can contact us on
ProductSafety@springernature.com

In case Publisher is established outside the EU,
the EU authorized representative is:
Springer Nature Customer Service Center GmbH
Europaplatz 3, 69115 Heidelberg, Germany

Printed by Libri Plureos GmbH
in Hamburg, Germany